STUDENT HELP TECHNOLOGY OVERVIEW

STUDENT COMPANION WEBSITE

BigIdeasMath.com

ONLINE PUPIL EDITION

INTERACTIVE GLOSSARY
English, Spanish, Haitian Creole

SKILLS REVIEW HANDBOOK

LESSON TUTORIALS

BIG IDEAS MATH 8®

FLORIDA EDITION

Ron Larson
Laurie Boswell

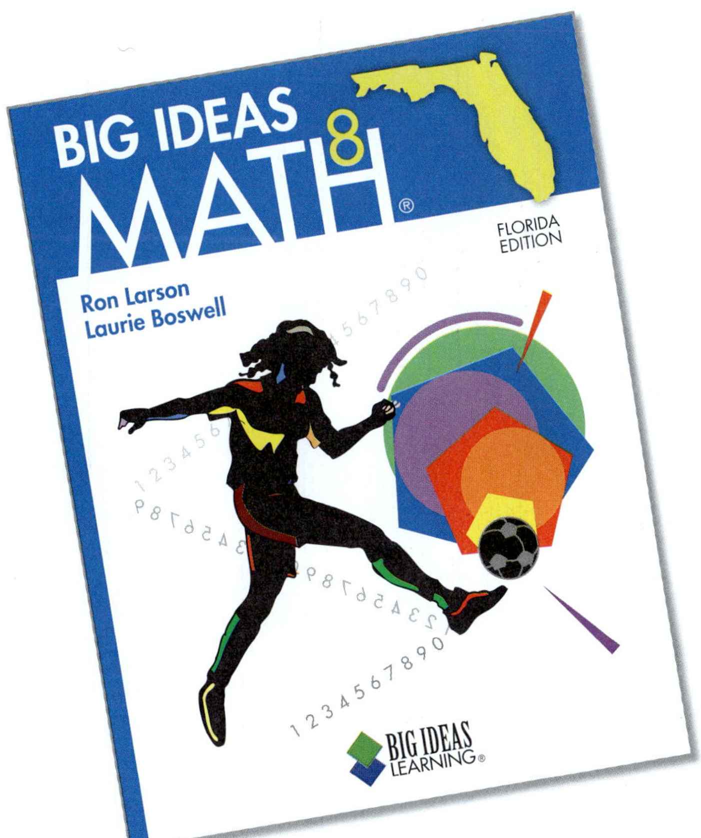

BIG IDEAS LEARNING®

Erie, Pennsylvania
BigIdeasLearning.com

BIG IDEAS LEARNING®

Big Ideas Learning, LLC
1762 Norcross Road
Erie, PA 16510-3838
USA

For product information and customer support, contact Big Ideas Learning at **1-877-552-7766** or visit us at ***BigIdeasLearning.com***.

Printed in the U.S.A.

ISBN 13: 978-1-60840-002-7
ISBN 10: 1-60840-002-6

3 4 5 6 7 8 9 10 WEB 13 12 11 10

AUTHORS

Ron Larson is a professor of mathematics at Penn State University in Erie, Pennsylvania where he has taught since receiving his Ph.D. in mathematics from the University of Colorado in 1970. Dr. Larson is well known as the lead author of a comprehensive program for mathematics that spans middle school, high school, and college courses. His high school and Advanced Placement books are published by Holt McDougal. Dr. Larson's numerous professional activities keep him in constant touch with the needs of students, teachers, and supervisors. Dr. Larson and Laurie Boswell began writing together in 1992. Since that time, they have authored over two dozen textbooks. In their collaboration, Dr. Larson is primarily responsible for the pupil edition and Laurie is primarily responsible for the teaching edition of the text.

Laurie Boswell is a mathematics teacher at the Riverside School in Lyndonville, Vermont. She is a recipient of the Presidential Award for Excellence in Mathematics Teaching. Laurie has taught math to students at all levels, elementary through college. In addition, Laurie was a Tandy Technology Scholar, and served on the NCTM Board of Directors from 2002 to 2005. She currently serves on the board of NCSM, and is a popular national speaker. Along with Dr. Larson, Laurie has co-authored numerous math programs.

ABOUT THE BOOK

This book is brand new! It is not a revision of a previously published book.

When Florida released its new mathematics standards in September 2007, we were delighted to see that a state has had the courage to follow the NCTM Focal Points guidelines for teaching a *Narrower and Deeper*™ middle school mathematics program.

- **DEEPER** Each section is designed for 2–3 day coverage.
- **DYNAMIC** Each section begins with a full class period of active learning.
- **DOABLE** Each section is accompanied by full student and teacher support.
- **DAZZLING** How else can we say this? This book puts the dazzle back in math!

Ron Larson   Laurie Boswell

TEACHER REVIEWERS

- Natasha Bell
  Math Department Head
  Parkway Middle School of the Arts
- Lisa Goldsmith
  Math Department Chair
  Broward County Public Schools
- Michelle Kadel
  Mathematics Instructor
  Tyrone Middle School
- Scott Lapsley
  Geometry and Pre-Algebra Instructor
  Pinellas County Schools
- Marcela Mansur
  Curriculum Specialist
  Walter C. Young Middle School
- Tammy Rush
  District Resource Teacher
  Hillsborough County Schools
- Patricia D. Seger
  Math Department Chair
  Bartow Middle School
- Elizabeth Wright
  Math Department Chair
  Bay Point Middle School
- Barbara Zorin
  Mathematics Honors Course Instructor
  Meadowlawn Middle School

STUDENT REVIEWERS

- Ashley Benovic
- Vanessa Bowser
- Sara Chinsky
- Kaitlyn Grimm
- Lakota Noble
- Norhan Omar
- Jack Puckett
- Abby Quinn
- Victoria Royal
- Madeline Su
- Lance Williams

CONSULTANTS

- Patsy Davis
  Educational Consultant
  Knoxville, Tennessee
- Deb Johnson
  Differentiated Instruction Consultant
  Missoula, Montana
- Mark Johnson
  Mathematics Assessment Consultant
  Raymond, New Hampshire
- Ryan Keating
  Special Education Advisor
  Gilbert, Arizona
- Michael McDowell
  Project-Based Instruction Specialist
  Scottsdale, Arizona
- Sean McKeighan
  Interdisciplinary Advisor
  Midland, Texas
- Bill Setzer
  Retired Mathematics Supervisor
  Roanoke County, Virginia
- Bonnie Spence
  Differentiated Instruction Consultant
  Missoula, Montana

Florida Standards for Grade 8

Chapter Coverage for Big and Supporting Ideas

1 2 3 4 5 6 7 8 9

Big Idea 1

Analyze and represent linear functions and solve linear equations and systems of linear equations.

CODE	BENCHMARK
MA.8.A.1.1	Create and interpret tables, graphs, and models to represent, analyze, and solve problems related to linear equations, including analysis of domain, range and the difference between discrete and continuous data.
MA.8.A.1.2	Interpret the slope and the x- and y-intercepts when graphing a linear equation for a real-world problem.
MA.8.A.1.3	Use tables, graphs, and models to represent, analyze, and solve real-world problems related to systems of linear equations.
MA.8.A.1.4	Identify the solution to a system of linear equations using graphs.
MA.8.A.1.5	Translate among verbal, tabular, graphical and algebraic representations of linear functions.
MA.8.A.1.6	Compare the graphs of linear and non-linear functions for real-world situations.

1 2 3 4 5 6 7 8 9

Big Idea 2

Analyze two- and three-dimensional figures by using distance and angle.

CODE	BENCHMARK
MA.8.G.2.1	Use similar triangles to solve problems that include height and distances.
MA.8.G.2.2	Classify and determine the measure of angles, including angles created when parallel lines are cut by transversals.
MA.8.G.2.3	Demonstrate that the sum of the angles in a triangle is 180-degrees and apply this fact to find unknown measures of angles, and the sum of angles in polygons.
MA.8.G.2.4	Validate and apply the Pythagorean Theorem to find distances in real-world situations or between points in the coordinate plane.

1 2 3 4 5 6 7 8 9

Big Idea 3

Analyze and summarize data sets.

CODE	BENCHMARK
MA.8.S.3.1	Select, organize, and construct appropriate data displays, including box and whisker plots, scatter plots, and lines of best fit to convey information and make conjectures about possible relationships.
MA.8.S.3.2	Determine and describe how changes in data values impact measures of central tendency.

1 2 3 4 5 6 7 8 9

Supporting Idea Algebra

CODE	BENCHMARK
MA.8.A.4.1	Solve literal equations for a specified variable.
MA.8.A.4.2	Solve and graph one- and two-step inequalities in one variable.

1 2 3 4 5 6 7 8 9

Supporting Idea Geometry and Measurement

CODE	BENCHMARK
MA.8.G.5.1	Compare, contrast, and convert units of measure between different measurement systems (US customary or metric (SI)) and dimensions including temperature, area, volume, and derived units to solve problems.

1 2 3 4 5 6 7 8 9

Supporting Idea Number and Operations

CODE	BENCHMARK
MA.8.A.6.1	Use exponents and scientific notation to write large and small numbers and vice versa and to solve problems.
MA.8.A.6.2	Make reasonable approximations of square roots and mathematical expressions that include square roots, and use them to estimate solutions to problems and to compare mathematical expressions involving real numbers and radical expressions.
MA.8.A.6.3	Simplify real number expressions using the laws of exponents.
MA.8.A.6.4	Perform operations on real numbers (including integer exponents, radicals, percents, scientific notation, absolute value, rational numbers, and irrational numbers) using multi-step and real-world problems.

Unit 1 Chapters 1–4 **Linear Functions and Systems of Linear Equations**

Solving Equations

"I love my math book. It has so many interesting examples and homework problems. I have always liked math, but I didn't know how it could be used. Now I have lots of ideas."

Graphing Linear Equations and Linear Systems

"I like starting each new lesson with a partner activity. I just moved to this school and the activities helped me make friends."

Writing Linear Equations and Linear Systems

4

Functions

"I love the cartoons. They are funny and they help me remember the math. I want to be a cartoonist some day."

Unit 2 Analyzing Figures Using Distance and Angles

Chapters 5 and 6

Angles and Similarity

Square Roots and the Pythagorean Theorem

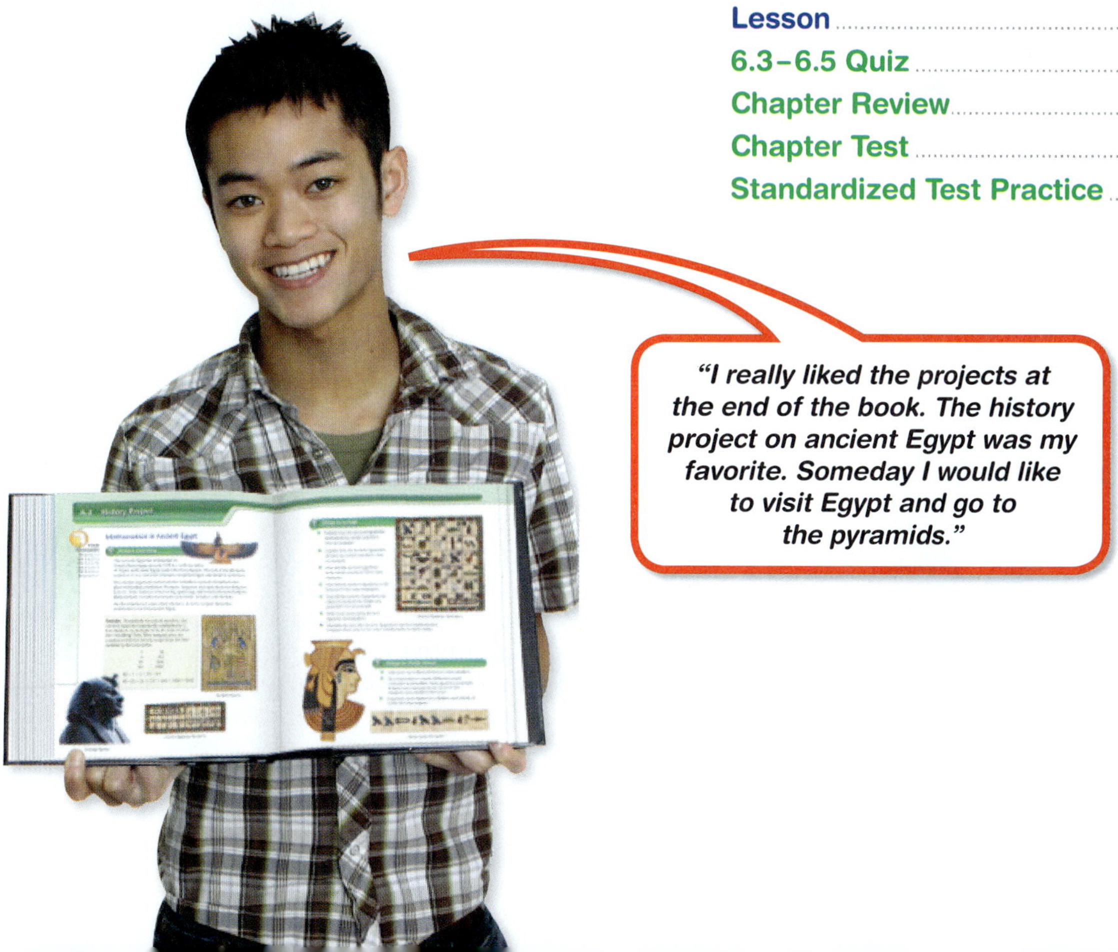

Unit 3 Chapter 7

Analyzing and Summarizing Data Sets

Data Analysis and Displays

"I like how the glossary in the book is part of the index. When I couldn't remember how a vocabulary word was defined, I could go to the index and find where the word was defined in the book."

Unit 4 Inequalities and Exponents

Chapters 8 and 9

Linear Inequalities

"I like the practice tests in the book. I get really nervous on tests. So, having a practice test to work on at home helped me to chill out when the real test came."

Exponents and Scientific Notation

"I like the review at the beginning of each chapter. This book has examples to help me remember things from last year. I don't like it when the review is just a list of questions."

Appendices A and B

Appendix A: My Big Ideas Projects

Appendix B: Financial Literacy

"I like the workbook (Record and Practice Journal). It saved me a lot of work to not have to copy all the questions and graphs."

How to Use Your Math Book

- Read the **Essential Question** in the activity.

  Work with a partner to decide **What Is Your Answer?**

  Now you are ready to do the **Practice** problems.

- Find the **Key Vocabulary** words, **highlighted in yellow**.

  Read their definitions. Study the concepts in each **Key Idea**.
  If you forget a definition, you can look it up online in the

  Multi-Language Glossary at BigIdeasMath.com.

- After you study each **EXAMPLE**, do the exercises in the **On Your Own**.

  Now You're Ready to do the exercises that correspond to the example.

  As you study, look for a **Study Tip** or a **Common Error**.

- The exercises are divided into 3 parts.

  Vocabulary and Concept Check

  Practice and Problem Solving

  Fair Game Review

  If an exercise has a 1 next to it, look back at Example 1 for help with that exercise.

  More help is available at **Check It Out** Lesson Tutorials BigIdeasMath.com.

- To help study for your test, use the following.

  Quiz **Study Help**

  Chapter Review **Chapter Test**

SCAVENGER HUNT

Use this *Scavenger Hunt* to find where things are in **Chapter 1**.

Newton

Find an **Essential Question**.

Find a **Practice**.

Find the definition of a **conversion factor**.

Find the **Key Idea** Division Property of Equality.
Use the index in the back of the book.

Find an **EXAMPLE** of solving a two-step equation.

Find an **On Your Own** about $-3x = 2x + 19$.

Find **Now You're Ready** Exercises 18–23.

Find a **Remember** about formulas.

Find a **Vocabulary** question about a literal equation.

Find a **Problem Solving** question about **ANTS**.
Find the answer in the back of the book.

Which examples **?** help with Exercise 8 on page 14?

Find 2 **Quiz**zes.

How many questions are on the **Chapter Test**?

Descartes

1 Solving Equations

1.1 Solving Simple Equations

1.2 Solving Multi-Step Equations

1.3 Solving Equations with Variables on Both Sides

1.4 Rewriting Equations and Formulas

1.5 Converting Units of Measure

"Dear Sir: Here is my suggestion for a good math problem."

"A box contains a total of 30 dog and cat treats. There are 5 times more dog treats than cat treats."

"How many of each type of treat are there?"

"Push faster, Descartes! According to the formula $R = D \div T$, the time needs to be 10 minutes or less to break our all-time speed record!"

What You Learned Before

Converting Measures (Grade 7)

You find a recipe for wheat germ bread.

WHEAT GERM BREAD

Preheat Oven to 220°C

Ingredients:

- $2\frac{3}{4}$ oz whole wheat grains
- 100 g wheat germ
- 400 g whole wheat flour
- $\frac{3}{4}$ tsp salt
- 340 mL water
- 60 mL orange juice
- 40 g honey
- $1\frac{2}{3}$ tsp yeast
- $\frac{1}{8}$ tsp cinnamon

Example 1 **How many cups of water do you need to make the recipe?**

$$340 \text{ mL} \times \frac{1 \text{ cup}}{237 \text{ mL}} \approx 1.4 \text{ cups}$$

You need about 1.4 cups of water to make the recipe.

Example 2 **Do you need more whole wheat grains or more whole wheat flour?**

$$400 \text{ grams} \times \frac{1 \text{ oz}}{28 \text{ grams}} \approx 14.3 \text{ oz of flour}$$

Because 14.3 ounces > $2\frac{3}{4}$ ounces, you need more whole wheat flour.

Adding and Subtracting Fractions (Grade 5)

Example 3 **How many teaspoons of spice (salt and cinnamon) are in the recipe?**

$$\frac{3}{4} + \frac{1}{8} = \frac{6}{8} + \frac{1}{8}$$
$$= \frac{7}{8}$$

So, there is $\frac{7}{8}$ teaspoon of spice.

Example 4 **How many more teaspoons of yeast than salt are in the recipe?**

$$1\frac{2}{3} - \frac{3}{4} = \frac{5}{3} - \frac{3}{4}$$
$$= \frac{20}{12} - \frac{9}{12}$$
$$= \frac{11}{12}$$

So, there is $\frac{11}{12}$ teaspoon more yeast than salt.

Try It Yourself

Use the recipe to answer the questions.

1. You have one-quarter cup of orange juice. Do you have enough to make the recipe? Explain.

2. You have $\frac{1}{8}$ ounce of whole wheat grains. How many more ounces do you need to make the recipe?

1.1 Solving Simple Equations

STATE STANDARDS
MA.8.A.1.1

Essential Question How can you use inductive reasoning to discover rules in mathematics? How can you test a rule?

1 ACTIVITY: Sum of the Angles of a Triangle

Work with a partner. Copy the triangles. Use a protractor to measure the angles of each triangle. Copy and complete the table to organize your results.

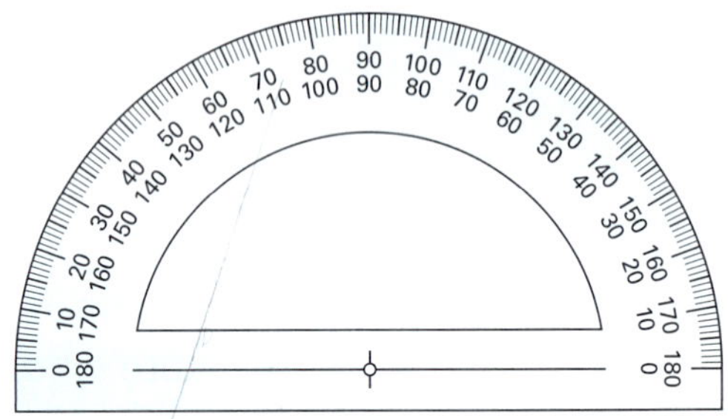

a.

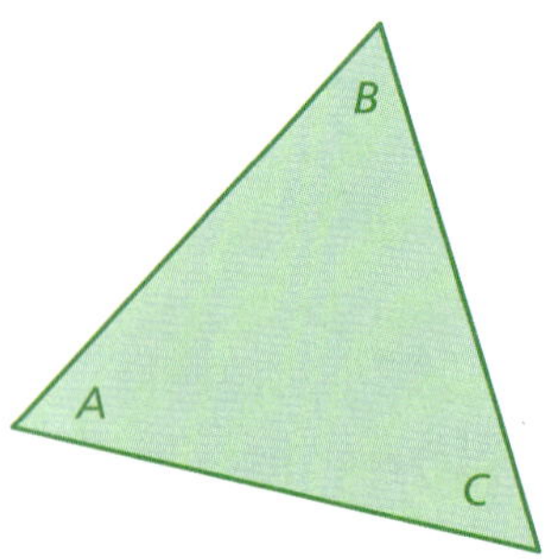

b.

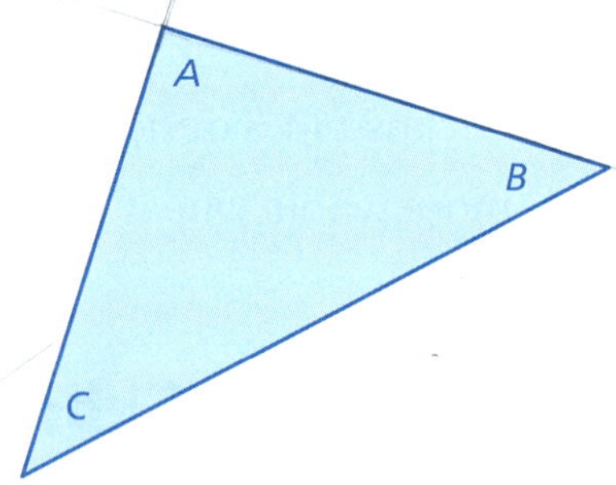

c.

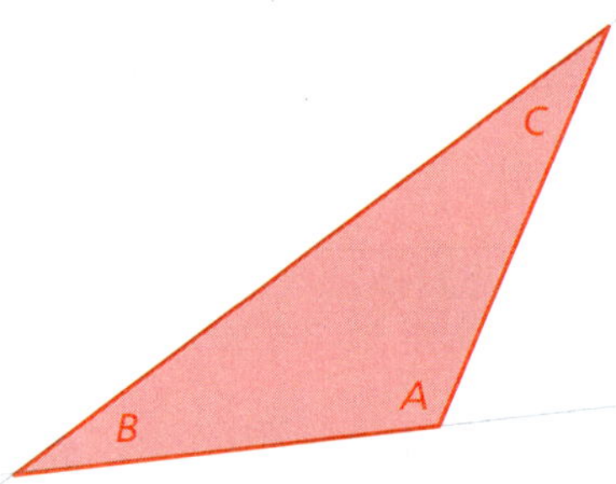

d.

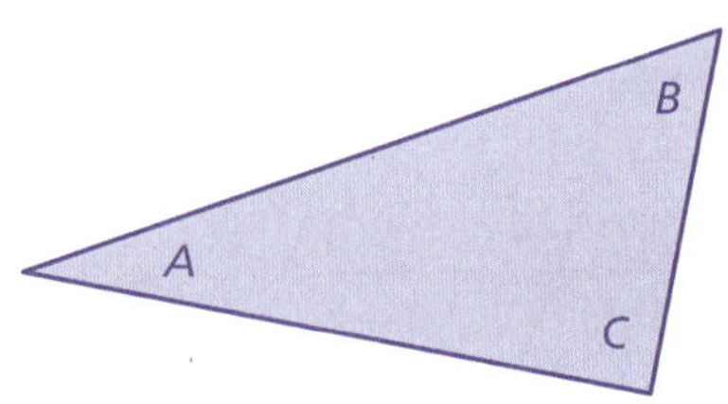

Triangle	Angle A (degrees)	Angle B (degrees)	Angle C (degrees)	$A + B + C$
a.				
b.				
c.				
d.				

2 ACTIVITY: Writing a Rule

Work with a partner. Use inductive reasoning to write and test a rule.

a. Use the completed table in Activity 1 to write a rule about the sum of the angle measures of a triangle.

b. **TEST YOUR RULE** Draw four triangles that are different from those in Activity 1. Measure the angles of each triangle. Organize your results in a table. Find the sum of the angle measures of each triangle.

3 ACTIVITY: Applying Your Rule

Work with a partner. Use the rule you wrote in Activity 2 to write an equation for each triangle. Then, solve the equation to find the value of *x*. Use a protractor to check the reasonableness of your answer.

a.

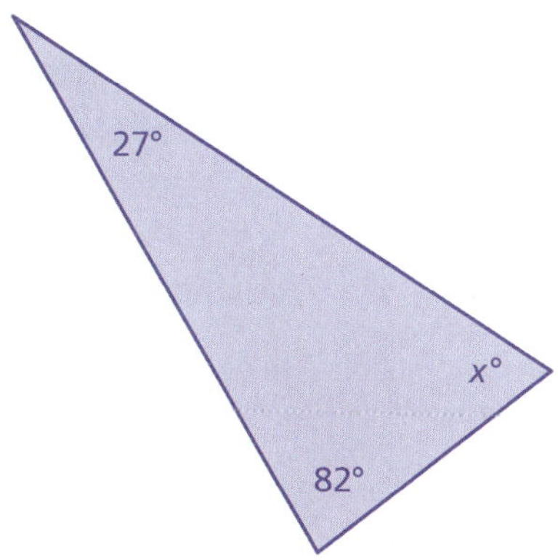

b.

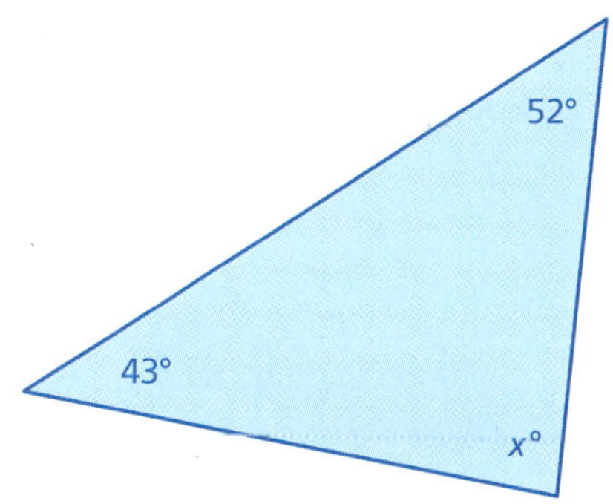

c.

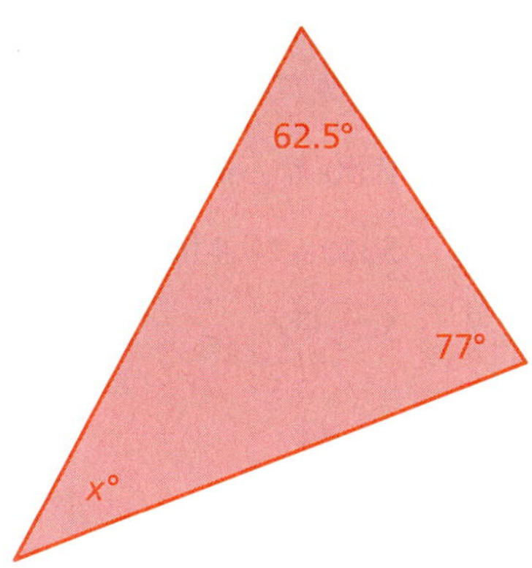

d.

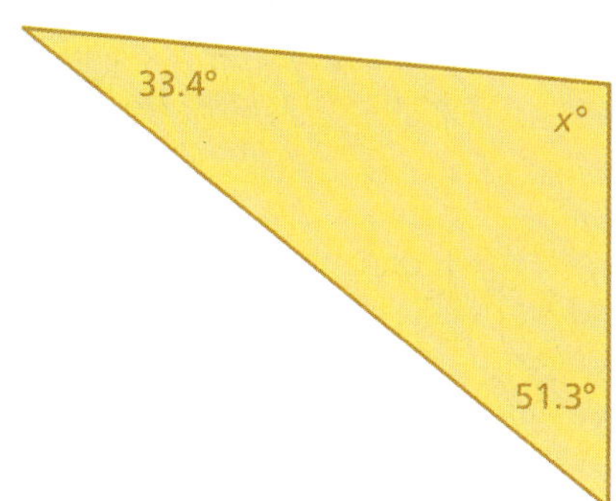

What Is Your Answer?

4. **IN YOUR OWN WORDS** How can you use inductive reasoning to discover rules in mathematics? How can you test a rule? How can you use a rule to solve problems in mathematics?

Practice Use what you learned about solving simple equations to complete Exercises 4–6 on page 7.

1.1 Lesson

Check It Out
Lesson Tutorials
BigIdeasMath.com

Remember

Addition and subtraction are inverse operations.

Key Ideas

Addition Property of Equality

Words Adding the same number to each side of an equation produces an equivalent equation.

Algebra If $a = b$, then $a + c = b + c$.

Subtraction Property of Equality

Words Subtracting the same number from each side of an equation produces an equivalent equation.

Algebra If $a = b$, then $a - c = b - c$.

EXAMPLE 1 Solving Equations Using Addition or Subtraction

a. Solve $x - 7 = -6$.

$x - 7 = -6$ Write the equation.

Undo the subtraction. → $\underline{+7} \quad \underline{+7}$ Add 7 to each side.

$x = 1$ Simplify.

The solution is $x = 1$.

Check

$x - 7 = -6$

$1 - 7 \stackrel{?}{=} -6$

$-6 = -6$ ✓

b. Solve $y + 3.4 = 0.5$.

$y + 3.4 = 0.5$ Write the equation.

Undo the addition. → $\underline{-3.4} \quad \underline{-3.4}$ Subtract 3.4 from each side.

$y = -2.9$ Simplify.

The solution is $y = -2.9$.

Check

$y + 3.4 = 0.5$

$-2.9 + 3.4 \stackrel{?}{=} 0.5$

$0.5 = 0.5$ ✓

c. Solve $h + 2\pi = 3\pi$.

$h + 2\pi = 3\pi$ Write the equation.

Undo the addition. → $\underline{-2\pi} \quad \underline{-2\pi}$ Subtract 2π from each side.

$h = \pi$ Simplify.

The solution is $h = \pi$.

On Your Own

Now You're Ready
Exercises 7–15

Solve the equation. Check your solution.

1. $b + 2 = -5$
2. $g - 1.7 = -0.9$
3. $-3 = k + 3$
4. $r - \pi = \pi$
5. $t - \frac{1}{4} = -\frac{3}{4}$
6. $5.6 + z = -8$

Remember

Multiplication and division are inverse operations.

Key Ideas

Multiplication Property of Equality

Words Multiplying each side of an equation by the same number produces an equivalent equation.

Algebra If $a = b$, then $a \cdot c = b \cdot c$.

Division Property of Equality

Words Dividing each side of an equation by the same number produces an equivalent equation.

Algebra If $a = b$, then $a \div c = b \div c$, $c \neq 0$.

EXAMPLE 2 Solving Equations Using Multiplication or Division

a. Solve $-\frac{3}{4}n = -2$.

$-\frac{3}{4}n = -2$ — Write the equation.

Use the reciprocal. → $-\frac{4}{3} \cdot \left(-\frac{3}{4}n\right) = -\frac{4}{3} \cdot -2$ — Multiply each side by $-\frac{4}{3}$, the reciprocal of $-\frac{3}{4}$.

$n = \frac{8}{3}$ — Simplify.

The solution is $n = \frac{8}{3}$.

b. Solve $\pi x = 3\pi$.

$\pi x = 3\pi$ — Write the equation.

Undo the multiplication. → $\frac{\pi x}{\pi} = \frac{3\pi}{\pi}$ — Divide each side by π.

$x = 3$ — Simplify.

The solution is $x = 3$.

Check

$\pi x = 3\pi$

$\pi(3) \stackrel{?}{=} 3\pi$

$3\pi = 3\pi$ ✓

On Your Own

Now You're Ready
Exercises 18–26

Solve the equation. Check your solution.

7. $\frac{y}{4} = -7$
8. $6\pi = \pi x$
9. $0.09w = 1.8$

EXAMPLE 3 Standardized Test Practice

What value of k makes the equation $k + 4 \div 0.2 = 5$ true?

Ⓐ -15 Ⓑ -5 Ⓒ -3 Ⓓ 1.5

$$
\begin{aligned}
k + 4 \div 0.2 &= 5 && \text{Write the equation.}\\
k + 20 &= 5 && \text{Divide 4 by 0.2.}\\
-20 \quad & \quad -20 && \text{Subtract 20 from each side.}\\
k &= -15 && \text{Simplify.}
\end{aligned}
$$

The correct answer is Ⓐ.

EXAMPLE 4 Real-Life Application

The melting point of bromine is -7°C.

The *melting point* of a solid is the temperature at which the solid becomes a liquid. The melting point of bromine is $\frac{1}{30}$ of the melting point of nitrogen. Write and solve an equation to find the melting point of nitrogen.

Words	The melting point of bromine	is	$\frac{1}{30}$	of	the melting point of nitrogen.
Variable	Let n be the melting point of nitrogen.				
Equation	-7	$=$	$\frac{1}{30}$	$\cdot$	n

$$
\begin{aligned}
-7 &= \frac{1}{30}n && \text{Write the equation.}\\
30 \cdot (-7) &= 30 \cdot \left(\frac{1}{30}n\right) && \text{Multiply each side by 30.}\\
-210 &= n && \text{Simplify.}
\end{aligned}
$$

The melting point of nitrogen is -210°C.

On Your Own

Now You're Ready
Exercises 33–38

10. Solve $p - 8 \div \frac{1}{2} = -3$.

11. Solve $q + |-10| = 2$.

12. The melting point of mercury is about $\frac{1}{4}$ of the melting point of krypton. The melting point of mercury is -39°C. Write and solve an equation to find the melting point of krypton.

1.1 Exercises

Check It Out
Help with Homework
BigIdeasMath.com

Vocabulary and Concept Check

1. **VOCABULARY** Which of the operations $+$, $-$, $\times$, and $\div$ are inverses of each other?
2. **VOCABULARY** Are the equations $3x = -9$ and $4x = -12$ equivalent? Explain.
3. **WHICH ONE DOESN'T BELONG?** Which equation does *not* belong with the other three? Explain your reasoning.

$x - 2 = 4$ $\quad$ $x - 3 = 6$ $\quad$ $x - 5 = 1$ $\quad$ $x - 6 = 0$

Practice and Problem Solving

Find the value of x. Use a protractor to check the reasonableness of your answer.

4. 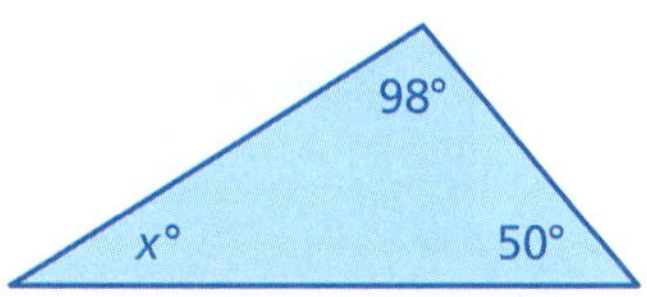

5. 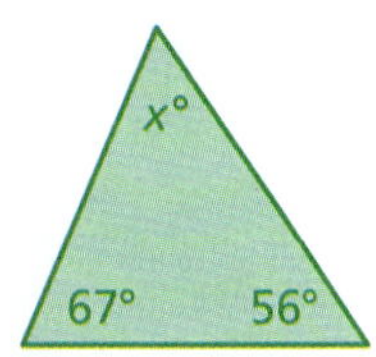

6. 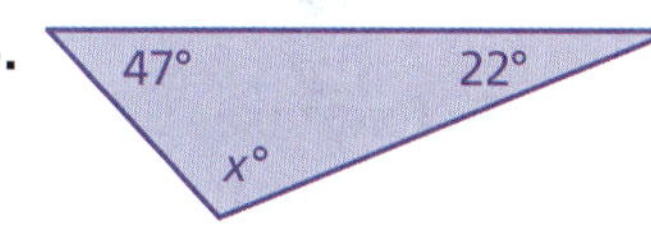

Solve the equation. Check your solution.

7. $x + 12 = 7$
8. $g - 16 = 8$
9. $-9 + p = 12$
10. $0.7 + y = -1.34$
11. $x - 8\pi = \pi$
12. $4\pi = w - 6\pi$
13. $\frac{5}{6} = \frac{1}{3} + d$
14. $\frac{3}{8} = r + \frac{2}{3}$
15. $n - 1.4 = -6.3$

16. **CONCERT** A discounted concert ticket is \$14.50 less than the original price p. You pay \$53 for a discounted ticket. Write and solve an equation to find the original price.

17. **BOWLING** Your friend's final bowling score is 105. Your final bowling score is 14 pins less than your friend's final score.
    a. Write and solve an equation to find your final score.
    b. Your friend made a spare in the tenth frame. Did you? Explain.

9	10	FINAL SCORE
8 – / 89	7 / 6 / 105	105
6 3 / 82	9	?

Solve the equation. Check your solution.

2 **18.** $7x = 35$

19. $4 = -0.8n$

20. $6 = -\frac{w}{8}$

21. $\frac{m}{\pi} = 7.3$

22. $-4.3g = 25.8$

23. $\frac{3}{2} = \frac{9}{10}k$

24. $-7.8x = -1.56$

25. $-2 = \frac{6}{7}p$

26. $3\pi d = 12\pi$

27. ERROR ANALYSIS Describe and correct the error in solving the equation.

✗
$-1.5 + k = 8.2$
$k = 8.2 + (-1.5)$
$k = 6.7$

28. TENNIS A gym teacher orders 42 tennis balls. Each package contains 3 tennis balls. Which of the following equations represents the number x of packages?

$x + 3 = 42$ $3x = 42$ $\frac{x}{3} = 42$ $x = \frac{3}{42}$

In Exercises 29–32, write and solve an equation to answer the question.

29. PARK You clean a community park for 6.5 hours. You earn $42.25. How much do you earn per hour?

30. SPACE SHUTTLE A space shuttle is scheduled to launch from Kennedy Space Center in 3.75 hours. What time is it now?

31. BANKING After earning interest, the balance of an account is $420. The new balance is $\frac{7}{6}$ of the original balance. How much interest was earned?

Tallest Coasters at Cedar Point	
Roller Coaster	**Height (feet)**
Top Thrill Dragster	420
Millennium Force	310
Magnum XL-200	205
Mantis	?

32. ROLLER COASTER Cedar Point amusement park has some of the tallest roller coasters in the United States. The Mantis is 165 feet shorter than the Millennium Force. What is the height of the Mantis?

Solve the equation. Check your solution.

3 **33.** $-3 = h + 8 \div 2$ **34.** $12 = w - |-7|$ **35.** $q + |6.4| = 9.6$

36. $d - 2.8 \div 0.2 = -14$ **37.** $\frac{8}{9} = x + \frac{1}{3}(7)$ **38.** $p - \frac{1}{4} \cdot 3 = -\frac{5}{6}$

39. CRITICAL THINKING Is the solution of $-2x = -15$ *greater than* or *less than* -15? Explain.

40. OPEN-ENDED Write a subtraction equation and a division equation that each has a solution of -2.

41. ANTS Some ant species can carry 50 times their body weight. It takes 32 ants to carry the cherry. About how much does each ant weigh?

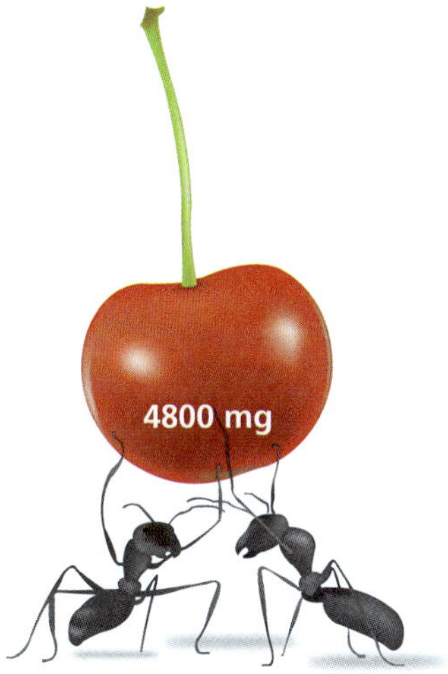

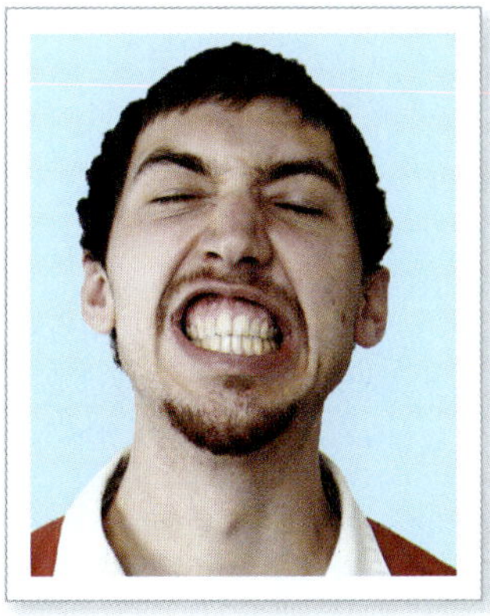

42. PICTURES One-fourth of the girls and one-eighth of the boys in an eighth grade retake their school pictures. The photographer retakes pictures for 16 girls and 7 boys. How many students are in the eighth grade?

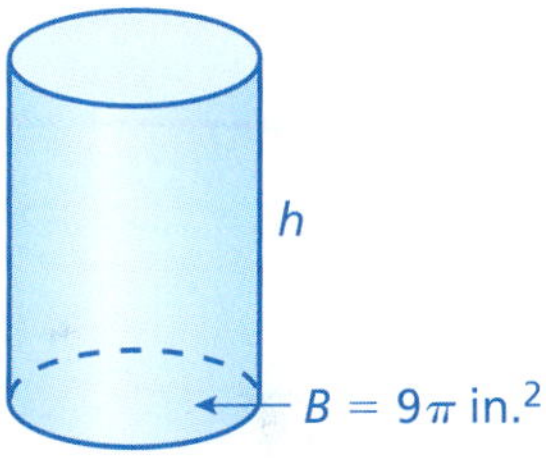

43. VOLUME The volume V of the cylinder is 72π cubic inches. Use the formula $V = Bh$ to find the height h of the cylinder.

44. Critical Thinking A neighbor pays you and two friends \$90 to paint her garage. The money is divided three ways in the ratio 2 : 3 : 5.

a. How much is each share?

b. What is one possible reason the money is not divided evenly?

Fair Game Review What you learned in previous grades & lessons

Simplify the expression. *(Skills Review Handbook)*

45. $2(x - 2) + 5x$ **46.** $0.4b - 3.2 + 1.2b$ **47.** $\frac{1}{4}g + 6g - \frac{2}{3}$

48. MULTIPLE CHOICE The temperature at 4 P.M. was -12°C. By 11 P.M. the temperature had dropped 14 degrees. What was the temperature at 11 P.M.? *(Skills Review Handbook)*

Ⓐ -26°C Ⓑ -2°C Ⓒ 2°C Ⓓ 26°C

1.2 Solving Multi-Step Equations

STATE STANDARDS
MA.8.A.1.1

Essential Question How can you solve a multi-step equation? How can you check the reasonableness of your solution?

1 ACTIVITY: Solving for the Angles of a Triangle

Work with a partner. Write an equation for each triangle. Solve the equation to find the value of the variable. Then find the angle measures of each triangle. Use a protractor to check the reasonableness of your answer.

a.

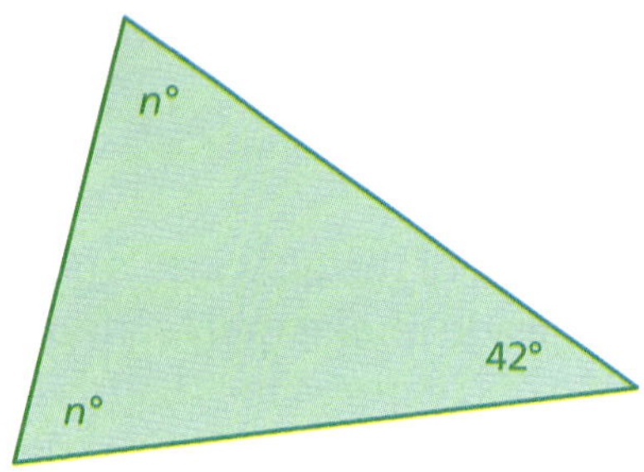

b.

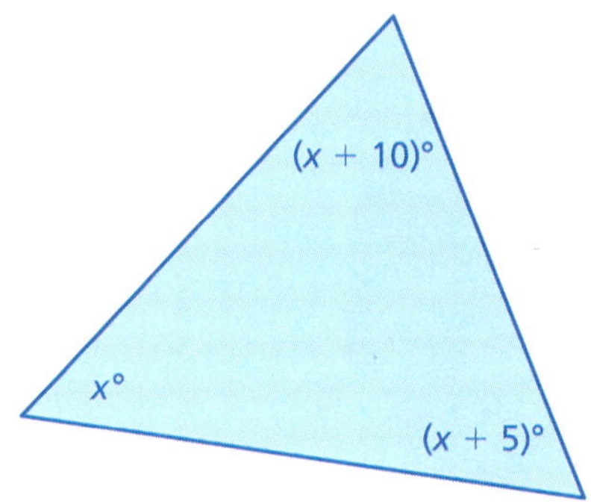

c.

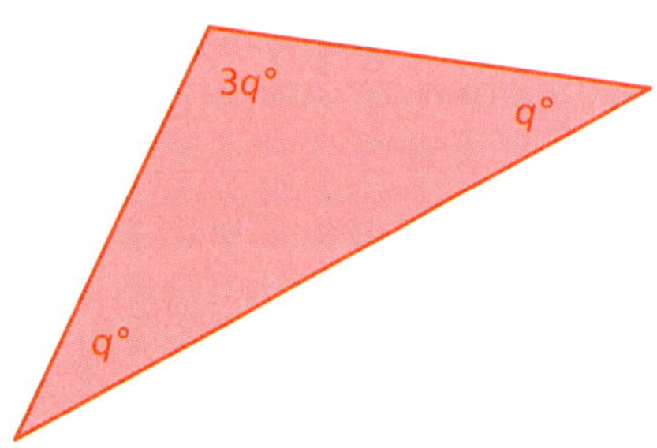

d.

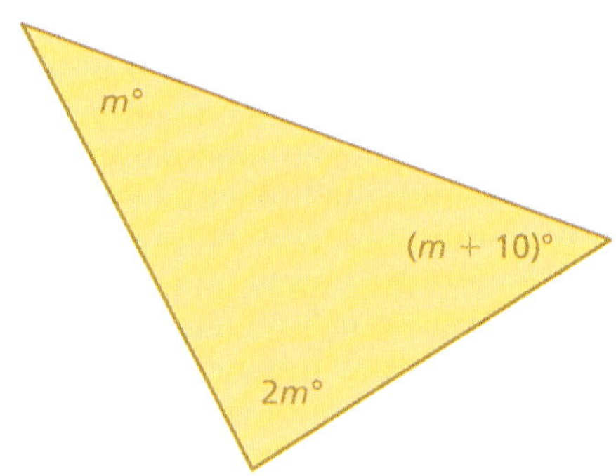

e.

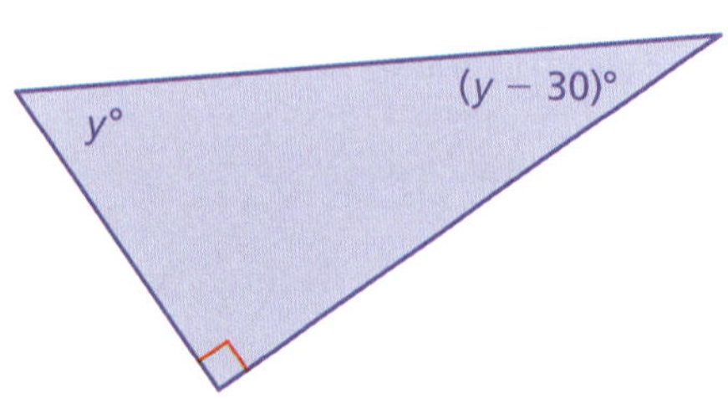

f.

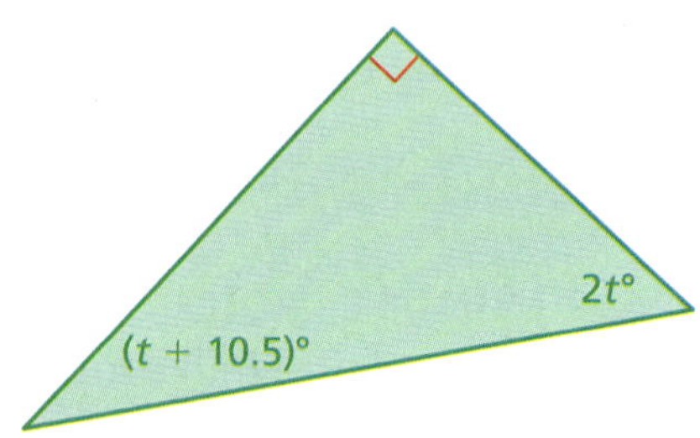

2 ACTIVITY: Problem-Solving Strategy

Work with a partner.

The six triangles form a rectangle.

Find the angle measures of each triangle. Use a protractor to check the reasonableness of your answers.

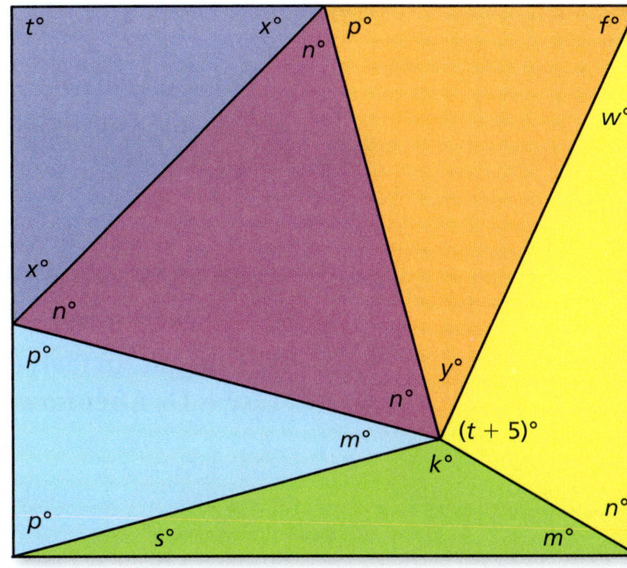

3 ACTIVITY: Puzzle

Work with a partner. A survey asked 200 people to name their favorite weekday. The results are shown in the circle graph.

a. How many degrees are in each part of the circle graph?

b. What percent of the people chose each day?

c. How many people chose each day?

d. Organize your results in a table.

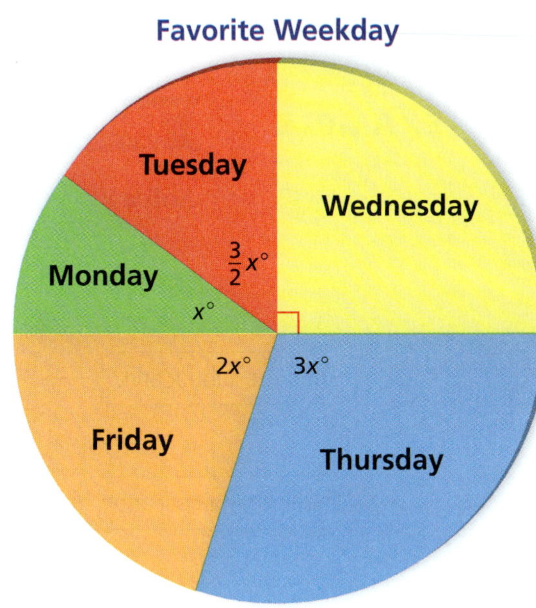

What Is Your Answer?

4. **IN YOUR OWN WORDS** How can you solve a multi-step equation? How can you check the reasonableness of your solution?

Practice Use what you learned about solving multi-step equations to complete Exercises 3–5 on page 14.

1.2 Lesson

Check It Out
Lesson Tutorials
BigIdeasMath.com

Key Idea

Solving Multi-Step Equations

To solve multi-step equations, use inverse operations to isolate the variable.

EXAMPLE 1 Solving a Two-Step Equation

The height (in feet) of a palm tree after x years is $1.5x + 15$. After how many years is the tree 24 feet tall?

	$1.5x + 15 = 24$	Write an equation.
Undo the addition.	$-15 \quad -15$	Subtract 15 from each side.
	$1.5x = 9$	Simplify.
Undo the multiplication.	$\dfrac{1.5x}{1.5} = \dfrac{9}{1.5}$	Divide each side by 1.5.
	$x = 6$	Simplify.

∴ The tree is 24 feet tall after 6 years.

EXAMPLE 2 Combining Like Terms to Solve an Equation

Solve $8x - 6x - 25 = -35$.

	$8x - 6x - 25 = -35$	Write the equation.
	$2x - 25 = -35$	Combine like terms.
Undo the subtraction.	$+25 \quad +25$	Add 25 to each side.
	$2x = -10$	Simplify.
Undo the multiplication.	$\dfrac{2x}{2} = \dfrac{-10}{2}$	Divide each side by 2.
	$x = -5$	Simplify.

∴ The solution is $x = -5$.

On Your Own

Now You're Ready
Exercises 6–9

Solve the equation. Check your solution.

1. $-3z + 1 = 7$
2. $\frac{1}{2}x - 9 = -25$
3. $-4n - 8n + 17 = 23$

EXAMPLE 3 Using the Distributive Property to Solve an Equation

Solve $2(1 - 5x) + 4 = -8$.

$2(1 - 5x) + 4 = -8$	Write the equation.
$2(1) - 2(5x) + 4 = -8$	Use Distributive Property.
$2 - 10x + 4 = -8$	Multiply.
$-10x + 6 = -8$	Combine like terms.
$\underline{-6} \quad \underline{-6}$	Subtract 6 from each side.
$-10x = -14$	Simplify.
$\dfrac{-10x}{-10} = \dfrac{-14}{-10}$	Divide each side by -10.
$x = 1.4$	Simplify.

Study Tip

Here is another way to solve the equation in Example 3.

$$2(1 - 5x) + 4 = -8$$
$$2(1 - 5x) = -12$$
$$1 - 5x = -6$$
$$-5x = -7$$
$$x = 1.4$$

EXAMPLE 4 Real-Life Application

Use the table to find the number of miles x you need to run on Friday so that the mean number of miles run per day is 1.5.

Day	Miles
Monday	2
Tuesday	0
Wednesday	1.5
Thursday	0
Friday	x

Write an equation using the definition of mean.

sum of the data → (numerator); number of values → (denominator)	$\dfrac{2 + 0 + 1.5 + 0 + x}{5} = 1.5$	Write the equation.
	$\dfrac{3.5 + x}{5} = 1.5$	Combine like terms.
Undo the division. →	$5 \cdot \dfrac{3.5 + x}{5} = 5 \cdot 1.5$	Multiply each side by 5.
	$3.5 + x = 7.5$	Simplify.
Undo the addition. →	$\underline{-3.5} \qquad \underline{-3.5}$	Subtract 3.5 from each side.
	$x = 4$	Simplify.

∴ You need to run 4 miles on Friday.

On Your Own

Now You're Ready
Exercises 10 and 11

Solve the equation. Check your solution.

4. $-3(x + 2) + 5x = -9$
5. $5 + 1.5(2d - 1) = 0.5$
6. You scored 88, 92, and 87 on three tests. Write and solve an equation to find the score you need on the fourth test so that your mean test score is 90.

1.2 Exercises

Check It Out
Help with Homework
BigIdeasMath.com

Vocabulary and Concept Check

1. **WRITING** Write the verbal statement as an equation. Then solve.

   2 more than 3 times a number is 17.

2. **OPEN-ENDED** Explain how to solve the equation $2(4x - 11) + 9 = 19$.

Practice and Problem Solving

Find the value of the variable. Then find the angle measures of the polygon. Use a protractor to check the reasonableness of your answer.

3. 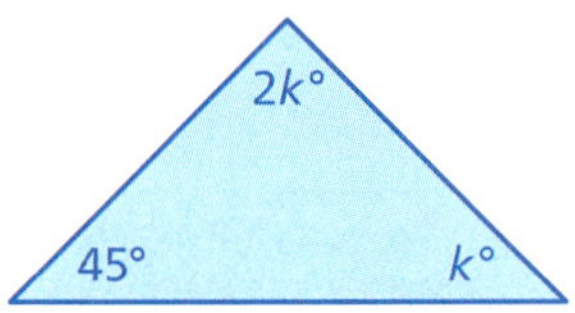

Sum of angle measures: 180°

4. 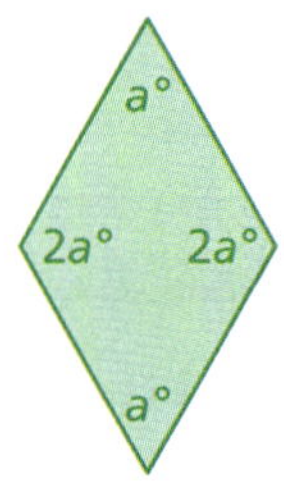

Sum of angle measures: 360°

5. 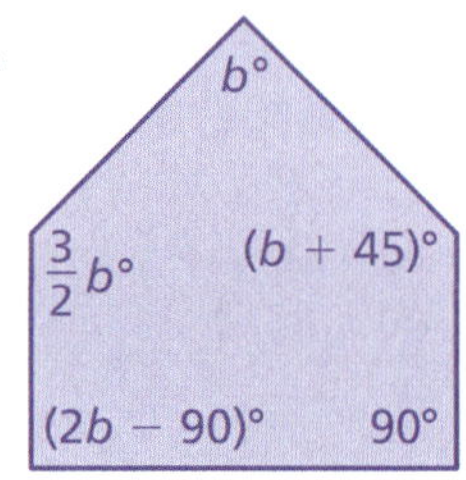

Sum of angle measures: 540°

Solve the equation. Check your solution.

① ② 6. $10x + 2 = 32$

7. $19 - 4c = 17$

8. $1.1x + 1.2x - 5.4 = -10$

9. $\frac{2}{3}h - \frac{1}{3}h + 11 = 8$

③ 10. $6(5 - 8v) + 12 = -54$

11. $21(2 - x) + 12x = 44$

12. **ERROR ANALYSIS** Describe and correct the error in solving the equation.

   ✗
   $$
   \begin{aligned}
   -2(7 - y) + 4 &= -4 \\
   -14 - 2y + 4 &= -4 \\
   -10 - 2y &= -4 \\
   -2y &= 6 \\
   y &= -3
   \end{aligned}
   $$

13. **WATCHES** The cost (in dollars) of making n watches is represented by $C = 15n + 85$. How many watches are made when the cost is \$385?

14. **HOUSE** The height of the house is 26 feet. What is the height x of each story?

In Exercises 15–17, write and solve an equation to answer the question.

15. POSTCARD The area of the postcard is 24 square inches. What is the width b of the message (in inches)?

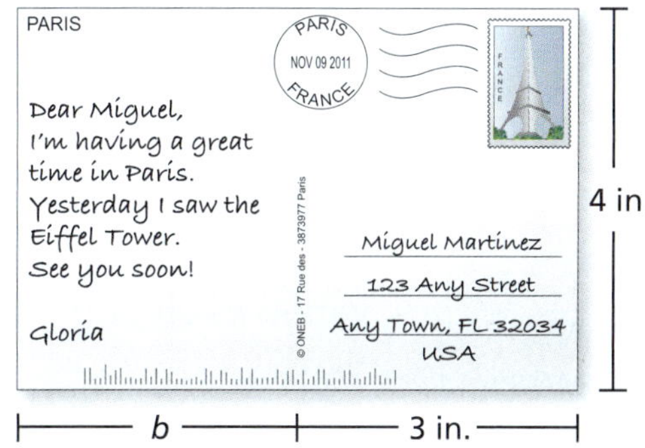

16. BREAKFAST You order two servings of pancakes and a fruit cup. The cost of the fruit cup is $1.50. You leave a 15% tip. Your total bill is $11.50. How much does one serving of pancakes cost?

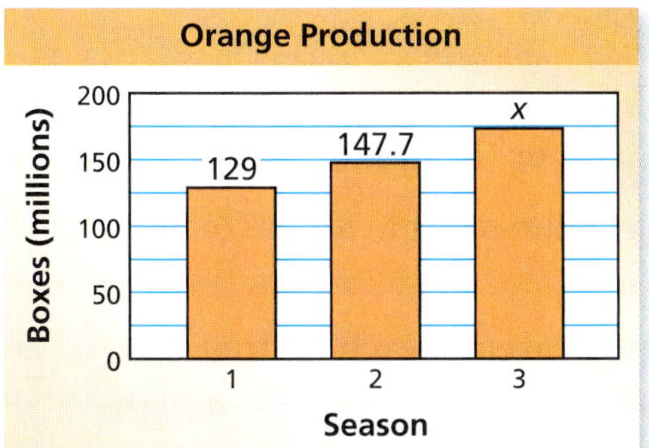

17. ORANGES How many boxes x of oranges must Florida produce in the third season to average 150 million boxes over the three seasons?

18. DIVING Olympic divers are scored by an international panel of judges. The highest and lowest scores are dropped. The total of the remaining scores is multiplied by the degree of difficulty of the dive. This product is multiplied by 0.6 to determine the final score.

a. A diver's final score is 77.7. What is the degree of difficulty of the dive?

Judge	Russia	China	Mexico	Germany	Italy	Japan	Brazil
Score	7.5	8.0	6.5	8.5	7.0	7.5	7.0

b. Critical Thinking The degree of difficulty of a dive is 4.0. The diver's final score is 97.2. Judges award half or whole points from 0 to 10. What scores could the judges have given the diver?

© Paul Slaughter, www.slaughterphoto.com
Greg Louganis diving at the 1984 Olympics

Fair Game Review What you learned in previous grades & lessons

Let $a = 3$ and $b = -2$. Copy and complete the statement using <, >, or =. *(Skills Review Handbook)*

19. $-5a$ ___ 4

20. 5 ___ $b + 7$

21. $a - 4$ ___ $10b + 8$

22. MULTIPLE CHOICE What value of x makes the equation $x + 5 = 2x$ true? *(Skills Review Handbook)*

Ⓐ -1 Ⓑ 0 Ⓒ 3 Ⓓ 5

1.3 Solving Equations with Variables on Both Sides

STATE STANDARDS
MA.8.A.1.1

Essential Question How can you solve an equation that has variables on both sides?

1 ACTIVITY: Perimeter and Area

Work with a partner. Each figure has the unusual property that the value of its perimeter (in feet) is equal to the value of its area (in square feet).

- **Write an equation (value of perimeter = value of area) for each figure.**
- **Solve each equation for *x*.**
- **Use the value of *x* to find the perimeter and area of each figure.**
- **Check your solution by comparing the value of the perimeter and the value of the area of each figure.**

a.

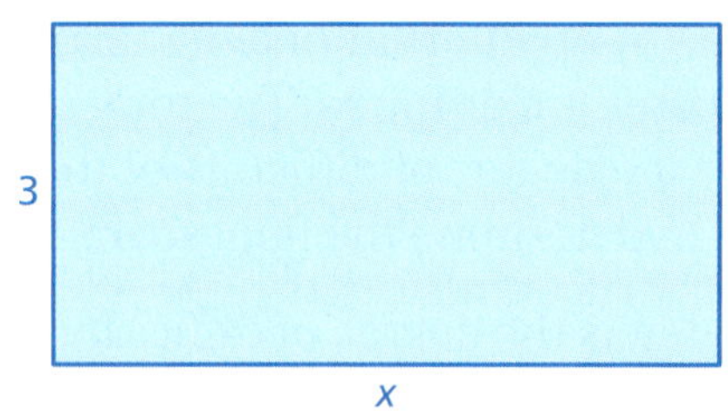

b.

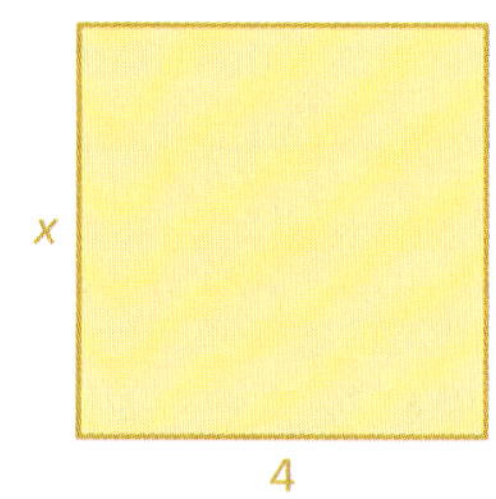

c.

d.

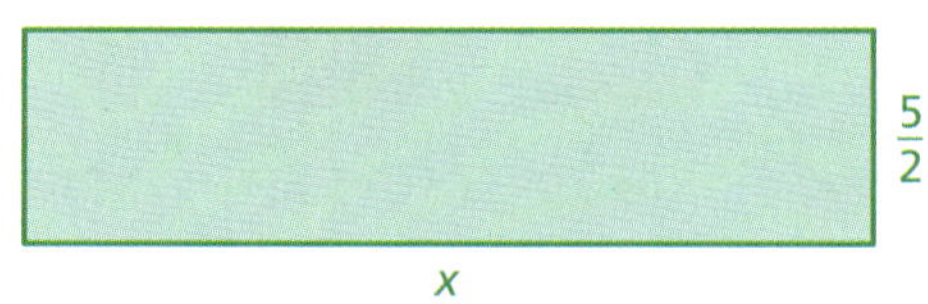

e.

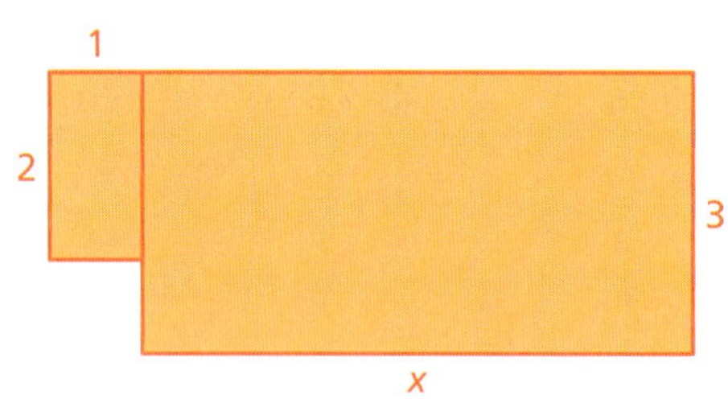

f.

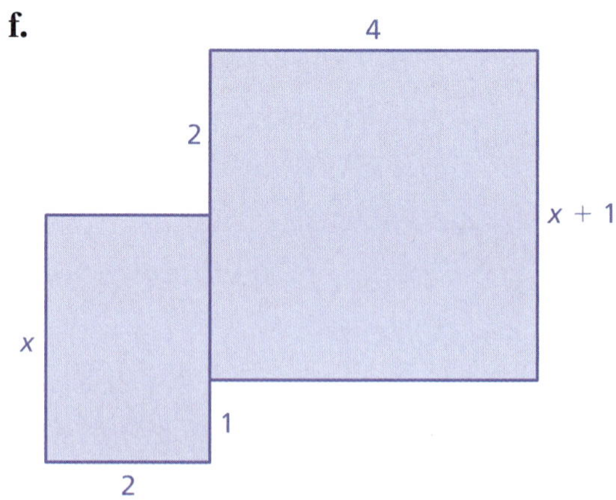

g.

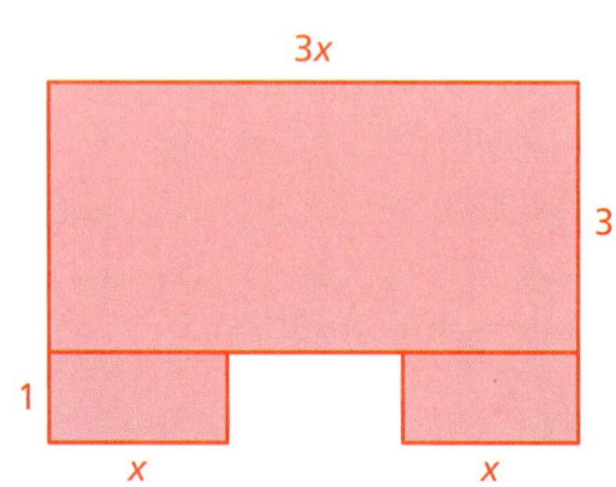

2 ACTIVITY: Surface Area and Volume

Work with a partner. Each solid has the unusual property that the value of its surface area (in square inches) is equal to the value of its volume (in cubic inches).

- **Write an equation (value of surface area = value of volume) for each figure.**
- **Solve each equation for x.**
- **Use the value of x to find the surface area and volume of each figure.**
- **Check your solution by comparing the value of the surface area and the value of the volume of each figure.**

a.

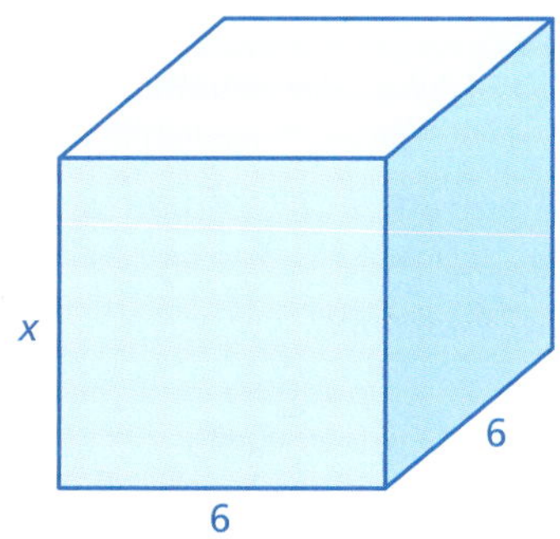

b.

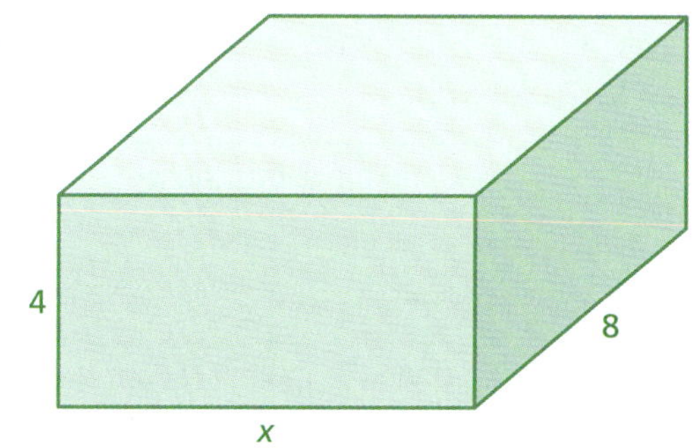

3 ACTIVITY: Puzzle

Work with a partner. The two triangles are similar. The perimeter of the larger triangle is 150% of the perimeter of the smaller triangle. Find the dimensions of each triangle.

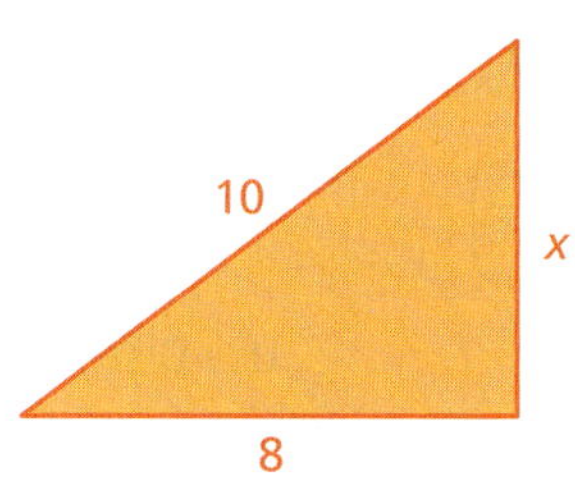

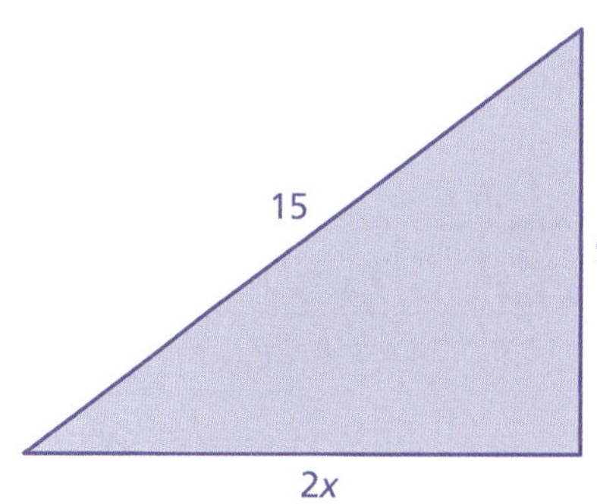

What Is Your Answer?

4. **IN YOUR OWN WORDS** How can you solve an equation that has variables on both sides? Write an equation that has variables on both sides. Solve the equation.

Practice Use what you learned about solving equations with variables on both sides to complete Exercises 3–5 on page 20.

1.3 Lesson

Check It Out
Lesson Tutorials
BigIdeasMath.com

Key Idea

Solving Equations with Variables on Both Sides

To solve equations with variables on both sides, collect the variable terms on one side and the constant terms on the other side.

EXAMPLE 1 Solving an Equation with Variables on Both Sides

Solve $15 - 2x = -7x$. Check your solution.

	$15 - 2x = -7x$	Write the equation.
Undo the subtraction. →	$+\,2x \quad +\,2x$	Add $2x$ to each side.
	$15 = -5x$	Simplify.
Undo the multiplication. →	$\frac{15}{-5} = \frac{-5x}{-5}$	Divide each side by -5.
	$-3 = x$	Simplify.

Check

$15 - 2x = -7x$

$15 - 2(-3) \stackrel{?}{=} -7(-3)$

$21 = 21$ ✓

The solution is $x = -3$.

EXAMPLE 2 Using the Distributive Property to Solve an Equation

Solve $-2(x - 5) = 6\left(2 - \frac{1}{2}x\right)$.

	$-2(x - 5) = 6\left(2 - \frac{1}{2}x\right)$	Write the equation.
	$-2x + 10 = 12 - 3x$	Use Distributive Property.
Undo the subtraction. →	$+\,3x \quad +\,3x$	Add $3x$ to each side.
	$x + 10 = 12$	Simplify.
Undo the addition. →	$-\,10 \quad -\,10$	Subtract 10 from each side.
	$x = 2$	Simplify.

The solution is $x = 2$.

On Your Own

Now You're Ready
Exercises 6–14

Solve the equation. Check your solution.

1. $-3x = 2x + 19$
2. $2.5y + 6 = 4.5y - 1$
3. $6(4 - z) = 2z$

EXAMPLE 3 Standardized Test Practice

The circles are congruent. What is the area of each circle?

Ⓐ 2 Ⓑ 4 Ⓒ 16π Ⓓ 64π

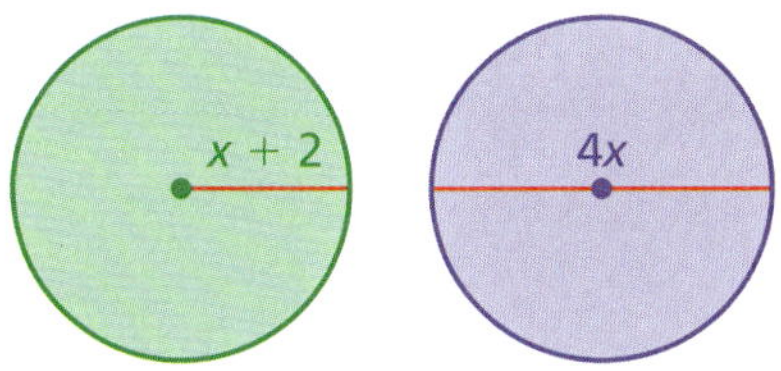

The circles are congruent, so the radius of each circle is the same.

$x + 2 = 2x$	Write an equation. The radius of the purple circle is $2x$.
$\underline{-x} \qquad \underline{-x}$	Subtract x from each side.
$2 = x$	Simplify.

∴ The area of each circle is $\pi r^2 = \pi(4)^2 = 16\pi$. So, the correct answer is Ⓒ.

EXAMPLE 4 Real-Life Application

A boat travels x miles per hour from Hillsborough Bay to the Sunshine Skyway Bridge. The boat travels 4 miles per hour faster on the return trip. How far does the boat travel each way?

The speed of the boat on the return trip is $(x + 4)$ miles per hour.

Distance from bay to bridge = Distance from bridge to bay

$2x = 1.5(x + 4)$	Write an equation.
$2x = 1.5x + 6$	Use Distributive Property.
$\underline{-1.5x} \qquad \underline{-1.5x}$	Subtract $1.5x$ from each side.
$0.5x = 6$	Simplify.
$\dfrac{0.5x}{0.5} = \dfrac{6}{0.5}$	Divide each side by 0.5.
$x = 12$	Simplify.

∴ The boat travels 12 miles per hour for 2 hours on the initial trip. So, it travels 24 miles each way.

On Your Own

4. **WHAT IF?** In Example 3, the diameter of the purple circle is $3x$. What is the area of each circle?

5. A boat travels x miles per hour from one island to another island in 2.5 hours. The boat travels 5 miles per hour faster on the return trip of 2 hours. What is the distance between the islands?

1.3 Exercises

Check It Out
Help with Homework
BigIdeasMath.com

Vocabulary and Concept Check

1. **WRITING** Is $x = 3$ a solution of the equation $3x - 5 = 4x - 9$? Explain.

2. **OPEN-ENDED** Write an equation that has variables on both sides and has a solution of -3.

Practice and Problem Solving

The value of the figure's surface area is equal to the value of the figure's volume. Find the value of *x*.

3. 

4. 

5. 

Solve the equation. Check your solution.

(1) (2)

6. $m - 4 = 2m$

7. $3k - 1 = 7k + 2$

8. $6.7x = 5.2x + 12.3$

9. $-24 - \frac{1}{8}p = \frac{3}{8}p$

10. $12(2w - 3) = 6w$

11. $2(n - 3) = 4n + 1$

12. $2(4z - 1) = 3(z + 2)$

13. $0.1x = 0.2(x + 2)$

14. $\frac{1}{6}d + \frac{2}{3} = \frac{1}{4}(d - 2)$

15. **ERROR ANALYSIS** Describe and correct the error in solving the equation.

✗
$$3x - 4 = 2x + 1$$
$$3x - 4 - 2x = 2x + 1 - 2x$$
$$x - 4 = 1$$
$$x - 4 + 4 = 1 - 4$$
$$x = -3$$

16. **TRAIL MIX** The equation $4.05p + 14.40 = 4.50(p + 3)$ represents the number p of pounds of peanuts you need to make trail mix. How many pounds of peanuts do you need for the trail mix?

17. **CARS** Write and solve an equation to find the number of miles you must drive to have the same cost for each of the car rentals.

$15 plus $0.50 per mile

$25 plus $0.25 per mile

A polygon is *regular* if each of its sides has the same length. Find the perimeter of the regular polygon.

18.

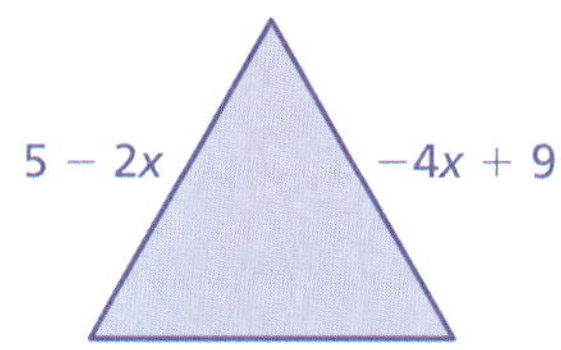

19.

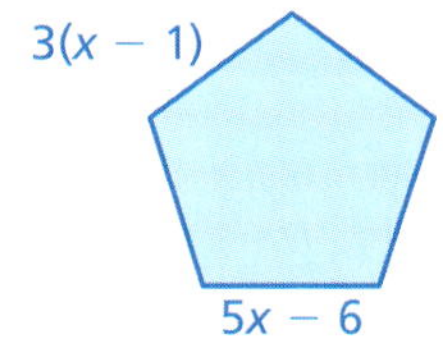

20.

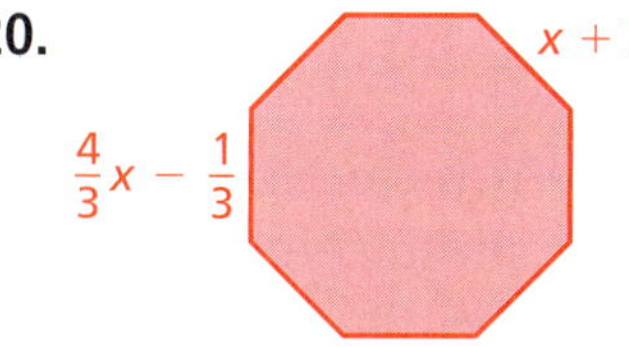

21. POSTAGE The cost of mailing a DVD in an envelope by express mail is equal to the cost of mailing a DVD in a box by priority mail. What is the weight of the DVD with its packing material? Round your answer to the nearest hundredth.

	Packing Material	Priority Mail	Express Mail
Box	\$2.25	\$2.50 per lb	\$8.50 per lb
Envelope	\$1.10	\$2.50 per lb	\$8.50 per lb

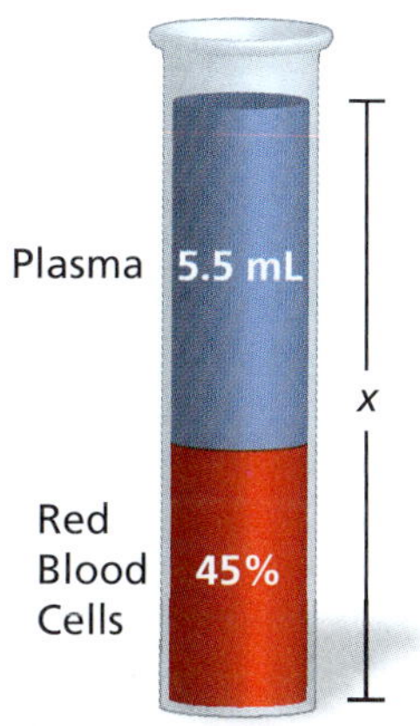

22. REASONING Would you solve the equation $0.25x + 7 = \frac{1}{3}x - 8$ using fractions or decimals? Explain.

23. BLOOD SAMPLE The amount of red blood cells in a blood sample is equal to the total amount in the sample minus the amount of plasma. What is the total amount x of blood drawn?

24. BEACHES In Florida, about 59% of the beaches are eroding and about 338 miles of beaches are not eroding. About how many miles of beaches are in Florida? Round your answer to the nearest whole number.

25. **Geometry** The perimeter of the square is equal to the perimeter of the triangle. What are the side lengths of each figure?

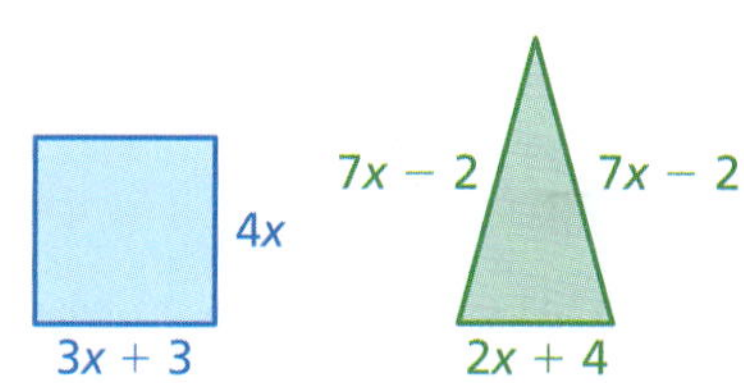

Fair Game Review What you learned in previous grades & lessons

Find the volume of the figure. Use 3.14 for π. *(Skills Review Handbook)*

26.

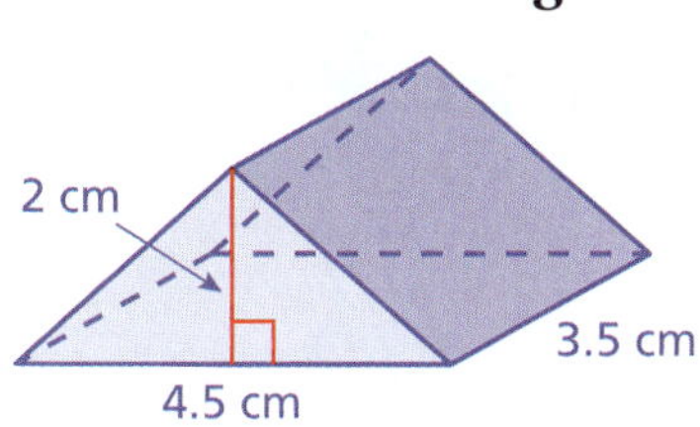

27.

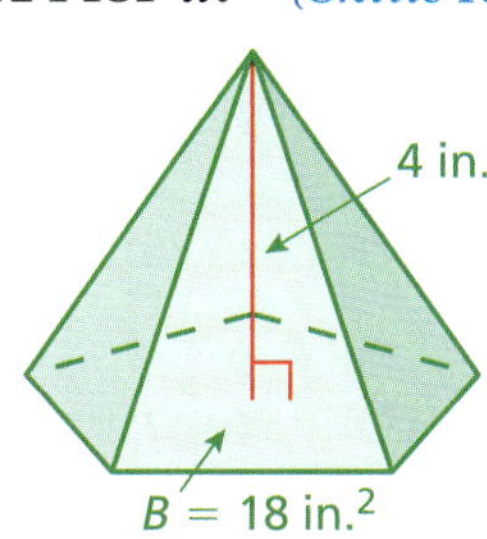

28.

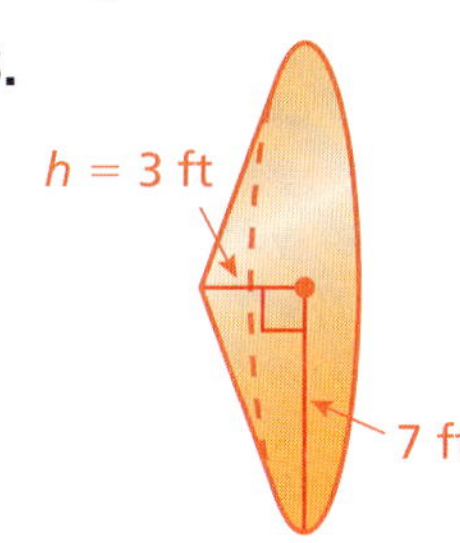

29. MULTIPLE CHOICE A car travels 480 miles on 15 gallons of gasoline. How many miles does the car travel per gallon? *(Section 1.1)*

Ⓐ 28 mi/gal Ⓑ 30 mi/gal Ⓒ 32 mi/gal Ⓓ 35 mi/gal

1 Study Help

Check It Out
Graphic Organizer
BigIdeasMath.com

You can use a **Y chart** to compare two topics. List differences in the branches and similarities in the base of the Y. Here is an example of a Y chart that compares solving simple equations to solving multi-step equations.

Solving Simple Equations

- You can solve the equation in one step.

Solving Multi-Step Equations

- You must use more than one step to solve the equation.
- Undo the operations in the reverse order of the order of operations.

Both

- As necessary, use the Addition, Subtraction, Multiplication, and Division Properties of Equality to solve for the variable.
- The variable can end up on either side of the equation.
- It is always a good idea to check your solution.

On Your Own

Make a Y chart to help you study and compare these topics.

1. solving equations with the variable on one side and solving equations with variables on both sides
2. solving multi-step equations and solving equations with variables on both sides

After you complete this chapter, make Y charts for the following topics.

3. solving multi-step equations and rewriting literal equations
4. converting meters to feet and converting feet to meters
5. converting one unit to another and converting one rate to another

"I made a Y chart to compare and contrast yours and Fluffy's characteristics."

1.1–1.3 Quiz

Check It Out
Progress Check
BigIdeasMath.com

Solve the equation. Check your solution. *(Section 1.1)*

1. $-\frac{1}{2} = y - 1$

2. $-3\pi + w = 2\pi$

3. $1.2m = 0.6$

Solve the equation. Check your solution. *(Section 1.2)*

4. $-4k + 17 = 1$

5. $\frac{1}{4}z + 8 = 12$

Find the value of *x*. Then find the angle measures of the polygon. *(Section 1.2)*

6.

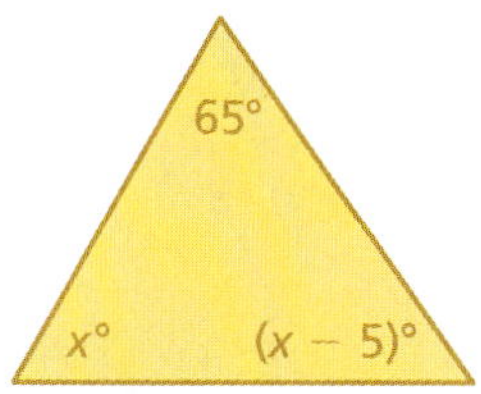

Sum of angle measures: 180°

7.

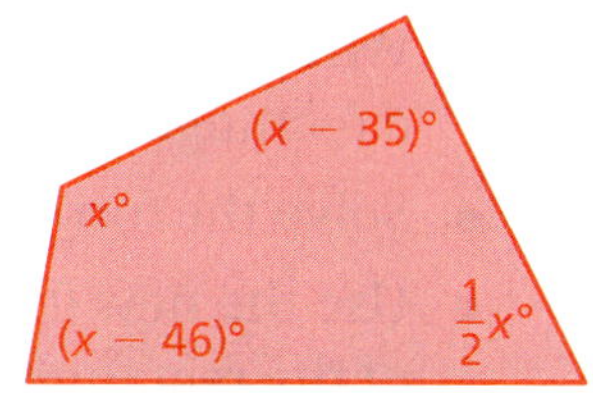

Sum of angle measures: 360°

Solve the equation. Check your solution. *(Section 1.3)*

8. $2(x + 4) = -5x + 1$

9. $\frac{1}{2}s = 4s - 21$

10. JEWELER The equation $P = 2.5m + 35$ represents the price P (in dollars) of a bracelet, where m is the cost of the materials (in dollars). The price of a bracelet is \$115. What is the cost of the materials? *(Section 1.2)*

11. PASTURE A 455-foot fence encloses a pasture. What is the length of each side of the pasture? *(Section 1.2)*

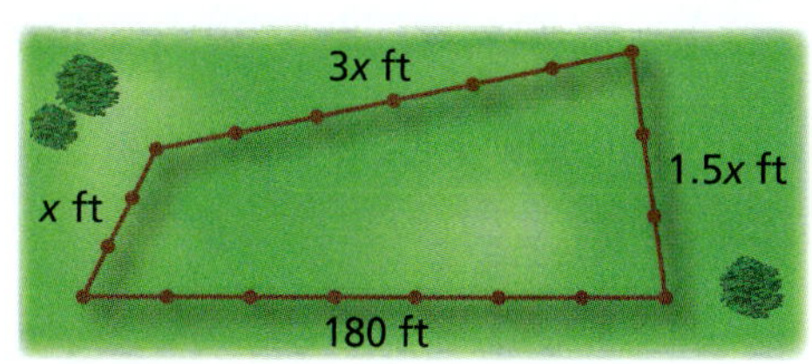

12. POSTERS A machine prints 230 movie posters each hour. Write and solve an equation to find the number of hours it takes the machine to print 1265 posters. *(Section 1.1)*

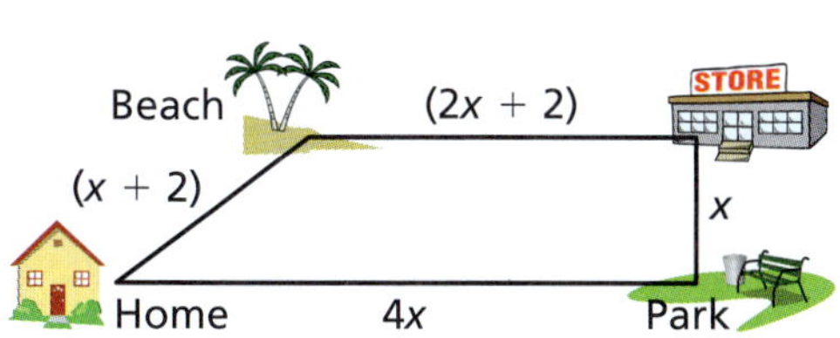

13. ROUTES From your home, the route to the store that passes the beach is 2 miles shorter than the route to the store that passes the park. What is the length of each route? *(Section 1.3)*

1.4 Rewriting Equations and Formulas

STATE STANDARDS
MA.8.A.4.1
MA.8.G.5.1

Essential Question How can you use a formula for one measurement to write a formula for a different measurement?

1 ACTIVITY: Using Perimeter and Area Formulas

Work with a partner.

a.
- Write a formula for the perimeter P of a rectangle.
- Solve the formula for w.
- Use the new formula to find the width of the rectangle.

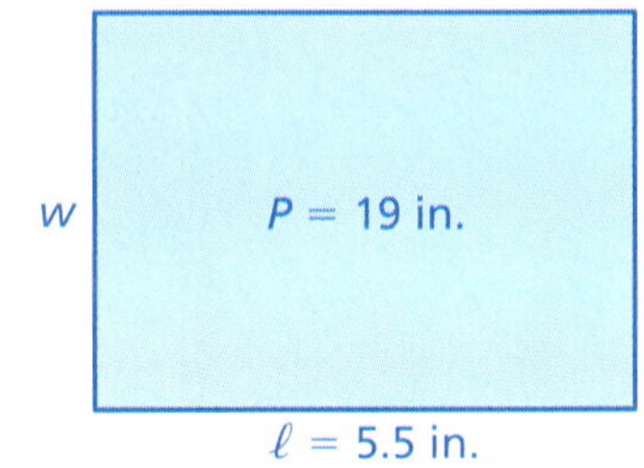

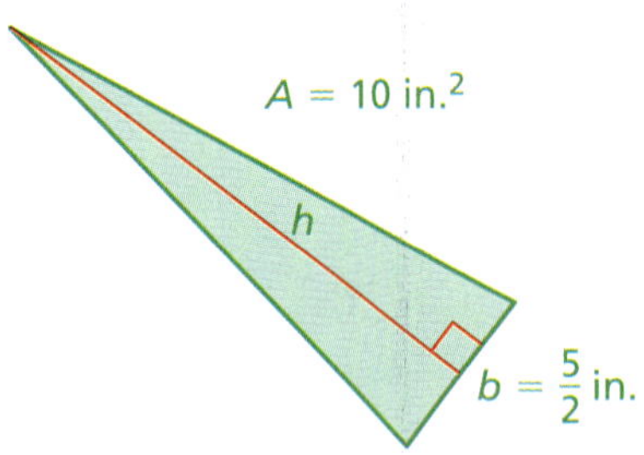

b.
- Write a formula for the area A of a triangle.
- Solve the formula for h.
- Use the new formula to find the height of the triangle.

c.
- Write a formula for the circumference C of a circle.
- Solve the formula for r.
- Use the new formula to find the radius of the circle.

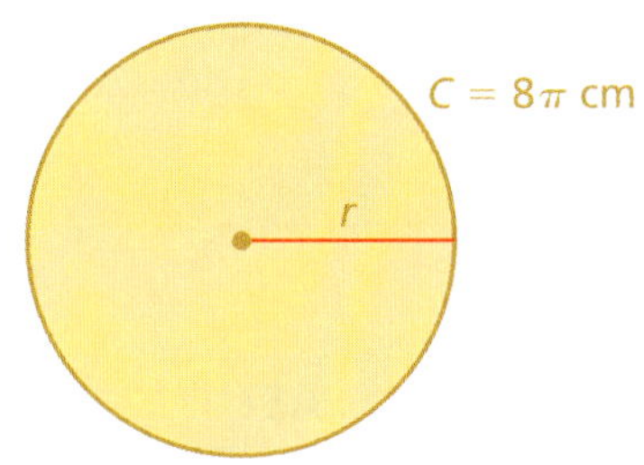

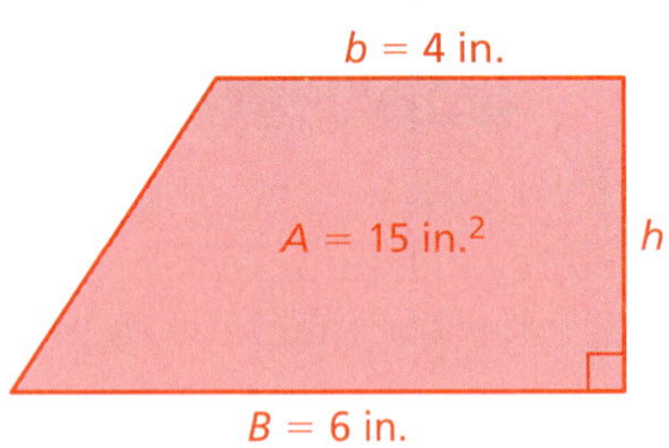

d.
- Write a formula for the area A of a trapezoid.
- Solve the formula for h.
- Use the new formula to find the height of the trapezoid.

e.
- Write a formula for the area A of a parallelogram.
- Solve the formula for h.
- Use the new formula to find the height of the parallelogram.

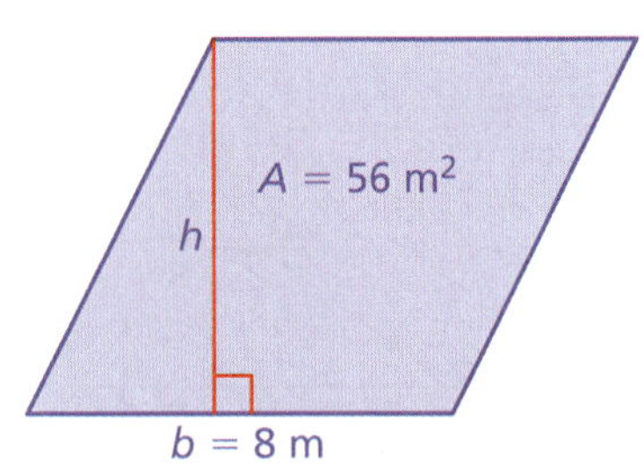

2 ACTIVITY: Using Volume Formulas

Work with a partner.

a. • Write a formula for the volume V of a prism.
• Solve the formula for h.
• Use the new formula to find the height of the prism.

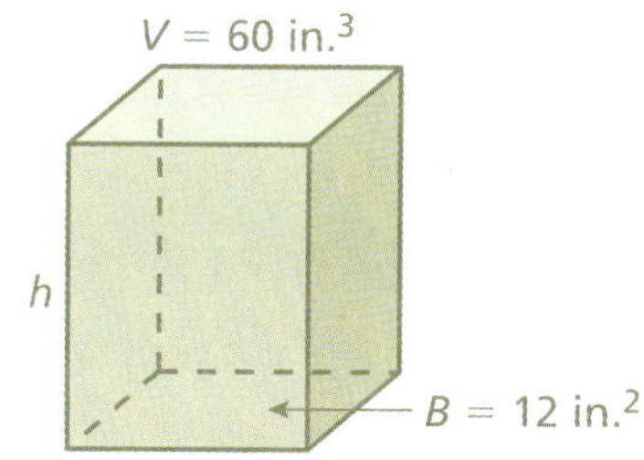

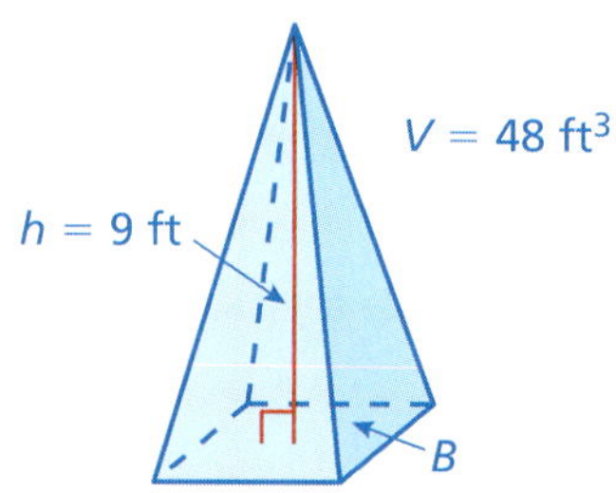

b. • Write a formula for the volume V of a pyramid.
• Solve the formula for B.
• Use the new formula to find the area of the base of the pyramid.

c. • Write a formula for the volume V of a cylinder.
• Solve the formula for B.
• Use the new formula to find the area of the base of the cylinder.

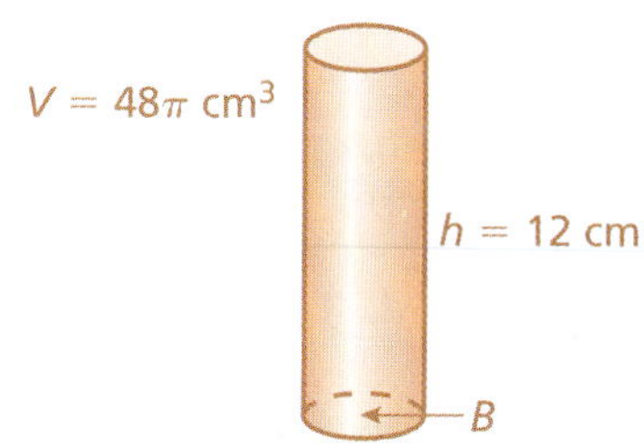

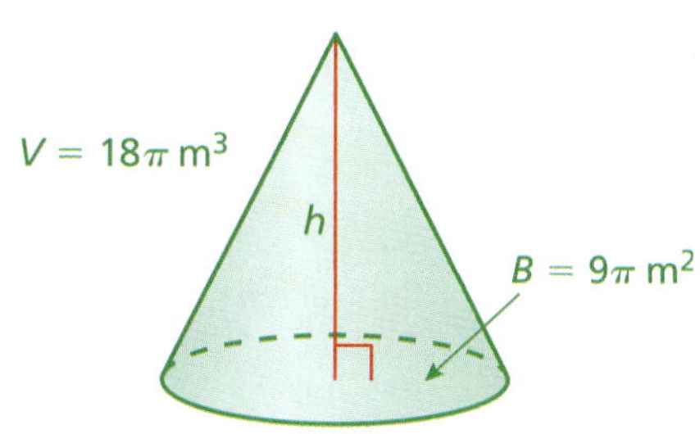

d. • Write a formula for the volume V of a cone.
• Solve the formula for h.
• Use the new formula to find the height of the cone.

What Is Your Answer?

3. **IN YOUR OWN WORDS** How can you use a formula for one measurement to write a formula for a different measurement? Give an example that is different from the examples on these two pages.

Practice Use what you learned about rewriting equations and formulas to complete Exercises 3 and 4 on page 28.

1.4 Lesson

Check It Out Lesson Tutorials BigIdeasMath.com

Key Vocabulary
literal equation, *p. 26*

An equation that has two or more variables is called a **literal equation**. To rewrite a literal equation, solve for one variable in terms of the other variable(s).

EXAMPLE 1 Rewriting an Equation

Solve the equation $2y + 5x = 6$ for y.

$2y + 5x = 6$ — Write the equation.

Undo the addition. → $2y + 5x - 5x = 6 - 5x$ — Subtract $5x$ from each side.

$2y = 6 - 5x$ — Simplify.

Undo the multiplication. → $\frac{2y}{2} = \frac{6 - 5x}{2}$ — Divide each side by 2.

$y = 3 - \frac{5}{2}x$ — Simplify.

On Your Own

Now You're Ready Exercises 5–10

Solve the equation for y.

1. $5y - x = 10$
2. $4x - 4y = 1$
3. $12 = 6x + 3y$

EXAMPLE 2 Rewriting a Formula

The formula for the surface area S of a cone is $S = \pi r^2 + \pi r\ell$. Solve the formula for the slant height ℓ.

Remember
A *formula* shows how one variable is related to one or more other variables. A formula is a type of literal equation.

$S = \pi r^2 + \pi r\ell$ — Write the equation.

$S - \pi r^2 = \pi r^2 - \pi r^2 + \pi r\ell$ — Subtract πr^2 from each side.

$S - \pi r^2 = \pi r\ell$ — Simplify.

$\frac{S - \pi r^2}{\pi r} = \frac{\pi r\ell}{\pi r}$ — Divide each side by πr.

$\frac{S - \pi r^2}{\pi r} = \ell$ — Simplify.

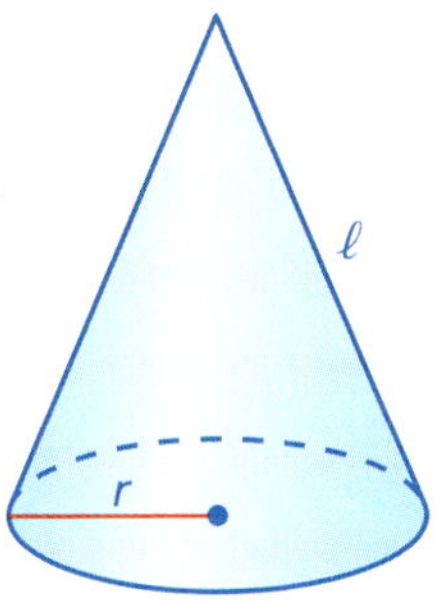

On Your Own

Now You're Ready Exercises 14–19

Solve the formula for the red variable.

4. Area of rectangle: $A = bh$
5. Simple interest: $I = Prt$
6. Surface area of cylinder: $S = 2\pi r^2 + 2\pi rh$

Key Idea

Temperature Conversion

A formula for converting from degrees Fahrenheit F to degrees Celsius C is

$$C = \frac{5}{9}(F - 32).$$

EXAMPLE 3 Rewriting the Temperature Formula

Solve the temperature formula for F.

$C = \frac{5}{9}(F - 32)$ — Write the temperature formula.

Use the reciprocal. → $\frac{9}{5} \cdot C = \frac{9}{5} \cdot \frac{5}{9}(F - 32)$ — Multiply each side by $\frac{9}{5}$, the reciprocal of $\frac{5}{9}$.

$\frac{9}{5}C = F - 32$ — Simplify.

Undo the subtraction. → $\frac{9}{5}C + 32 = F - 32 + 32$ — Add 32 to each side.

$\frac{9}{5}C + 32 = F$ — Simplify.

∴ The rewritten formula is $F = \frac{9}{5}C + 32$.

EXAMPLE 4 Real-Life Application

Which has the greater temperature?

Convert the Celsius temperature of lightning to Fahrenheit.

$F = \frac{9}{5}C + 32$ — Write the rewritten formula from Example 3.

$= \frac{9}{5}(30{,}000) + 32$ — Substitute 30,000 for C.

$= 54{,}032$ — Simplify.

∴ Because 54,032 °F is greater than 11,000 °F, lightning has the greater temperature.

On Your Own

7. Lake Kissimmee has a water temperature of 85 °F.
Lake Okeechobee has a water temperature of 30 °C.
Which lake has the greater water temperature? Explain.

1.4 Exercises

Check It Out
Help with Homework
BigIdeasMath.com

Vocabulary and Concept Check

1. **VOCABULARY** Is $-2x = \frac{3}{8}$ a literal equation? Explain.

2. **DIFFERENT WORDS, SAME QUESTION** Which is different? Find "both" answers.

Solve $4x - 2y = 6$ for y.	Solve $6 = 4x - 2y$ for y.
Solve $4x - 2y = 6$ for y in terms of x.	Solve $4x - 2y = 6$ for x in terms of y.

Practice and Problem Solving

3. **a.** Write a formula for the area A of a triangle.
   b. Solve the formula for b.
   c. Use the new formula to find the base of the triangle.

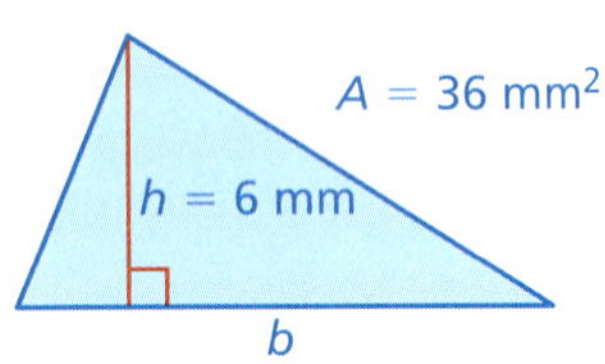

4. **a.** Write a formula for the volume V of a prism.
   b. Solve the formula for B.
   c. Use the new formula to find the area of the base of the prism.

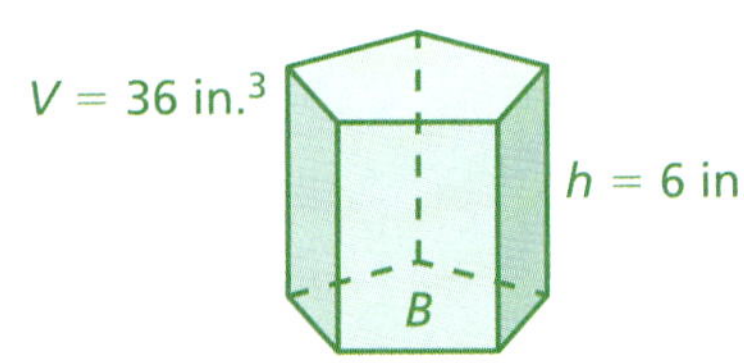

Solve the equation for y.

1 **5.** $\frac{1}{3}x + y = 4$ **6.** $3x + \frac{1}{5}y = 7$ **7.** $6 = 4x + 9y$

8. $\pi = 7x - 2y$ **9.** $4.2x - 1.4y = 2.1$ **10.** $6y - 1.5x = 8$

11. **ERROR ANALYSIS** Describe and correct the error in rewriting the equation.

$$2x - y = 5$$
$$y = -2x + 5$$

12. **TEMPERATURE** The formula $K = C + 273.15$ converts temperatures from Celsius C to Kelvin K.
    a. Solve the formula for C.
    b. Convert 300 K to Celsius.

13. **INTEREST** The formula for simple interest is $I = Prt$.
    a. Solve the formula for t.
    b. Use the new formula to find the value of t in the table.

I	\$75
P	\$500
r	5%
t	

Solve the equation for the red variable.

14. $d = rt$ **15.** $e = mc^2$ **16.** $R - C = P$

17. $A = \frac{1}{2}\pi w^2 + 2\ell w$ **18.** $B = 3\frac{V}{h}$ **19.** $g = \frac{1}{6}(w + 40)$

20. WRITING Why is it useful to rewrite a formula in terms of another variable?

21. TEMPERATURE The formula $K = \frac{5}{9}(F - 32) + 273.15$ converts temperatures from Fahrenheit F to Kelvin K.

a. Solve the formula for F.

b. The freezing point of water is 273.15 Kelvin. What is this temperature in Fahrenheit?

c. The temperature of dry ice is −78.5 °C. Which is colder, dry ice or liquid nitrogen?

Navy Pier Ferris Wheel

22. FERRIS WHEEL The Navy Pier Ferris Wheel in Chicago has a circumference that is 56% of the circumference of the first Ferris wheel built in 1893.

a. What is the radius of the Navy Pier Ferris Wheel?

b. What was the radius of the first Ferris wheel?

c. The first Ferris wheel took 9 minutes to make a complete revolution. How fast was the wheel moving?

23. Geometry The formula for the volume of a sphere is $V = \frac{4}{3}\pi r^3$. Solve the formula for r^3. Use guess, check, and revise to find the radius of the sphere.

$V = 381.51$ in.^{3} $\vdash r \dashv$

Fair Game Review What you learned in previous grades & lessons

Multiply. *(Skills Review Handbook)*

24. $5 \times \frac{3}{4}$ **25.** $2.4 \times \frac{8}{3}$ **26.** $\frac{1}{4} \times \frac{3}{2} \times \frac{8}{9}$ **27.** $25 \times \frac{3}{5} \times \frac{1}{12}$

28. MULTIPLE CHOICE Which of the following is not equivalent to $\frac{3}{4}$? *(Skills Review Handbook)*

Ⓐ 0.75 Ⓑ 3 : 4 Ⓒ 75% Ⓓ 4 : 3

1.5 Converting Units of Measure

STATE STANDARDS
MA.8.G.5.1

Essential Question How can you convert from one measurement system to another?

1 ACTIVITY: Converting Units of Measure

Work with a partner. Copy and complete the table. Describe the pattern in the completed table.

		Perimeter, in. to ft ratio	Area, in.^2 to ft^2 ratio
Sample: a.	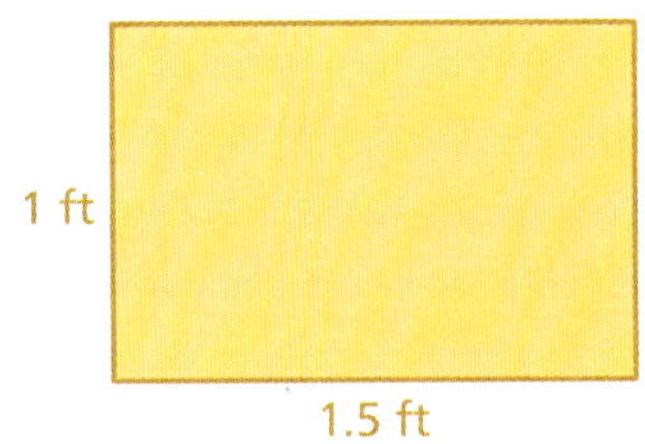	$\frac{60 \text{ in.}}{5 \text{ ft}} = \frac{12 \text{ in.}}{1 \text{ ft}}$	$\frac{216 \text{ in.}^2}{1.5 \text{ ft}^2} = \frac{144 \text{ in.}^2}{1 \text{ ft}^2}$
b.	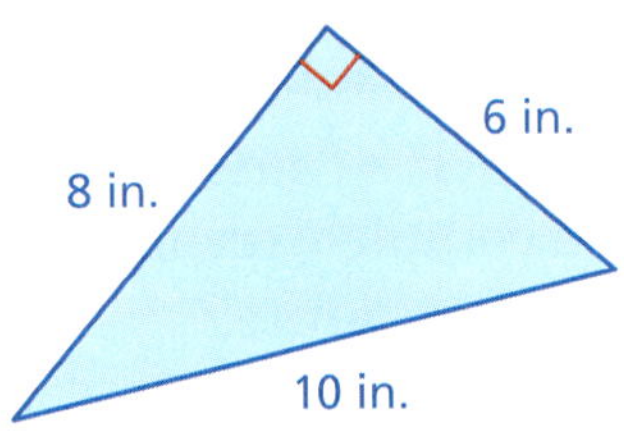	$\frac{\square \text{ in.}}{\square \text{ ft}} = \square$	$\frac{\square \text{ in.}^2}{\square \text{ ft}^2} = \square$
c.	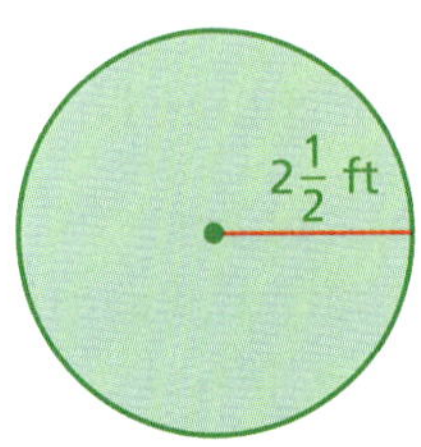	$\frac{\square \text{ in.}}{\square \text{ ft}} = \square$	$\frac{\square \text{ in.}^2}{\square \text{ ft}^2} = \square$
d.	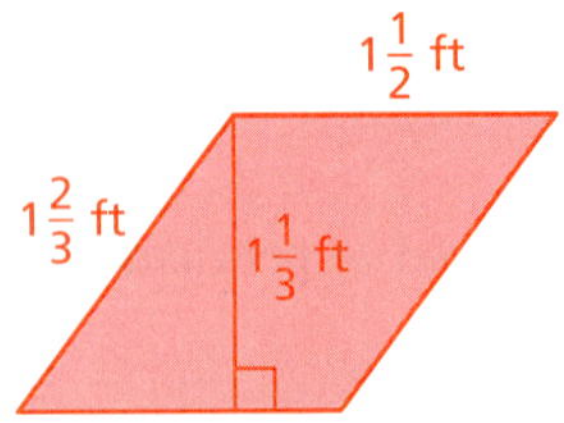	$\frac{\square \text{ in.}}{\square \text{ ft}} = \square$	$\frac{\square \text{ in.}^2}{\square \text{ ft}^2} = \square$
e.	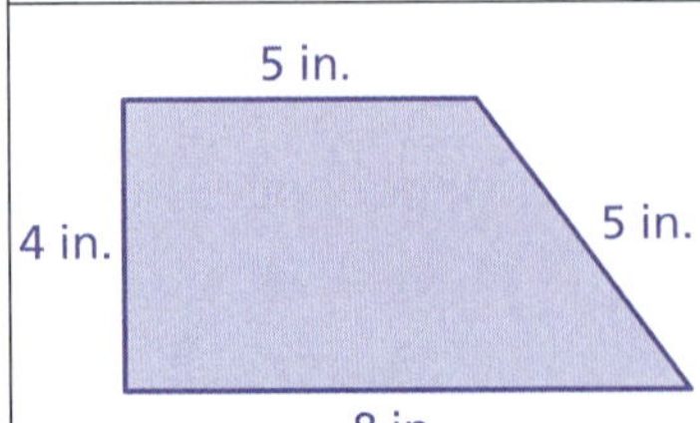	$\frac{\square \text{ in.}}{\square \text{ ft}} = \square$	$\frac{\square \text{ in.}^2}{\square \text{ ft}^2} = \square$

2 ACTIVITY: Comparing Units of Measure

Work with a partner. Name the units for each pair of "rulers".

a.

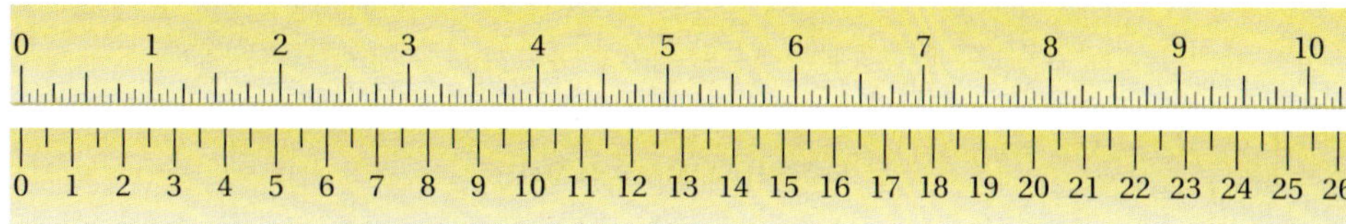

b.

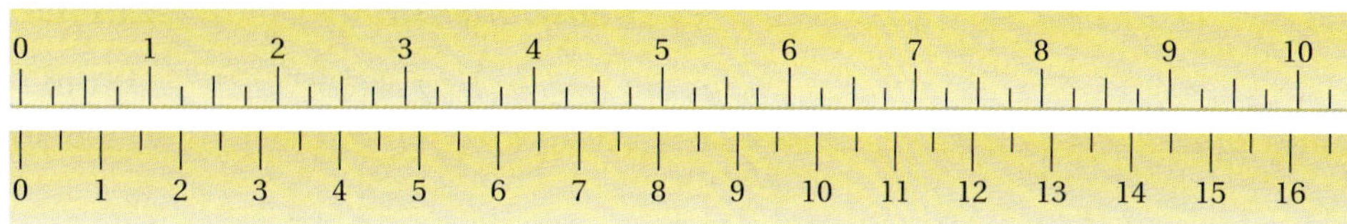

c.

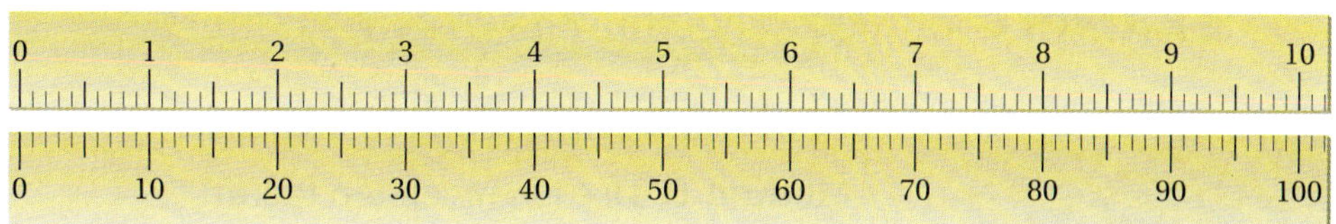

3 ACTIVITY: Puzzle

Who is correct, Fred or Sam? Explain your reasoning.

John said, "We left camp this morning, and walked 1 mile due south. Then, we saw a polar bear and turned due east and ran 1 kilometer. Finally, we turned due north and walked 1 mile and ended back at camp."

Fred said, "That is not possible!"

Sam explained, "Yes it is. And I know exactly where the camp was."

What Is Your Answer?

4. **IN YOUR OWN WORDS** How can you convert from one measurement system to another? The examples on these two pages are measurements of length and area. Describe a conversion between two types of temperature units.

Practice Use what you learned about converting units of measure to complete Exercises 4–6 on page 35.

1.5 Lesson

Check It Out
Lesson Tutorials
BigIdeasMath.com

Key Vocabulary
conversion factor, *p. 32*

To convert between customary and metric units, multiply by one or more *conversion factors*.

Key Idea

Conversion Factor

A **conversion factor** is a rate that equals 1.

	Relationship	*Conversion factors*
Example	$1 \text{ m} \approx 3.28 \text{ ft}$	$\frac{1 \text{ m}}{3.28 \text{ ft}}$ and $\frac{3.28 \text{ ft}}{1 \text{ m}}$

EXAMPLE 1 Converting Between Systems

Convert 20 centimeters to inches.

Method 1: Use a conversion factor.

1 in. ≈ 2.54 cm

$$20 \text{ cm} \cdot \frac{1 \text{ in.}}{2.54 \text{ cm}} \approx 7.87 \text{ in.}$$

So, 20 centimeters is about 7.87 inches.

Method 2: Use a proportion.

Let x be the number of inches equivalent to 20 centimeters.

$$\frac{1}{2.54} \approx \frac{x}{20}$$ (inches / centimeters) Write a proportion.

$$20 \cdot \frac{1}{2.54} \approx 20 \cdot \frac{x}{20}$$ Multiply each side by 20.

$$7.87 \approx x$$ Simplify.

So, 20 centimeters is about 7.87 inches.

On Your Own

Now You're Ready
Exercises 7–15

Copy and complete the statement.

1. $10 \text{ qt} \approx$ ___ L
2. $4 \text{ km} \approx$ ___ mi
3. $18 \text{ in.} \approx$ ___ cm
4. $84 \text{ lb} \approx$ ___ kg

Multi-Language Glossary at BigIdeasMath.com.

EXAMPLE 2 Converting a Rate: Changing One Unit

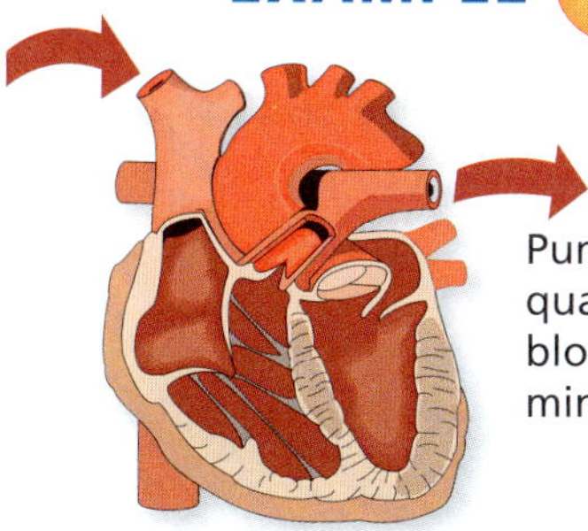

Convert the pumping rate of the human heart to liters per minute.

1 qt ≈ 0.95 L

$$\frac{5\ \cancel{qt}}{1\ min} \cdot \frac{0.95\ L}{1\ \cancel{qt}} \approx \frac{4.75\ L}{1\ min}$$

∴ The rate of 5 quarts per minute is about 4.75 liters per minute.

EXAMPLE 3 Converting a Speed: Changing Both Units

Convert the speed of the zip liner to feet per second.

$$\frac{15\ \cancel{mi}}{1\ \cancel{h}}\left(\frac{5280\ ft}{1\ \cancel{mi}}\right)\left(\frac{1\ \cancel{h}}{3600\ sec}\right) = \frac{15 \cdot 5280\ ft}{3600\ sec}$$

$$= \frac{79{,}200\ ft}{3600\ sec}$$

$$= \frac{22\ ft}{1\ sec}$$

1 mi = 5280 ft

1 h = 3600 sec

∴ The speed of the zip liner is 22 feet per second.

Study Tip

Here is another way to convert the rate in Example 3.

- Write the rate as $15\ \frac{\text{miles}}{\text{hour}}$.
- Substitute 5280 feet for miles and 3600 seconds for hour.

On Your Own

Now You're Ready
Exercises 18–23

5. An oil tanker is leaking oil at a rate of 300 gallons per minute. Convert this rate to gallons per second.
6. A tennis ball travels at a speed of 120 miles per hour. Convert this rate to feet per second.
7. A kite boarder travels at a speed of 10 meters per second. Convert this rate to kilometers per minute.

Key Idea

Converting Units for Area or Volume

To convert units for area, multiply the area by the *square* of the conversion factor.

To convert units for volume, multiply the volume by the *cube* of the conversion factor.

EXAMPLE 4 Converting Units for Area

The painting *Fracture* by Benedict Gibson has an area of 2880 square inches. What is the area of the painting in square feet?

> **Remember**
> Area is measured in *square units*. Volume is measured in *cubic units*.

1 ft = 12 in.

$$2880 \text{ in.}^2 = 2880 \text{ in.}^2 \cdot \left(\frac{1 \text{ ft}}{12 \text{ in.}}\right)^2$$

$$= 2880 \cancel{\text{in.}^2} \cdot \frac{1 \text{ ft}^2}{144 \cancel{\text{in.}^2}}$$

$$= \frac{2880}{144} \text{ ft}^2$$

$$= 20 \text{ ft}^2$$

The area of the painting is 20 square feet.

EXAMPLE 5 Converting Units for Volume

What is the volume of the cylinder in cubic centimeters?

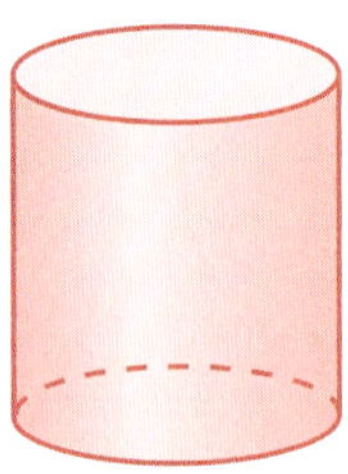

Volume = 80 m^3

1 m = 100 cm

$$80 \text{ m}^3 = 80 \text{ m}^3 \cdot \left(\frac{100 \text{ cm}}{1 \text{ m}}\right)^3$$

$$= 80 \cancel{\text{m}^3} \cdot \frac{1{,}000{,}000 \text{ cm}^3}{1 \cancel{\text{m}^3}}$$

$$= 80{,}000{,}000 \text{ cm}^3$$

The volume is 80,000,000 cubic centimeters.

On Your Own

Now You're Ready
Exercises 30–35

8. The painting *Busy Market* by Haitian painter Frantz Petion has an area of 6 square feet. What is the area of the painting in square inches?

9. The volume of a pyramid is 50 cubic centimeters. What is the volume of the pyramid in cubic millimeters?

1.5 Exercises

Check It Out
Help with Homework
BigIdeasMath.com

Vocabulary and Concept Check

1. **VOCABULARY** Is $\frac{10\text{ mm}}{1\text{ cm}}$ a conversion factor? Explain.

2. **WRITING** Describe how to convert 2 liters per hour to milliliters per second.

3. **WHICH ONE DOESN'T BELONG?** Which measurement does *not* belong with the other three? Explain your reasoning.

100 in.	254 cm	6.25 ft	2.54 m

Practice and Problem Solving

Find the perimeter in feet and in yards.

4. 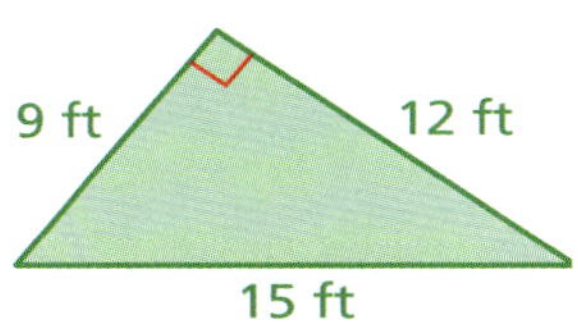

5. 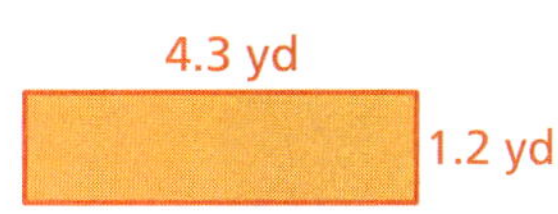

6. 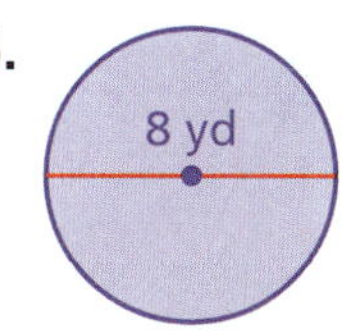

Copy and complete the statement.

1 7. 12 L $\approx$ ___ qt

8. 14 m $\approx$ ___ ft

9. 4 ft $\approx$ ___ m

10. 64 lb $\approx$ ___ kg

11. 0.3 km $\approx$ ___ mi

12. 75 in. $\approx$ ___ cm

13. 17 kg $\approx$ ___ lb

14. 15 cm $\approx$ ___ in.

15. 9 mi $\approx$ ___ km

16. **ERROR ANALYSIS** Describe and correct the error in converting the units.

$$8\text{ L} \approx 8\text{ L} \cdot \frac{0.95\text{ qt}}{1\text{ L}} = 8\cancel{\text{L}} \cdot \frac{0.95\text{ qt}}{1\cancel{\text{L}}} = 7.6\text{ qt}$$

17. **BRIDGE** The Sunshine Skyway Bridge connects St. Petersburg and Bradenton across Tampa Bay.

   a. How high above the water is the roadway in meters?

   b. The bridge spans 29,040 feet. What is the length in kilometers?

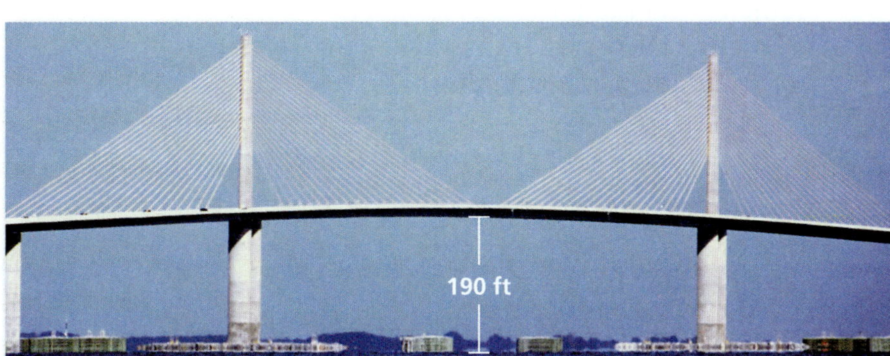

Copy and complete the statement.

(2) (3) **18.** $\dfrac{13\text{ km}}{\text{h}} \approx \dfrac{\square\text{ mi}}{\text{h}}$ **19.** $\dfrac{22\text{ L}}{\text{min}} = \dfrac{\square\text{ L}}{\text{h}}$ **20.** $\dfrac{63\text{ mi}}{\text{h}} = \dfrac{\square\text{ mi}}{\text{sec}}$

21. $\dfrac{3\text{ km}}{\text{min}} \approx \dfrac{\square\text{ mi}}{\text{h}}$ **22.** $\dfrac{17\text{ gal}}{\text{h}} \approx \dfrac{\square\text{ qt}}{\text{min}}$ **23.** $\dfrac{6\text{ cm}}{\text{min}} = \dfrac{\square\text{ m}}{\text{sec}}$

24. **SNAIL** What is the speed of the snail in kilometers per hour?

25. **BLOOD DRIVE** A donor gives blood at a rate of 0.125 pint per minute. What is the rate in milliliters per second?

26. **POSTER** A poster of your favorite band has a width of 15 inches. You have a space on your wall that has a width of 1.2 feet. Will the poster fit? Explain.

27. **ROME** Ancient Romans used the *talent* and the *mina* as measures of weight. How many minas are in 100 pounds?

28. **FUEL EFFICIENCY** The fuel efficiency standard for cars in Japan is 20 kilometers per liter. The fuel efficiency standard for cars in the United States is 28 miles per gallon. Which country has a greater fuel efficiency standard?

29. **BIRDS** The table shows the flying speeds of several birds.

a. Which bird is the fastest? Which is the slowest?

b. The peregrine falcon has a dive speed of 322 kilometers per hour. Is the dive speed of the peregrine falcon faster than the flying speed of any of the birds? Explain.

Bird	Speed
Spine-tailed swift	2843.2 m/min
Spur-winged goose	129.1 ft/sec
Eider duck	31.3 m/sec
Mallard	65 mi/h

Copy and complete the statement.

(4) **30.** $4\text{ yd}^2 = \square\text{ ft}^2$ **31.** $0.00125\text{ mi}^2 = \square\text{ ft}^2$ **32.** $30\text{ mm}^2 = \square\text{ cm}^2$

(5) **33.** $3\text{ km}^3 = \square\text{ m}^3$ **34.** $2\text{ ft}^3 = \square\text{ in.}^3$ **35.** $420\text{ cm}^3 = \square\text{ m}^3$

36. **FIREWOOD** The volume of a cord of firewood is 128 cubic feet. What is the volume of a cord of firewood in cubic yards? Round your answer to the nearest hundredth.

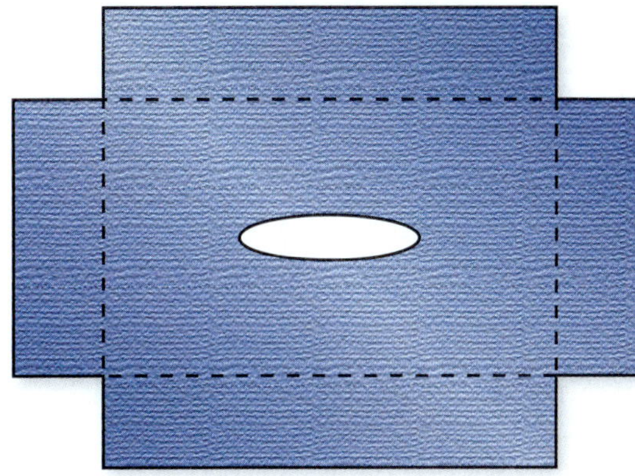

1 cm : 2 in.

37. **FABRIC COVER** The pattern shows the dimensions of a fabric cover for a tissue box.

 a. Use the pattern and a ruler to estimate the volume of a tissue box.

 b. The volume of a tissue is about 0.864 cubic inch. About how many tissues are in a box?

38. **PROJECT** The table shows the currencies of four countries.

 a. **RESEARCH** Use the Internet to find the exchange rates for the currencies listed in the table.

 b. How much of each currency would you receive in exchange for $20?

Country	Currency	Value in Dollars
United States	Dollar	$1
Japan	Yen	
Spain	Euro	
Great Britain	Pound	

39. **SHAMPOO** Your shampoo bottle is 80% full. The total volume of the bottle is 565 cubic centimeters. How much shampoo have you used? Write your answer in cubic millimeters.

40. **Critical Thinking** You make Floating Island Punch for a party.

 a. Your punch bowl holds 6 liters. Will the punch fit into the bowl? Explain.

 b. One milliliter is equal to 1 cubic centimeter. Can you store the punch in a container with a capacity of 3000 cubic centimeters?

Recipe for: Floating Island Punch
From: Mom

2 cups water	2 cups sugar
1 L ginger ale	1 L carbonated water
1 pt orange sherbet	4 cups ice
$1\frac{1}{2}$ cups frozen lemonade concentrate	$1\frac{1}{2}$ cups frozen orange juice concentrate

Fair Game Review What you learned in previous grades & lessons

Plot the ordered pair in a coordinate plane. *(Skills Review Handbook)*

41. (1, 2)
42. (0, −3)
43. (−6, −8)
44. (−5, 7)

45. **MULTIPLE CHOICE** Which equation shows direct variation? *(Skills Review Handbook)*

 (A) $y = 2x + 1$ (B) $y = \frac{1}{3}x$ (C) $4 = xy$ (D) $y = 2x - 1$

1.4–1.5 Quiz

Check It Out
Progress Check
BigIdeasMath.com

Solve the equation for *y*. *(Section 1.4)*

1. $6x - 3y = 9$

2. $8 = 2y - 10x$

Solve the formula for the red variable. *(Section 1.4)*

3. Volume of a cylinder: $V = \pi r^2 h$

4. Area of a trapezoid: $A = \frac{1}{2}h(b + B)$

Copy and complete the statement. *(Section 1.5)*

5. $30 \text{ cm} \approx$ ▭ in.

6. $0.7 \text{ km} \approx$ ▭ mi

7. $15 \text{ L} \approx$ ▭ qt

8. $3000 \text{ cm}^2 =$ ▭ m^2

9. $45 \text{ in.}^3 =$ ▭ ft^3

10. $50 \text{ yd}^3 =$ ▭ ft^3

11. TEMPERATURE In which city is the water temperature higher? *(Section 1.4)*

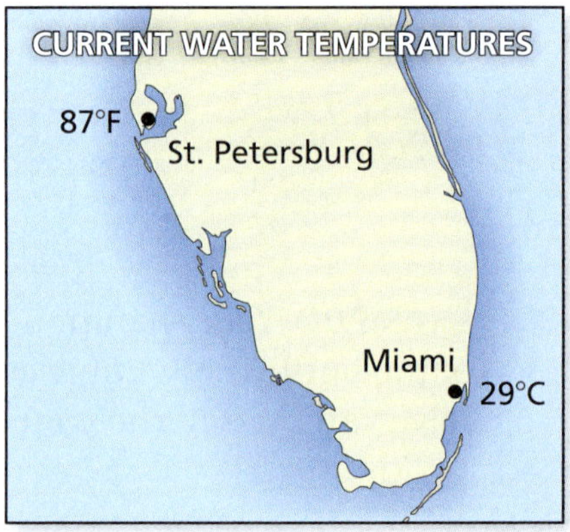

12. INTEREST The formula for simple interest I is $I = Prt$. Solve the formula for the interest rate r. What is the interest rate r if the principal P is \$1500, the time t is 2 years, and the interest earned I is \$900? *(Section 1.4)*

13. RIVER The Loxahatchee River Canoe Trail is 8 miles long. How long is the trail in kilometers? *(Section 1.5)*

14. MEDICINE A grain is a measure of weight equal to 0.06 gram. An aspirin tablet weighs 5 grains. How many milligrams does the tablet weigh? *(Section 1.5)*

15. HAWK What is the speed of the hawk in kilometers per hour? *(Section 1.5)*

1 Chapter Review

Check It Out
Vocabulary Help
BigIdeasMath.com

Review Key Vocabulary

literal equation, *p. 26*

conversion factor, *p. 32*

Review Examples and Exercises

1.1 Solving Simple Equations *(pp. 2–9)*

The *boiling point* of a liquid is the temperature at which the liquid becomes a gas. The boiling point of mercury is about $\frac{41}{200}$ of the boiling point of lead. Write and solve an equation to find the boiling point of lead.

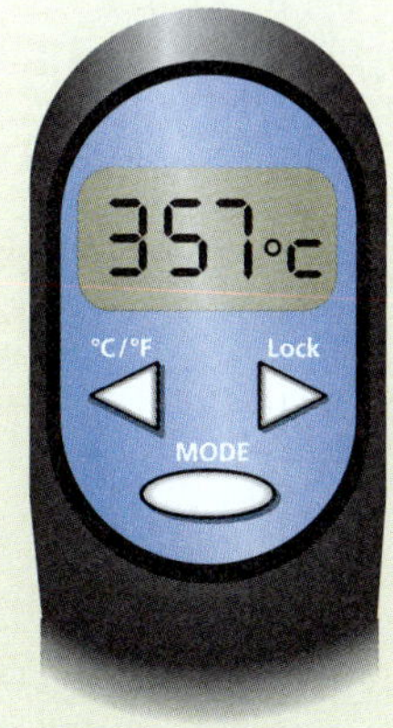

Mercury 357°C

Let x be the boiling point of lead.

$\frac{41}{200}x = 357$ — Write the equation.

$\frac{200}{41} \cdot \left(\frac{41}{200}x\right) = \frac{200}{41} \cdot 357$ — Multiply each side by $\frac{200}{41}$.

$x \approx 1741$ — Simplify.

The boiling point of lead is about 1741°C.

Exercises

Solve the equation. Check your solution.

1. $y + 8 = -11$
2. $3.2 = -0.4n$
3. $-\frac{t}{4} = -3\pi$

1.2 Solving Multi-Step Equations *(pp. 10–15)*

Solve $-14x + 28 + 6x = -44$.

$-14x + 28 + 6x = -44$ — Write the equation.

$-8x + 28 = -44$ — Combine like terms.

$\underline{-28} \quad \underline{-28}$ — Subtract 28 from each side.

$-8x = -72$ — Simplify.

$\frac{-8x}{-8} = \frac{-72}{-8}$ — Divide each side by −8.

$x = 9$ — Simplify.

The solution is $x = 9$.

Exercises

Find the value of x. Then find the angle measures of the polygon.

4.

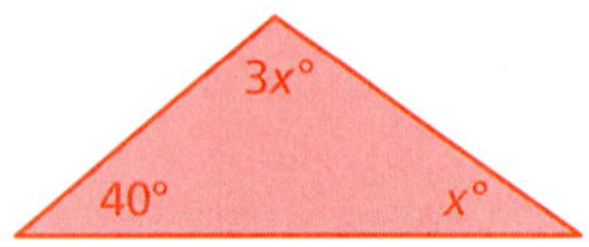

Sum of angle measures: 180°

5.

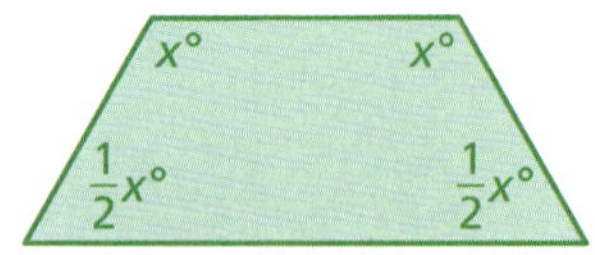

Sum of angle measures: 360°

6.

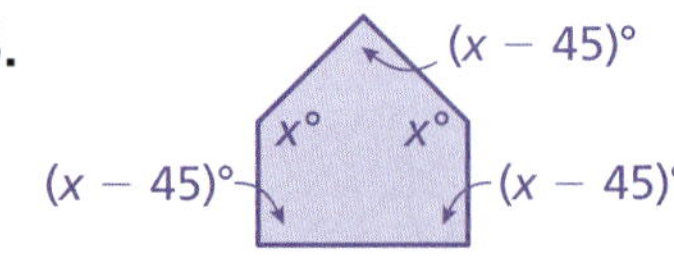

Sum of angle measures: 540°

1.3 Solving Equations with Variables on Both Sides *(pp. 16–21)*

Solve $3(x - 4) = -2(4 - x)$.

$3(x - 4) = -2(4 - x)$	Write the equation.
$3x - 12 = -8 + 2x$	Use Distributive Property.
$-2x \qquad\qquad -2x$	Subtract $2x$ from each side.
$x - 12 = -8$	Simplify.
$+12 \qquad +12$	Add 12 to each side.
$x = 4$	Simplify.

The solution is $x = 4$.

Exercises

Solve the equation. Check your solution.

7. $5m - 1 = 4m + 5$

8. $3(5p - 3) = 5(p - 1)$

9. $\frac{2}{5}n + \frac{1}{10} = \frac{1}{2}(n + 4)$

1.4 Rewriting Equations and Formulas *(pp. 24–29)*

The equation for a line in slope-intercept form is $y = mx + b$. Solve the equation for x.

$y = mx + b$	Write the equation.
$y - b = mx + b - b$	Subtract b from each side.
$y - b = mx$	Simplify.
$\frac{y - b}{m} = \frac{mx}{m}$	Divide each side by m.
$\frac{y - b}{m} = x$	Simplify.

So, $x = \frac{y - b}{m}$.

Exercises

10. **a.** The formula $F = \frac{9}{5}(K - 273.15) + 32$ converts a temperature from Kelvin K to Fahrenheit F. Solve the formula for K.

b. Convert 240°F to Kelvin K. Round your answer to the nearest hundredth.

11. **a.** Write the formula for the area A of a trapezoid.

b. Solve the formula for h.

c. Use the new formula to find the height h of the trapezoid.

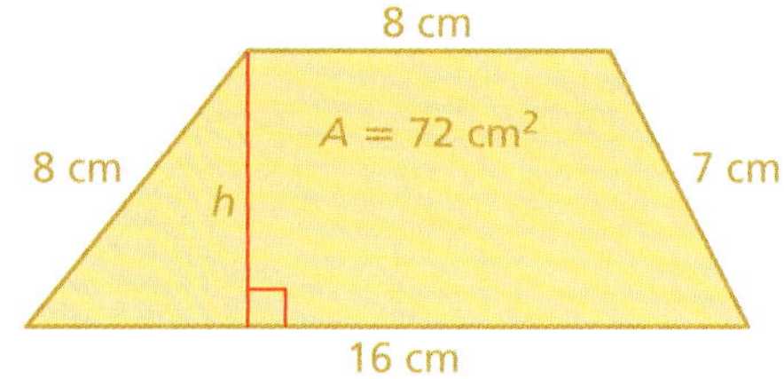

1.5 Converting Units of Measure (pp. 30–37)

Convert the speed of the train to miles per hour.

$$\frac{5 \text{ km}}{1 \text{ min}}\left(\frac{60 \text{ min}}{1 \text{ h}}\right)\left(\frac{0.6 \text{ mi}}{1 \text{ km}}\right)$$

60 min = 1 h   0.6 mi ≈ 1 km

$$\frac{5 \cancel{\text{km}}}{1 \cancel{\text{min}}}\left(\frac{60 \cancel{\text{min}}}{1 \text{ h}}\right)\left(\frac{0.6 \text{ mi}}{1 \cancel{\text{km}}}\right) = \frac{5 \cdot 60 \cdot 0.6 \text{ mi}}{1 \text{ h}}$$

$$= \frac{180 \text{ mi}}{1 \text{ h}}$$

∴ The speed of the train is 180 miles per hour.

Exercises

Copy and complete the statement.

12. 17 in. ≈ ___ cm   **13.** 80 L ≈ ___ qt   **14.** 4800 ft ≈ ___ km

15. **BASEBALL** A baseball pitch is clocked at 93 miles per hour. Convert this rate to feet per second.

16. **POOL** A community pool is filling at a rate of 40 liters per minute. Convert this rate to gallons per hour.

17. **GEOMETRY** What is the volume of the cube in cubic meters?

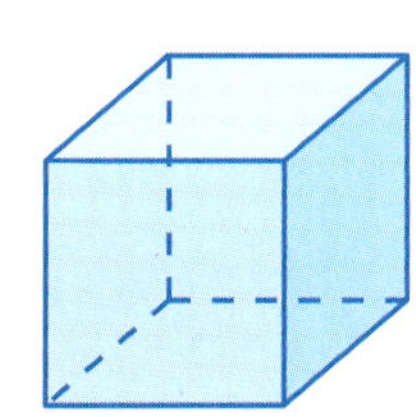

Volume = 34,300 cm^3

1 Chapter Test

Solve the equation. Check your solution.

1. $4 + y = 9.5$
2. $x - 3\pi = 5\pi$
3. $3.8n - 13 = 1.4n + 5$

Find the value of x. Then find the angle measures of the polygon.

4.

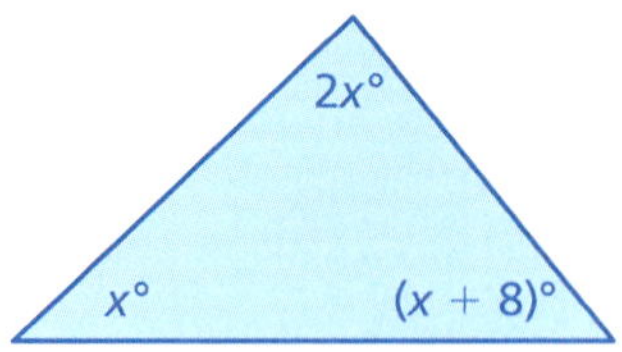

Sum of angle measures: 180°

5.

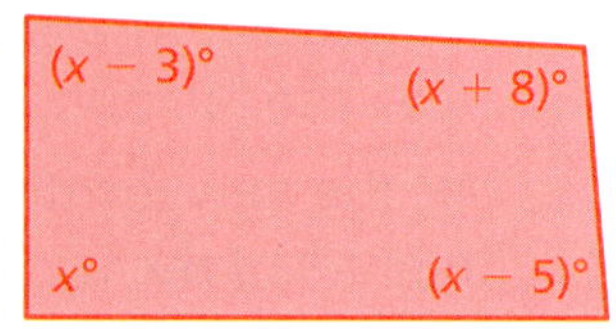

Sum of angle measures: 360°

Solve the equation for y.

6. $1.2x - 4y = 28$
7. $0.5 = 0.4y - 0.25x$

Solve the formula for the red variable.

8. Perimeter of a rectangle: $P = 2\ell + 2w$
9. Distance formula: $d = rt$

Copy and complete the statement.

10. 27 in. ≈ ___ cm
11. 14 mi ≈ ___ km
12. 12 L ≈ ___ qt

13. **BASKETBALL** Your basketball team wins a game by 13 points. The opposing team scores 72 points. Write and solve an equation to find your team's score.

14. **AQUARIUM** You want to buy the aquarium that has the greater volume. Which of the aquariums should you buy? Explain.

15. **AIRPORT** Runway 17 at an airport is 5020 feet long. What is the length of the runway in meters? Round your answer to the nearest whole number.

16. **JOBS** Your profit for mowing lawns this week is \$24. You are paid \$8 per hour and you paid \$40 for gas for the lawnmower. How many hours did you work this week?

1 Standardized Test Practice

Test-Taking Strategy
Solve Directly or Eliminate Choices

"You can eliminate A and D. Then, solve directly to determine that the correct answer is B."

1. Which value of x makes the equation true?

$$4x = 32$$

A. 8

B. 28

C. 36

D. 128

2. A taxi ride costs \$3 plus \$2 for each mile driven. When you rode in a taxi, the total cost was \$39. This can be modeled by the equation below, where m represents the number of miles driven.

$$2m + 3 = 39$$

How long was your taxi ride?

F. 72 mi

G. 34 mi

H. 21 mi

I. 18 mi

3. One fluid ounce (fl oz) contains 6 teaspoons. You add $\frac{1}{4}$ teaspoon of vanilla each time you make hot chocolate. How many times can you make hot chocolate using the bottle of vanilla shown?

A. 6

B. 24

C. 40

D. 96

4. A bicyclist is riding at a speed of 900 feet per minute. How many yards does the bicyclist ride in 1 second?

5. The formula below relates distance, rate, and time.

$$d = rt$$

Solve this formula for t.

F. $t = dr$

G. $t = \frac{d}{r}$

H. $t = d - r$

I. $t = \frac{r}{d}$

6. What could be the first step to solve the equation shown below?

$$3x + 5 = 2(x + 7)$$

A. Combine $3x$ and 5.

B. Multiply x by 2 and 7 by 2.

C. Subtract x from $3x$.

D. Subtract 5 from 7.

7. You work as a sales representative. You earn \$400 per week plus 5% of your total sales for the week.

Think Solve Explain

Part A Last week, you had total sales of \$5000. Find your total earnings. Show your work.

Part B One week, you earned \$1350. Let s represent your total sales that week. Write an equation that could be used to find s.

Part C Using your equation from Part B, find s. Show all steps clearly.

8. In ten years, Maria will be 39 years old. Let m represent Maria's age today. Which equation can be used to find m?

F. $m = 39 + 10$

G. $m - 10 = 39$

H. $m + 10 = 39$

I. $10m = 39$

9. Which value of y makes the equation below true?

$$3y + 8 = 7y + 11$$

A. -4.75

B. -0.75

C. 0.75

D. 4.75

10. The equation below is used to convert a Fahrenheit temperature F to its equivalent Celsius temperature C.

$$C = \frac{5}{9}(F - 32)$$

Which formula can be used to convert a Celsius temperature to its equivalent Fahrenheit temperature?

F. $F = \frac{5}{9}(C - 32)$

G. $F = \frac{9}{5}(C + 32)$

H. $F = \frac{9}{5}C + \frac{32}{5}$

I. $F = \frac{9}{5}C + 32$

11. You have already saved \$35 for a new cell phone. You need \$175 in all. You think you can save \$10 per week. At this rate, how many more weeks will you need to save money before you can buy the new cell phone?

12. The cube shown below has edge lengths of 1 foot. What is the volume of the cube?

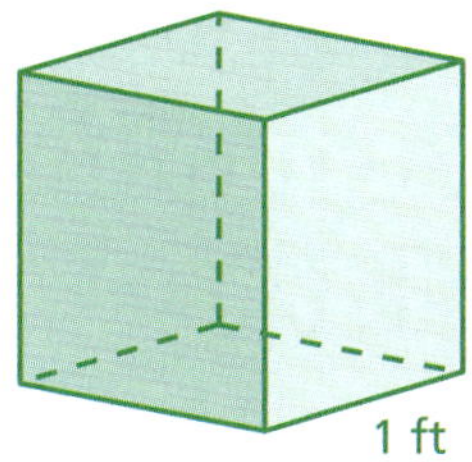

A. 12 in.^3

B. 36 in.^3

C. 144 in.^3

D. 1728 in.^3

13. Which value of x makes the equation below true?

$$6(x - 3) = 4x - 7$$

F. -5.5

G. -2

H. 1.1

I. 5.5

14. The drawing below shows equal weights on two sides of a balance scale.

What can you conclude from the drawing?

A. A mug weighs one-third as much as a trophy.

B. A mug weighs one-half as much as a trophy.

C. A mug weighs twice as much as a trophy.

D. A mug weighs three times as much as a trophy.

2 Graphing Linear Equations and Linear Systems

"Okay Descartes, stand on the *y*-axis and try to intercept the pass when I throw."

"Here's an easy example of a line with a slope of 1."

"You eat one mouse treat the first day. Two treats the second day. And so on. Get it?"

What You Learned Before

"I estimate that we are on a slope of about −0.625. What do you think?"

Evaluating Expressions Using Order of Operations (Grade 6)

Example 1 **Evaluate $2xy + 3(x + y)$ when $x = 4$ and $y = 7$.**

$2xy + 3(x + y) = 2(4)(7) + 3(4 + 7)$ Substitute 4 for x and 7 for y.

$= 8(7) + 3(4 + 7)$ Use order of operations.

$= 56 + 3(11)$ Simplify.

$= 56 + 33$ Multiply.

$= 89$ Add.

Try It Yourself

Evaluate the expression when $a = \frac{1}{4}$ and $b = 6$.

1. $-8ab$ **2.** $16a^2 - 4b$ **3.** $\frac{5b}{32a^2}$ **4.** $12a + (b - a - 4)$

Plotting Points (Grade 7)

Example 2 **Write the ordered pair that corresponds to Point U.**

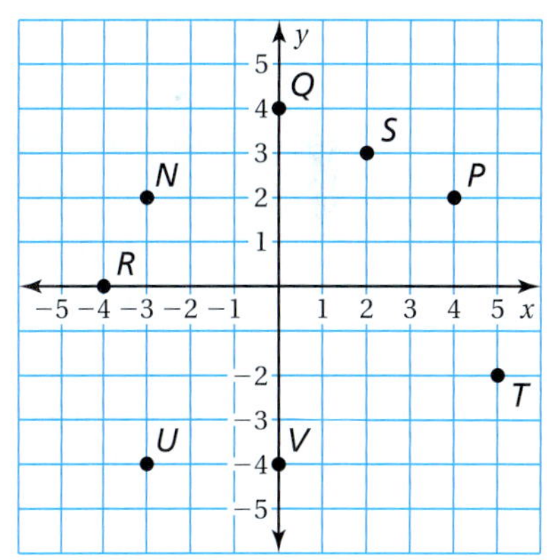

Point U is 3 units to the left of the origin and 4 units down. So, the x-coordinate is -3 and the y-coordinate is -4.

The ordered pair $(-3, -4)$ corresponds to Point U.

Example 3 **Which point is located at $(5, -2)$?**

Start at the origin. Move 5 units right and 2 units down.

Point T is located at $(5, -2)$.

Try It Yourself

Use the graph to answer the question.

5. Write the ordered pair that corresponds to Point Q.

6. Write the ordered pair that corresponds to Point P.

7. Which point is located at $(-4, 0)$?

8. Which point is located in Quadrant II?

2.1 Graphing Linear Equations

STATE STANDARDS
MA.8.A.1.1

Essential Question How can you recognize a linear equation? How can you draw its graph?

1 ACTIVITY: Graphing a Linear Equation

Work with a partner.

a. Use the equation $y = \frac{1}{2}x + 1$ to complete the table. (Choose any two x-values and find the y-values.)

	Solution Points	
x		
$y = \frac{1}{2}x + 1$		

b. Write the two ordered pairs given by the table. These are called **solution points** of the equation.

c. Plot the two solution points. Draw a line *exactly* through the two points.

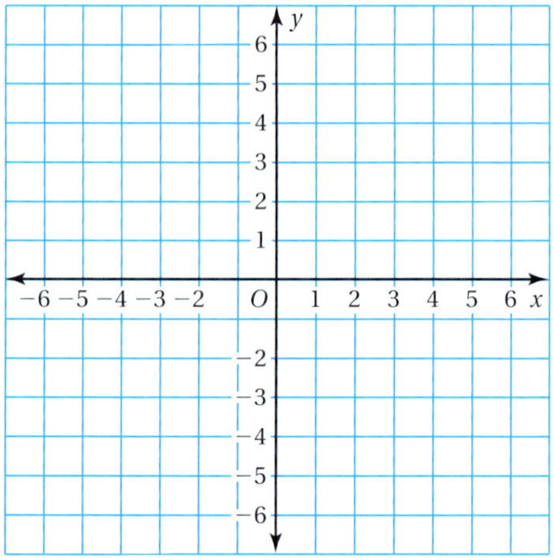

d. Find a different point on the line. Check that this point is a solution point of the equation $y = \frac{1}{2}x + 1$.

e. **GENERALIZE** Do you think it is true that *any* point on the line is a solution point of the equation $y = \frac{1}{2}x + 1$? Explain.

f. Choose five additional x-values for the table. (Choose positive and negative x-values.) Plot the five corresponding solution points. Does each point lie on the line?

	Solution Points				
x					
$y = \frac{1}{2}x + 1$					

g. **GENERALIZE** Do you think it is true that *any* solution point of the equation $y = \frac{1}{2}x + 1$ is a point on the line? Explain.

h. **THE MEANING OF A WORD** Why is $y = ax + b$ called a *linear equation*?

2 HISTORY: Analytic Geometry

René Descartes (1596–1650)

René Descartes was a French philosopher, scientist, and mathematician.

Up until the time of Descartes, *algebra* and *geometry* were separate fields of mathematics. Descartes's invention of the coordinate plane was of huge importance to mathematics. For the first time, people could "see" solutions of equations. No longer did people have to work with algebra from a purely symbolic point of view.

Descartes's combination of geometry and algebra is called *analytic* (or algebraic) *geometry*. One of the main discoveries in analytic geometry is that all of the important types of graphs (lines, parabolas, circles, ellipses, and so on) can be represented by simple algebraic equations.

Within a few dozen years, other mathematicians were able to discover all of *calculus*, a field of mathematics that is of great value in business, science, and engineering.

In this book, you will study lines. In Algebra 1 and Algebra 2, you will study many other types of equations.

Line: $y = ax + b$

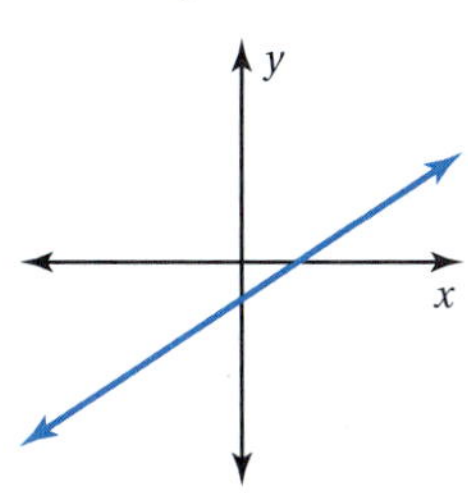

Parabola: $y = ax^2 + b$

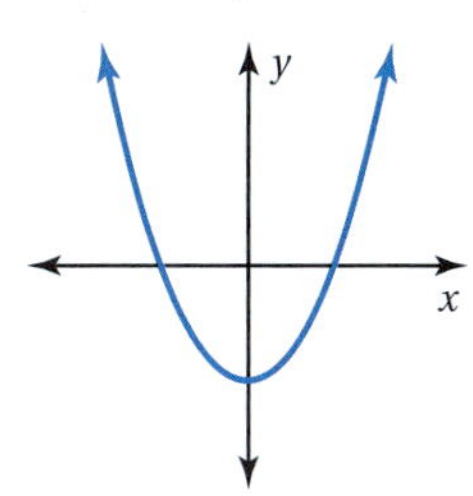

Circle: $x^2 + y^2 = r^2$

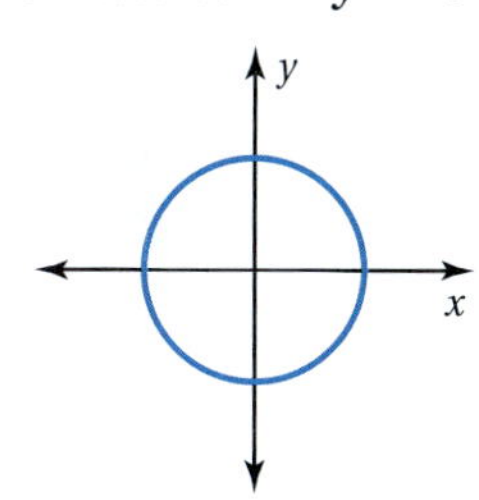

What Is Your Answer?

3. **IN YOUR OWN WORDS** How can you recognize a linear equation? How can you draw its graph? Write an equation that is linear. Write an equation that is *not* linear.

4. Are you a visual learner? Most people can learn mathematics more easily when they see "pictures" of the mathematics. Why do you think Descartes's invention was important to mathematics?

Practice Use what you learned about graphing linear equations to complete Exercises 3 and 4 on page 52.

2.1 Lesson

Check It Out
Lesson Tutorials
BigIdeasMath.com

Key Vocabulary
linear equation, *p. 50*
solution of a linear equation, *p. 50*

Remember
An ordered pair (x, y) is used to locate a point in a coordinate plane.

Key Idea

Linear Equations

A **linear equation** is an equation whose graph is a line. The points on the line are **solutions** of the equation.

You can use a graph to show the solutions of a linear equation. The graph below is for the equation $y = x + 1$.

x	y	(x, y)
-1	0	$(-1, 0)$
0	1	$(0, 1)$
2	3	$(2, 3)$

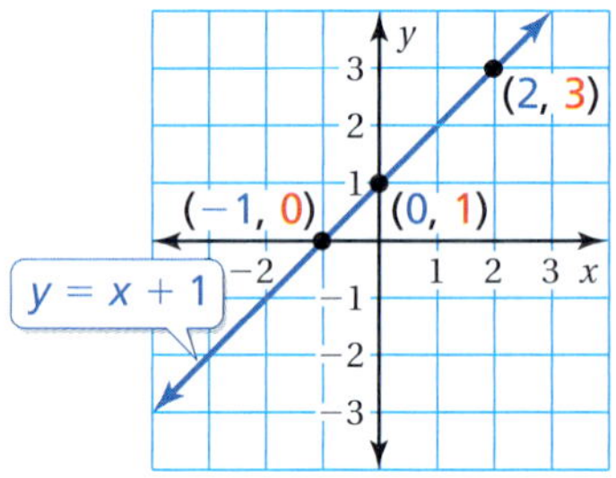

EXAMPLE 1 Graphing a Linear Equation

Graph $y = -2x + 1$.

Step 1: Make a table of values.

x	$y = -2x + 1$	y	(x, y)
-1	$y = -2(-1) + 1$	3	$(-1, 3)$
0	$y = -2(0) + 1$	1	$(0, 1)$
2	$y = -2(2) + 1$	-3	$(2, -3)$

Step 2: Plot the ordered pairs.

Step 3: Draw a line through the points.

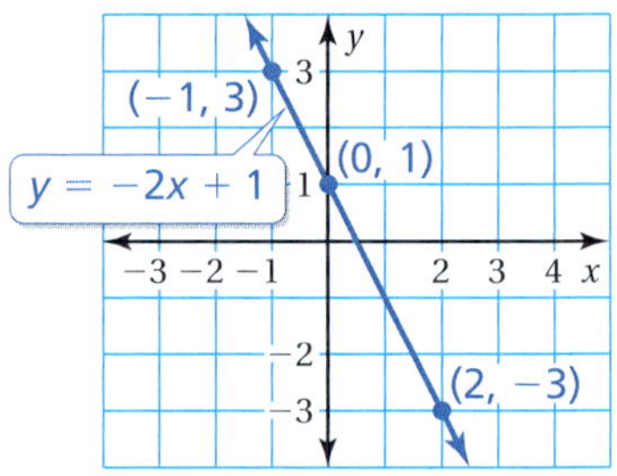

Key Idea

Graphing a Horizontal Line

The graph of $y = a$ is a horizontal line passing through $(0, a)$.

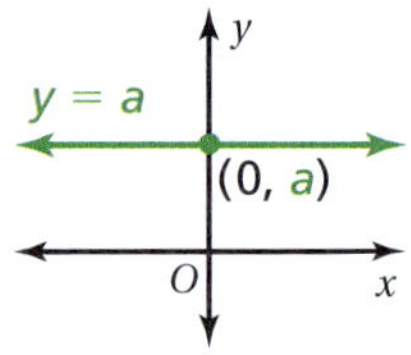

Multi-Language Glossary at BigIdeasMath.com.

EXAMPLE 2 Graphing a Horizontal Line

Graph $y = -3$.

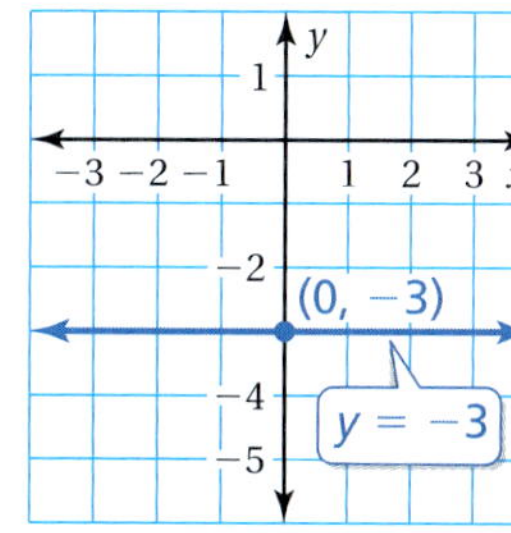

The graph of $y = -3$ is a horizontal line passing through $(0, -3)$.

Plot $(0, -3)$. Draw a horizontal line through the point.

On Your Own

Now You're Ready
Exercises 5–13

Graph the linear equation.

1. $y = 3x$
2. $y = -\frac{1}{2}x + 2$
3. $y = \pi$
4. $y = -1.5$

EXAMPLE 3 Real-Life Application

The wind speed y (in miles per hour) of a tropical storm is $y = 2x + 66$, where x is the number of hours after the storm enters the Gulf of Mexico.

a. Graph the equation.

b. When does the storm become a hurricane?

A tropical storm becomes a hurricane when wind speeds are at least 74 miles per hour.

a. Make a table of values.

x	$y = 2x + 66$	y	(x, y)
0	$y = 2(0) + 66$	66	(0, 66)
1	$y = 2(1) + 66$	68	(1, 68)
2	$y = 2(2) + 66$	70	(2, 70)
3	$y = 2(3) + 66$	72	(3, 72)

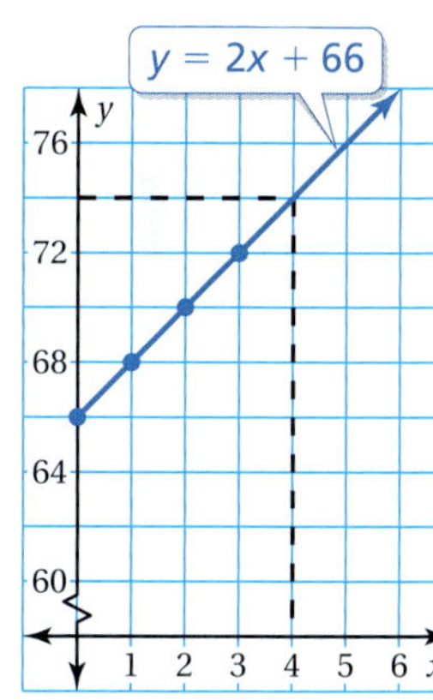

Plot the ordered pairs and draw a line through the points.

b. From the graph, you can see that $y = 74$ when $x = 4$.

So, the storm becomes a hurricane 4 hours after it enters the Gulf of Mexico.

On Your Own

5. **WHAT IF?** In Example 3, the wind speed of the storm is $y = 1.5x + 62$. When does the storm become a hurricane?

2.1 Exercises

Check It Out
Help with Homework
BigIdeasMath.com

Vocabulary and Concept Check

1. **VOCABULARY** What type of graph represents the solutions of the equation $y = 2x + 3$?

2. **WHICH ONE DOESN'T BELONG?** Which equation does *not* belong with the other three? Explain your reasoning.

$y = 0.5x - 0.2$ $\qquad 4x + 3 = y \qquad y = x^2 + 6 \qquad \frac{3}{4}x + \frac{1}{3} = y$

Practice and Problem Solving

Copy and complete the table. Plot the two solution points and draw a line *exactly* through the two points. Find a different solution point on the line.

3.

x		
$y = 3x - 1$		

4.

x		
$y = \frac{1}{3}x + 2$		

Graph the linear equation.

1 2

5. $y = -5x$
6. $y = \frac{1}{4}x$
7. $y = 5$
8. $y = x - 3$
9. $y = -7x - 1$
10. $y = -\frac{x}{3} + 4$
11. $y = \frac{3}{4}x - \frac{1}{2}$
12. $y = -\frac{2}{3}$
13. $y = 6.75$

14. **ERROR ANALYSIS** Describe and correct the error in graphing the equation.

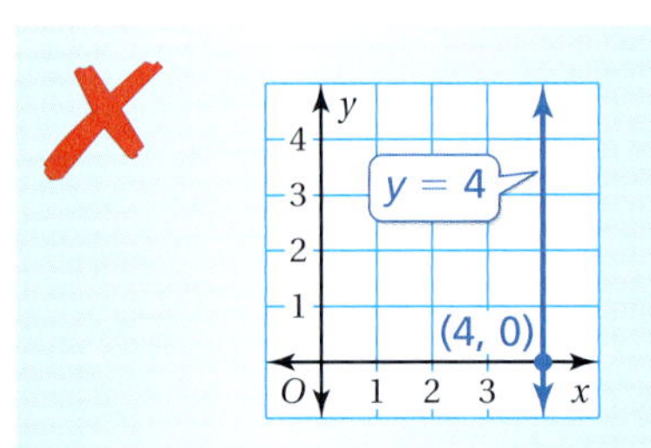

15. **MESSAGING** You sign up for an unlimited text messaging plan for your cell phone. The equation $y = 20$ represents the cost y (in dollars) for sending x text messages. Graph the equation.

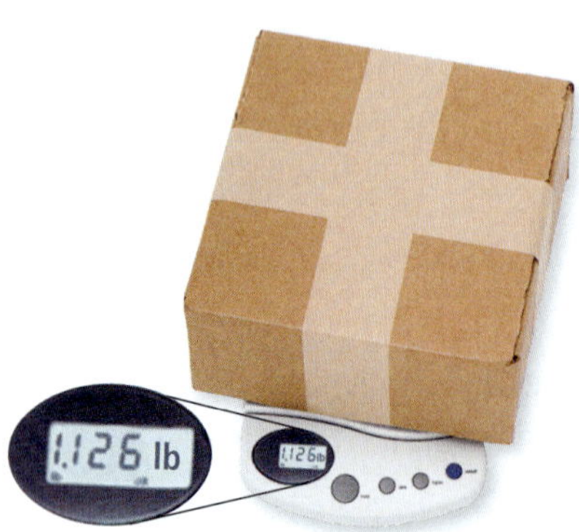

16. **MAIL** The equation $y = 2x + 3$ represents the cost y (in dollars) of mailing a package that weighs x pounds.

   a. Graph the equation.

   b. Use the graph to estimate how much it costs to mail the package.

   c. Use the equation to find exactly how much it costs to mail the package.

Solve for y. Then graph the equation.

17. $y - 3x = 1$

18. $5x + 2y = 4$

19. $-\frac{1}{3}y + 4x = 3$

20. $x + 0.5y = 1.5$

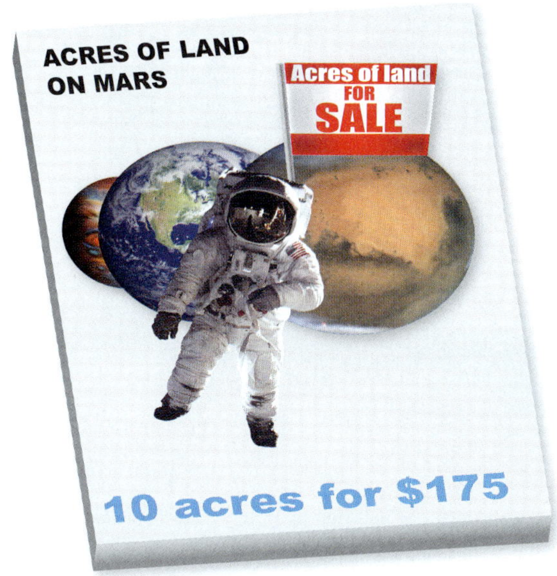

21. SAVINGS You have $100 in your savings account and plan to deposit $12.50 each month.

a. Write and graph a linear equation that represents the balance in your account.

b. How many months will it take you to save enough money to buy 10 acres of land on Mars?

22. CAMERA One second of video on your digital camera uses the same amount of memory as two pictures. Your camera can store 250 pictures.

a. Write and graph a linear equation that represents the number y of pictures your camera can store if you take x seconds of video.

b. How many pictures can your camera store after you take the video shown?

23. SEA LEVEL Along the coast of Florida, the sea level is rising about 3 millimeters per year.

a. Write and graph a linear equation that represents how much sea level rises over a period of time.

b. How many millimeters has sea level risen since you were born?

24. Geometry The sum S of the measures of the angles of a polygon is $S = (n - 2) \cdot 180°$, where n is the number of sides of the polygon. Plot four points (n, S) that satisfy the equation. Do the points lie on a line? Explain your reasoning.

Fair Game Review What you learned in previous grades & lessons

Write the ordered pair corresponding to the point. *(Skills Review Handbook)*

25. Point A

26. Point B

27. Point C

28. Point D

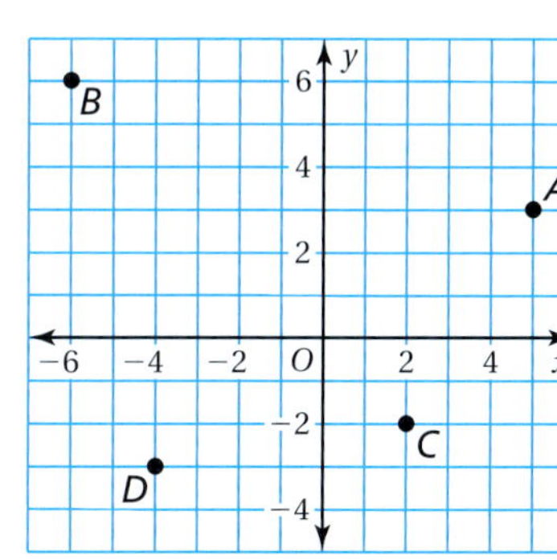

29. MULTIPLE CHOICE A debate team has 15 female members. The ratio of females to males is 3 : 2. How many males are on the debate team? *(Skills Review Handbook)*

Ⓐ 6 Ⓑ 10 Ⓒ 22 Ⓓ 25

2.2 Slope of a Line

STATE STANDARDS
MA.8.A.1.2

Essential Question

How can the slope of a line be used to describe the line?

Slope is the rate of change between any two points on a line. It is the measure of the *steepness* of the line.

To find the slope of a line, find the ratio of the change in y (vertical change) to the change in x (horizontal change).

$$\text{slope} = \frac{\text{change in } y}{\text{change in } x}$$

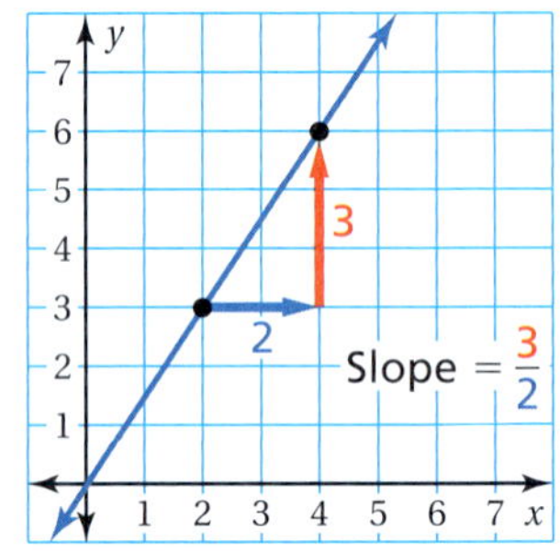

1 ACTIVITY: Finding the Slope of a Line

Work with a partner. Find the slope of each line using two methods.

Method 1: Use the two black points. ●

Method 2: Use the two pink points. ●

Do you get the same slope using each method?

a.

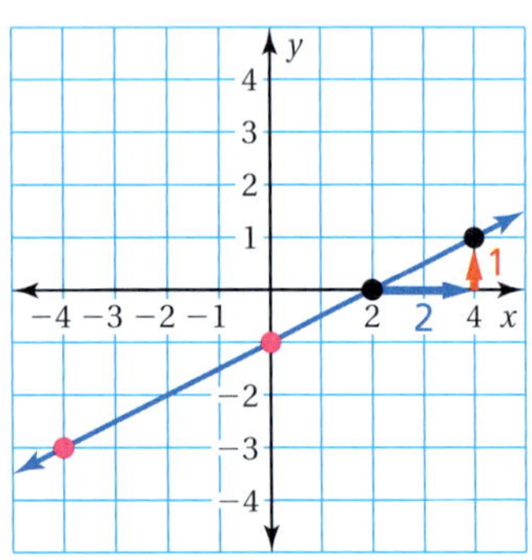

b.

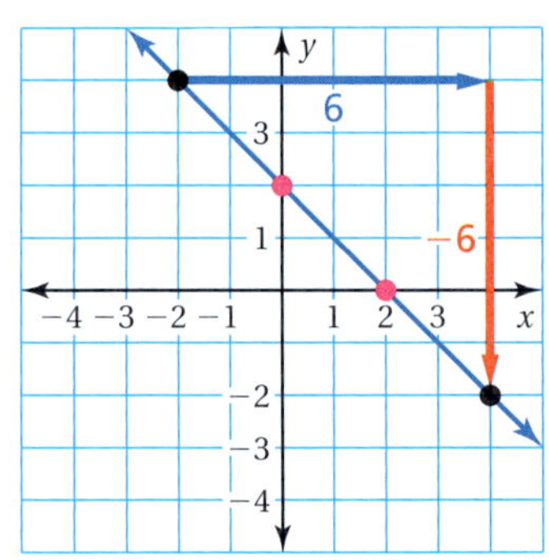

c.

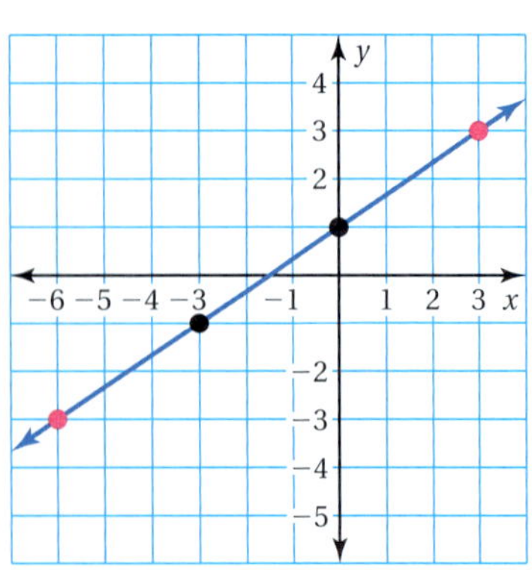

d.

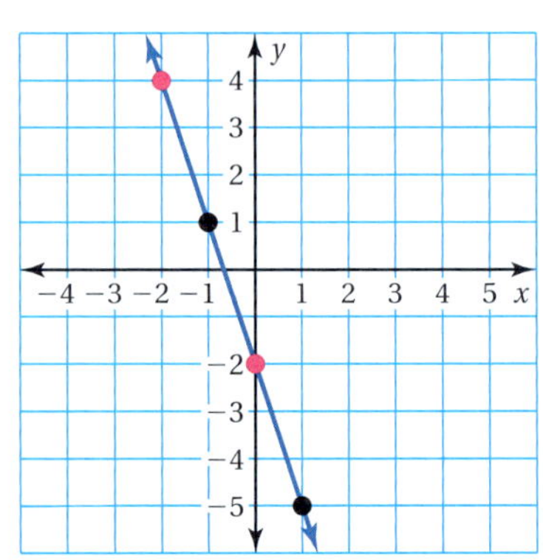

2 ACTIVITY: Drawing Lines with Given Slopes

Work with a partner.

- **Draw a line through the black point using the given slope.**
- **Draw a line through the pink point using the given slope.**
- **What do you notice about the two lines?**

a. Slope $= 2$

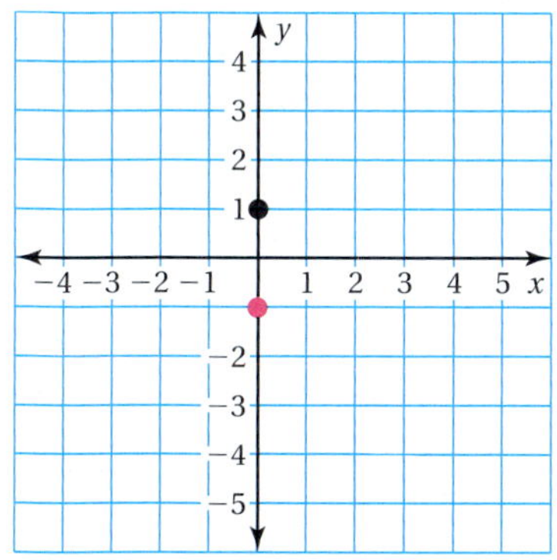

b. Slope $= -\frac{1}{2}$

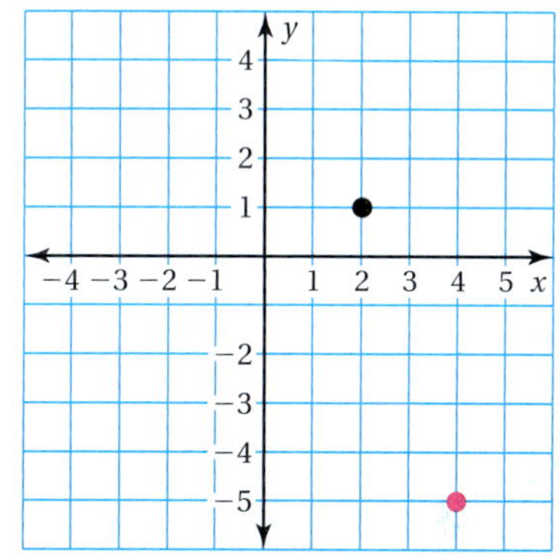

c. Slope $= \frac{3}{4}$

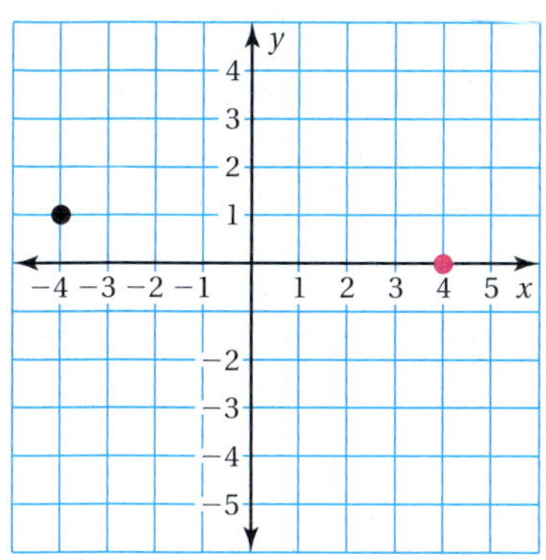

d. Slope $= -2$

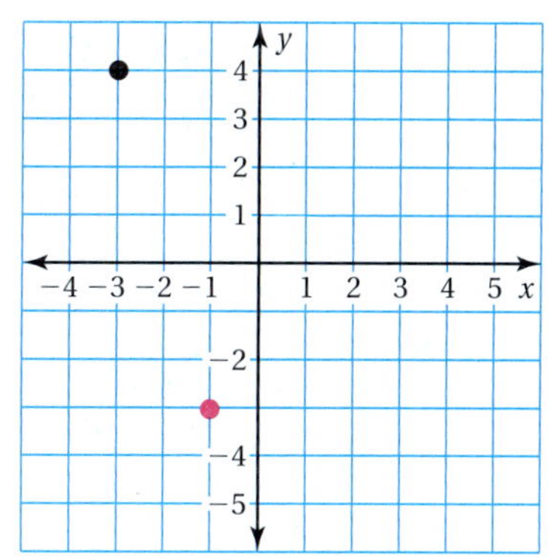

What Is Your Answer?

3. **IN YOUR OWN WORDS** How can the slope of a line be used to describe the line?

a. Draw three lines that have positive slopes.

b. Draw three lines that have negative slopes.

4. Line A has a slope of 1. Line B has a slope of 2. Compare the slopes of the lines. Illustrate your comparison.

5. Line C has a slope of -1. Line D has a slope of -2. Compare the slopes of the lines. Illustrate your comparison.

Practice Use what you learned about the slope of a line to complete Exercises 4–6 on page 59.

2.2 Lesson

Check It Out
Lesson Tutorials
BigIdeasMath.com

Key Vocabulary
slope, *p. 56*
rise, *p. 56*
run, *p. 56*

Key Idea

Slope

The **slope** of a line is a ratio of the change in y (the **rise**) to the change in x (the **run**) between any two points on the line.

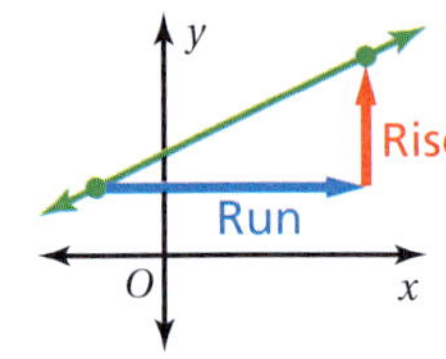

$$\text{slope} = \frac{\text{change in } y}{\text{change in } x} = \frac{\text{rise}}{\text{run}}$$

Positive slope

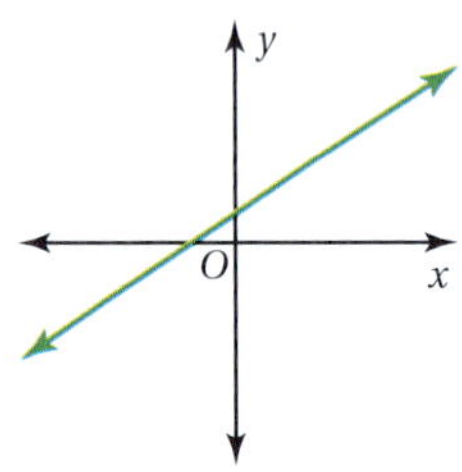

The line rises from left to right.

Negative slope

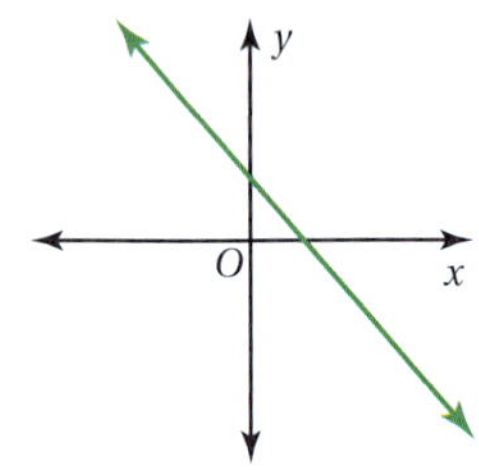

The line falls from left to right.

EXAMPLE 1 Finding the Slope of a Line

Tell whether the slope of the line is *positive* or *negative*. Then find the slope.

a.

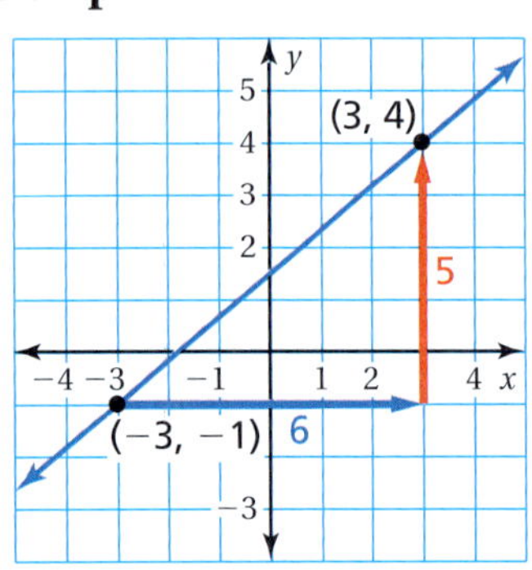

The line rises from left to right. So, the slope is positive.

$$\text{slope} = \frac{\text{rise}}{\text{run}}$$

$$= \frac{5}{6}$$

The slope is $\frac{5}{6}$.

b.

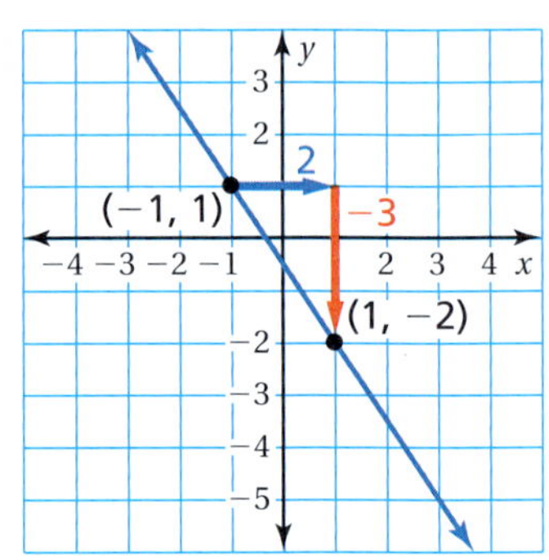

The line falls from left to right. So, the slope is negative.

$$\text{slope} = \frac{\text{rise}}{\text{run}}$$

$$= \frac{-3}{2}, \text{ or } -\frac{3}{2}$$

The slope is $-\frac{3}{2}$.

Multi-Language Glossary at BigIdeasMath.com.

EXAMPLE 2 Finding the Slope of a Horizontal Line

Study Tip

The slope of a horizontal line is always 0.

Find the slope of the line.

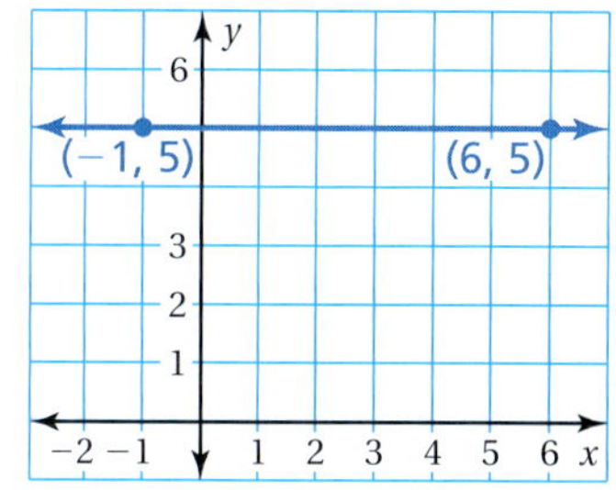

The line is not rising or falling. So, the rise is 0.

$$\text{slope} = \frac{\text{rise}}{\text{run}}$$

$$= \frac{0}{7}$$

The slope is 0.

On Your Own

Now You're Ready Exercises 7–12

Find the slope of the line.

1. 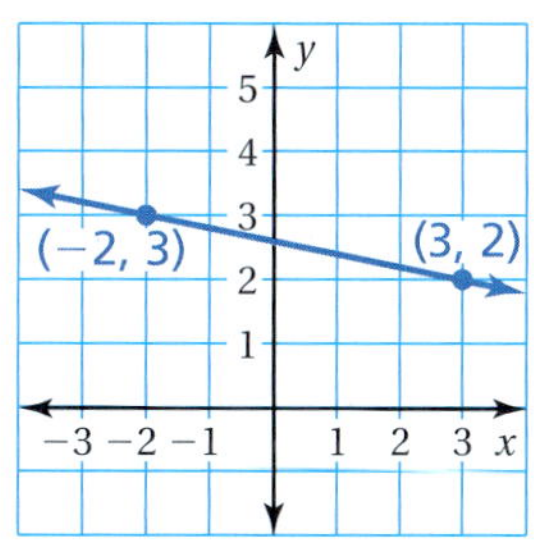

2. 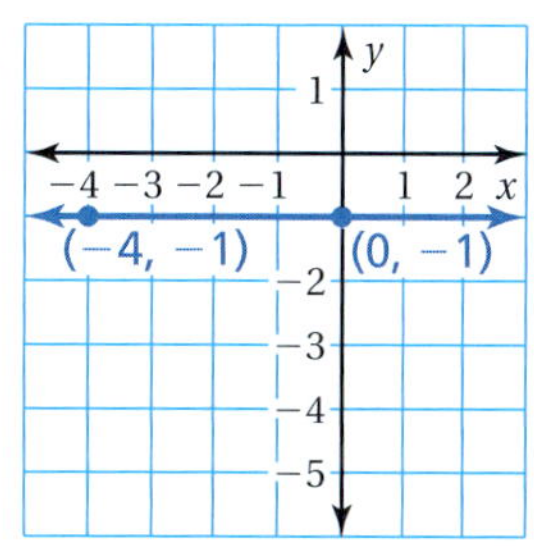

3. 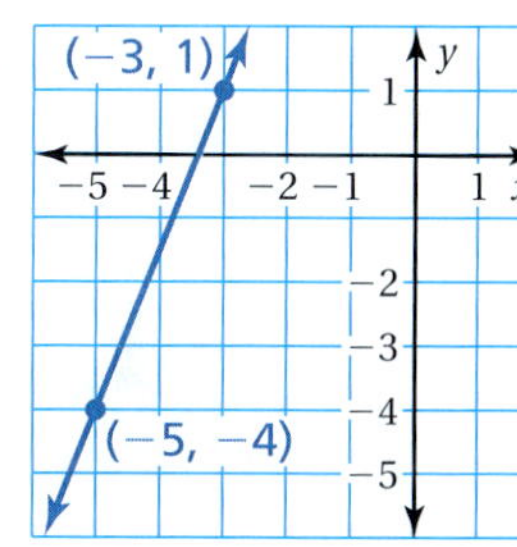

EXAMPLE 3 Finding Slope from a Table

The points in the table lie on a line. Find the slope of the line. Then draw its graph.

x	1	4	7	10
y	8	6	4	2

Choose any two points from the table. Then find the change in y and the change in x.

Use the points (1, 8) and (4, 6).

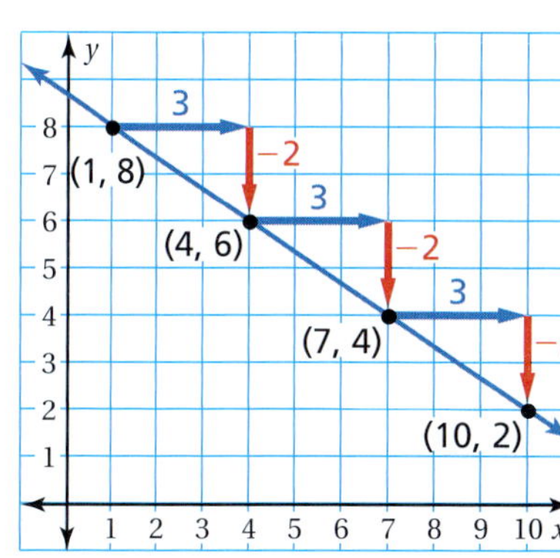

$$\text{slope} = \frac{\text{change in } y}{\text{change in } x}$$

$$= \frac{6 - 8}{4 - 1}$$

$$= \frac{-2}{3}$$

The slope is $-\frac{2}{3}$.

On Your Own

Now You're Ready
Exercises 15–18

The points in the table lie on a line. Find the slope of the line. Then draw its graph.

4.

x	1	3	5	7
y	2	5	8	11

5.

x	−3	−2	−1	0
y	6	4	2	0

Key Idea

Parallel Lines and Slopes

Two lines in the same plane that do not intersect are parallel lines. Two lines with the same slope are parallel.

EXAMPLE 4 Finding Parallel Lines

Which two lines are parallel? Explain.

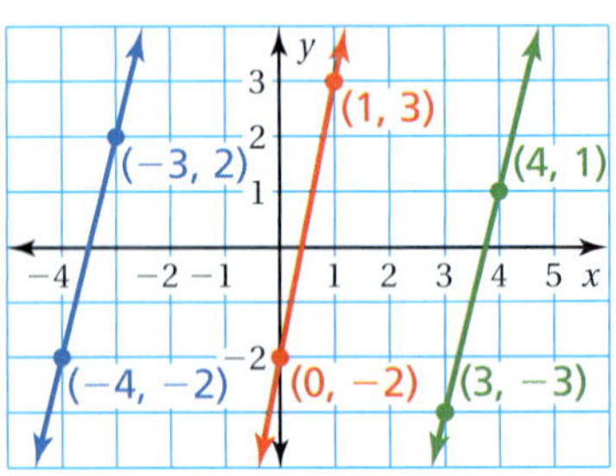

Find the slope of each line.

Blue Line	*Red Line*	*Green Line*
$\text{slope} = \dfrac{\text{rise}}{\text{run}}$	$\text{slope} = \dfrac{\text{rise}}{\text{run}}$	$\text{slope} = \dfrac{\text{rise}}{\text{run}}$
$= \dfrac{4}{1}$	$= \dfrac{5}{1}$	$= \dfrac{4}{1}$
$= 4$	$= 5$	$= 4$

The slope of the blue and green lines is 4. The slope of the red line is 5.

The blue and green lines have the same slope, so they are parallel.

On Your Own

Now You're Ready
Exercises 21 and 22

6. Which two lines are parallel? Explain.

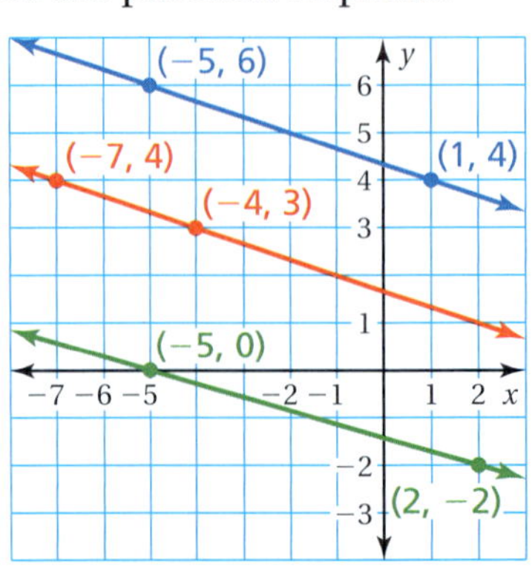

2.2 Exercises

Check It Out
Help with Homework
BigIdeasMath.com

Vocabulary and Concept Check

1. **CRITICAL THINKING** Refer to the graph.

   a. Which lines have positive slopes?

   b. Which line has the steepest slope?

   c. Are any two of the lines parallel? Explain.

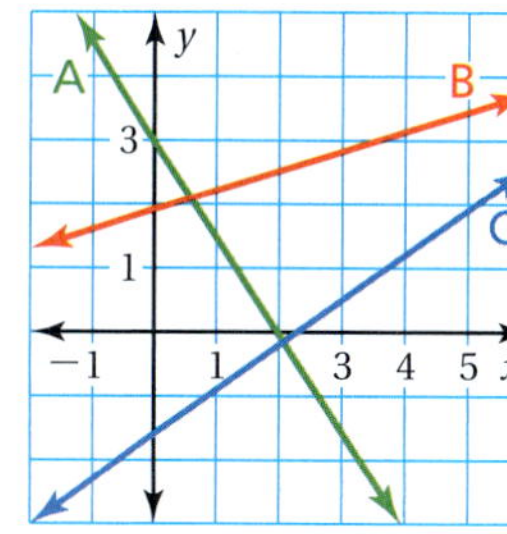

2. **OPEN-ENDED** Describe a real-life situation that involves slope.

3. **REASONING** The slope of a line is 0. What do you know about the line?

Practice and Problem Solving

Draw a line through each point using the given slope. What do you notice about the two lines?

4. Slope $= 1$

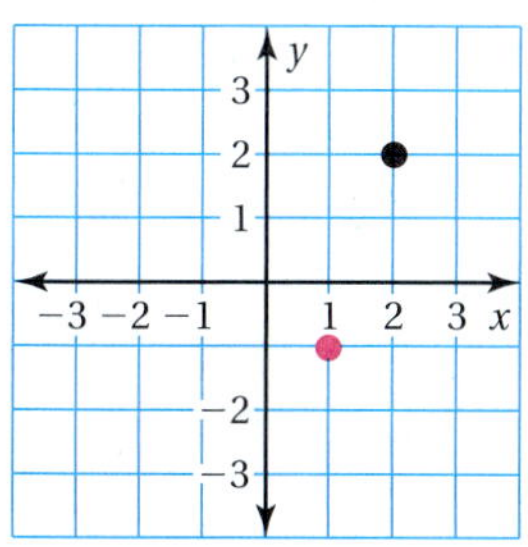

5. Slope $= -3$

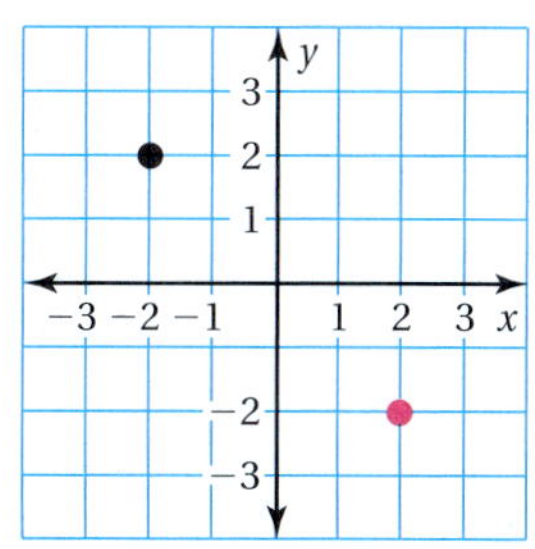

6. Slope $= \frac{1}{4}$

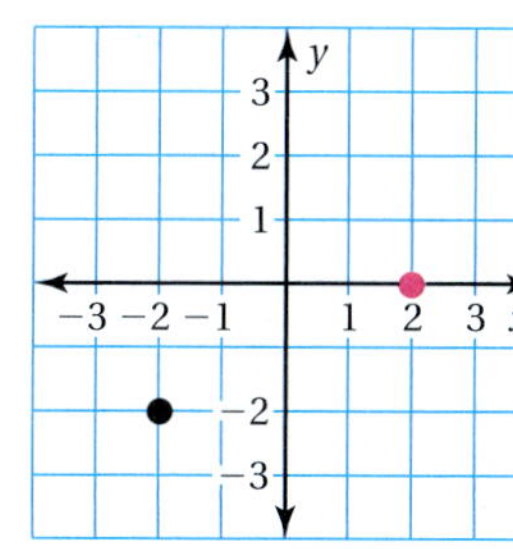

Find the slope of the line.

1 2 7.

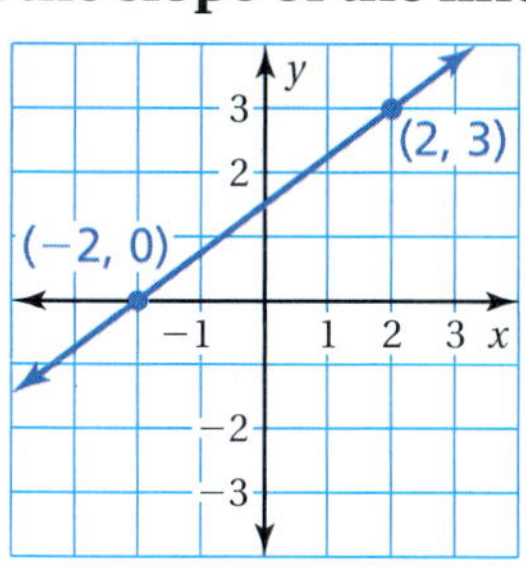

8.

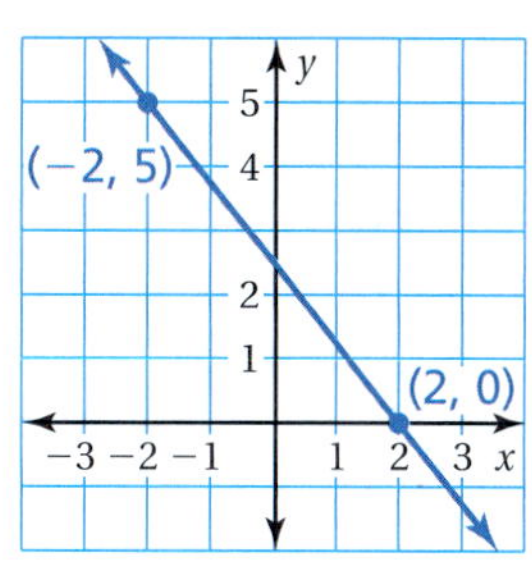

9.

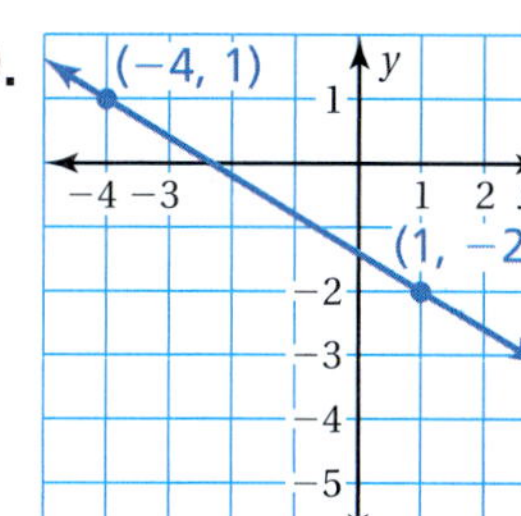

10.

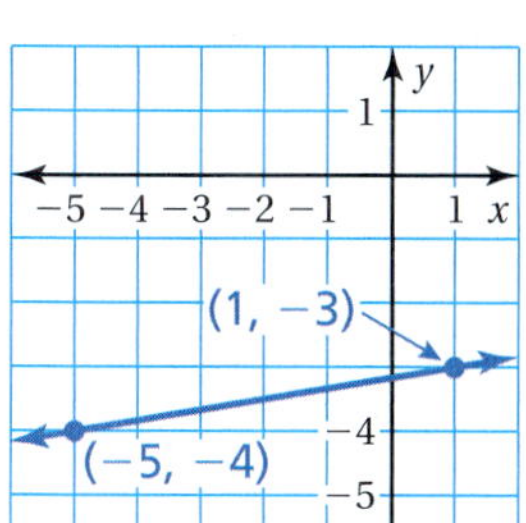

11.

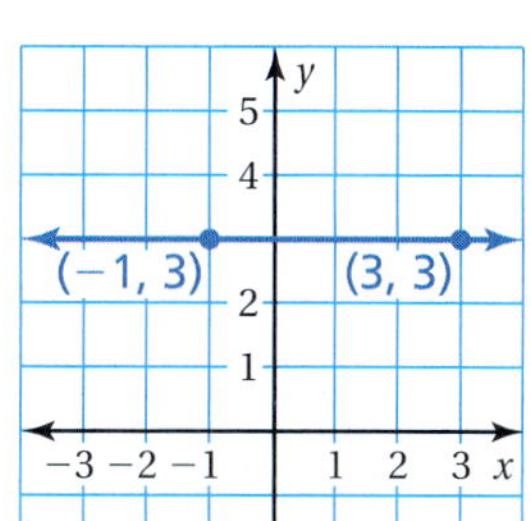

12.

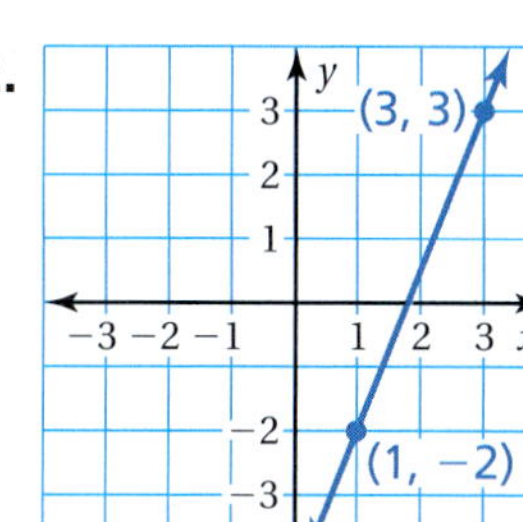

13. **ERROR ANALYSIS** Describe and correct the error in finding the slope of the line.

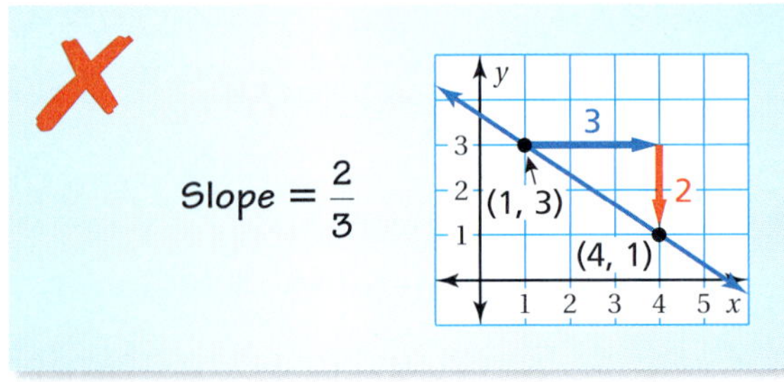

14. **CRITICAL THINKING** Is it more difficult to walk up the ramp or the hill? Explain.

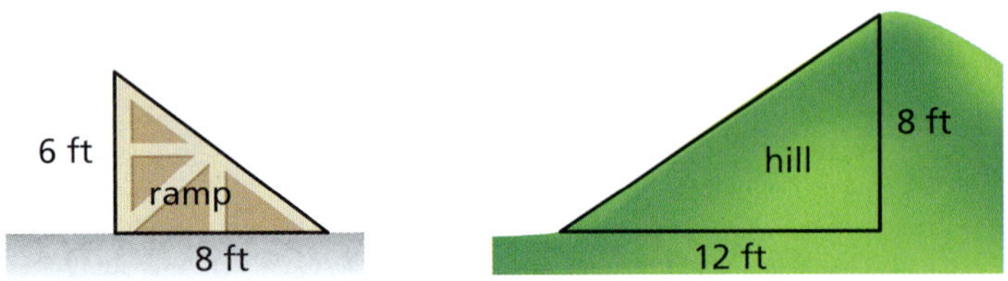

The points in the table lie on a line. Find the slope of the line. Then draw its graph.

③ 15.

x	1	3	5	7
y	2	10	18	26

16.

x	−3	2	7	12
y	0	2	4	6

17.

x	−6	−2	2	6
y	8	5	2	−1

18.

x	−8	−2	4	10
y	8	1	−6	−13

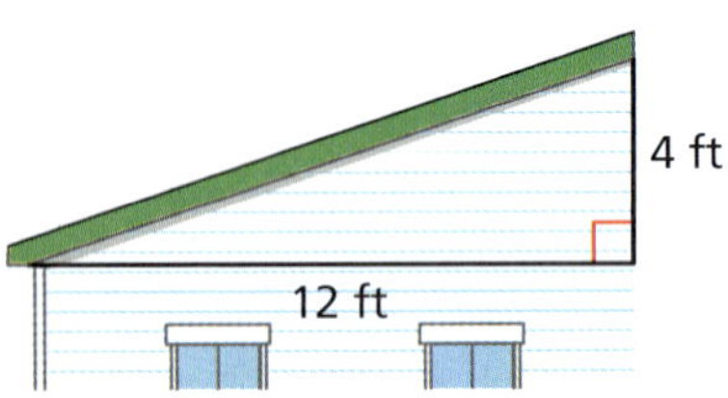

19. **PITCH** Carpenters refer to the slope of a roof as the *pitch* of the roof. Find the pitch of the roof.

20. **PROJECT** The guidelines for a wheelchair ramp suggest that the ratio of the rise to the run be no greater than 1 : 12.

 a. Find a wheelchair ramp in your school or neighborhood. Measure its slope. Does the ramp follow the guidelines?

 b. Design a wheelchair ramp that provides access to a building with a front door that is 2.5 feet higher than the sidewalk. Illustrate your design.

Which two lines are parallel? Explain.

④ 21.

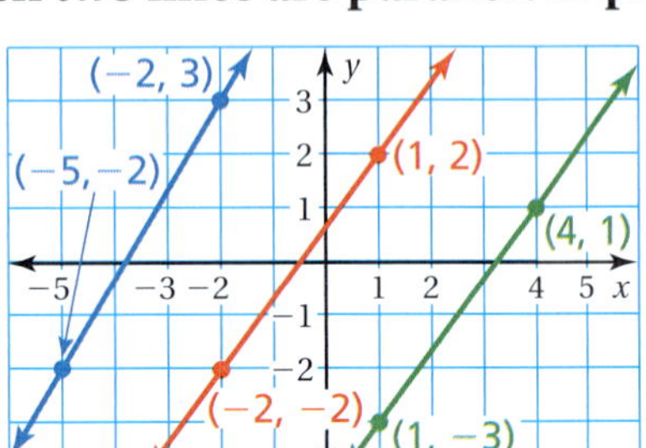

22.

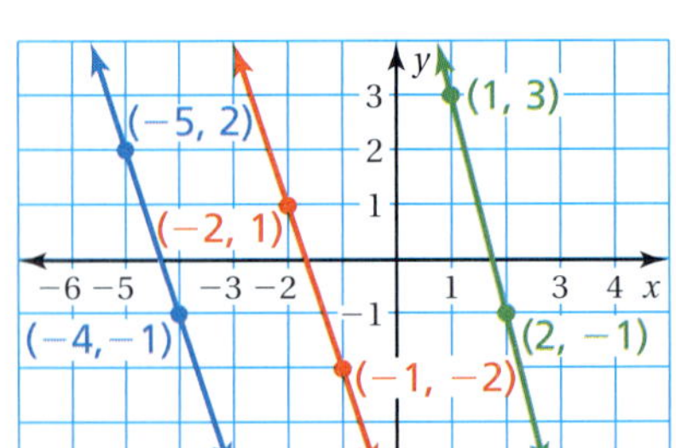

Tell whether the quadrilateral is a parallelogram. Explain.

23.

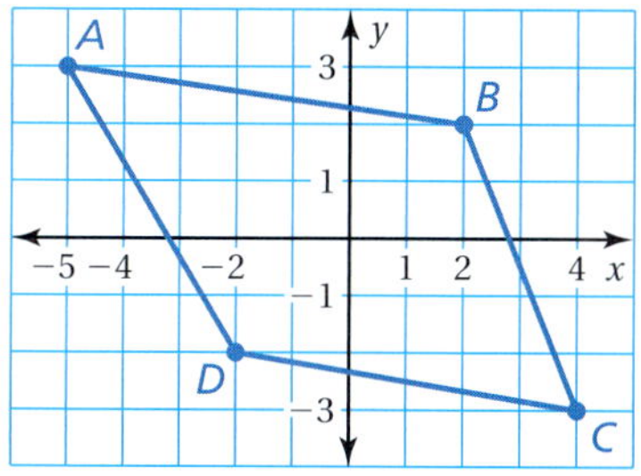

24.

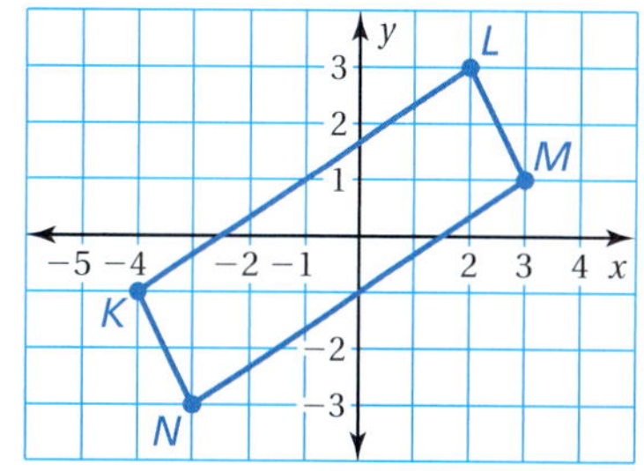

25. **TURNPIKE** The graph shows the cost of traveling by car on the Florida Turnpike.

a. Find the slope of the line.

b. Explain the meaning of the slope as a rate of change.

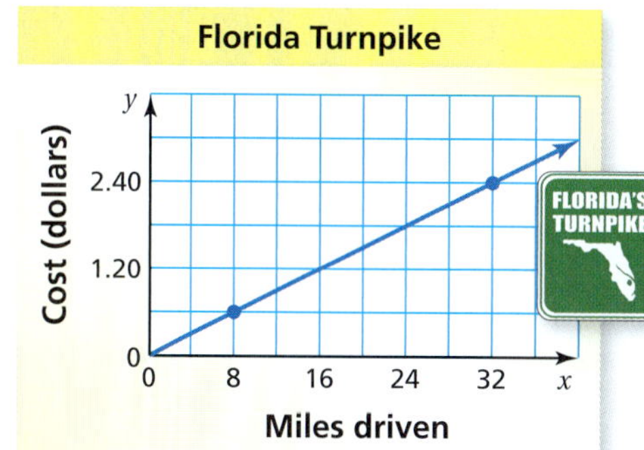

26. **BOAT RAMP** Which is steeper: the boat ramp or a road with a 12% grade? Explain. (*Note:* Road grade is the vertical increase divided by the horizontal distance.)

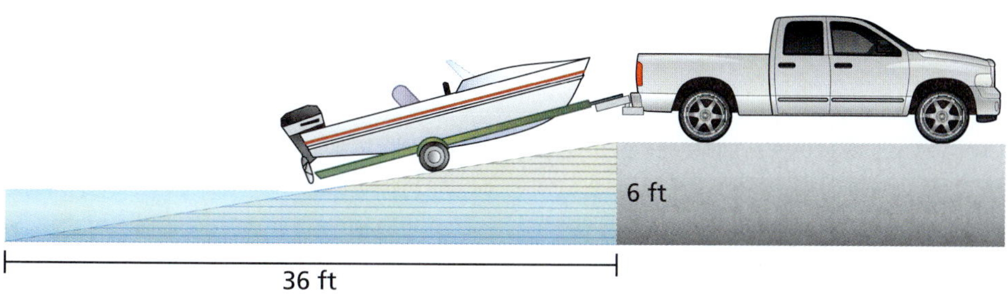

27. Critical Thinking The top and bottom of the slide are parallel to the ground.

a. What is the slope of the main portion of the slide?

b. How does the slope change if the bottom of the slide is only 12 inches above the ground? Is the slide steeper? Explain.

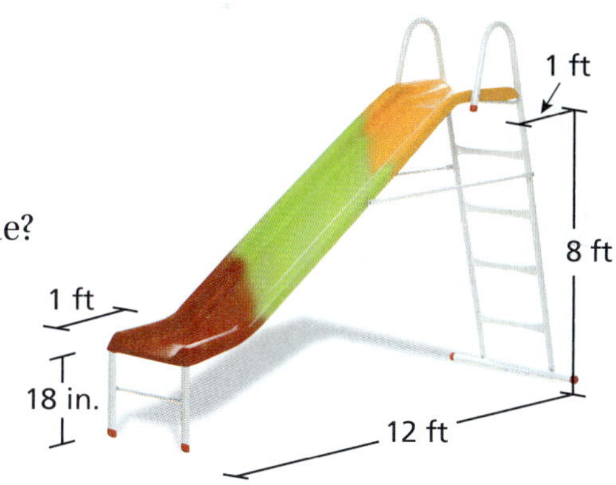

Fair Game Review What you learned in previous grades & lessons

Graph the linear equation. *(Section 2.1)*

28. $y = -\frac{1}{2}x$

29. $y = 3x - \frac{3}{4}$

30. $y = -\frac{x}{3} - \frac{3}{2}$

31. **MULTIPLE CHOICE** What is the prime factorization of 84? *(Skills Review Handbook)*

Ⓐ $2 \times 3 \times 7$ Ⓑ $2^2 \times 3 \times 7$ Ⓒ $2 \times 3^2 \times 7$ Ⓓ $2^2 \times 21$

2.3 Graphing Linear Equations in Slope-Intercept Form

STATE STANDARDS
MA.8.A.1.1
MA.8.A.1.2

Essential Question How can you describe the graph of the equation $y = mx + b$?

1 ACTIVITY: Finding Slopes and y-Intercepts

Work with a partner.

- **Graph the equation.**
- **Find the slope of the line.**
- **Find the point where the line crosses the y-axis.**

a. $y = -\frac{1}{2}x + 1$

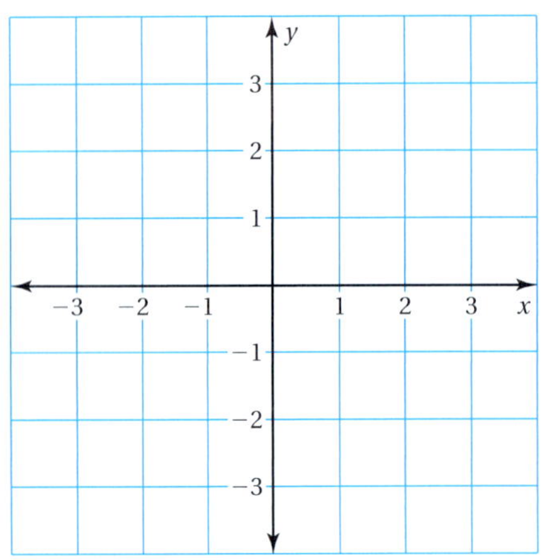

b. $y = -x + 2$

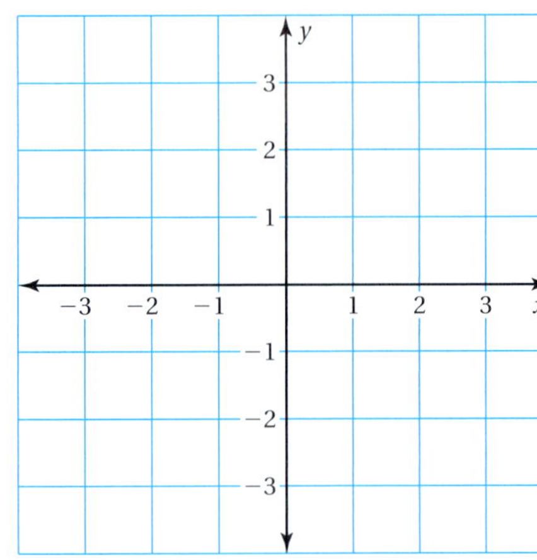

c. $y = -x - 2$

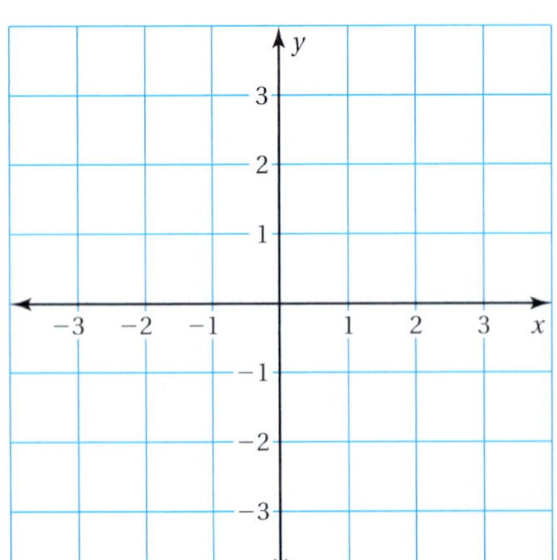

d. $y = \frac{1}{2}x + 1$

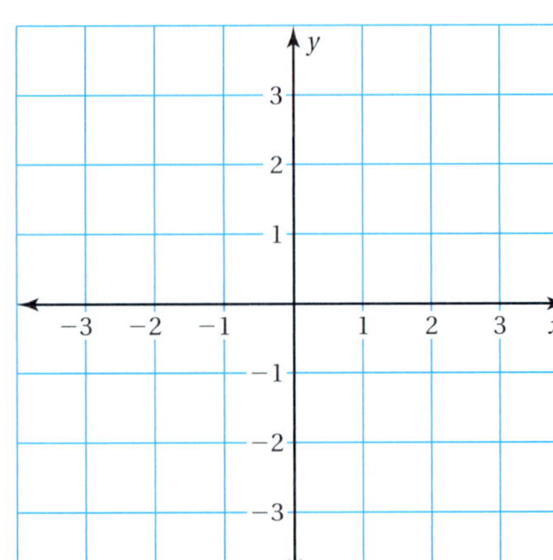

Inductive Reasoning

Work with a partner. Graph each equation. Then copy and complete the table.

	Equation	Description of Graph	Slope of Graph	Point of Intersection with *y*-axis
1a	**2.** $y = -\frac{1}{2}x + 1$	Line	$-\frac{1}{2}$	(0, 1)
1b	**3.** $y = -x + 2$			
1c	**4.** $y = -x - 2$			
1d	**5.** $y = \frac{1}{2}x + 1$			
	6. $y = x + 2$			
	7. $y = x - 2$			
	8. $y = \frac{1}{2}x - 1$			
	9. $y = -\frac{1}{2}x - 1$			
	10. $y = 3x + 2$			
	11. $y = 3x - 2$			
	12. $y = -2x + 3$			

What Is Your Answer?

13. **IN YOUR OWN WORDS** How can you describe the graph of the equation $y = mx + b$?

a. How does the value of m affect the graph of the equation?

b. How does the value of b affect the graph of the equation?

c. Check your answers to parts (a) and (b) with three equations that are not in the table.

14. Why is $y = mx + b$ called the "slope-intercept" form of the equation of a line?

Practice Use what you learned about graphing linear equations in slope-intercept form to complete Exercises 4–6 on page 66.

Check It Out
Lesson Tutorials
BigIdeasMath.com

Key Vocabulary
x-intercept, *p. 64*
y-intercept, *p. 64*
slope-intercept form, *p. 64*

Key Ideas

Intercepts

The ***x*-intercept** of a line is the x-coordinate of the point where the line crosses the x-axis. It occurs when $y = 0$.

The ***y*-intercept** of a line is the y-coordinate of the point where the line crosses the y-axis. It occurs when $x = 0$.

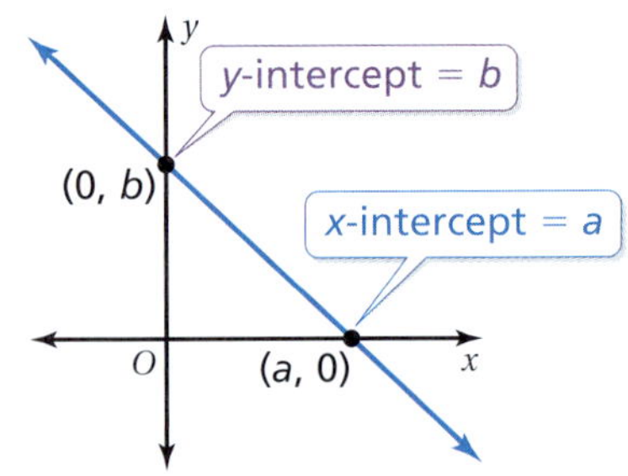

Slope-Intercept Form

Words An equation written in the form $y = mx + b$ is in **slope-intercept form**. The slope of the line is m and the y-intercept of the line is b.

Algebra $y = mx + b$ — m: slope; b: y-intercept

EXAMPLE 1 Identifying Slopes and y-Intercepts

Find the slope and y-intercept of the graph of each linear equation.

a. $y = -4x - 2$

$y = -4x + (-2)$ Write in slope-intercept form.

The slope is -4 and the y-intercept is -2.

b. $y - 5 = \frac{3}{2}x$

$y = \frac{3}{2}x + 5$ Add 5 to each side.

The slope is $\frac{3}{2}$ and the y-intercept is 5.

On Your Own

Now You're Ready
Exercises 7–15

Find the slope and y-intercept of the graph of the linear equation.

1. $y = 3x - 7$
2. $y - 1 = -\frac{2}{3}x$

Multi-Language Glossary at BigIdeasMath.com.

EXAMPLE 2 Graphing a Linear Equation in Slope-Intercept Form

Graph $y = -3x + 3$. Identify the x-intercept.

Step 1: Find the slope and y-intercept.

$$y = -3x + 3$$

slope: -3 y-intercept: 3

Step 2: The y-intercept is 3. So, plot (0, 3).

Step 3: Use the slope to find another point and draw the line.

$$\text{slope} = \frac{\text{rise}}{\text{run}} = \frac{-3}{1}$$

Plot the point that is 1 unit right and 3 units down from (0, 3). Draw a line through the two points.

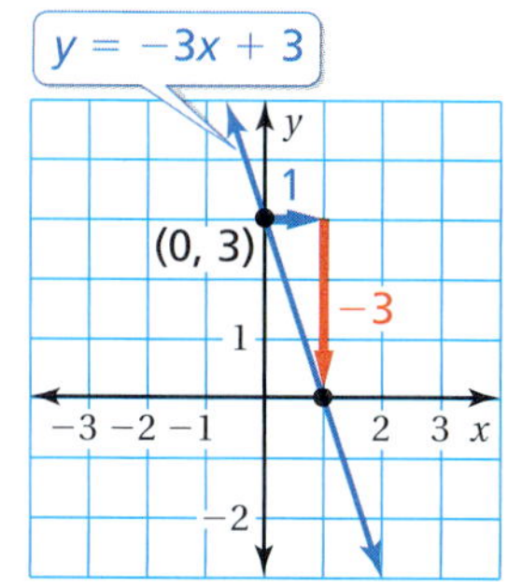

The line crosses the x-axis at (1, 0). So, the x-intercept is 1.

Study Tip

You can check the x-intercept by substituting $y = 0$ in the equation and solving for x.

$$y = -3x + 3$$
$$0 = -3x + 3$$
$$-3 = -3x$$
$$1 = x$$

EXAMPLE 3 Real-Life Application

The cost y (in dollars) of taking a taxi x miles is $y = 2.5x + 2$. (a) Graph the equation. (b) Interpret the y-intercept and slope.

a. The slope of the line is $2.5 = \frac{5}{2}$. Use the slope and y-intercept to graph the equation.

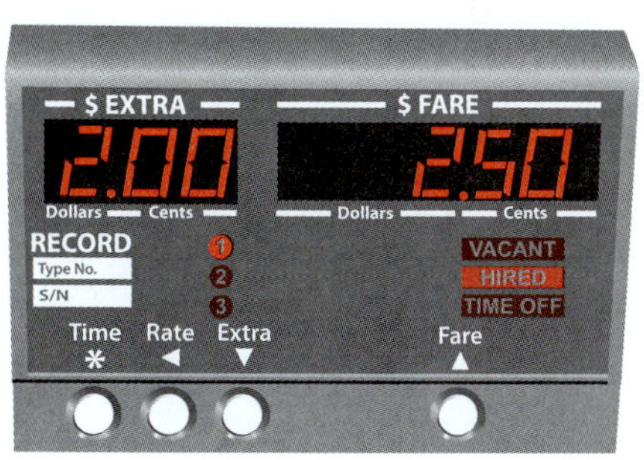

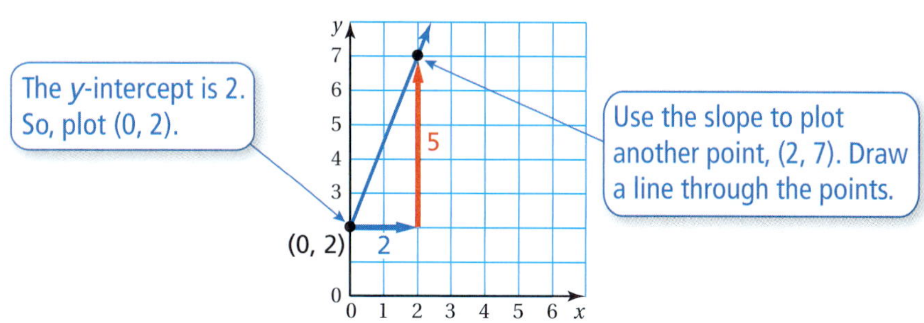

b. The slope is 2.5. So, the cost per mile is \$2.50. The y-intercept is 2. So, there is an initial fee of \$2 to take the taxi.

On Your Own

Now You're Ready
Exercises 18–23

Graph the linear equation. Identify the x-intercept.

3. $y = x - 4$

4. $y = -\frac{1}{2}x + 1$

5. In Example 3, the cost y (in dollars) of taking a different taxi x miles is $y = 2x + 1.5$. Interpret the y-intercept and slope.

2.3 Exercises

Check It Out
Help with Homework
BigIdeasMath.com

Vocabulary and Concept Check

1. **VOCABULARY** How can you find the x-intercept of the graph of $2x + 3y = 6$?
2. **CRITICAL THINKING** Is the equation $y = 3x$ in slope-intercept form? Explain.
3. **OPEN-ENDED** Describe a real-life situation that can be modeled by a linear equation. Write the equation. Interpret the y-intercept and slope.

Practice and Problem Solving

Match the equation with its graph. Identify the slope and y-intercept.

4. $y = 2x + 1$
5. $y = \frac{1}{3}x - 2$
6. $y = -\frac{2}{3}x + 1$

A.

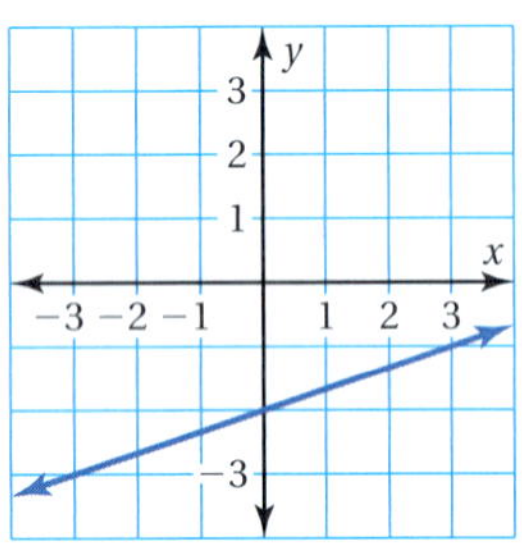

B.

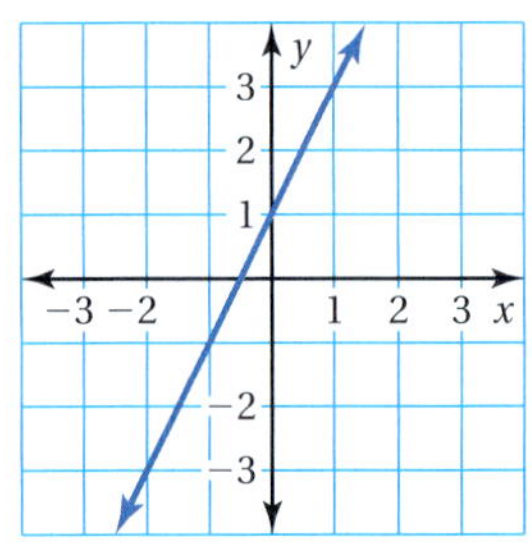

C.

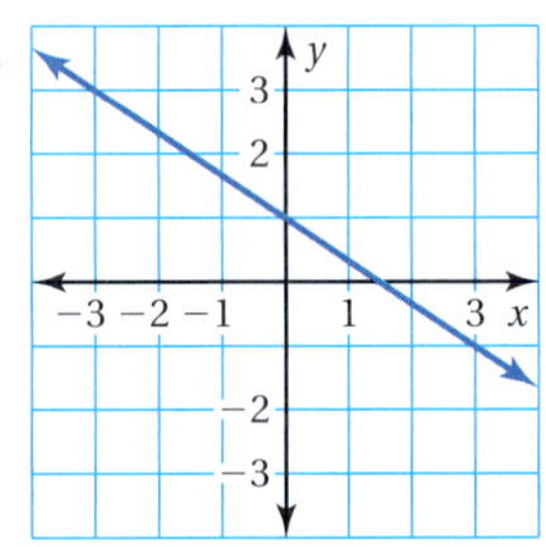

Find the slope and y-intercept of the graph of the linear equation.

1 7. $y = 4x - 5$
8. $y = -7x + 12$
9. $y = -\frac{4}{5}x - 2$
10. $y = 2.25x + 3$
11. $y + 1 = \frac{4}{3}x$
12. $y - 6 = \frac{3}{8}x$
13. $y - 3.5 = -2x$
14. $y + 5 = -\frac{1}{2}x$
15. $y = 1.5x + 11$

16. **ERROR ANALYSIS** Describe and correct the error in finding the slope and y-intercept of the graph of the linear equation.

> ✗ $y = 4x - 3$
> The slope is 4 and the y-intercept is 3.

17. **SKYDIVING** A skydiver parachutes to the ground. The height y (in feet) of the skydiver after x seconds is $y = -10x + 3000$.
   a. Graph the equation.
   b. Interpret the x-intercept and slope.

Graph the linear equation. Identify the x-intercept.

2 **18.** $y = \frac{1}{5}x + 3$ **19.** $y = 6x - 7$ **20.** $y = -\frac{8}{3}x + 9$

21. $y = -1.4x - 1$ **22.** $y + 9 = -3x$ **23.** $y - 4 = -\frac{3}{5}x$

24. PHONES The cost y (in dollars) of making a long distance phone call for x minutes is $y = 0.25x + 2$.

a. Graph the equation.

b. Interpret the slope and y-intercept.

25. APPLES Write a linear equation that models the cost y of picking x pounds of apples. Graph the equation.

26. ELEVATOR The basement of a building is 40 feet below ground level. The elevator rises at a rate of 5 feet per second. You enter the elevator in the basement. Write an equation that represents the height y (in feet) of the elevator after x seconds. Graph the equation.

27. BONUS You work in an electronics store. You earn a fixed amount of $35 per day, plus a 15% bonus on the merchandise you sell. Write an equation that models the amount y (in dollars) you earn for selling x dollars of merchandise in one day. Graph the equation.

28. Critical Thinking Six friends create a website. The website earns money by selling banner ads. The site has five banner ads. It costs $120 a month to operate the website.

a. A banner ad earns $0.005 per click. Write a linear equation that represents the monthly income y (in dollars) for x clicks.

b. Draw a graph of the equation in part (a). On the graph, label the number of clicks needed for the friends to start making a profit.

Fair Game Review What you learned in previous grades & lessons

Solve the equation for y. *(Section 1.4)*

29. $y - 2x = 3$ **30.** $4x + 5y = 13$ **31.** $2x - 3y = 6$ **32.** $7x + 4y = 8$

33. MULTIPLE CHOICE Which point is a solution of the equation $3x - 8y = 11$? *(Section 2.1)*

Ⓐ $(1, 1)$ Ⓑ $(1, -1)$ Ⓒ $(-1, 1)$ Ⓓ $(-1, -1)$

2.4 Graphing Linear Equations in Standard Form

STATE STANDARDS
MA.8.A.1.1
MA.8.A.1.2

Essential Question How can you describe the graph of the equation $ax + by = c$?

1 ACTIVITY: Using a Table to Plot Points

Work with a partner. You sold a total of \$16 worth of tickets to a school concert. You lost track of how many of each type of ticket you sold.

$$\frac{\$4}{\text{Adult}} \cdot \text{Number of Adult Tickets} + \frac{\$2}{\text{Child}} \cdot \text{Number of Child Tickets} = \$16$$

a. Let x represent the number of adult tickets.
Let y represent the number of child tickets.
Write an equation that relates x and y.

b. Copy and complete the table showing the different combinations of tickets you might have sold.

Number of Adult Tickets, x					
Number of Child Tickets, y					

c. Plot the points from the table. Describe the pattern formed by the points.

d. If you remember how many adult tickets you sold, can you determine how many child tickets you sold? Explain your reasoning.

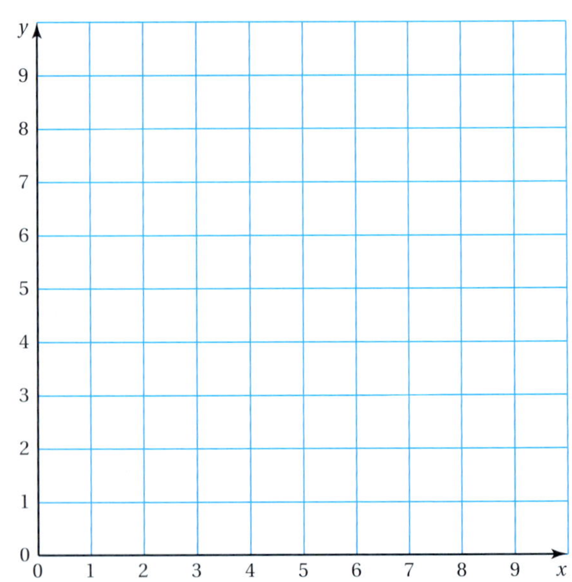

2 ACTIVITY: Rewriting an Equation

Work with a partner. You sold a total of $16 worth of cheese. You forgot how many pounds of each type of cheese you sold.

$$\frac{\$4}{\text{lb}} \cdot \text{Pounds of Swiss} + \frac{\$2}{\text{lb}} \cdot \text{Pounds of Cheddar} = \$16$$

a. Let x represent the number of pounds of Swiss cheese.
Let y represent the number of pounds of Cheddar cheese.
Write an equation that relates x and y.

b. Write the equation in slope-intercept form. Then graph the equation.

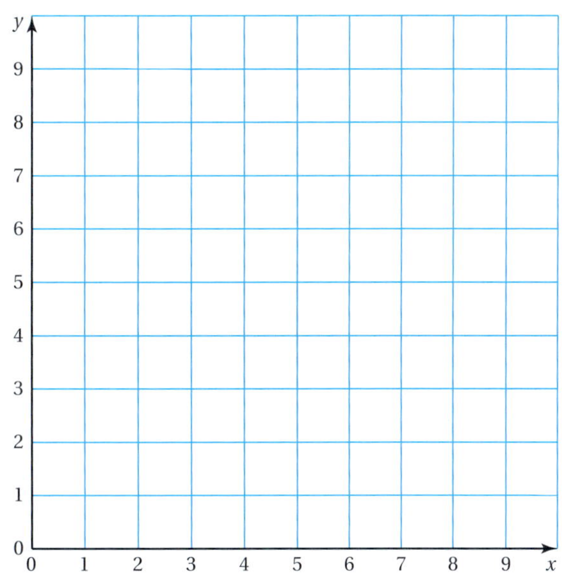

What Is Your Answer?

3. **IN YOUR OWN WORDS** How can you describe the graph of the equation $ax + by = c$?
4. Activities 1 and 2 show two different methods for graphing $ax + by = c$. Describe the two methods. Which method do you prefer? Explain.
5. Write a real-life problem that is similar to those shown in Activities 1 and 2.

Practice Use what you learned about graphing linear equations in standard form to complete Exercises 3 and 4 on page 72.

2.4 Lesson

Check It Out
Lesson Tutorials
BigIdeasMath.com

Key Vocabulary
standard form, *p. 70*

Study Tip
Any linear equation can be written in standard form.

Key Idea

Standard Form of a Linear Equation

The **standard form** of a linear equation is

$$ax + by = c$$

where a and b are not both zero.

EXAMPLE 1 Graphing a Linear Equation in Standard Form

Graph $-2x + 3y = -6$.

Step 1: Write the equation in slope-intercept form.

$$-2x + 3y = -6 \qquad \text{Write the equation.}$$

$$3y = 2x - 6 \qquad \text{Add } 2x \text{ to each side.}$$

$$y = \frac{2}{3}x - 2 \qquad \text{Divide each side by 3.}$$

Step 2: Use the slope and y-intercept to graph the equation.

$$y = \frac{2}{3}x + (-2)$$

slope: $\frac{2}{3}$; y-intercept: -2

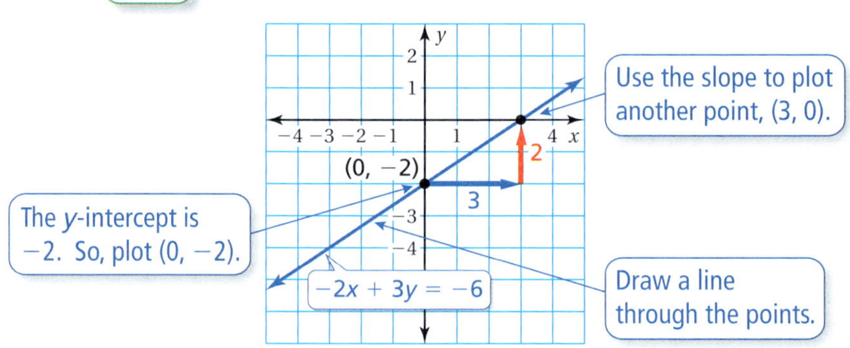

On Your Own

Now You're Ready
Exercises 5–10

Graph the linear equation.

1. $x + y = -2$
2. $-\frac{1}{2}x + 2y = 6$
3. $-\frac{2}{3}x + y = 0$
4. $2x + y = 5$

Multi-Language Glossary at BigIdeasMath.com.

EXAMPLE 2 Graphing a Linear Equation in Standard Form

Graph $x + 3y = -3$ using intercepts.

Step 1: To find the x-intercept, substitute 0 for y.

$$x + 3y = -3$$
$$x + 3(0) = -3$$
$$x = -3$$

To find the y-intercept, substitute 0 for x.

$$x + 3y = -3$$
$$0 + 3y = -3$$
$$y = -1$$

Step 2: Graph the equation.

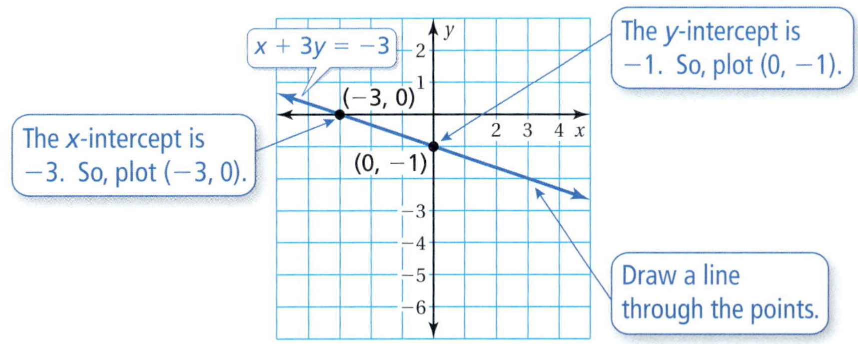

EXAMPLE 3 Real-Life Application

You have \$6 to spend on apples and bananas. (a) Graph the equation $1.5x + 0.6y = 6$, where x is the number of pounds of apples and y is the number of pounds of bananas. (b) Interpret the intercepts.

a. Find the intercepts and graph the equation.

x-intercept	**y-intercept**
$1.5x + 0.6y = 6$	$1.5x + 0.6y = 6$
$1.5x + 0.6(0) = 6$	$1.5(0) + 0.6y = 6$
$x = 4$	$y = 10$

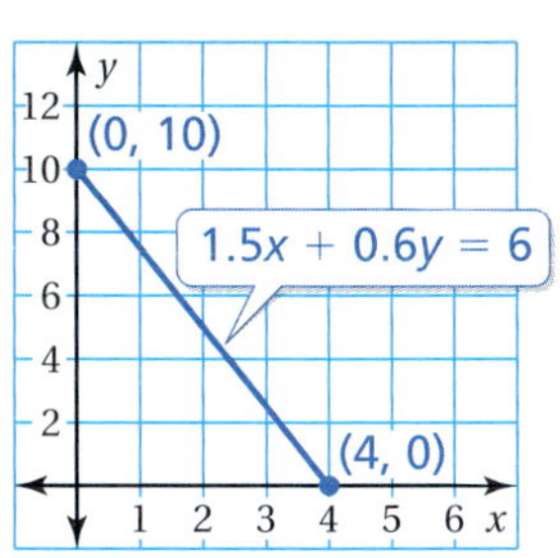

b. The x-intercept shows that you can buy 4 pounds of apples if you don't buy any bananas. The y-intercept shows that you can buy 10 pounds of bananas if you don't buy any apples.

On Your Own

Now You're Ready Exercises 16–18

Graph the linear equation using intercepts.

5. $2x - y = 8$

6. $x + 3y = 6$

7. **WHAT IF?** In Example 3, you buy y pounds of oranges instead of bananas. Oranges cost \$1.20 per pound. Graph the equation $1.5x + 1.2y = 6$. Interpret the intercepts.

2.4 Exercises

Check It Out
Help with Homework
BigIdeasMath.com

Vocabulary and Concept Check

1. **VOCABULARY** Is the equation $y = -2x + 5$ in standard form? Explain.

2. **REASONING** Does the graph represent a linear equation? Explain.

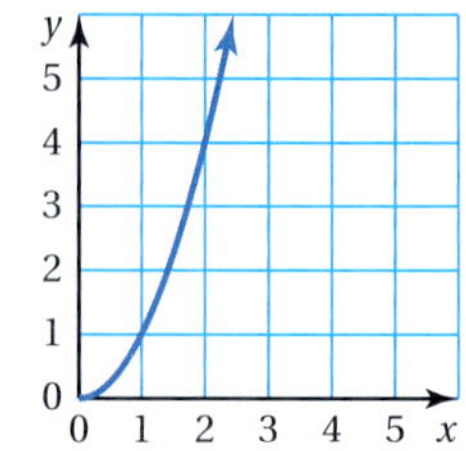

Practice and Problem Solving

Define two variables for the verbal model. Write an equation in slope-intercept form that relates the variables. Graph the equation.

3. $\dfrac{\$2.00}{\text{pound}} \cdot$ Pounds of peaches $+ \dfrac{\$1.50}{\text{pound}} \cdot$ Pounds of apples $= \$15$

4. $\dfrac{16\text{ miles}}{\text{hour}} \cdot$ Hours biked $+ \dfrac{2\text{ miles}}{\text{hour}} \cdot$ Hours walked $=$ 32 miles

Write the linear equation in slope-intercept form.

1 5. $2x + y = 17$

6. $5x - y = \frac{1}{4}$

7. $-\frac{1}{2}x + y = 10$

Graph the linear equation.

8. $-18x + 9y = 72$

9. $16x - 4y = 2$

10. $\frac{1}{4}x + \frac{3}{4}y = 1$

Use the graph to find the x- and y-intercepts.

11.

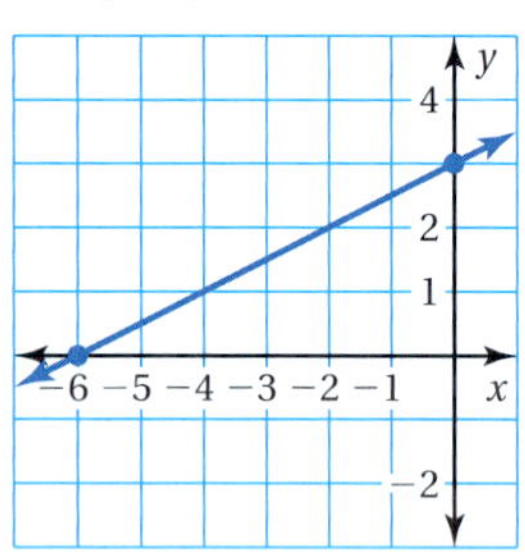

12.

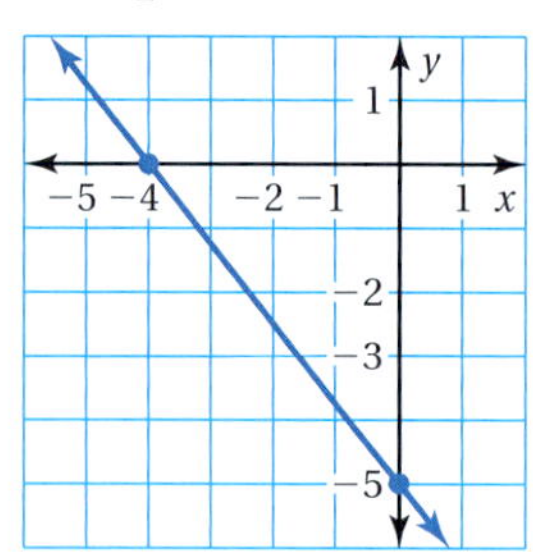

13.

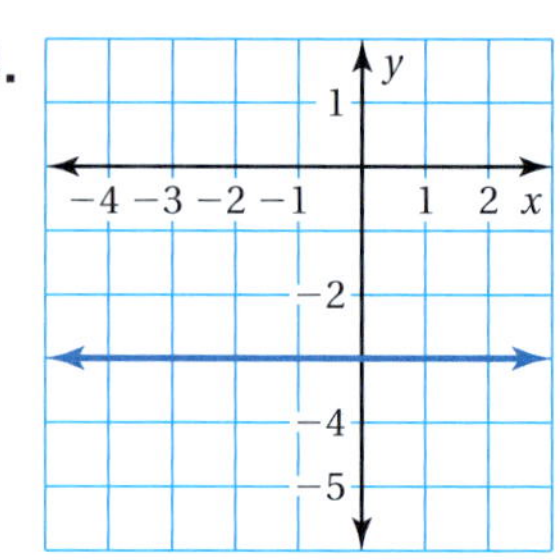

14. **ERROR ANALYSIS** Describe and correct the error in finding the x-intercept.

$$-2x + 3y = 12$$
$$-2(0) + 3y = 12$$
$$3y = 12$$
$$y = 4$$

15. **BRACELET** A charm bracelet costs \$65, plus \$25 for each charm.

 a. Write an equation in standard form that represents the total cost of the bracelet.

 b. How much does the bracelet shown cost?

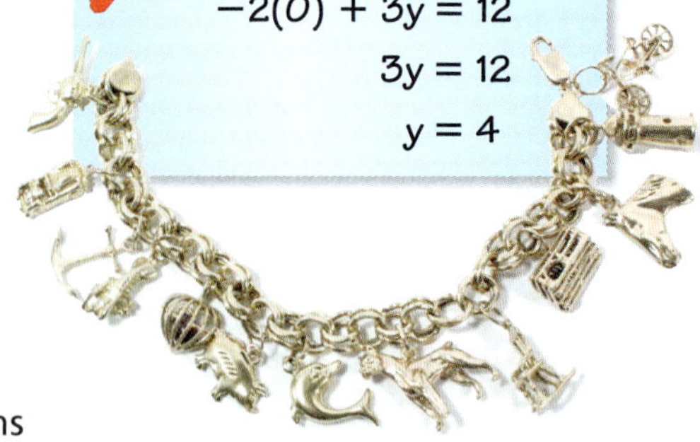

Graph the linear equation using intercepts.

2 **16.** $3x - 4y = -12$ **17.** $2x + y = 8$ **18.** $\frac{1}{3}x - \frac{1}{6}y = -\frac{2}{3}$

19. **SHOPPING** The amount of money you spend on x CDs and y DVDs is given by the equation $14x + 18y = 126$. Find the intercepts and graph the equation.

20. **SCUBA** Five friends go scuba diving off Key Largo. They rent a boat for x days and scuba gear for y days. The total spent is $1000.

a. Write an equation in standard form that represents the situation.

b. Graph the equation and interpret the intercepts.

21. **WAGES** You work at a restaurant as a host and a server. You earn $9.45 for each hour you work as a host and $7.65 for each hour you work as a server.

a. Write an equation in standard form that models your earnings.

b. Graph the equation.

Basic Information
Pay to the Order of: .... John Doe
# of hours worked as .... host: x
# of hours worked as .... server: y
Earnings for this pay .... period: $160.65

22. **REASONING** Does the graph of every linear equation have an x-intercept? Explain your reasoning. Include an example.

23. Critical Thinking For a house call, a veterinarian charges $70, plus $40 an hour.

a. Write an equation that represents the total fee y charged by the veterinarian for a visit lasting x hours.

b. Find the x-intercept. Will this point appear on the graph of the equation? Explain your reasoning.

c. Graph the equation.

Fair Game Review What you learned in previous grades & lessons

Copy and complete the table of values. *(Skills Review Handbook)*

24.

x	-2	-1	0	1	2
$2x + 5$					

25.

x	-2	-1	0	1	2
$-5 - 3x$					

26. **MULTIPLE CHOICE** Which value of x makes the equation $4x - 12 = 3x - 9$ true? *(Section 1.3)*

Ⓐ -1 Ⓑ 0 Ⓒ 1 Ⓓ 3

Check It Out
Graphic Organizer
BigIdeasMath.com

You can use a **process diagram** to show the steps involved in a procedure. Here is an example of a process diagram for graphing a linear equation.

Graphing a linear equation	Example: Graph $y = 2x - 1$
↓ Make a table of values.	↓ (table below)
↓ Plot the ordered pairs.	↓ 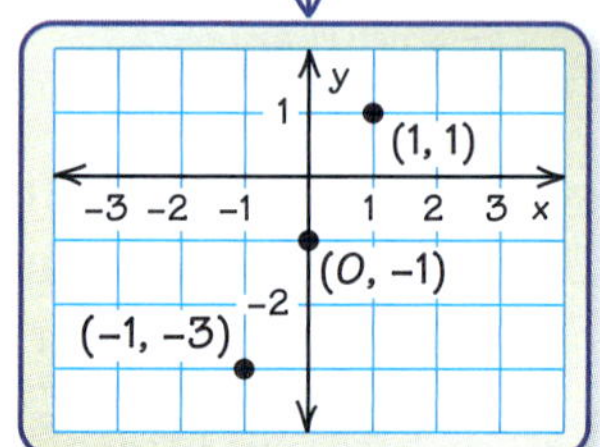
↓ Draw a line through the points.	↓ 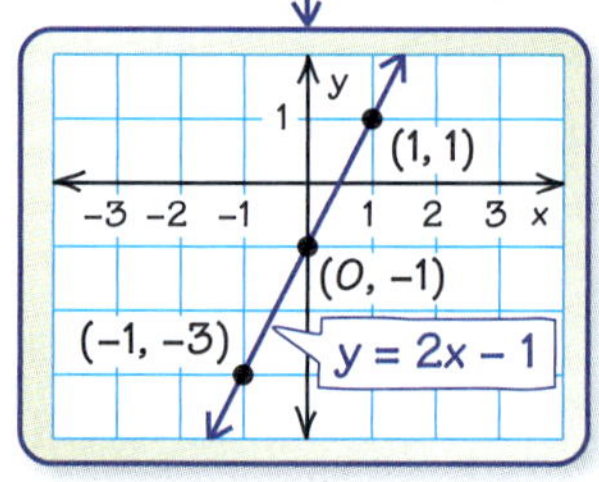

x	−1	0	1
y	−3	−1	1

On Your Own

Make a process diagram with an example to help you study these topics.

1. finding the slope of a line
2. graphing a linear equation using
   a. slope and y-intercept
   b. x- and y-intercepts

After you complete this chapter, make process diagrams for the following topics.

3. solving a linear system
   a. using a table
   b. using a graph
   c. algebraically
4. finding the number of solutions of a linear system
5. solving an equation by graphing

"Here is a process diagram with suggestions for what to do if a hyena knocks on your door."

2.1–2.4 Quiz

Check It Out
Progress Check
BigIdeasMath.com

Graph the linear equation using a table. *(Section 2.1)*

1. $y = -12x$ **2.** $y = -x + 8$ **3.** $y = \frac{x}{3} - 4$ **4.** $y = 3.5$

Find the slope of the line. *(Section 2.2)*

5.

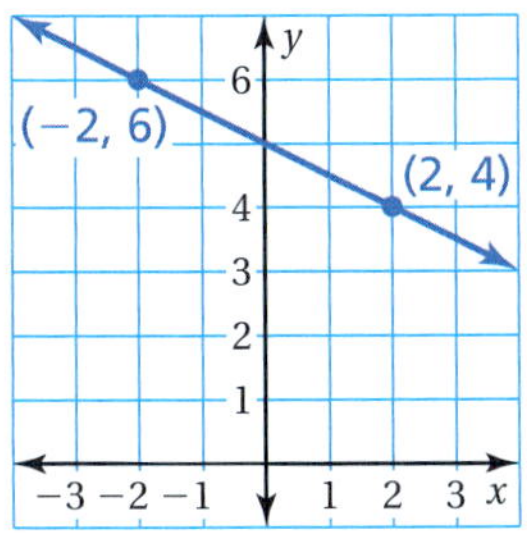

6.

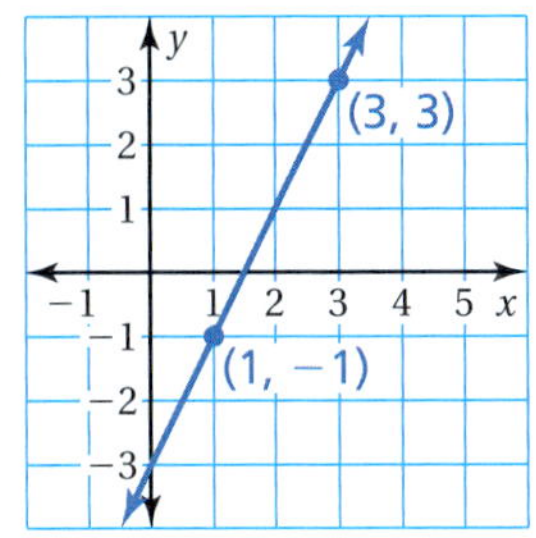

Find the slope and *y*-intercept of the graph of the equation. *(Section 2.3)*

7. $y = \frac{1}{4}x - 8$ **8.** $y = -x + 3$

Find the *x*- and *y*-intercepts of the graph of the equation. *(Section 2.4)*

9. $3x - 2y = 12$ **10.** $x + 5y = 15$

11. **BARBEQUE** The equation $3x + 2y = 30$ represents the amount of money your family spends on x pounds of beef and y pounds of chicken for a barbeque. Graph the equation and interpret the intercepts. *(Section 2.4)*

12. **BANKING** A bank charges \$3 each time you use an out-of-network ATM. At the beginning of the month, you have \$1500 in your bank account. You withdraw \$60 from your bank account each time you use an out-of-network ATM. Write and graph a linear equation that represents the balance in your account after you use an out-of-network ATM x times. *(Section 2.1)*

13. **STATE FAIR** Write a linear equation that models the cost y of one person going on x rides at the fair. Graph the equation. *(Section 2.3)*

14. **PAINTING** You used \$90 worth of paint for a school float. *(Section 2.4)*

a. Graph the equation $18x + 15y = 90$, where x is the number of gallons of blue paint and y is the number of gallons of white paint.

b. Interpret the intercepts.

2.5 Systems of Linear Equations

STATE STANDARDS
MA.8.A.1.3
MA.8.A.1.4

Essential Question How can you solve a system of linear equations?

1 ACTIVITY: Writing a System of Linear Equations

Work with a partner.

Your family starts a bed-and-breakfast in your home. You spend \$500 fixing up a bedroom to rent. Your cost for food and utilities is \$10 per night. Your family charges \$60 per night to rent the bedroom.

a. Write an equation that represents your costs.

Cost, C (in dollars)	=	\$10 per night	•	Number of nights, x	+	\$500

b. Write an equation that represents your revenue (income).

Revenue, R (in dollars)	=	\$60 per night	•	Number of nights, x

c. A set of two (or more) linear equations is called a **system of linear equations**. Write the system of linear equations for this problem.

2 ACTIVITY: Using a Table to Solve a System

Use the cost and revenue equations from Activity 1 to find how many nights you need to rent the bedroom before you recover the cost of fixing up the bedroom. This is the *break-even point* for your business.

a. Copy and complete the table.

x	0	1	2	3	4	5	6	7	8	9	10	11
C												
R												

b. How many nights do you need to rent the bedroom before you break even?

3 ACTIVITY: Using a Graph to Solve a System

a. Graph the cost equation from Activity 1.

b. In the same coordinate plane, graph the revenue equation from Activity 1.

c. Find the point of intersection of the two graphs. The x-value of this point is the number of nights you need to rent the bedroom to break even.

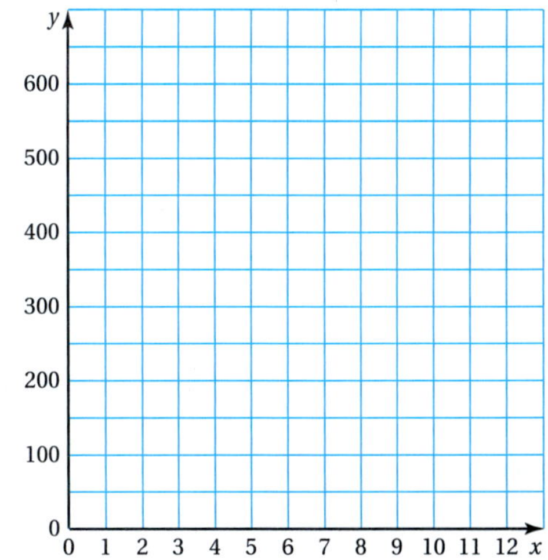

4 ACTIVITY: Using an Equation to Solve a System

a. Write the cost equation from Activity 1.

$C =$ ______

b. Write the revenue equation from Activity 1.

$R =$ ______

c. The break-even point occurs when $C = R$. Set the expression for C equal to the expression for R. You should obtain an equation with x on both sides. Solve this equation for x. The solution is your break-even point.

d. Did you obtain the same break-even point in Activities 2, 3, and 4? If not, check your work. The break-even point should be the same in all three activities.

What Is Your Answer?

5. **IN YOUR OWN WORDS** How can you solve a system of linear equations?

6. When solving a system of linear equations, explain why it is a good idea to use two different ways to find the solution.

Practice Use what you learned about systems of linear equations to complete Exercises 3 and 4 on page 80.

2.5 Lesson

Check It Out Lesson Tutorials BigIdeasMath.com

Key Vocabulary
system of linear equations, *p. 78*
solution of a system of linear equations, *p. 78*

A **system of linear equations** is a set of two or more linear equations in the same variables. A **solution of a system of linear equations** in two variables is an ordered pair that makes each equation true.

EXAMPLE 1 Solving a System of Linear Equations Using a Table

Solve the system. $y = x - 5$ Equation 1
$y = -3x + 7$ Equation 2

Reading
A system of linear equations is also called a *linear system*.

Step 1: Make a table of values.

Step 2: Find an x-value that gives the same y-value for both equations.

x	0	1	2	3
$y = x - 5$	−5	−4	−3	−2
$y = -3x + 7$	7	4	1	−2

The solution is (3, −2).

EXAMPLE 2 Solving a System of Linear Equations Using a Graph

Solve the system. $y = 2x + 3$ Equation 1
$y = -x + 6$ Equation 2

Step 1: Graph each equation.

Step 2: Find the point of intersection. The graphs appear to intersect at (1, 5).

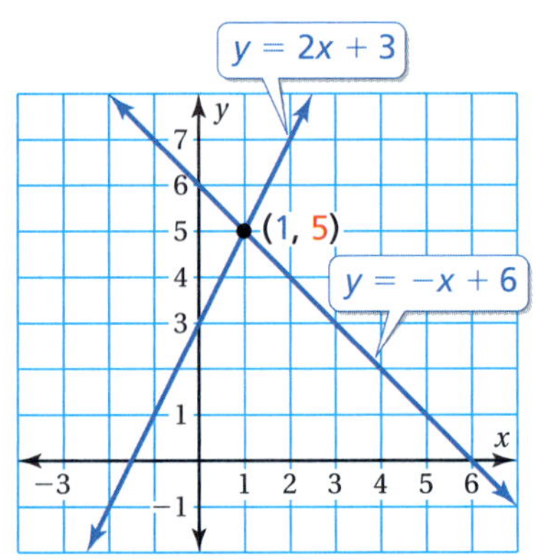

Step 3: Check your solution.

Equation 1

$y = 2x + 3$

$5 \stackrel{?}{=} 2(1) + 3$

$5 = 5$ ✓

Equation 2

$y = -x + 6$

$5 \stackrel{?}{=} -1 + 6$

$5 = 5$ ✓

The solution is (1, 5).

On Your Own

Now You're Ready Exercises 5–7 and 10–12

Solve the system of linear equations using a table and using a graph.

1. $y = x - 1$
   $y = -x + 3$
2. $y = -5x + 14$
   $y = x - 10$
3. $y = x$
   $y = 2x + 1$

Multi-Language Glossary at BigIdeasMath.com.

Key Idea

Solving a System of Linear Equations Algebraically

Step 1 Solve both equations for one of the variables.

Step 2 Set the expressions equal to each other and solve for the variable.

Step 3 Substitute back into one of the original equations and solve for the other variable.

EXAMPLE 3 Solving a System of Linear Equations Algebraically

A middle school yearbook committee has 35 members. There are 7 more girls than boys. Use the models to write a system of linear equations. Then solve the system to find the number of boys x and the number of girls y.

Number of boys, x + Number of girls, y = 35

Number of girls, y = Number of boys, x + 7

The system is $x + y = 35$ and $y = x + 7$.

Step 1: Solve $x + y = 35$ for y.

$y = 35 - x$ Subtract x from each side.

Step 2: Set the expressions equal to each other and solve for x.

$35 - x = x + 7$ Set expressions equal to each other.

$28 = 2x$ Subtract 7 from each side. Add x to each side.

$14 = x$ Divide each side by 2.

Step 3: Substitute $x = 14$ into one of the original equations and solve for y.

$y = x + 7$ Write one of the original equations.

$= 14 + 7$ Substitute 14 for x.

$= 21$ Add.

There are 14 boys and 21 girls on the yearbook committee.

Study Tip

Be sure to check your solutions.

On Your Own

Now You're Ready
Exercises 13–15

4. **WHAT IF?** In Example 3, the yearbook committee has 45 members. Use the models to write a system of linear equations. Then solve the system to find the number of boys x and the number of girls y.

2.5 Exercises

Check It Out
Help with Homework
BigIdeasMath.com

Vocabulary and Concept Check

1. **VOCABULARY** Do the equations $4a - 3b = 5$ and $7b + 2a = -8$ form a system of linear equations? Explain.

2. **REASONING** Can a point in Quadrant II be a break-even point for a system? Explain.

Practice and Problem Solving

Use the table to find the break-even point. Check your solution.

3. $C = 15x + 150$

   $R = 45x$

x	0	1	2	3	4	5	6
C							
R							

4. $C = 24x + 80$

   $R = 44x$

x	0	1	2	3	4	5	6
C							
R							

Solve the system of linear equations using a table.

1 5. $y = x + 4$

   $y = 3x - 1$

6. $y = 1.5x - 2$

   $y = -x + 13$

7. $y = \frac{2}{3}x - 3$

   $y = -2x + 5$

8. **ERROR ANALYSIS** Describe and correct the error in solving the system of linear equations.

x	0	1	2	3
$y = -2x - 1$	−1	−3	−5	−7
$y = x - 7$	−7	−6	−5	−4

The solution is (−5, −5).

9. **CARRIAGE RIDES** The cost C (in dollars) for the care and maintenance of a horse and carriage is $C = 15x + 2000$, where x is the number of rides.

   a. Write an equation for the revenue R in terms of the number of rides.

   b. How many rides are needed for the business to break even?

Solve the system of linear equations using a graph.

2 **10.** $y = 2x + 9$
$y = 6 - x$

11. $y = -x - 4$
$y = \frac{3}{5}x + 4$

12. $y = 2x + 5$
$y = \frac{1}{2}x - 1$

Solve the system of linear equations algebraically.

3 **13.** $x + y = 27$
$y = x + 3$

14. $y - x = 17$
$y = 4x + 2$

15. $x - y = 7$
$0.5x + y = 5$

16. **HOMEWORK** You have 42 math and science problems for homework. You have 10 more math problems than science problems. Use the model to write a system of linear equations. How many problems do you have in each subject?

Number of math problems, x + Number of science problems, y = 42

Number of science problems, y = Number of math problems, x − 10

17. **CANOEING** You and your friend are canoeing on the Withlacoochee River Canoe Trail. Your position on the trail y (in miles) is represented by $y = 3.5x + 28$, where x is in hours. Your friend's position is represented by $y = 2x + 37$.

a. How long will it take you to catch up with your friend?

b. How far will you each have traveled when you catch up with your friend?

18. Critical Thinking You buy x bottles of face paint and y brushes at two stores. The amounts you spend are represented by $10x + 7.5y = 42.5$ and $8x + 6y = 34$. How many bottles of face paint and brushes did you buy?

Fair Game Review What you learned in previous grades & lessons

Decide whether the two equations are equivalent. *(Section 1.2 and Section 1.3)*

19. $4n + 1 = n - 8$
$3n = -9$

20. $2a + 6 = 12$
$a + 3 = 6$

21. $7v - \frac{3}{2} = 5$
$14v - 3 = 15$

22. **MULTIPLE CHOICE** Which line has the same slope as $y = \frac{1}{2}x - 3$? *(Section 2.3)*

Ⓐ $y = -2x + 4$ Ⓑ $y = 2x + 3$ Ⓒ $y - 2x = 5$ Ⓓ $2y - x = 7$

2.6 Special Systems of Linear Equations

STATE STANDARDS
MA.8.A.1.3
MA.8.A.1.4

Essential Question Can a system of linear equations have no solution? Can a system of linear equations have many solutions?

1 ACTIVITY: Writing a System of Linear Equations

Work with a partner.

Your cousin is 3 years older than you. Your ages can be represented by two linear equations.

$y = t$ Your age

$y = t + 3$ Your cousin's age

a. Graph both equations in the same coordinate plane.

b. What is the vertical distance between the two graphs? What does this distance represent?

c. Do the two graphs intersect? If not, what does this mean in terms of your age and your cousin's age?

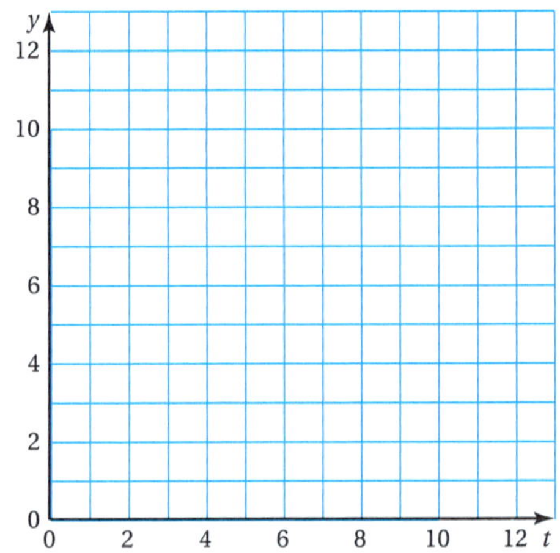

2 ACTIVITY: Using a Table to Solve a System

Work with a partner. You invest \$500 for equipment to make dog backpacks. Each backpack costs you \$15 for materials. You sell each backpack for \$15.

a. Copy and complete the table for your cost C and your revenue R.

x	0	1	2	3	4	5	6	7	8	9	10
C											
R											

b. When will your company break even? What is wrong?

3 ACTIVITY: Using a Graph to Solve a Puzzle

Work with a partner. Let x and y be two numbers. Here are two clues about the values of x and y.

	Words	Equation
Clue 1:	y is 4 more than twice the value of x.	$y = 2x + 4$
Clue 2:	The difference of $3y$ and $6x$ is 12.	$3y - 6x = 12$

a. Graph both equations in the same coordinate plane.

b. Do the two lines intersect? Explain.

c. What is the solution of the puzzle?

d. Use the equation $y = 2x + 4$ to complete the table.

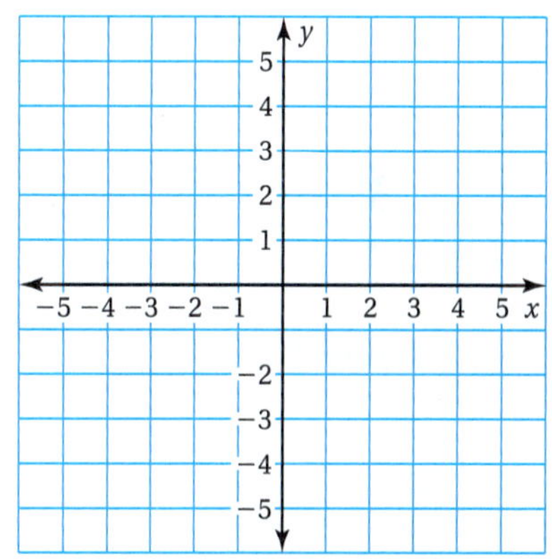

x	0	1	2	3	4	5	6	7	8	9	10
y											

e. Does each solution in the table satisfy *both* clues?

f. What can you conclude? How many solutions does the puzzle have? How can you describe them?

What Is Your Answer?

4. IN YOUR OWN WORDS Can a system of linear equations have no solution? Can a system of linear equations have many solutions? Give examples to support your answers.

Practice Use what you learned about special systems of linear equations to complete Exercises 4 and 5 on page 86.

2.6 Lesson

Check It Out
Lesson Tutorials
BigIdeasMath.com

EXAMPLE 1 Solving a Special System of Linear Equations

Solve the system.

$\mathbf{y = 3x + 1}$ Equation 1

$\mathbf{y = 3x - 5}$ Equation 2

Graph each equation.

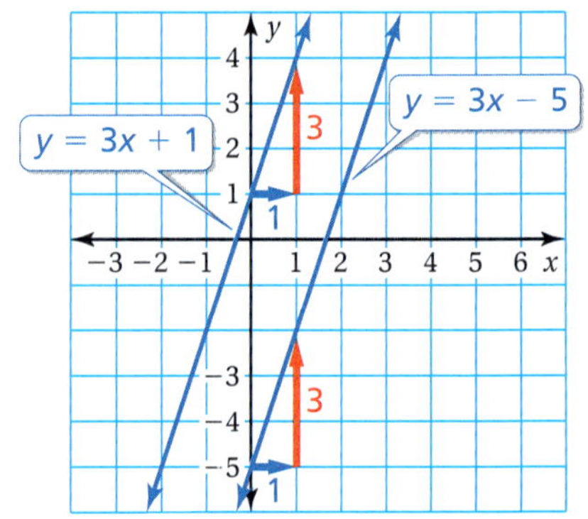

The lines have the same slope and different y-intercepts. So, the lines are parallel.

Because parallel lines do not intersect, there is no point that is a solution of both equations.

∴ So, the system of linear equations has no solution.

EXAMPLE 2 Solving a Special System of Linear Equations

Solve the system.

$\mathbf{y = -2x + 4}$ Equation 1

$\mathbf{4x + 2y = 8}$ Equation 2

Write $4x + 2y = 8$ in slope-intercept form.

$4x + 2y = 8$ Write the equation.

$2y = -4x + 8$ Subtract $4x$ from each side.

$y = -2x + 4$ Divide each side by 2.

The equations are the same.

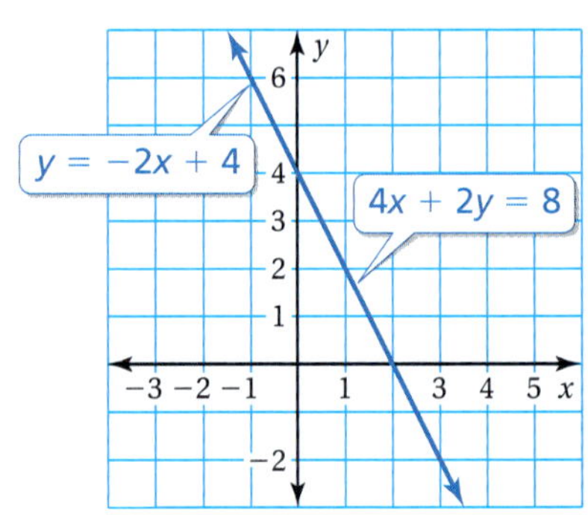

The solution of the system is all the points on the line $y = -2x + 4$.

∴ So, the system of linear equations has infinitely many solutions.

On Your Own

Now You're Ready
Exercises 6–11

Solve the system of linear equations.

1. $-4x + 4y = 8$
$y = x + 2$

2. $y = -5x - 2$
$5x + y = 3$

3. $x + y = 6$
$y = -x$

EXAMPLE 3 Solving a Special System of Linear Equations

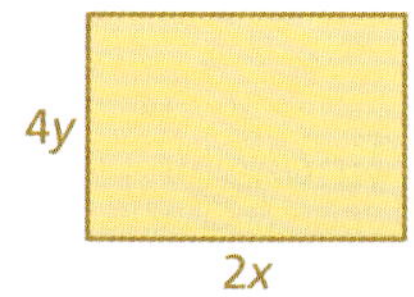

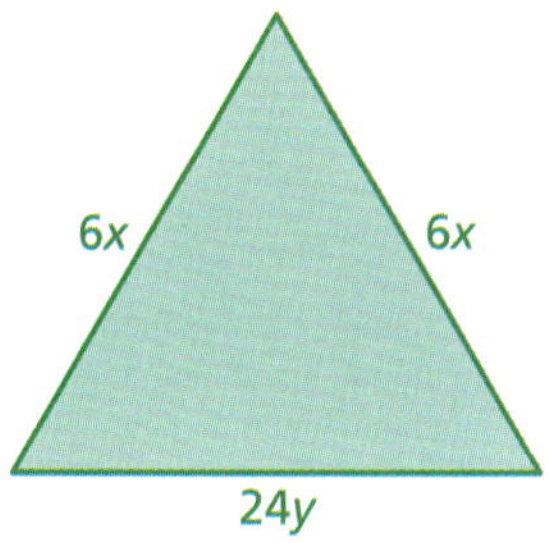

The perimeter of the rectangle is 36 units. The perimeter of the triangle is 108 units. Write and solve a system of linear equations to find the values of x and y.

Perimeter of rectangle		*Perimeter of triangle*	
$2(2x) + 2(4y) = 36$		$6x + 6x + 24y = 108$	
$4x + 8y = 36$	Equation 1	$12x + 24y = 108$	Equation 2

The system of linear equations is $4x + 8y = 36$ and $12x + 24y = 108$.

Write both equations in slope-intercept form.

$4x + 8y = 36$	$12x + 24y = 108$
$8y = -4x + 36$	$24y = -12x + 108$
$y = -\frac{1}{2}x + \frac{9}{2}$	$y = -\frac{1}{2}x + \frac{9}{2}$

The equations are the same. So, the solution of the system is all the points on the line $y = -\frac{1}{2}x + \frac{9}{2}$.

∴ The system of linear equations has infinitely many solutions.

On Your Own

4. **WHAT IF?** What happens to the solution in Example 3 if the perimeter of the rectangle is 54 units? Explain.

Summary

Solutions of Systems of Linear Equations

A system of linear equations can have *one solution, no solution,* or *infinitely many solutions.*

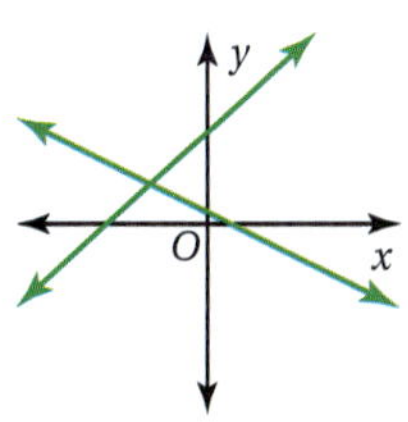

One solution

The lines intersect.

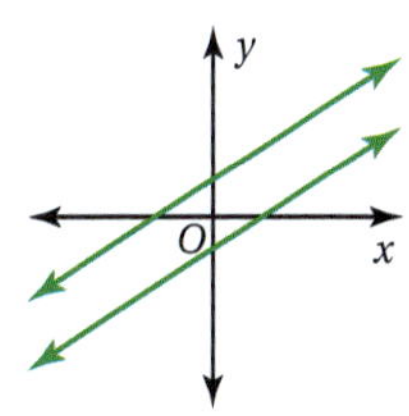

No solution

The lines are parallel.

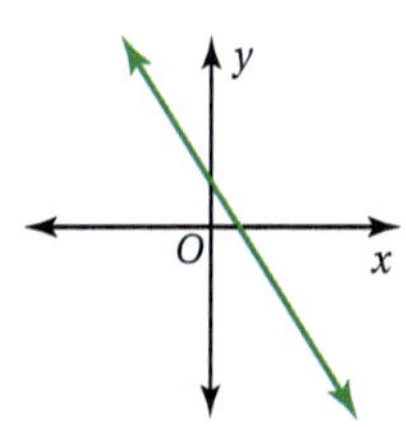

Infinitely many solutions

The lines are the same.

2.6 Exercises

Check It Out
Help with Homework
BigIdeasMath.com

Vocabulary and Concept Check

1. **VOCABULARY** What is the difference between the graph of a system of linear equations that has *no solution* and the graph of a system of linear equations that has *infinitely many solutions*?

2. **NUMBER SENSE** Determine the number of solutions of the system of linear equations without writing the equations in slope-intercept form. Explain your reasoning.

$$2x + y = 5$$
$$4x + 2y = 10$$

3. **REASONING** One equation in a system of linear equations has a slope of -3. The other equation has a slope of 4. How many solutions does the system have? Explain.

Practice and Problem Solving

Let x and y be two numbers. Find the solution of the puzzle.

4. y is $\frac{1}{3}$ more than 4 times the value of x.

   The difference of $3y$ and $12x$ is 1.

5. $\frac{1}{2}$ of x plus 3 is equal to y.

   x is 6 more than twice the value of y.

Solve the system of linear equations.

(1) (2) 6. $-6x + 3y = 18$
   $y = 2x - 2$

7. $y = -\frac{1}{6}x + 5$
   $x + 6y = 30$

8. $-x + 2y = -3$
   $9x - 3y = -3$

9. $3x + 2y = 0$
   $y = x - 5$

10. $y = \frac{4}{9}x + \frac{1}{3}$
    $-4x + 9y = 3$

11. $y = -6x + 8$
    $12x + 2y = -8$

12. **ERROR ANALYSIS** Describe and correct the error in solving the system of linear equations.

$$y = -2x + 6$$
$$8x + 4y = 24$$

13. **PIG RACE** In a pig race, your pig gets a head start of 3 feet running at a rate of 2 feet per second. Your friend's pig is running at a rate of 2 feet per second. A system of linear equations that represents this situation is $y = 2x + 3$ and $y = 2x$. Will your friend's pig catch up to your pig? Explain.

Solve the system of linear equations.

14. $y + 4.6x = 5.8x + 0.4$
$-4.8x + 4y = 1.6$

15. $y = \frac{\pi}{3}x + \pi$
$-\pi x + 3y = -6\pi$

16. $-2x + y = 1.3$
$2(0.5x - y) = 4.6$

$4x + 8y = 64$
$8x + 16y = 128$

17. MONEY You and a friend both work two different jobs. The system of linear equations represents the total earnings for x hours worked at the first job and y hours worked at the second job. Your friend earns twice as much as you.

a. One week, both of you work 4 hours at the first job. How many hours do you and your friend work at the second job?

b. Both of you work the same number of hours at the second job. Compare the number of hours you each work at the first job.

18. BOAT RACE The Mug Race is a timed race that is 42 miles long from Palatka to Jacksonville along the St. Johns River.

a. Which boat left the starting point first?

b. Compare the speeds of the boats.

c. Estimate how long it takes each boat to finish the race.

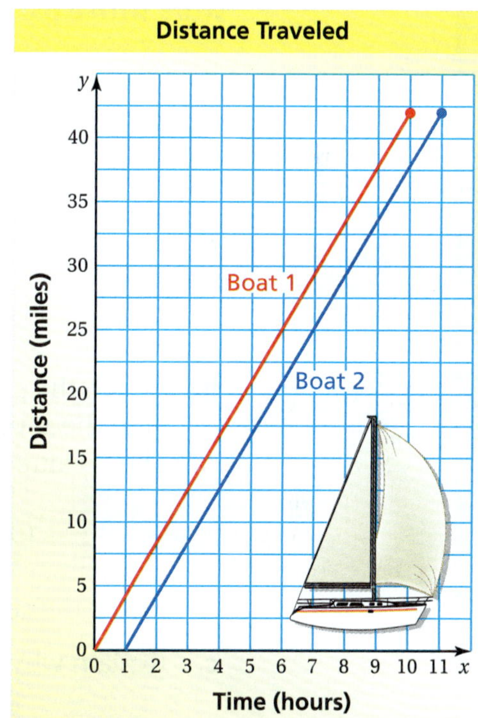

19. Critical Thinking One equation in a system of linear equations is $y = 3x - 1$.

a. Write a second equation so that $(-2, -7)$ is the only solution of the system.

b. Write a second equation so that the system has no solution.

c. Write a second equation so that the system has infinitely many solutions.

Fair Game Review What you learned in previous grades & lessons

Solve the equation. Check your solution. *(Section 1.3)*

20. $3x - 5 = 2x + 1$

21. $-3(x - 1) = -8x - 12$

22. $\frac{1}{2}x + 4 = \frac{3}{4}x + 6$

23. MULTIPLE CHOICE What is the slope of the line represented by the points in the table? *(Section 2.2)*

x	0	4	8	12
y	2	5	8	11

Ⓐ $-\frac{3}{4}$ Ⓑ $\frac{3}{4}$ Ⓒ $\frac{4}{3}$ Ⓓ 2

2.7 Solving Equations by Graphing

STATE STANDARDS
MA.8.A.1.1
MA.8.A.1.3
MA.8.A.1.4

Essential Question How can you use a system of linear equations to solve an equation that has variables on both sides?

You learned how to use algebra to solve equations with variables on both sides. Another way is by using a system of linear equations.

1 ACTIVITY: Solving a System of Linear Equations

Work with a partner. Find the solution of $2x - 1 = -\frac{1}{2}x + 4$.

a. Use the left side of the equation to write one linear equation. Then, use the right side to write another linear equation.

$$2x - 1 = -\frac{1}{2}x + 4$$

$y = 2x - 1$ $\qquad$ $y = -\frac{1}{2}x + 4$

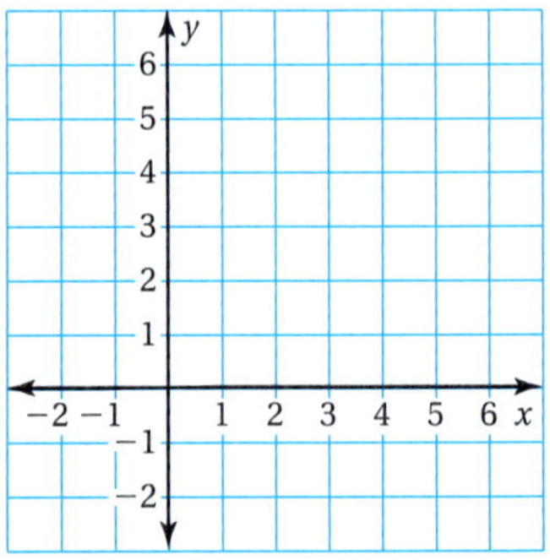

b. Sketch the graphs of the two linear equations. Find the x-value of the point of intersection. The x-value is the solution of

$$2x - 1 = -\frac{1}{2}x + 4.$$

Check the solution.

c. Explain why this "graphical method" works.

2 ACTIVITY: Using a Graphing Calculator

Use a graphing calculator to graph the two linear equations.

$$y = 2x - 1$$

$$y = -\frac{1}{2}x + 4$$

The steps used to enter the equations depend on the calculator model that you have.

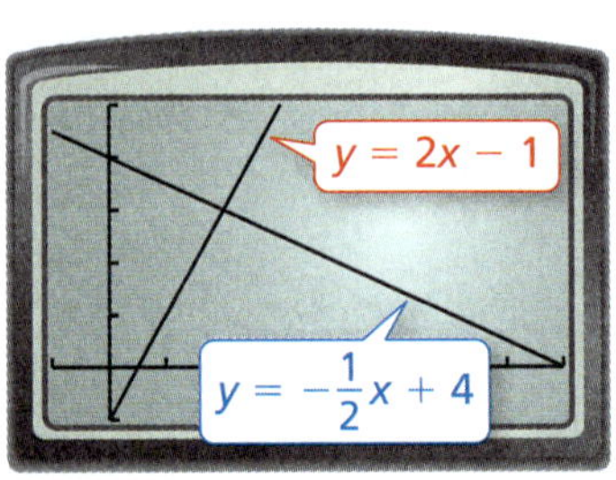

3 ACTIVITY: Using a System of Linear Equations

Work with a partner. Solve each equation using two methods.

- **Method 1:** Use an algebraic method.
- **Method 2:** Use a graphical method.
- Is the solution the same using both methods?

a. $\frac{1}{2}x + 4 = -\frac{1}{4}x + 1$

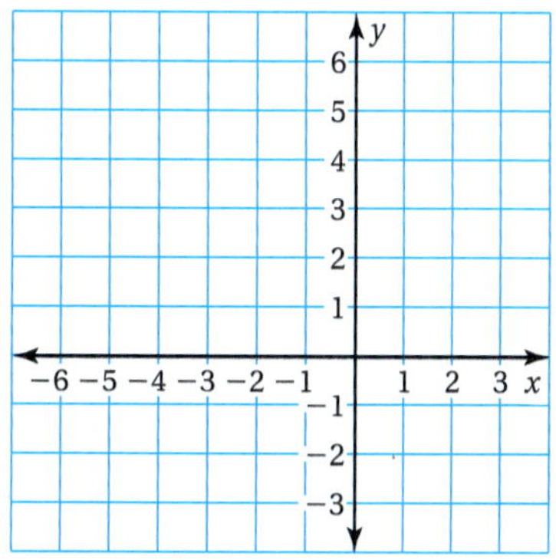

b. $\frac{2}{3}x + 4 = \frac{1}{3}x + 3$

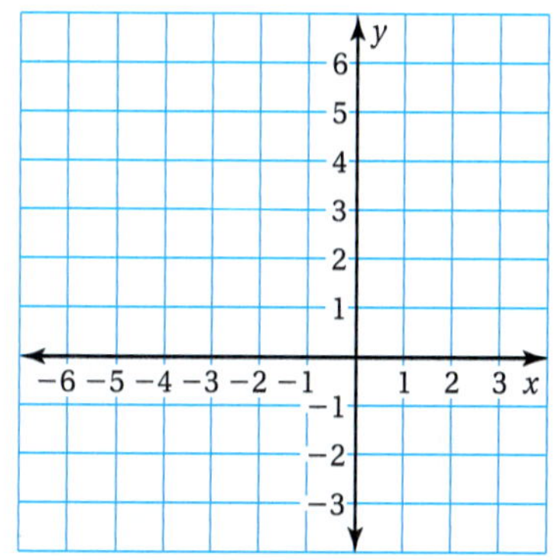

c. $-\frac{2}{3}x - 1 = \frac{1}{3}x - 4$

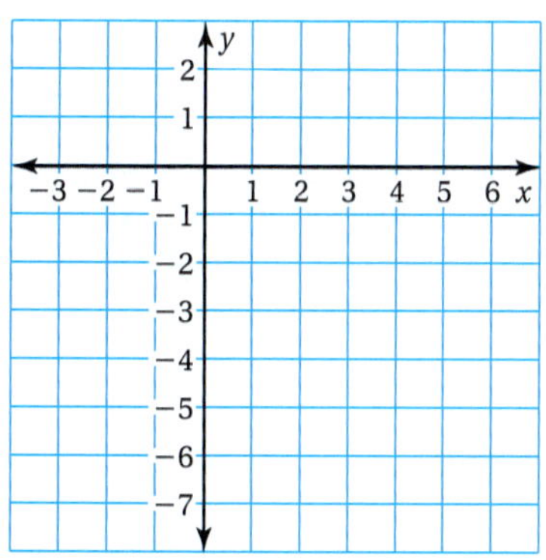

d. $\frac{4}{5}x + \frac{7}{5} = 3x - 3$

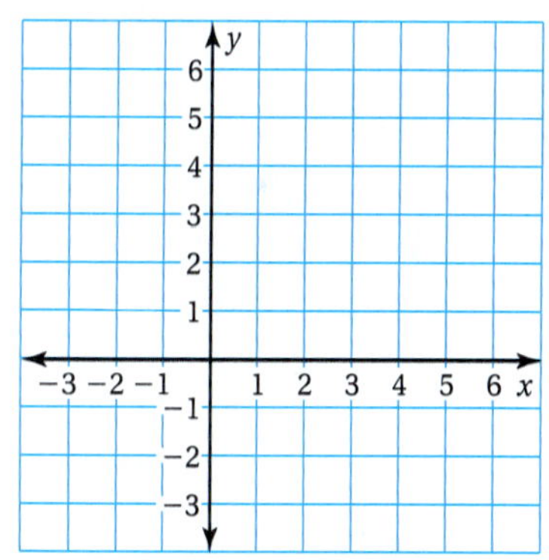

What Is Your Answer?

4. **IN YOUR OWN WORDS** How can you use a system of linear equations to solve an equation that has variables on both sides? Give an example that is different from those in Activities 1 and 3.

5. Describe three ways in which René Descartes's invention of the coordinate plane allows you to solve algebraic problems graphically.

Practice Use what you learned about solving equations by graphing to complete Exercises 3–5 on page 92.

2.7 Lesson

Key Idea

Solving Equations Using Graphs

Step 1: To solve the equation $ax + b = cx + d$, write two linear equations.

$$ax + b = cx + d$$

$y = ax + b$ and $y = cx + d$

Step 2: Graph the system of linear equations. The x-value of the solution of the system of linear equations is the solution of the equation $ax + b = cx + d$.

EXAMPLE 1 Solving an Equation Using a Graph

Solve $x - 2 = -\frac{1}{2}x + 1$ using a graph. Check your solution.

Step 1: Write a system of linear equations using each side of the equation.

$$x - 2 = -\frac{1}{2}x + 1$$

$y = x - 2$ $\qquad$ $y = -\frac{1}{2}x + 1$

Step 2: Graph the system.

$$y = x - 2$$

$$y = -\frac{1}{2}x + 1$$

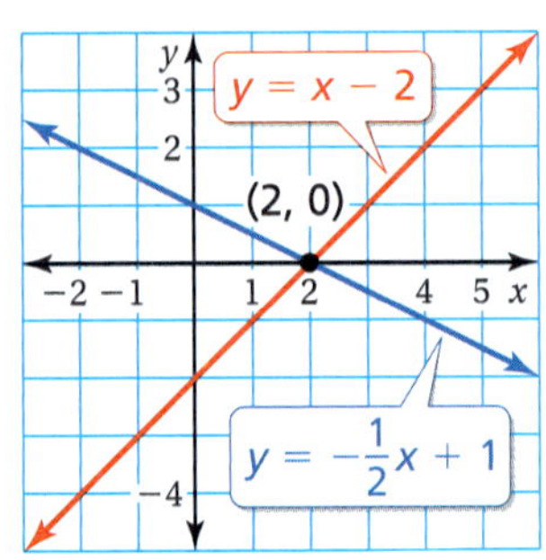

Check

$$x - 2 = -\frac{1}{2}x + 1$$

$$2 - 2 \stackrel{?}{=} -\frac{1}{2}(2) + 1$$

$$0 = 0 \checkmark$$

The graphs intersect at (2, 0). So, the solution is $x = 2$.

On Your Own

Now You're Ready
Exercises 6 and 7

Use a graph to solve the equation. Check your solution.

1. $\frac{1}{3}x = x + 8$

2. $1.5x + 2 = 11 - 3x$

EXAMPLE 2 Real-Life Application

Plant A grows 0.6 inch per month. Plant B grows twice as fast.

a. **Use the model to write an equation.**

b. **After how many months x are the plants the same height?**

Growth rate	•	Months, x	+	Original height	=	Growth rate	•	Months, x	+	Original height

a. The equation is $0.6x + 12 = 1.2x + 9$.

b. Write a system of linear equations using each side of the equation. Then graph the system.

$$0.6x + 12 = 1.2x + 9$$

$y = 0.6x + 12$ $\qquad$ $y = 1.2x + 9$

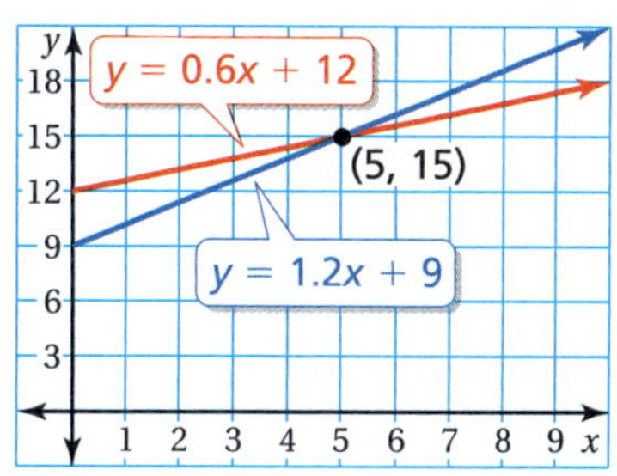

The solution of the system is $(x, y) = (5, 15)$. So, the plants are both 15 inches tall after 5 months.

On Your Own

3. Using the graph in Example 2, is the statement below true? Explain.

The system of linear equations $y = 0.6x + 12$ *and* $y = 1.2x + 9$ *has one solution.*

4. **WHAT IF?** In Example 2, the growth rate of Plant A is 0.5 inch per month. After how many months x are the plants the same height?

2.7 Exercises

Check It Out
Help with Homework
BigIdeasMath.com

Vocabulary and Concept Check

1. **CRITICAL THINKING** Would you rather solve the equation $x - \frac{4}{5} = -x + \frac{6}{5}$ using an algebraic method or a graphical method? Explain.

2. **DIFFERENT WORDS, SAME QUESTION** Which is different? Find "both" answers.

What is the solution of the equation $x - 3 = -\frac{1}{3}x + 5$?	What is the x-value of the solution of the linear system $y + 3 = x$ and $y + \frac{1}{3}x = 5$?
What is the y-coordinate of the intersection of $y = x - 3$ and $y = -\frac{1}{3}x + 5$?	What is the x-coordinate of the intersection of $y = x - 3$ and $y = -\frac{1}{3}x + 5$?

Practice and Problem Solving

Solve the equation algebraically and graphically.

3. $\frac{1}{3}x - 2 = -\frac{1}{6}x + 1$

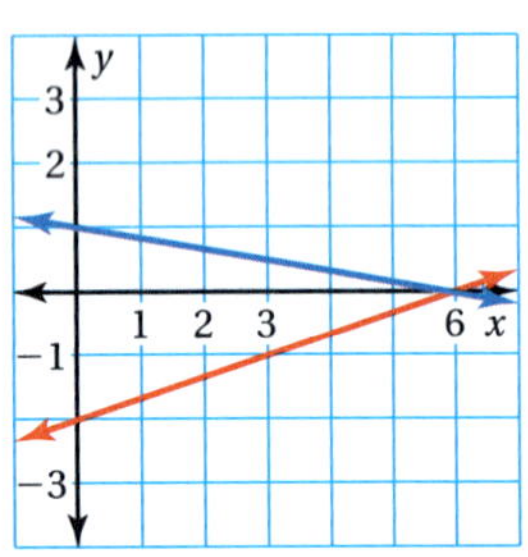

4. $-\frac{3}{4}x + \frac{5}{4} = x + 3$

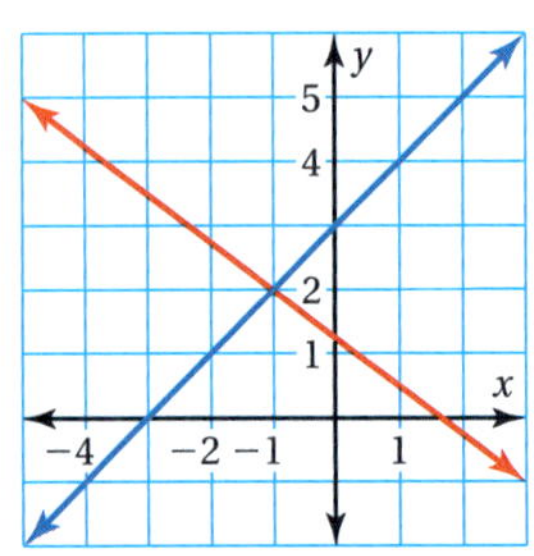

5. $\frac{5}{8}x - \frac{3}{4} = \frac{1}{3}x + 1$

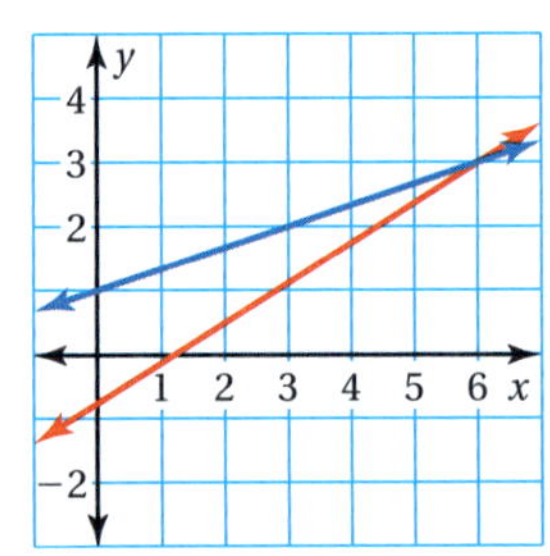

Use a graph to solve the equation. Check your solution.

1 6. $\frac{2}{5}x - 2 = -x + 12$

7. $-\frac{5}{6}x + \frac{1}{2} = -x + 1$

8. **ERROR ANALYSIS** Describe and correct the error in solving the equation $2x + 4 = -\frac{7}{2}x + 11$.

✗

$y = 2x + 4$

$y = -\frac{7}{2}x + 11$

There is no solution.

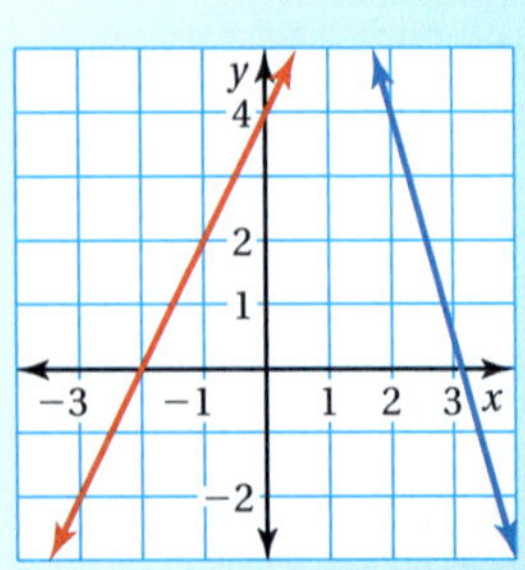

9. **KARAOKE** One night at karaoke, you sang $3x + 2$ songs. The next night, you sang $4x$ songs. Is it possible that you sang the same number of songs each night? Explain.

Use a graph to solve the equation. Check your solution.

10. $2.5x + 3 = 4x - 3$ **11.** $-1.4x + 1 = 1.6x - 5$ **12.** $0.7x - 1.2 = -1.4x - 7.5$

13. **CRITICAL THINKING** What happens when you use a graphical method to solve $\frac{1}{3}x - 5 = \frac{1}{3}x + 8$? Does an algebraic method give the same result?

14. **HIKING** You hike uphill at a rate of 200 feet per minute. Your friend hikes downhill on the same trail at a rate of 250 feet per minute. How long will it be until you meet?

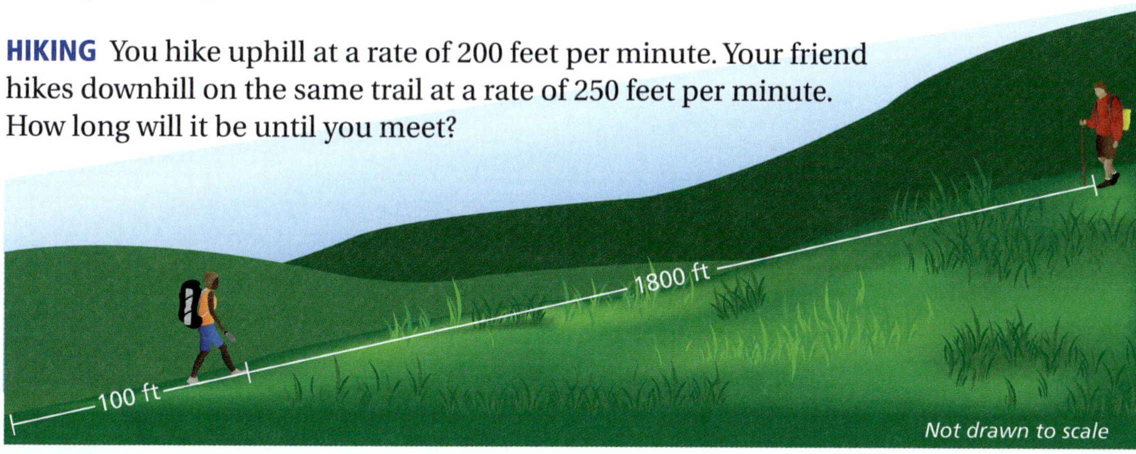

15. **SOCCER** A soccer team played four more home games and three-fourths as many away games this year than last year. The team played the same number of games each season. How many away games did the team play last year?

Last Year	
Home	**Away**
11	x

16. **Geometry** Candle A burns at an average rate of 11 cubic centimeters per hour. Candle B burns at an average rate of 18 cubic centimeters per hour. Do the candles ever have the same volume? Explain.

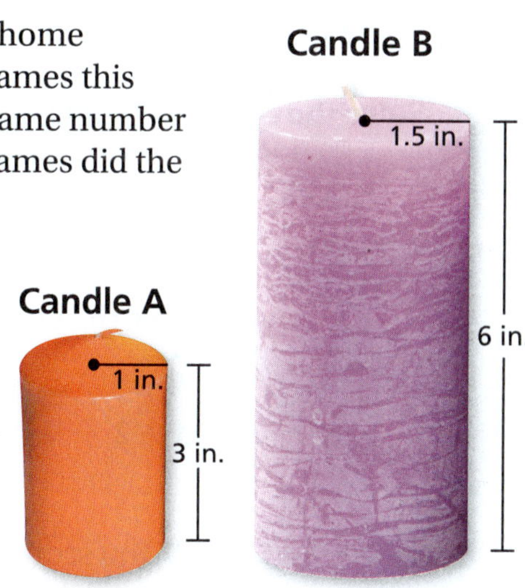

Fair Game Review What you learned in previous grades & lessons

Find the y-intercept of the graph of the linear equation. *(Section 2.3)*

17. $y = 3x + 4$ **18.** $y = -\frac{2}{3}x + 6$ **19.** $y = 2.4x - 3$ **20.** $y = 2x - \pi$

21. **MULTIPLE CHOICE** Which of the following is the slope of the line? *(Section 2.2)*

Ⓐ $-\frac{6}{7}$ Ⓑ $-\frac{7}{6}$

Ⓒ $\frac{6}{7}$ Ⓓ $\frac{7}{6}$

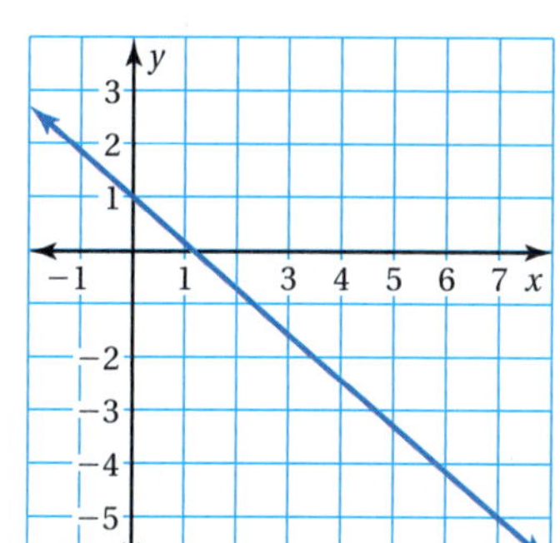

2.5–2.7 Quiz

Check It Out
Progress Check
BigIdeasMath.com

Solve the system of linear equations algebraically. *(Section 2.5)*

1. $y = 4 - x$
$y = x - 4$

2. $y = \frac{x}{2} + 10$
$y = 4x - 4$

Use the table to find the break-even point. Check your solution. *(Section 2.5)*

3. $C = 10x + 180$
$R = 46x$

x	0	1	2	3	4	5	6
C							
R							

Solve the system of linear equations using a graph. *(Section 2.5)*

4. $y = 3x + 2$
$y = x + 4$

5. $y = -3x - 1$
$y = -2x + 5$

Solve the system of linear equations using any method. *(Section 2.6)*

6. $y = 2x - 3$
$y - 6x = -9$

7. $y = 4x + 8$
$2y - 8x = 18$

Use a graph to solve the equation. Check your solution. *(Section 2.7)*

8. $\frac{1}{4}x - 4 = \frac{3}{4}x + 2$

9. $8x - 14 = -2x - 4$

10. **BASKETBALL** You score 24 points in a basketball game. You make 9 shots. How many three-point shots and two-point shots do you make? *(Section 2.5)*

Number of three-point shots, x + Number of two-point shots, y = 9

Value of a three-point shot • Number of three-point shots, x + Value of a two-point shot • Number of two-point shots, y = 24

11. **BICYCLE** One day, you ride $2x + 5$ kilometers on your bicycle. The next day, you ride $3x$ kilometers. Is it possible that you rode the same distance each day? Explain. *(Section 2.7)*

12. **TEMPERATURE** Two students write the expressions $\frac{1}{2}x + 49$ and $2x - 5$ to represent today's high temperature (in degrees Fahrenheit). What is today's high temperature? *(Section 2.7)*

2 Chapter Review

Check It Out
Vocabulary Help
BigIdeasMath.com

Review Key Vocabulary

linear equation *p. 50*
solution of a linear equation, *p. 50*
slope, *p. 56*
rise, *p. 56*
run, *p. 56*
x-intercept, *p. 64*
y-intercept, *p. 64*
slope-intercept form, *p. 64*
standard form, *p. 70*
system of linear equations, *p. 78*
solution of a system of linear equations, *p. 78*

Review Examples and Exercises

2.1 Graphing Linear Equations *(pp. 48–53)*

Graph $y = 3x - 1$.

Step 1: Make a table of values.

x	$y = 3x - 1$	y	(x, y)
-2	$y = 3(-2) - 1$	-7	$(-2, -7)$
-1	$y = 3(-1) - 1$	-4	$(-1, -4)$
0	$y = 3(0) - 1$	-1	$(0, -1)$
1	$y = 3(1) - 1$	2	$(1, 2)$

Step 2: Plot the ordered pairs.

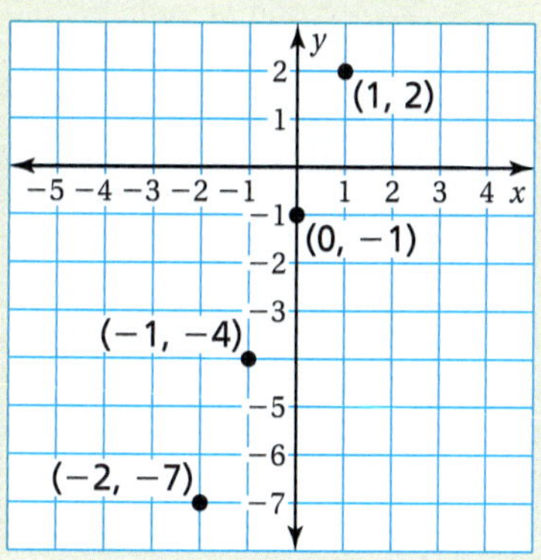

Step 3: Draw a line through the points.

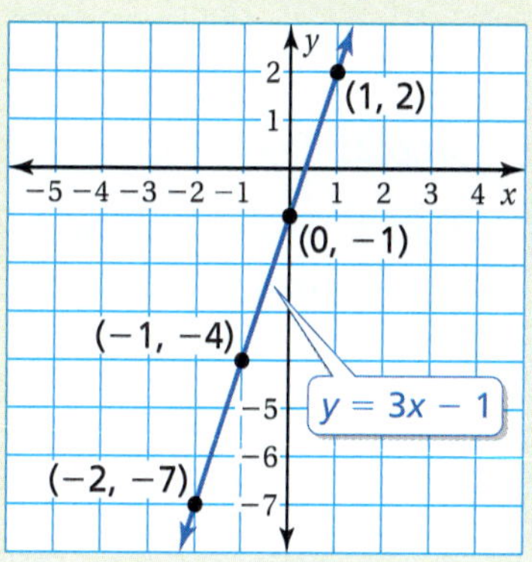

Exercises

Graph the linear equation.

1. $y = \frac{3}{5}x$
2. $y = -2$
3. $y = 9 - x$
4. $y = 1$
5. $y = \frac{2}{3}x + 2$
6. $y = 1 + x$

2.2 Slope of a Line (pp. 54–61)

Which two lines are parallel? Explain.

Find the slope of each line.

Red Line $\frac{-2}{6} = -\frac{1}{3}$

Blue Line $\frac{-2}{4} = -\frac{1}{2}$

Green Line $\frac{-3}{9} = -\frac{1}{3}$

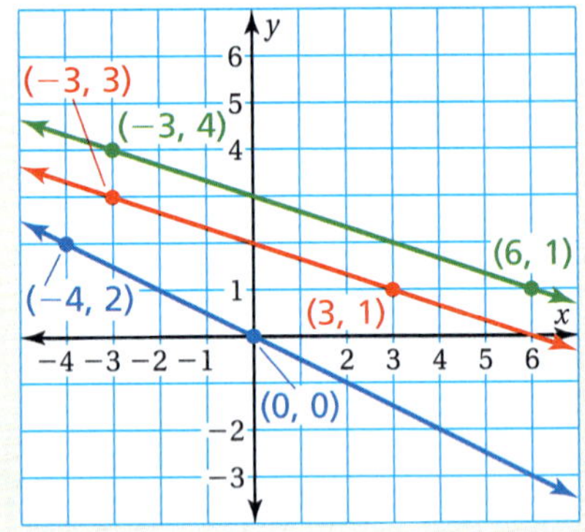

∴ The red and green lines have the same slope, so they are parallel.

Exercises

The points in the table lie on a line. Find the slope of the line. Then draw its graph.

7.

x	0	1	2	3
y	−1	0	1	2

8.

x	−2	0	2	4
y	3	4	5	6

2.3 Graphing Linear Equations in Slope-Intercept Form (pp. 62–67)

Graph $y = 0.5x - 3$. Identify the x-intercept.

Step 1: Find the slope and y-intercept.

$$y = 0.5x + (-3)$$

slope: 0.5; y-intercept: -3

Step 2: The y-intercept is -3. So, plot $(0, -3)$.

Step 3: Use the slope to find another point and draw the line.

$$\text{slope} = \frac{\text{rise}}{\text{run}} = \frac{1}{2}$$

Plot the point that is 2 units right and 1 unit up from (0, –3). Draw a line through the two points.

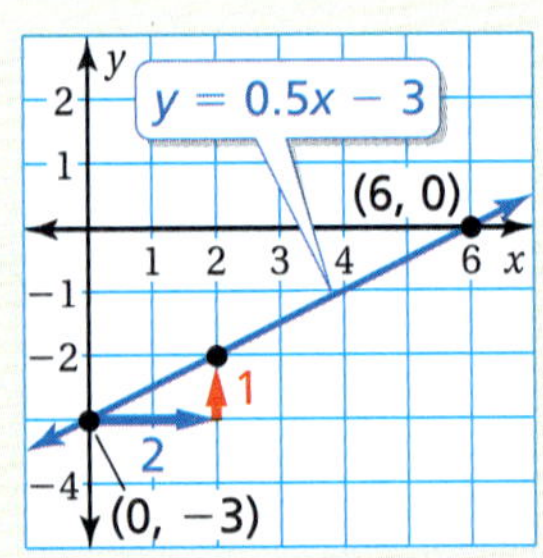

∴ The line crosses the x-axis at (6, 0). So, the x-intercept is 6.

Exercises

Graph the linear equation. Identify the x-intercept.

9. $y = 2x - 6$ **10.** $y = -4x + 8$ **11.** $y = -x - 8$

2.4 Graphing Linear Equations in Standard Form *(pp. 68–73)*

Graph $8x + 4y = 16$.

Step 1: Write the equation in slope-intercept form.

$$8x + 4y = 16 \quad \text{Write the equation.}$$

$$4y = -8x + 16 \quad \text{Subtract } 8x \text{ from each side.}$$

$$y = -2x + 4 \quad \text{Divide each side by 4.}$$

Step 2: Use the slope and y-intercept to plot two points.

$$y = -2x + 4$$

slope: -2 ; y-intercept: 4

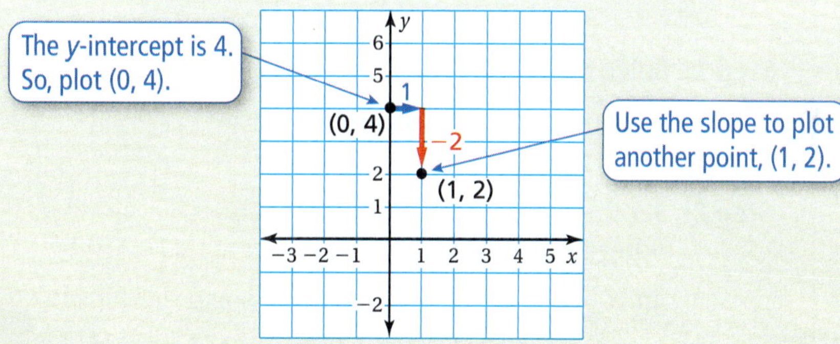

Step 3: Draw a line through the points.

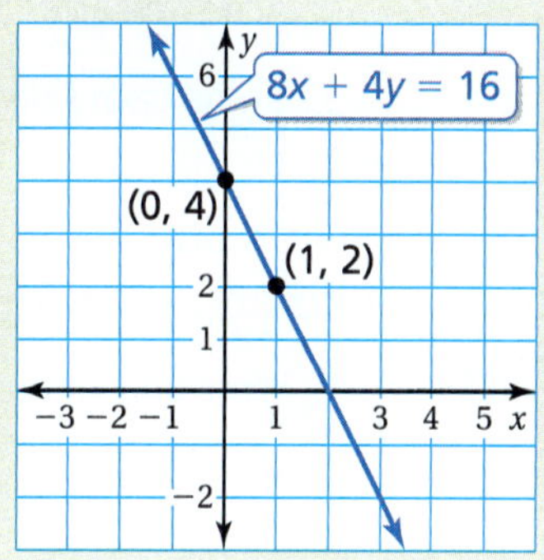

Exercises

Graph the linear equation.

12. $\frac{1}{4}x + y = 3$

13. $-4x + 2y = 8$

14. $x + 5y = 10$

15. $-\frac{1}{2}x + \frac{1}{8}y = \frac{3}{4}$

2.5 Systems of Linear Equations (pp. 76–81)

A middle school science club has 30 members. There are 12 more boys than girls. Use the models to write a system of linear equations. Then solve the system to find the number of boys x and the number of girls y.

Number of boys, x + Number of girls, y = 30

Number of boys, x = Number of girls, y + 12

The system is $x + y = 30$ and $x = y + 12$.

Step 1: Solve $x + y = 30$ for x.

$x = 30 - y$ Subtract y from each side.

Step 2: Set the expressions equal to each other and solve for y.

$30 - y = y + 12$ Set expressions equal to each other.

$18 = 2y$ Subtract 12 from each side. Add y to each side.

$9 = y$ Divide each side by 2.

Step 3: Substitute $y = 9$ into one of the original equations and solve for x.

$x = y + 12$ Write one of the original equations.

$= 9 + 12$ Substitute 9 for y.

$= 21$ Add.

There are 21 boys and 9 girls in the science club.

Exercises

Solve the system of linear equations algebraically.

16. $x + y = 20$
$y = 2x - 1$

17. $x - y = 3$
$x + 2y = -6$

18. $2x + y = 8$
$3x + 2y = 30$

2.6 Special Systems of Linear Equations (pp. 82–87)

Solve the system.

$$y = x - 2 \quad \text{Equation 1}$$

$$-3x + 3y = -6 \quad \text{Equation 2}$$

Write $-3x + 3y = -6$ in slope-intercept form.

$-3x + 3y = -6$ — Write the equation.

$3y = 3x - 6$ — Add $3x$ to both sides.

$y = x - 2$ — Divide each side by 3.

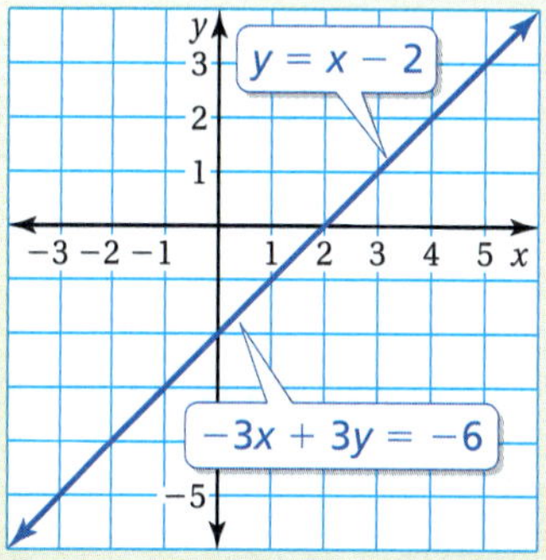

The equations are the same. The solution of the system is all the points on the line $y = x - 2$.

So, the system of linear equations has infinitely many solutions.

Exercises

19. Solve the system $y = -3x + 2$ and $6x + 2y = 10$.

2.7 Solving Equations by Graphing (pp. 88–93)

Solve $-\frac{2}{5}x - 10 = x + 4$ using a graph.

Step 1: Write a system of linear equations using each side of the equation.

$$-\frac{2}{5}x - 10 = x + 4$$

$y = -\frac{2}{5}x - 10$ (left side) $\quad y = x + 4$ (right side)

Step 2: Graph the system.

$$y = -\frac{2}{5}x - 10$$

$$y = x + 4$$

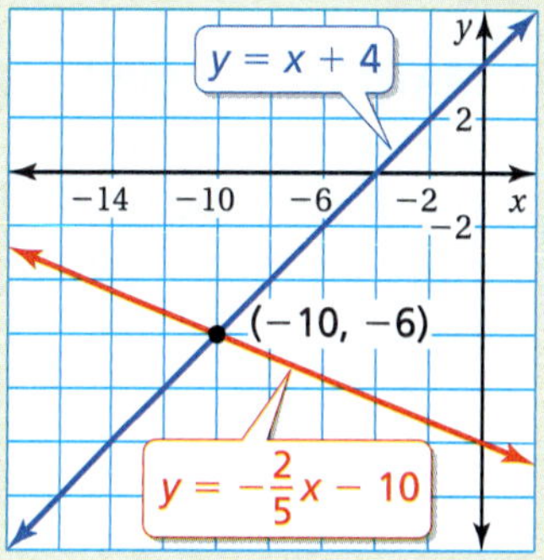

The graphs intersect at $(-10, -6)$.
So, the solution is $x = -10$.

Exercises

20. Use a graph to solve the equation $6x + 3 = 3x - 3$.

2 Chapter Test

Check It Out
Test Practice
BigIdeasMath.com

Find the slope and y-intercept of the graph of the linear equation.

1. $y = 6x - 5$

2. $y = 20x + 15$

3. $y = -5x - 16$

4. $y - 1 = 3x + 8.4$

5. $y + 4.3 = 0.1x$

6. $-\frac{1}{2}x + 2y = 7$

Graph the linear equation.

7. $y = 2x + 4$

8. $y = -\frac{1}{2}x - 5$

9. $-3x + 6y = 12$

10. Which two lines are parallel? Explain.

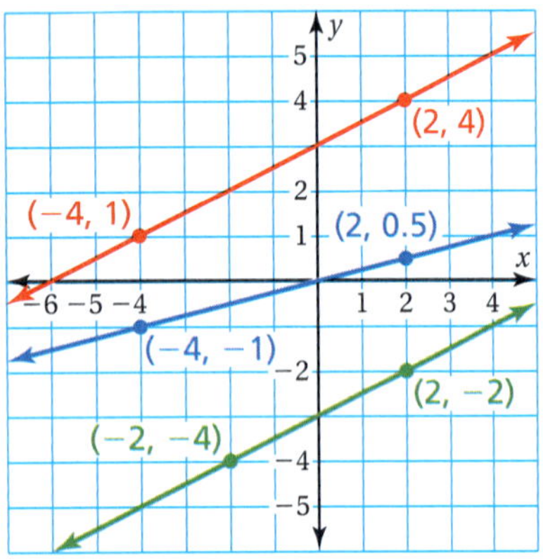

11. The points in the table lie on a line. Find the slope of the line. Then draw its graph.

x	y
−1	−4
0	−1
1	2
2	5

12. Solve the system of linear equations using any method.

$$2x - y = 16$$
$$y = 12 + x$$

13. Use a graph to solve the equation. Check your solution.

$$2x + 6 = 4x - 12$$

14. MATH TEST A math class has a test today. There are 30 problems on the test. The test has two types of problems: multiple choice problems and word problems. The multiple choice problems are worth 2 points and the word problems are worth 5 points. The teacher says there are a total of 75 points possible. How many multiple choice and word problems are on the test?

Number of multiple choice problems, x + Number of word problems, y = 30

Point value of multiple choice problems • Number of multiple choice problems, x + Point value of word problems • Number of word problems, y = 75

2 Standardized Test Practice

Test-Taking Strategy
Estimate the Answer

"Using **estimation** you can see that there are about 300 hairs. So, it's got to be C."

1. The graph below shows the value of United States dollars compared to Guatemalan quetzals.

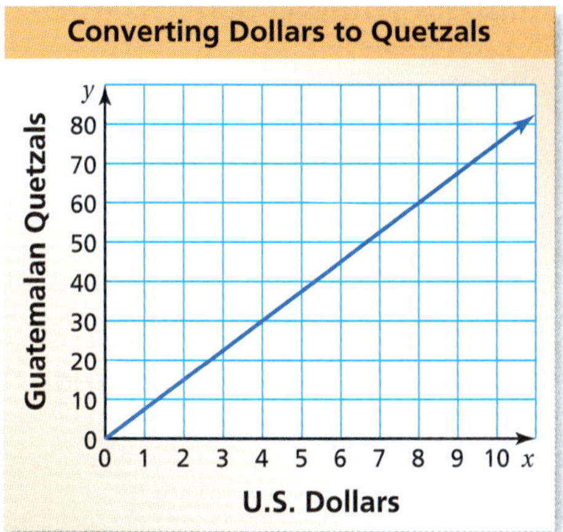

What is the value of 60 quetzals?

A. \$6 **C.** \$8

B. \$7 **D.** \$9

2. Which equation matches the line shown in the graph?

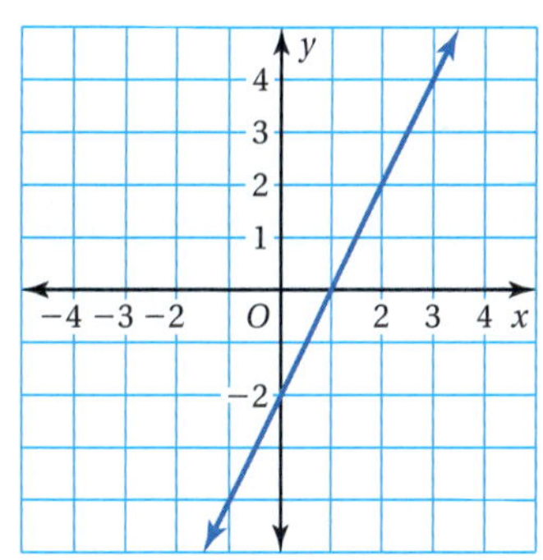

F. $y = 2x - 2$

G. $y = 2x + 1$

H. $y = x - 2$

I. $y = x + 1$

3. A faucet releases 6 quarts of water per minute. How many gallons of water will the faucet release in one hour?

A. 45 gal **C.** 360 gal

B. 90 gal **D.** 1440 gal

4. The equation $6x - 5y = 14$ is written in standard form. Which point lies on the graph of this equation?

F. $(-4, -1)$ **H.** $(-1, -4)$

G. $(-2, 4)$ **I.** $(4, -2)$

5. A system of two linear equations has no solutions. What can you conclude about the graphs of the two equations?

   A. The lines have the same slope and the same y-intercept.

   B. The lines have the same slope and different y-intercepts.

   C. The lines have different slopes and the same y-intercept.

   D. The lines have different slopes and different y-intercepts.

6. A cell phone plan costs \$10 per month plus \$0.10 for each minute used. Last month, you spent \$18.50 using this plan. This can be modeled by the equation below, where m represents the number of minutes used.

$$0.1m + 10 = 18.5$$

   How many minutes did you use last month?

   F. 8.4 min

   G. 85 min

   H. 185 min

   I. 285 min

7. What is the slope of the line that passes through the points $(2, -2)$ and $(8, 1)$?

8. It costs \$40 to rent a car for one day. In addition, the rental agency charges you for each mile driven as shown in the graph.

Think
Solve
Explain

   Part A Determine the slope of the line joining the points on the graph.

   Part B Explain what the slope represents.

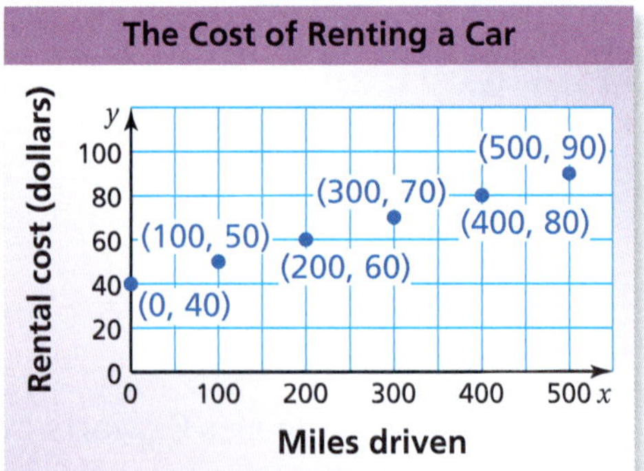

9. Which point is a solution of the system of equations shown below?

$$x + 3y = 10$$

$$x = 2y - 5$$

   A. $(1, 3)$

   B. $(3, 1)$

   C. $(55, -15)$

   D. $(-35, -15)$

10. Which line has a slope of 0?

F. 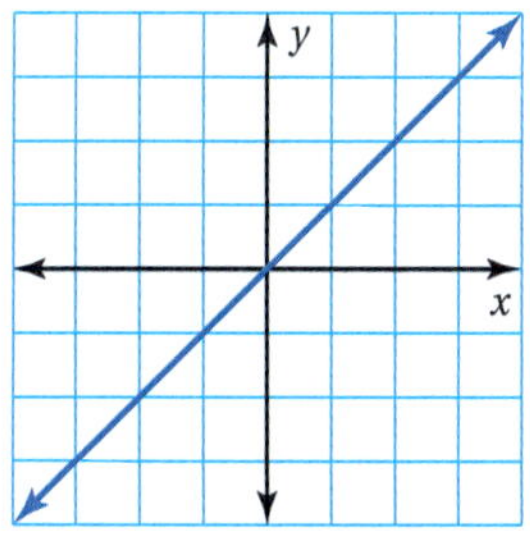

H. 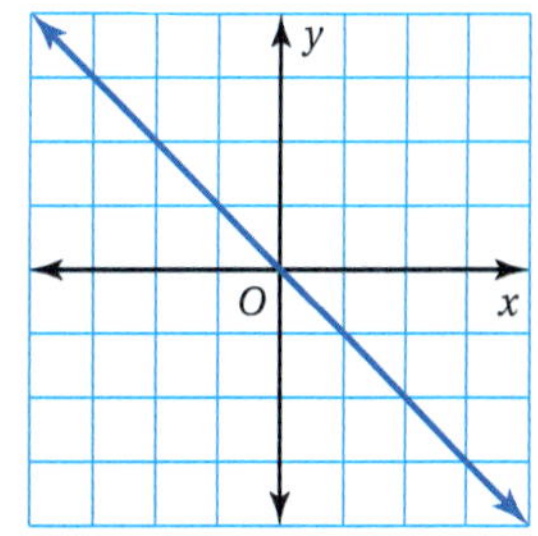

G. 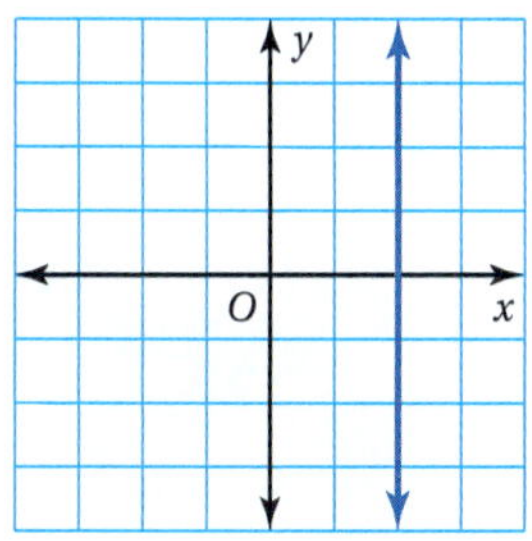

I. 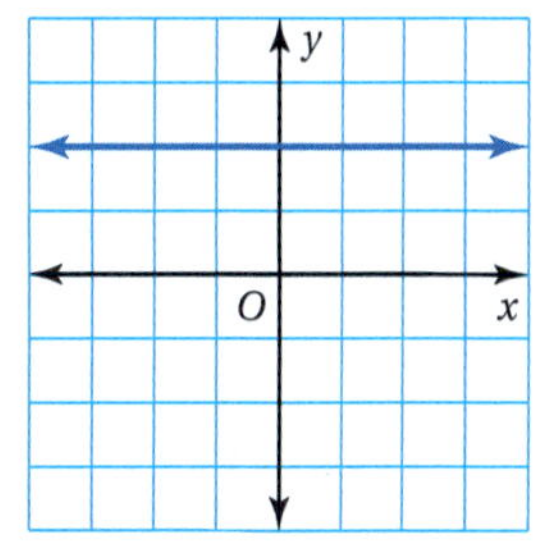

11. Solve the formula $K = 3M - 7$ for M.

A. $M = K + 7$

B. $M = \dfrac{K+7}{3}$

C. $M = \dfrac{K}{3} + 7$

D. $M = \dfrac{K-7}{3}$

12. The linear equation $5x + 2y = 10$ is written in standard form. What is the slope of the graph of this equation?

F. 5

G. 2.5

H. -2.5

I. -5

13. A package of breakfast cereal is labeled "750 g." This cereal is shipped in cartons that hold 24 packages. What is the total mass, in kilograms, of the breakfast cereal in 10 cartons?

14. A line has a slope of 4 and passes through the point (a, b). Which point must also lie on this line?

A. $(a, b + 4)$

B. $(2a, 8b)$

C. $(a + 1, b + 4)$

D. $(2a, 5b)$

3 Writing Linear Equations and Linear Systems

What You Learned Before

"Hold your tail a bit lower. We're trying to model a slope of 2."

Multiplying and Dividing Fractions (Grade 6)

A container of apple juice contains 16 cups.

Example 1 You drink $\frac{3}{8}$ of the juice. How many cups of juice did you drink?

$$16 \cdot \frac{3}{8} = \frac{16}{1} \cdot \frac{3}{8}$$

$$= \frac{\overset{2}{\cancel{16}} \cdot 3}{1 \cdot \underset{1}{\cancel{8}}}$$

$$= 6$$

You drank 6 cups of juice.

Example 2 A serving of juice is $\frac{4}{5}$ cup. How many servings are in the container?

$$16 \div \frac{4}{5} = \frac{16}{1} \div \frac{4}{5}$$

$$= \frac{16}{1} \cdot \frac{5}{4}$$

$$= \frac{\overset{4}{\cancel{16}} \cdot 5}{1 \cdot \underset{1}{\cancel{4}}}$$

$$= 20$$

There are 20 servings in the container.

Try It Yourself

Evaluate the expression.

1. $\frac{7}{15} \cdot \frac{2}{3}$
2. $\frac{5}{6} \cdot \frac{8}{9}$
3. $\frac{1}{7} \div \frac{8}{21}$
4. $\frac{9}{20} \div \frac{3}{4}$

Using Percents (Grade 7)

Example 3 Sales tax is 6%. What is the sales tax on an item that costs $43?

What is 6% of 43?

$43 \cdot 0.06 = 2.58$

The sales tax is $2.58.

Example 4 Your bill at a restaurant is $21.25. What is the amount of a 20% tip on the bill?

What is 20% of $21.25?

$21.25 \cdot 0.2 = 4.25$

The amount of the tip is $4.25.

Try It Yourself

5. Sales tax is 6%. What is the sales tax on an item that costs $55?
6. Your bill at a restaurant is $12.30. What is the amount of a 15% tip on the bill?

3.1 Writing Equations in Slope-Intercept Form

STATE STANDARDS
MA.8.A.1.1
MA.8.A.1.2

Essential Question How can you write an equation of a line when you are given the slope and *y*-intercept of the line?

1 ACTIVITY: Writing Equations of Lines

Work with a partner.

- **Find the slope of each line.**
- **Find the *y*-intercept of each line.**
- **Write an equation for each line.**
- **What do the three lines have in common?**

a.

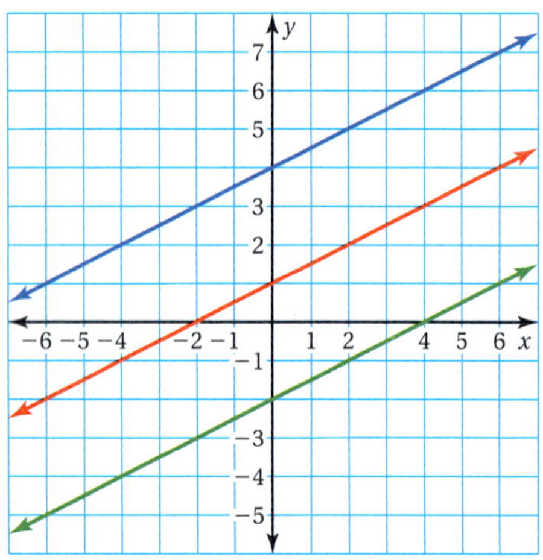

b.

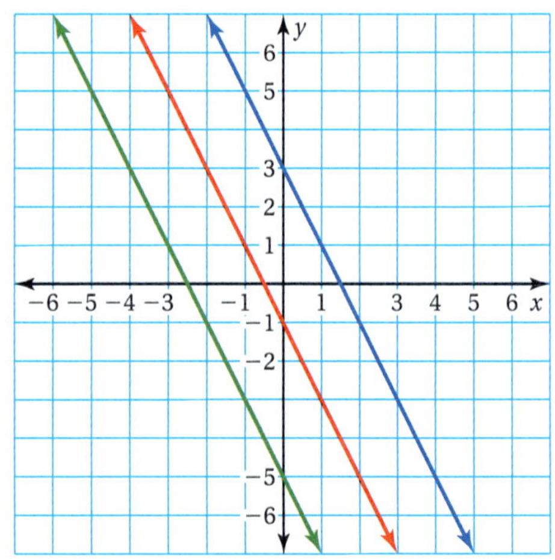

c.

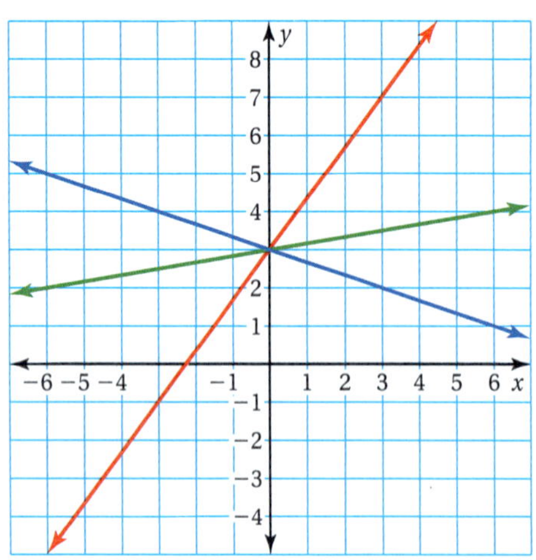

d.

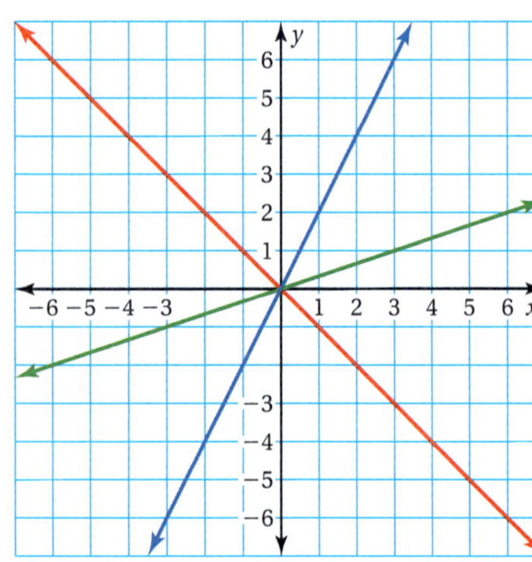

2 ACTIVITY: Describing a Parallelogram

Work with a partner.

- **Find the area of each parallelogram.**
- **Write an equation for each side of each parallelogram.**
- **What do you notice about the slopes of the opposite sides of each parallelogram?**

a.

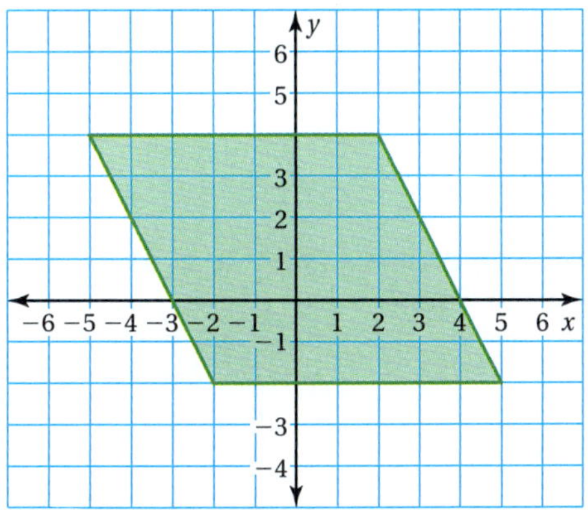

b.

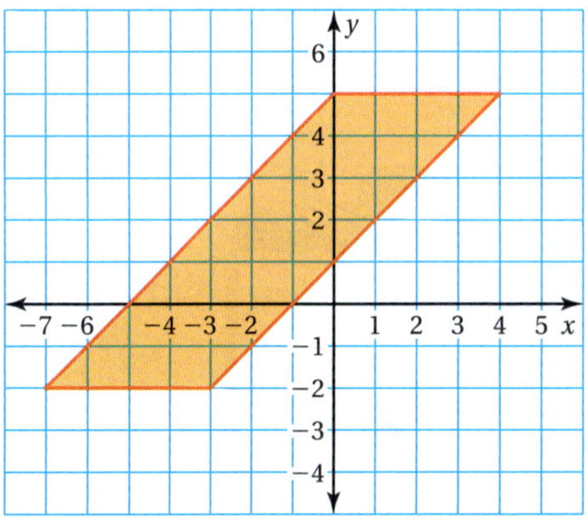

3 ACTIVITY: Interpreting the Slope and *y*-Intercept

Work with a partner. The graph shows a trip taken by a car where *t* is the time (in hours) and *y* is the distance (in miles) from Miami.

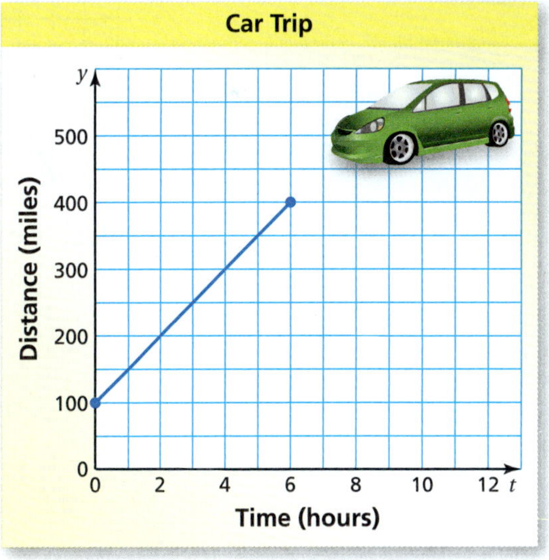

a. How far from Miami was the car at the beginning of the trip?

b. What was the car's speed?

c. How long did the trip last?

d. How far from Miami was the car at the end of the trip?

What Is Your Answer?

4. IN YOUR OWN WORDS How can you write an equation of a line when you are given the slope and y-intercept of the line? Give an example that is different from those in Activities 1, 2, and 3.

Practice Use what you learned about writing equations in slope-intercept form to complete Exercises 3 and 4 on page 110.

3.1 Lesson

Check It Out
Lesson Tutorials
BigIdeasMath.com

EXAMPLE 1 Writing Equations in Slope-Intercept Form

Write an equation of the line in slope-intercept form.

a.

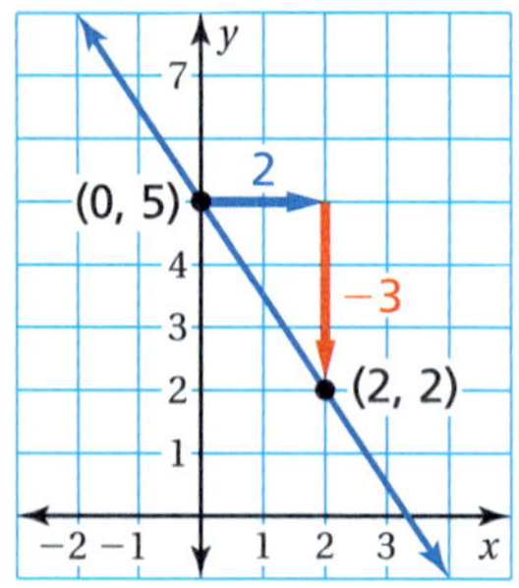

Find the slope and y-intercept.

$$\text{slope} = \frac{\text{rise}}{\text{run}} = \frac{-3}{2} = -\frac{3}{2}$$

Because the line crosses the y-axis at $(0, 5)$, the y-intercept is 5.

slope ↓ y-intercept ↓

∴ So, the equation is $y = -\frac{3}{2}x + 5$.

Study Tip

After writing an equation, check that the given points are solutions of the equation.

b.

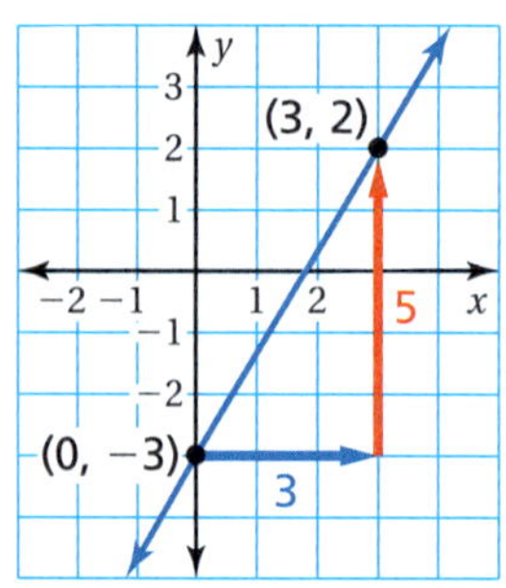

Find the slope and y-intercept.

$$\text{slope} = \frac{\text{rise}}{\text{run}} = \frac{5}{3}$$

Because the line crosses the y-axis at $(0, -3)$, the y-intercept is -3.

slope ↓ y-intercept ↓

∴ So, the equation is $y = \frac{5}{3}x + (-3)$, or $y = \frac{5}{3}x - 3$.

On Your Own

Now You're Ready
Exercises 5–10

Write an equation of the line in slope-intercept form.

1.

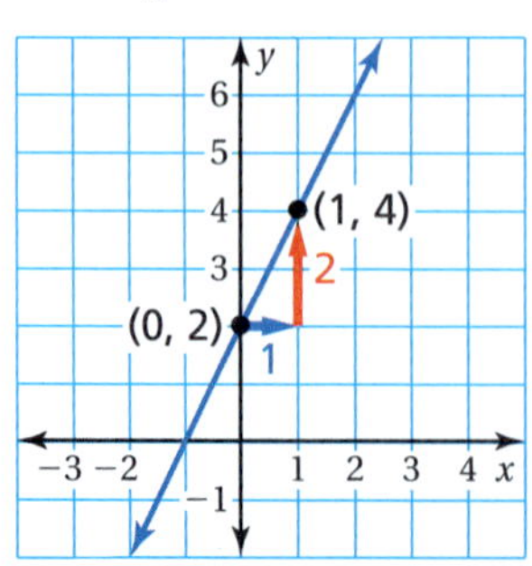

2.

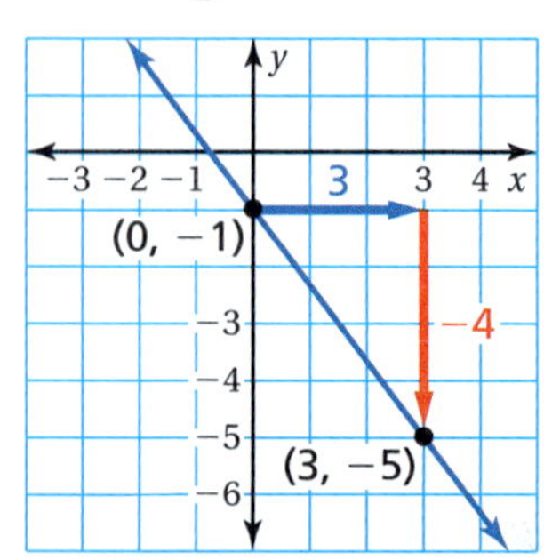

EXAMPLE 2 Standardized Test Practice

Which equation is shown in the graph?

Ⓐ $y = -4$ Ⓑ $y = -3$

Ⓒ $y = 0$ Ⓓ $y = -3x$

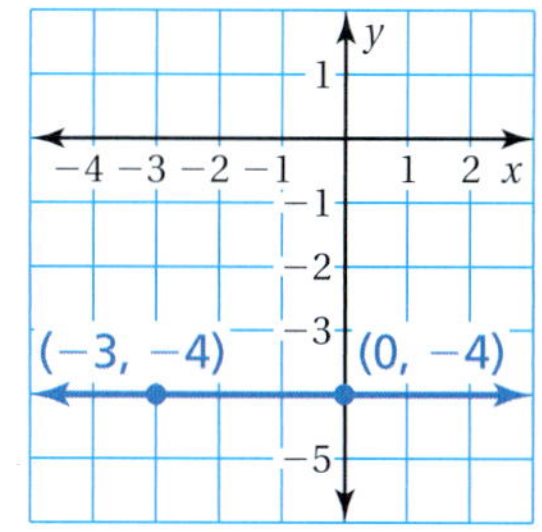

Remember

The graph of $y = a$ is a horizontal line that passes through $(0, a)$.

Find the slope and y-intercept.

The line is horizontal, so the rise is 0.

$$\text{slope} = \frac{\text{rise}}{\text{run}} = \frac{0}{3} = 0$$

Because the line crosses the y-axis at $(0, -4)$, the y-intercept is -4.

∴ So, the equation is $y = 0x + (-4)$, or $y = -4$. The correct answer is Ⓐ.

EXAMPLE 3 Real-Life Application

The Florida Department of Transportation plans to use a 42-foot wide tunnel boring machine to dig a 3900-foot long tunnel connecting Watson Island and Port of Miami by 2014.

The graph shows the distance remaining to complete a tunnel. (a) Write an equation that represents the distance y (in feet) remaining after x months. (b) How much time does it take to complete the tunnel?

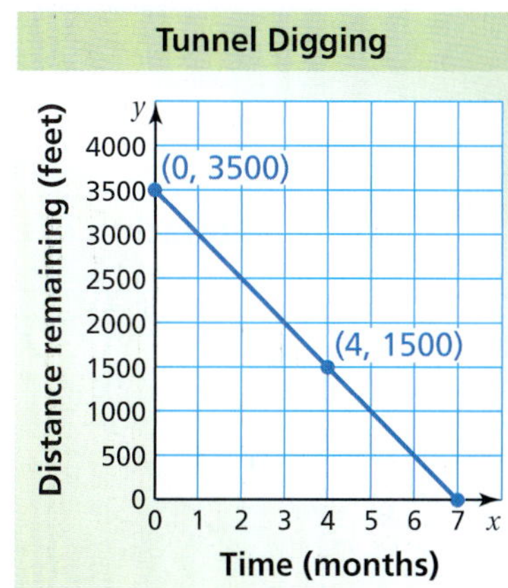

a. Find the slope and y-intercept.

$$\text{slope} = \frac{\text{rise}}{\text{run}} = \frac{-2000}{4} = -500$$

Because the line crosses the y-axis at $(0, 3500)$, the y-intercept is 3500.

∴ So, the equation is $y = -500x + 3500$.

b. The tunnel is complete when the distance remaining is 0 feet. So, find the value of x when $y = 0$.

$y = -500x + 3500$	Write the equation.
$0 = -500x + 3500$	Substitute 0 for y.
$-3500 = -500x$	Subtract 3500 from each side.
$7 = x$	Solve for x.

∴ It takes 7 months to complete the tunnel.

On Your Own

Now You're Ready Exercises 13–15

3. Write an equation of the line that passes through $(0, 5)$ and $(4, 5)$.

4. **WHAT IF?** In Example 3, the points are $(0, 3500)$ and $(5, 1500)$. How long does it take to complete the tunnel?

3.1 Exercises

Check It Out
Help with Homework
BigIdeasMath.com

Vocabulary and Concept Check

1. **WRITING** Explain how to find the slope of a line given the intercepts of the line.
2. **WRITING** Explain how to write an equation of a line using its graph.

Practice and Problem Solving

Write an equation for each side of the figure.

3. 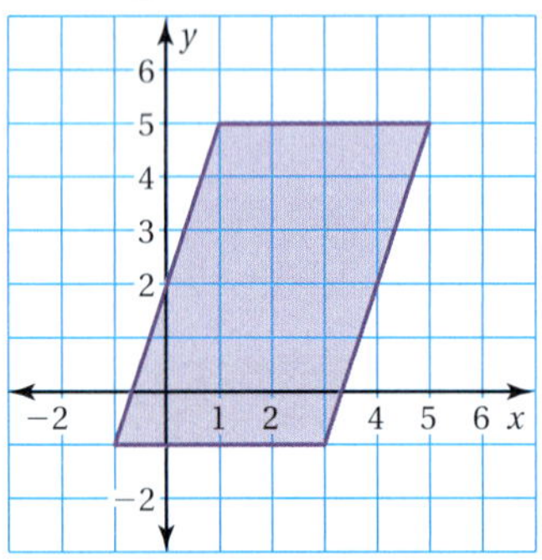

4. 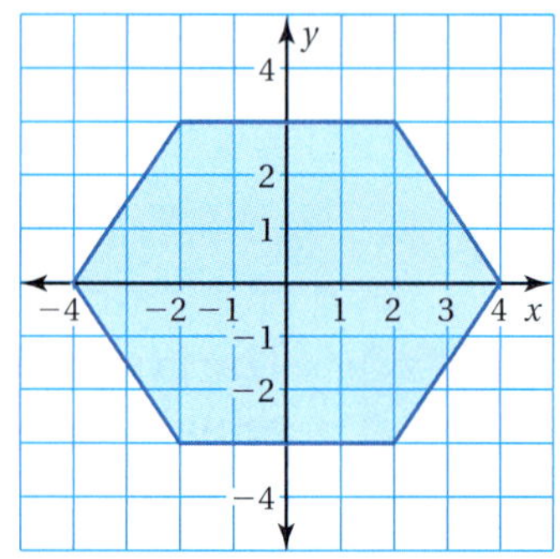

Write an equation of the line in slope-intercept form.

1 5. 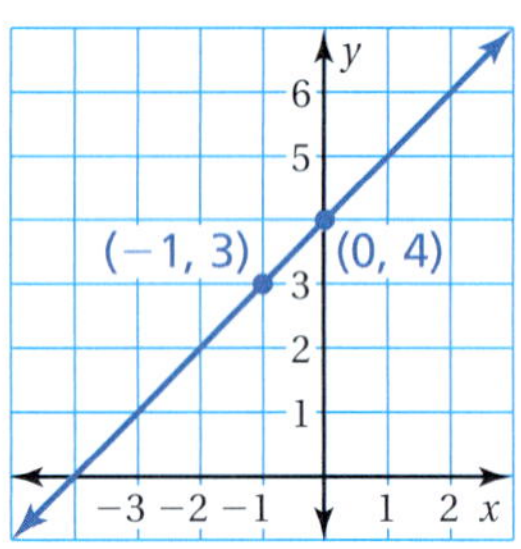

6. 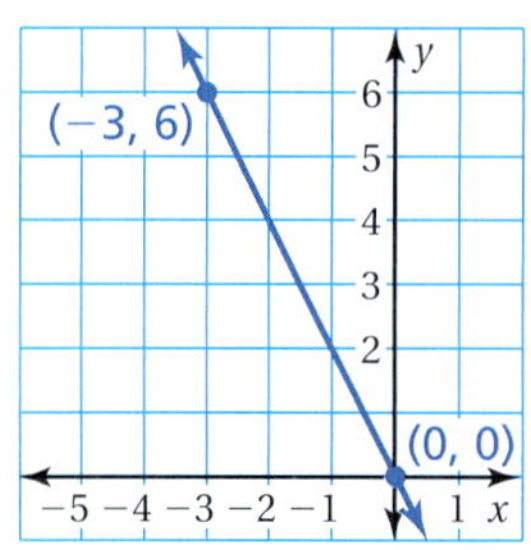

7. 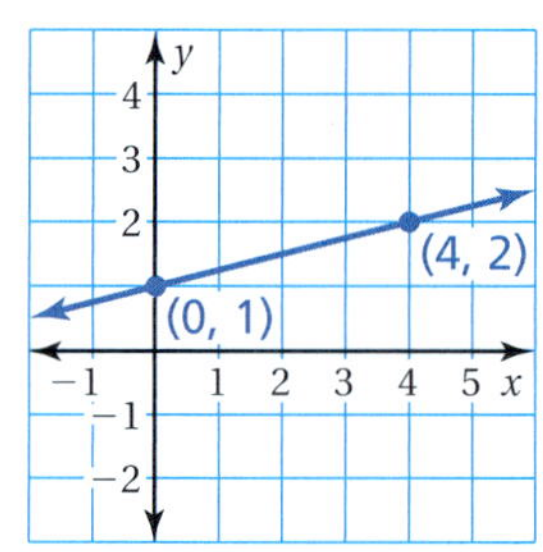

8. 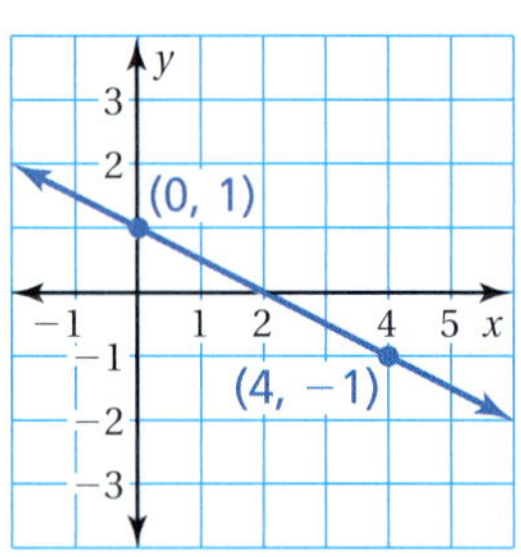

9. 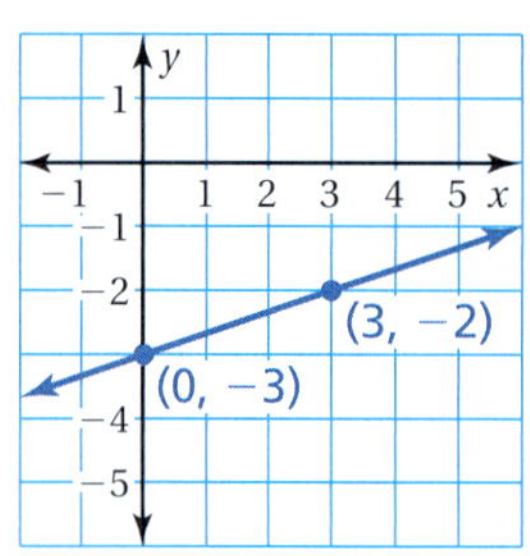

10. 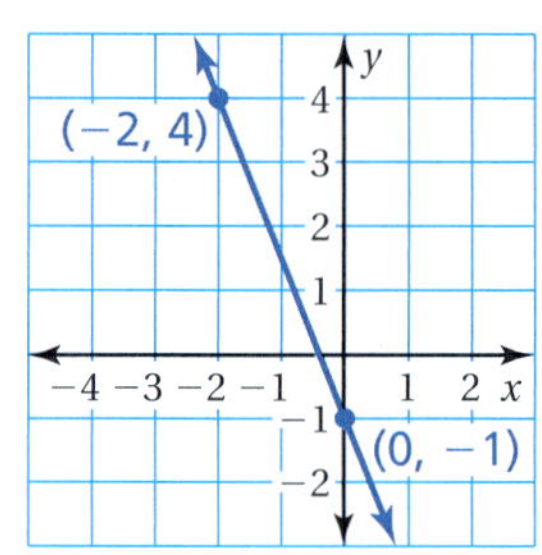

11. **ERROR ANALYSIS** Describe and correct the error in writing the equation of the line.

$y = \frac{1}{2}x + 4$

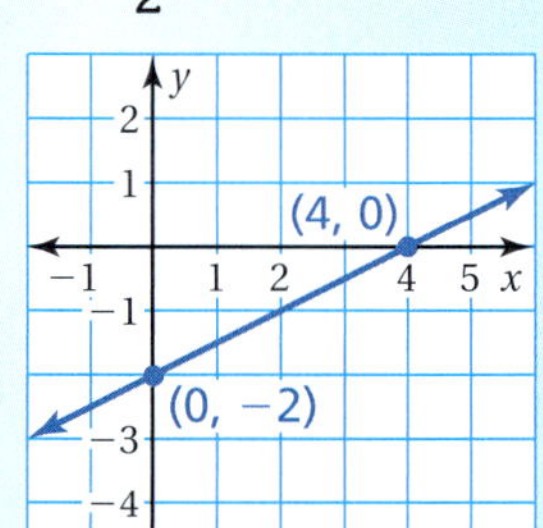

12. **BOA** A boa constrictor is 18 inches long at birth and grows 8 inches per year. Write an equation that represents the length y (in feet) of a boa constrictor that is x years old.

Write an equation of the line that passes through the points.

2 **13.** (2, 5), (0, 5) **14.** (−3, 0), (0, 0) **15.** (0, −2), (4, −2)

16. WALKATHON One of your friends gives you \$10 for a charity walkathon. Another friend gives you an amount per mile. After 5 miles, you have raised \$13.50 total. Write an equation that represents the amount y of money you have raised after x miles.

17. BRAKING TIME During each second of braking, an automobile slows by about 10 miles per hour.

a. Plot the points (0, 60) and (6, 0). What do the points represent?

b. Draw a line through the points. What does the line represent?

c. Write an equation of the line.

18. PAPER You have 500 sheets of notebook paper. After 1 week, you have 72% of the sheets left. You use the same number of sheets each week. Write an equation that represents the number y of pages remaining after x weeks.

19. Critical Thinking The palm tree on the left is 10 years old. The palm tree on the right is 8 years old. The trees grow at the same rate.

a. Estimate the height y (in feet) of each tree.

b. Plot the two points (x, y), where x is the age of each tree and y is the height of each tree.

c. What is the rate of growth of the trees?

d. Write an equation that represents the height of a palm tree in terms of its age.

Fair Game Review What you learned in previous grades & lessons

Plot the ordered pair in a coordinate plane. *(Skills Review Handbook)*

20. (1, 4) **21.** (−1, −2) **22.** (0, 1) **23.** (2, 7)

24. MULTIPLE CHOICE Which of the following statements is true? *(Section 2.3)*

Ⓐ The x-intercept is 5.

Ⓑ The x-intercept is −2.

Ⓒ The y-intercept is 5.

Ⓓ The y-intercept is −2.

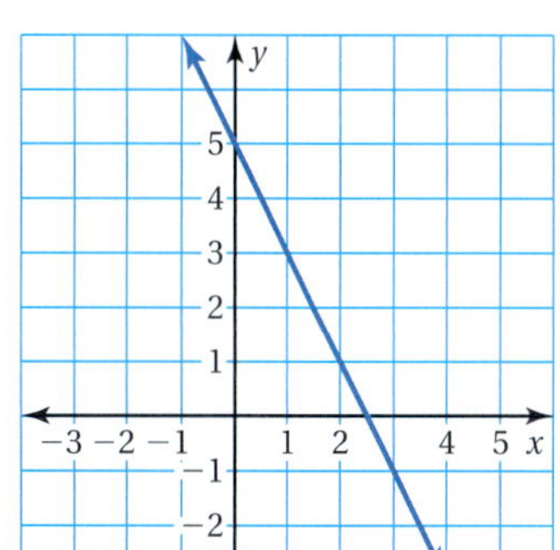

3.2 Writing Equations Using a Slope and a Point

STATE STANDARDS
MA.8.A.1.1
MA.8.A.1.2

Essential Question How can you write an equation of a line when you are given the slope and a point on the line?

1 ACTIVITY: Writing Equations of Lines

Work with a partner.

- **Sketch the line that has the given slope and passes through the given point.**
- **Find the *y*-intercept of the line.**
- **Write an equation of the line.**

a. $m = -2$

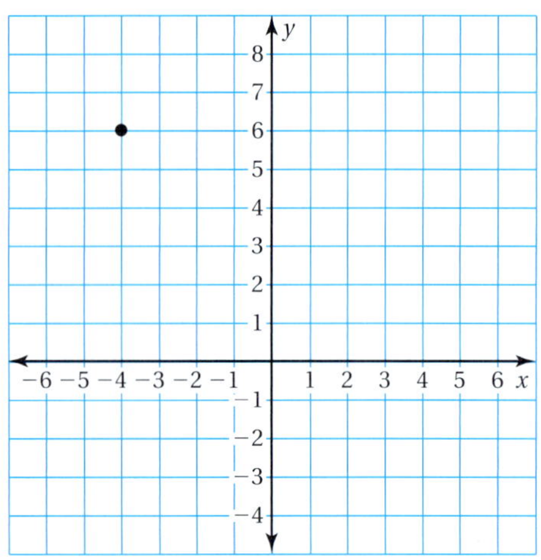

b. $m = \frac{1}{3}$

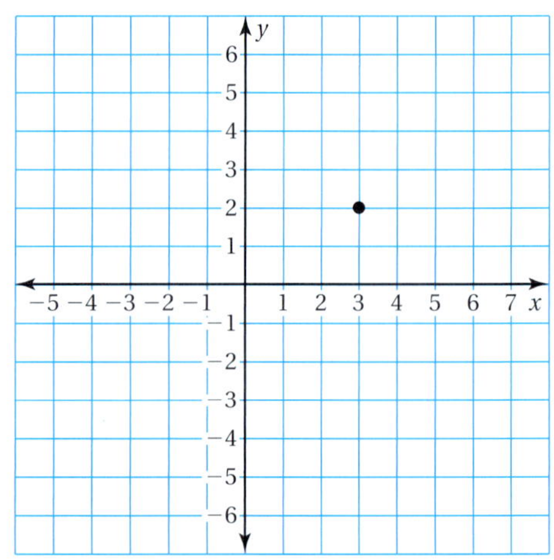

c. $m = -\frac{2}{3}$

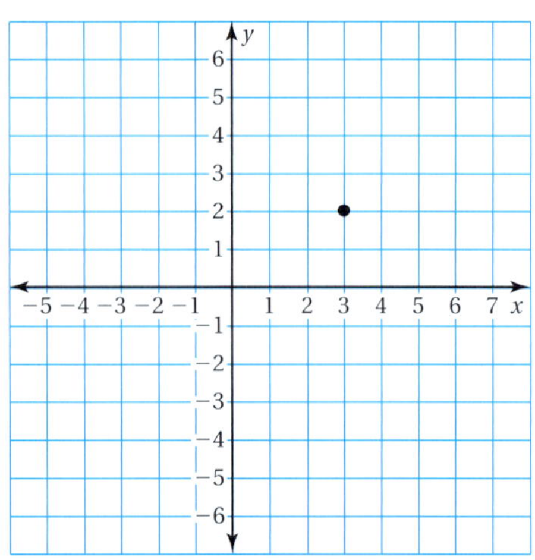

d. $m = \frac{5}{2}$

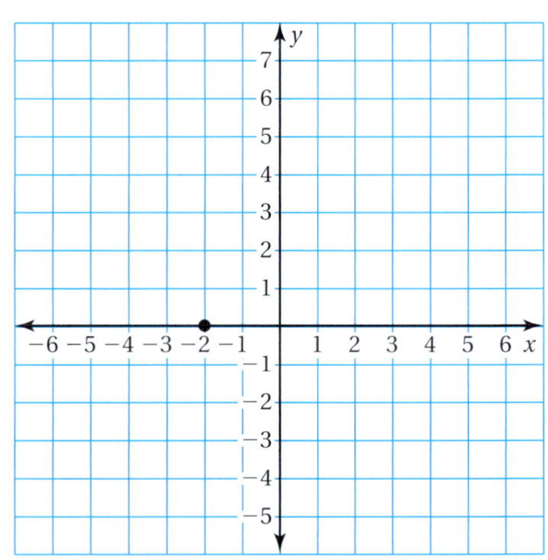

2 ACTIVITY: Writing Linear Equations

Work with a partner.

a. For 4 months, you have saved \$25 a month. You now have \$175 in your savings account.

- Draw a graph that shows the balance in your account after t months.
- Write an equation that represents the balance A after t months.

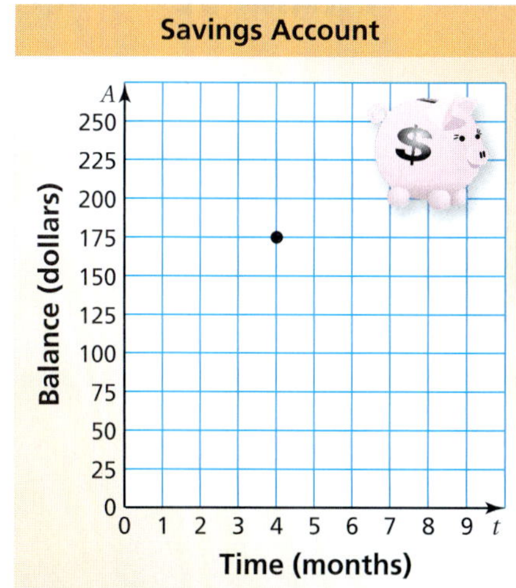

b. For 4 months, you have withdrawn \$25 a month from your savings account. Your account balance is now \$75.

- Draw a graph that shows the balance in your account after t months.
- Write an equation that represents the balance A after t months.

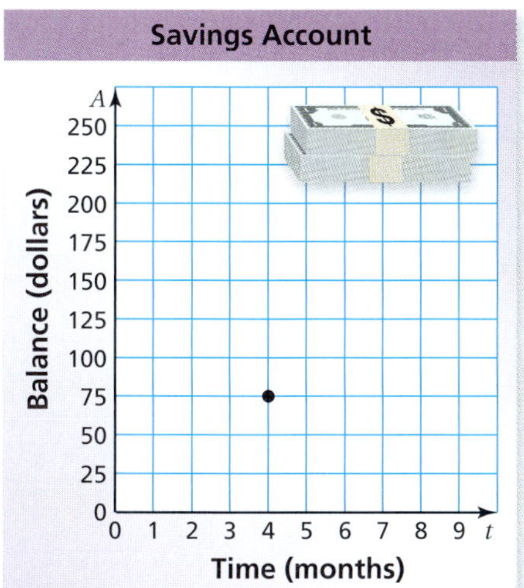

c. For 6 years, the population of a town has grown by 5000 people per year. The population is now 70,000.

- Draw a graph that shows the population after t years.
- Write an equation that represents the population P after t years.

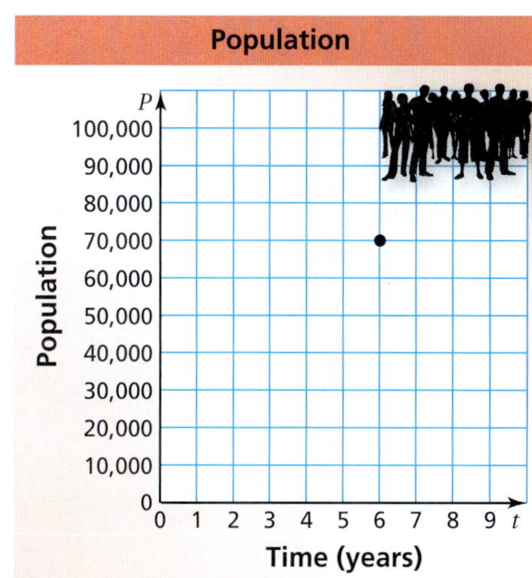

What Is Your Answer?

3. **IN YOUR OWN WORDS** How can you write an equation of a line when you are given the slope and a point on the line? Give an example that is different from those in Activities 1 and 2.

Practice Use what you learned about writing equations using a slope and a point to complete Exercises 3–5 on page 116.

Check It Out
Lesson Tutorials
BigIdeasMath.com

EXAMPLE 1 Writing Equations Using a Slope and a Point

Write an equation of the line with the given slope that passes through the given point.

a. $m = \frac{2}{3}$; $(-6, 1)$

Use a graph to find the y-intercept.

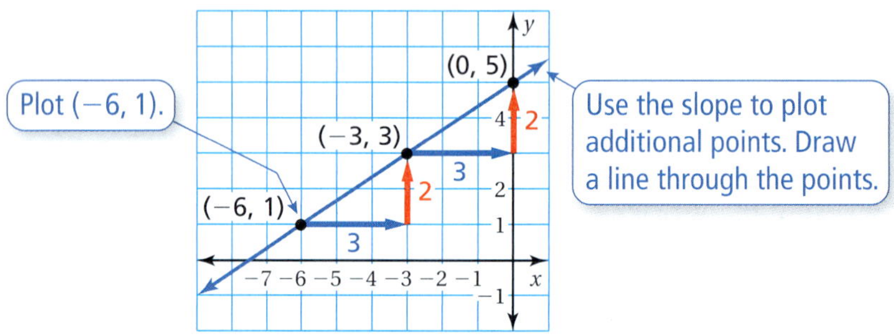

Because the line crosses the y-axis at $(0, 5)$, the y-intercept is 5.

So, the equation is $y = \frac{2}{3}x + 5$.

Check

Check that $(-6, 1)$ is a solution of the equation.

$y = \frac{2}{3}x + 5$

$1 \stackrel{?}{=} \frac{2}{3}(-6) + 5$

$1 = 1$ ✓

b. $m = -3$; $(1, -4)$

Use a graph to find the y-intercept.

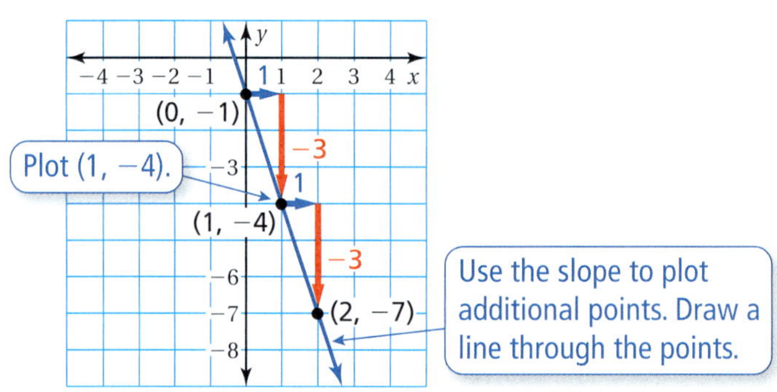

Because the line crosses the y-axis at $(0, -1)$, the y-intercept is -1.

So, the equation is $y = -3x + (-1)$, or $y = -3x - 1$.

Check

Check that $(1, -4)$ is a solution of the equation.

$y = -3x - 1$

$-4 \stackrel{?}{=} -3(1) - 1$

$-4 = -4$ ✓

On Your Own

Now You're Ready
Exercises 6–11

Write an equation of the line with the given slope that passes through the given point.

1. $m = 1$; $(2, 0)$

2. $m = -\frac{1}{2}$; $(2, 3)$

EXAMPLE 2 Real-Life Application

You finish parasailing and are being pulled back to the boat. After 2 seconds, you are 25 feet above the boat. (a) Write an equation that represents the height y (in feet) above the boat after x seconds. (b) At what height were you parasailing? (c) When do you reach the boat?

a. You are being pulled down at the rate of 10 feet per second. So, the slope is -10. You are 25 feet above the boat after 2 seconds. So, the line passes through $(2, 25)$.

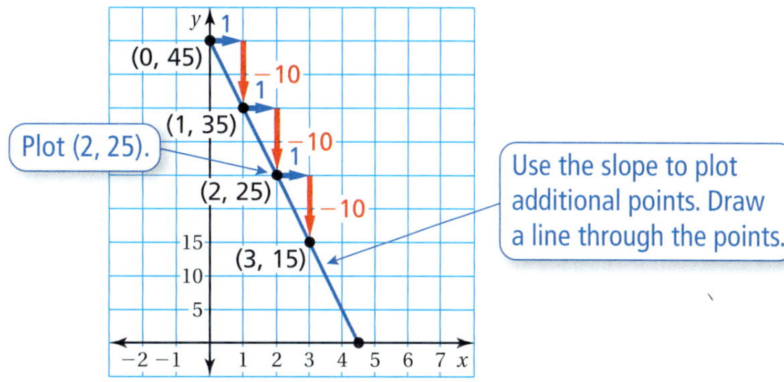

Because the line crosses the y-axis at $(0, 45)$, the y-intercept is 45.

So, the equation is $y = -10x + 45$.

Check

Check that (2, 25) is a solution of the equation.

$y = -10x + 45$	Write the equation.
$25 \stackrel{?}{=} -10(2) + 45$	Substitute.
$25 = 25$ ✓	Simplify.

b. You start descending when $x = 0$. The y-intercept is 45. So, you were parasailing at a height of 45 feet.

c. You reach the boat when $y = 0$.

$y = -10x + 45$	Write the equation.
$0 = -10x + 45$	Substitute 0 for y.
$-45 = -10x$	Subtract 45 from each side.
$4.5 = x$	Solve for x.

You reach the boat after 4.5 seconds.

On Your Own

3. **WHAT IF?** In Example 2, you are 35 feet above the boat after 2 seconds. When do you reach the boat?

3.2 Exercises

Check It Out
Help with Homework
BigIdeasMath.com

Vocabulary and Concept Check

1. **WRITING** What information do you need to write an equation of a line?

2. **WRITING** Describe how to write an equation of a line using its slope and a point on the line.

Practice and Problem Solving

Write an equation of the line with the given slope that passes through the given point.

3. $m = \frac{1}{2}$

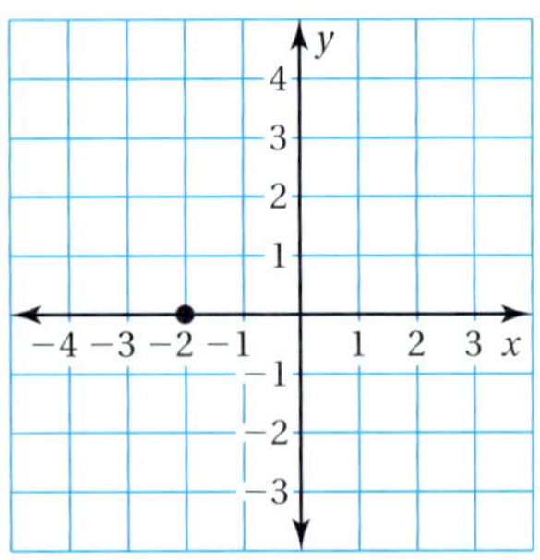

4. $m = -\frac{3}{4}$

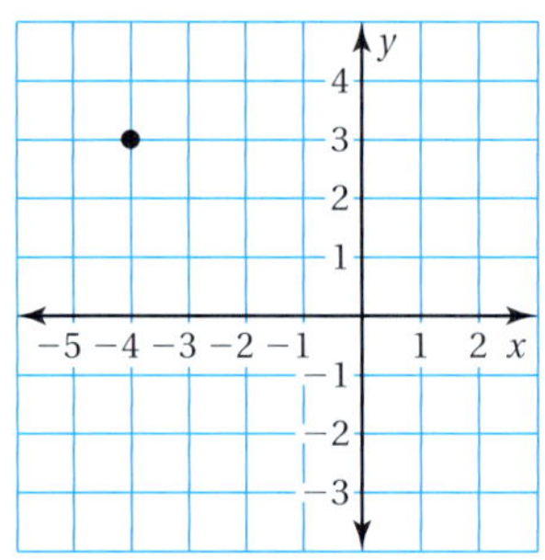

5. $m = -3$

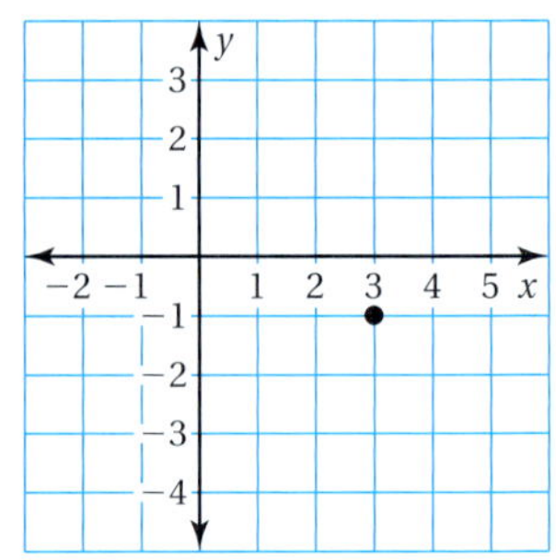

1 6. $m = -\frac{2}{3}$; $(3, 0)$

7. $m = \frac{3}{4}$; $(4, 8)$

8. $m = 4$; $(1, -3)$

9. $m = -\frac{1}{7}$; $(7, -5)$

10. $m = \frac{5}{3}$; $(3, 3)$

11. $m = -2$; $(-1, -4)$

12. **ERROR ANALYSIS** Describe and correct the error in writing an equation of the line with a slope of $\frac{1}{3}$ that passes through the point $(6, 4)$.

✗ $y = \frac{1}{3}x + 4$

13. **CHEMISTRY** At 0°C, the volume of a gas is 22 liters. For each degree the temperature T (in degrees Celsius) increases, the volume V (in liters) of the gas increases by $\frac{2}{25}$. Write an equation that represents the volume of the gas in terms of the temperature.

14. **CARS** After it is purchased, the value of a new car decreases \$4000 each year. After 3 years, the car is worth \$18,000.

   a. Write an equation that represents the value V (in dollars) of the car x years after it is purchased.

   b. What was the original value of the car?

15. **CRICKETS** According to Dolbear's Law, you can predict the temperature T (in degrees Fahrenheit) by counting the number x of chirps made by a snowy tree cricket in 1 minute. For each chirp the cricket makes in 1 minute, the temperature rises 0.25 degree.

   a. A cricket chirps 40 times in 1 minute when the temperature is 50°F. Write an equation that represents the temperature in terms of the number of chirps in 1 minute.

   b. You count 100 chirps in 1 minute. What is the temperature?

   c. The temperature is 96°F. How many chirps would you expect the cricket to make?

16. **AIRBOATS** You rent an airboat. The total cost includes a flat fee plus an hourly fee.

   a. After 4 hours the total cost is \$140. Write an equation that represents the total cost y after x hours.

   b. Interpret the y-intercept.

17. **Critical Thinking** Bone mineral density is a measure of the strength of bones. The average bone mineral density of a female astronaut who has never been in space is 2.9 grams per square centimeter. For the first three years she spends in space, her bone density decreases by 0.03 grams per square centimeter per month.

   a. Write an equation that represents the bone mineral density y of a female astronaut in terms of the number x of months she spends in space.

   b. What is her bone mineral density after 2 years and 6 months in space?

   c. Explain why the amount of time an astronaut can spend in space is limited.

Fair Game Review What you learned in previous grades & lessons

18. Plot the ordered pairs in the same coordinate plane. *(Skills Review Handbook)*

   $(2, 5), (-3, -6), (0, 7), (-5, 0), (-8, 9)$

19. **MULTIPLE CHOICE** What is the y-intercept of the equation $5x - 2y = 28$? *(Section 2.4)*

   Ⓐ $-\frac{5}{2}$ Ⓑ -14 Ⓒ $\frac{5}{2}$ Ⓓ 5.6

3.3 Writing Equations Using Two Points

STATE STANDARDS
MA.8.A.1.1
MA.8.A.1.2

Essential Question How can you write an equation of a line when you are given two points on the line?

1 ACTIVITY: Writing Equations of Lines

Work with a partner.

- **Sketch the line that passes through the given points.**
- **Find the slope and *y*-intercept of the line.**
- **Write an equation of the line.**

a.

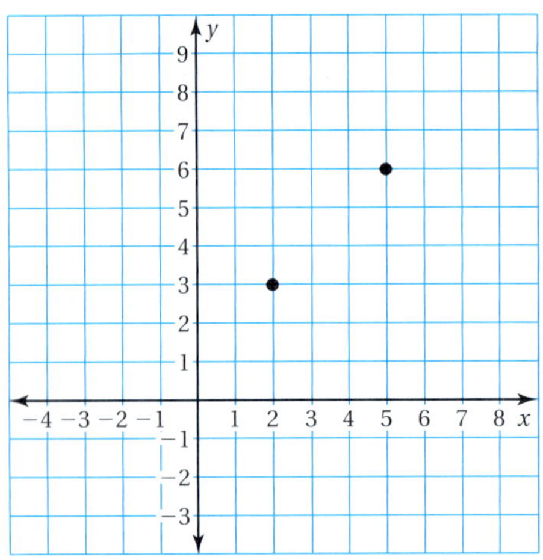

b.

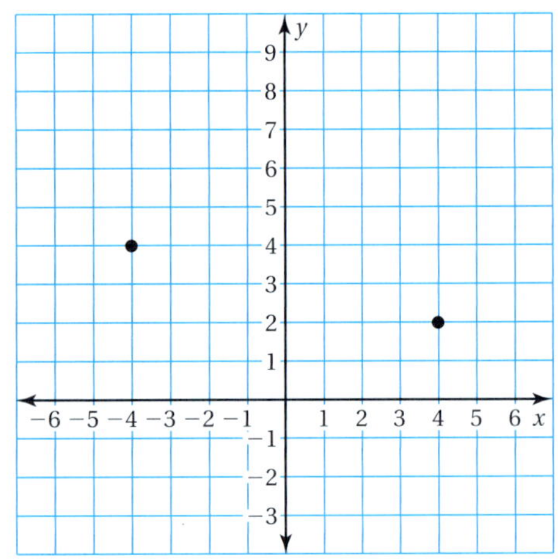

c.

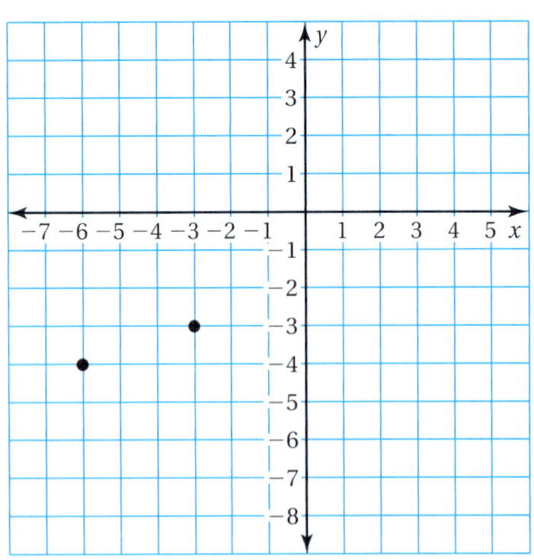

d.

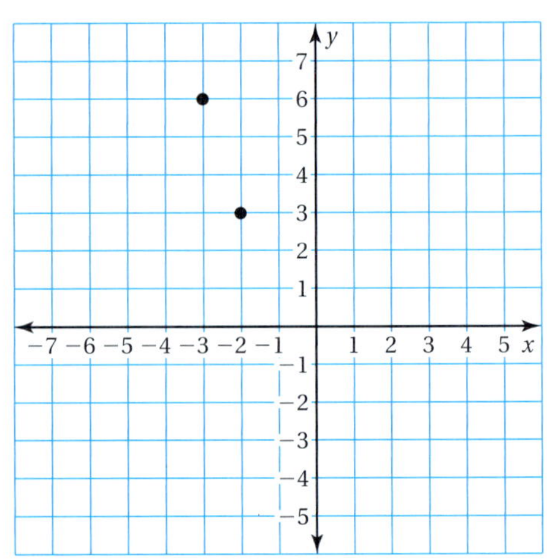

2 ACTIVITY: Writing and Using Linear Equations

Work with a partner.

a. You are rising in a hot air balloon. After 1 minute, you are 200 feet above the ground. After 4 minutes, you are 800 feet above the ground.

- Write an equation for the height h in terms of the time t.
- Use your equation to find the height of the balloon after 5 minutes.

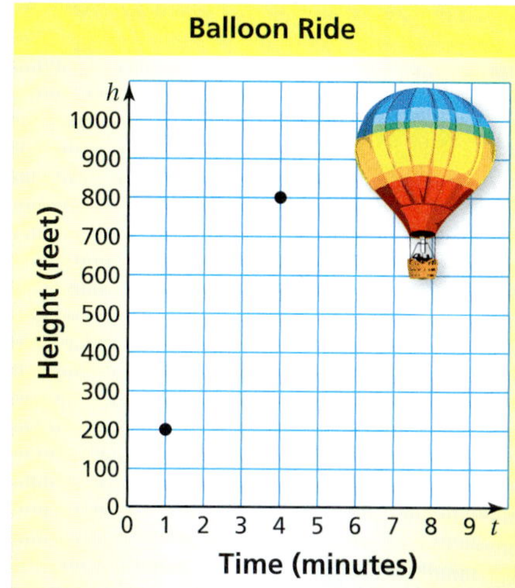

b. After 5 minutes, the hot air balloon starts to descend. After 6 minutes, you are 200 feet above the ground.

- Write an equation for the height h in terms of the time t.
- Use your equation to estimate when the balloon lands on the ground.

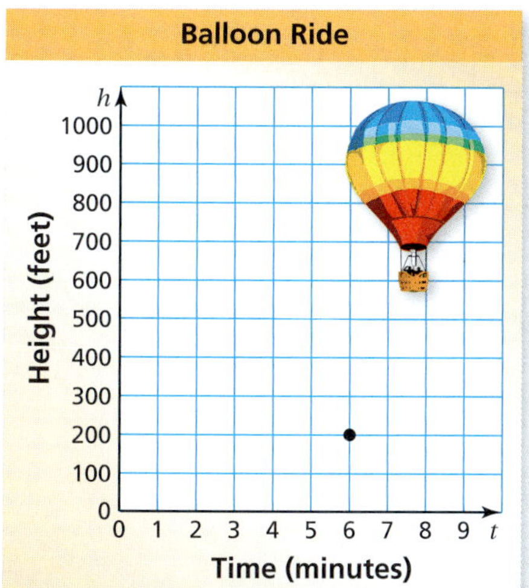

c. You are on a roller coaster. After 3 seconds, you are 190 feet above the ground and have reached maximum speed. One second later, you are 95 feet above the ground.

- Write an equation for the height h in terms of the time t.
- When will you reach ground level?

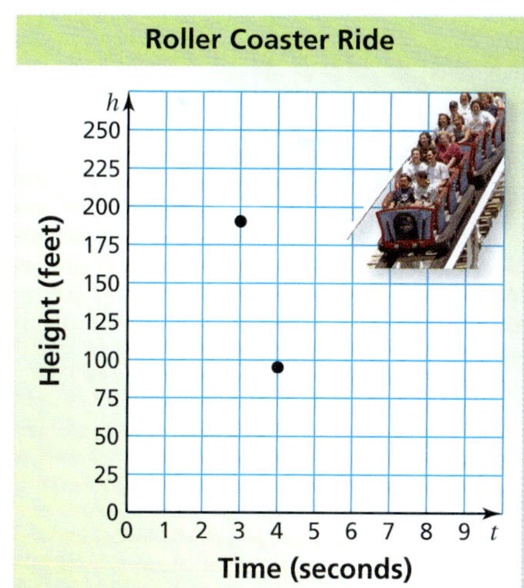

What Is Your Answer?

3. **IN YOUR OWN WORDS** How can you write an equation of a line when you are given two points on the line? Give an example that is different from those in Activities 1 and 2.

Practice Use what you learned about writing equations using two points to complete Exercises 3–5 on page 122.

3.3 Lesson

Check It Out
Lesson Tutorials
BigIdeasMath.com

EXAMPLE 1 Writing Equations Using Two Points

Write an equation of the line that passes through the points.

a. (−6, 6), (−3, 4)

Use a graph to find the slope and *y*-intercept.

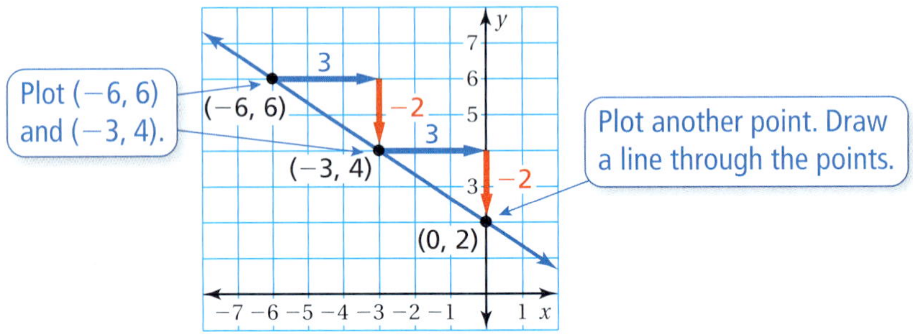

$$\text{slope} = \frac{\text{rise}}{\text{run}} = \frac{-2}{3} = -\frac{2}{3}$$

Because the line crosses the *y*-axis at (0, 2), the *y*-intercept is 2.

So, the equation is $y = -\frac{2}{3}x + 2$.

b. (−2, −4), (1, −1)

Use a graph to find the slope and *y*-intercept.

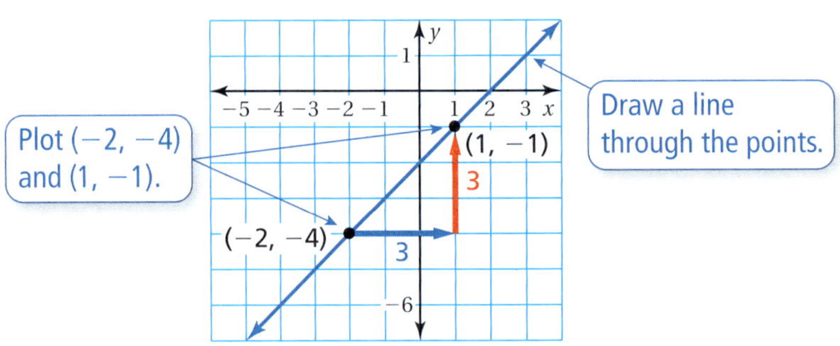

$$\text{slope} = \frac{\text{rise}}{\text{run}} = \frac{3}{3} = 1$$

Because the line crosses the *y*-axis at (0, −2), the *y*-intercept is −2.

So, the equation is $y = 1x + (-2)$, or $y = x - 2$.

Study Tip

After writing an equation, check that the given points are solutions of the equation.

On Your Own

Now You're Ready
Exercises 6–14

Write an equation of the line that passes through the points.

1. (2, 3), (4, 4)

2. (−1, 2), (1, −4)

EXAMPLE 2 Standardized Test Practice

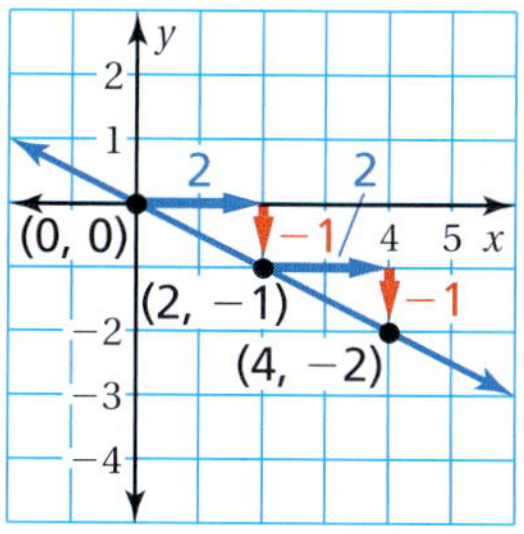

The graph of which equation passes through (2, −1) and (4, −2)?

Ⓐ $y = -\frac{1}{2}x$ Ⓑ $y = \frac{1}{2}x$

Ⓒ $y = -2x$ Ⓓ $y = 2x$

Graph the line through the points. Find the slope and y-intercept.

$$\text{slope} = \frac{\text{rise}}{\text{run}} = \frac{-1}{2} = -\frac{1}{2}$$

Because the line crosses the y-axis at (0, 0), the y-intercept is 0.

∴ So, the equation is $y = -\frac{1}{2}x + 0$, or $y = -\frac{1}{2}x$.

The correct answer is Ⓐ.

EXAMPLE 3 Real-Life Application

22.5 oz

A 2-week old kitten weighs 9 ounces. Two weeks later, it weighs 15 ounces. (a) Write an equation to represent the weight y (in ounces) of the kitten x weeks after birth. (b) How old is the kitten in the photo?

a. The kitten weighs 9 ounces after 2 weeks and 15 ounces after 4 weeks. So, graph the line that passes through (2, 9) and (4, 15).

$$\text{slope} = \frac{\text{rise}}{\text{run}} = \frac{6}{2} = 3$$

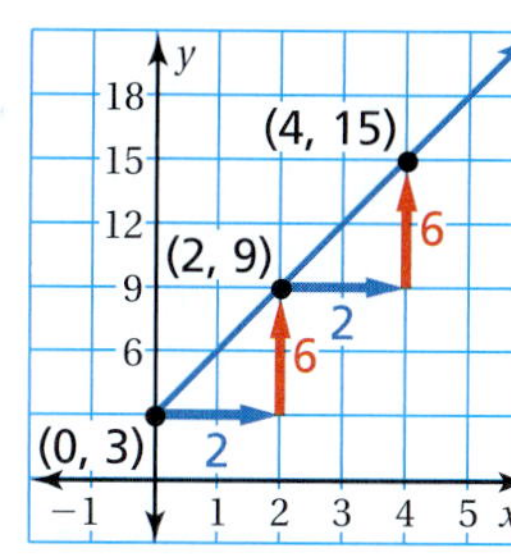

Because the line crosses the y-axis at (0, 3), the y-intercept is 3.

∴ So, the equation is $y = 3x + 3$.

b. Find the value of x when $y = 22.5$.

$y = 3x + 3$	Write the equation.
$22.5 = 3x + 3$	Substitute 22.5 for y.
$19.5 = 3x$	Subtract 3 from each side.
$6.5 = x$	Solve for x.

∴ The kitten in the photo is 6.5 weeks old.

On Your Own

3. The graph of which equation in Example 2 passes through (−2, 4) and (−1, 2)?

4. A 3-week old kitten weighs 12 ounces. Two weeks later, it weighs 18 ounces. How old is the kitten when it weighs 27 ounces?

3.3 Exercises

Vocabulary and Concept Check

1. **WRITING** Describe how to write an equation of a line using two points on the line.

2. **WHICH ONE DOESN'T BELONG?** Which pair of points does *not* belong with the other three? Explain your reasoning.

(0, 1), (2, 3)	(1, 2), (4, 5)	(2, 3), (5, 6)	(1, 2), (4, 6)

Practice and Problem Solving

Find the slope and y-intercept of the line that passes through the points. Then write an equation of the line.

3. 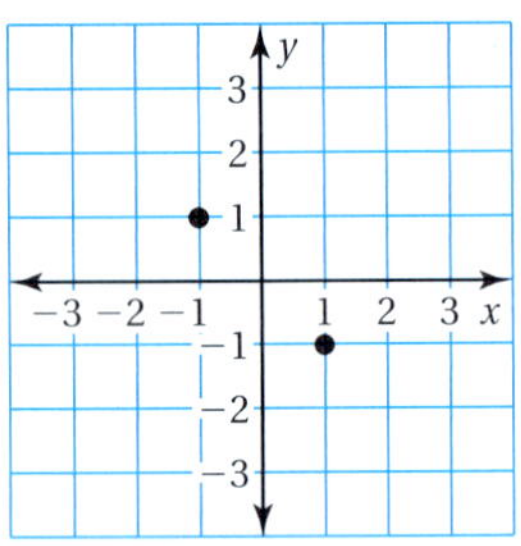

4. 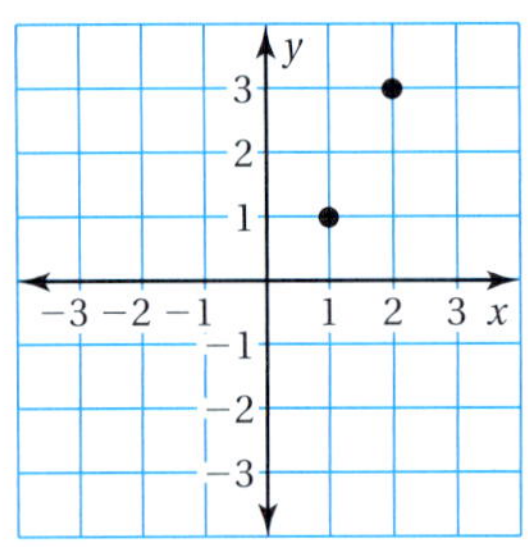

5. 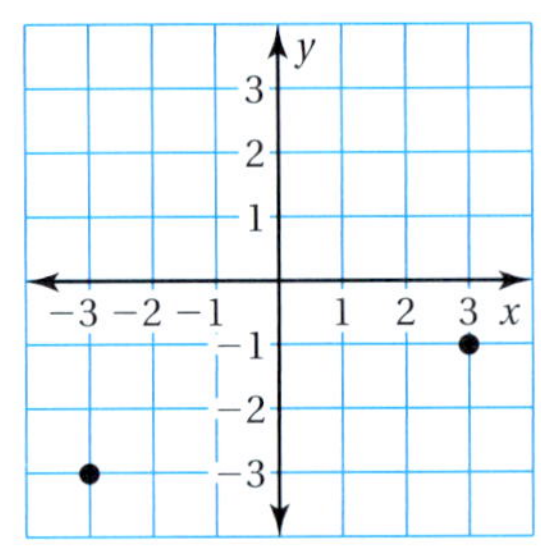

Write an equation of the line that passes through the points.

① 6. $(-1, -1), (1, 5)$
7. $(2, 4), (3, 6)$
8. $(-2, 3), (2, 7)$
9. $(4, 1), (8, 2)$
10. $(-9, 5), (-3, 3)$
11. $(1, 2), (-2, -1)$
12. $(-5, 2), (5, -2)$
13. $(2, -7), (8, 2)$
14. $(1, -2), (3, -8)$

15. **ERROR ANALYSIS** Describe and correct the error in finding the equation of the line that passes through $(-1, -6)$ and $(3, 2)$.

$$\text{slope} = \frac{\text{rise}}{\text{run}} = \frac{8}{4} = 2$$

The y-intercept is $(0, -4)$.

The equation is $y = -4x + 2$.

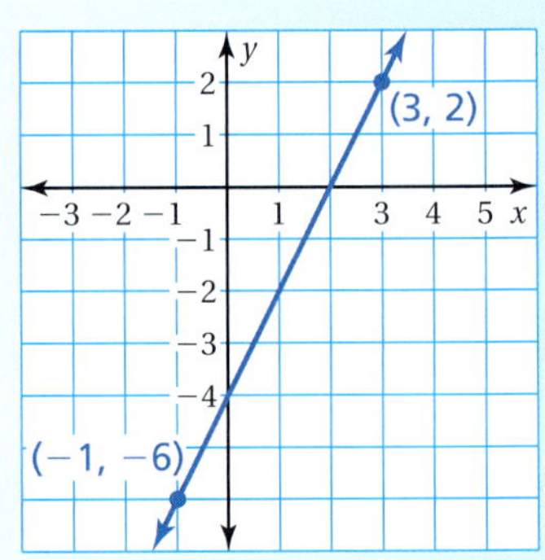

16. **JET SKI** It costs \$175 to rent a jet ski for 2 hours. It costs \$300 to rent a jet ski for 4 hours. Write an equation that represents the cost y (in dollars) of renting a jet ski for x hours.

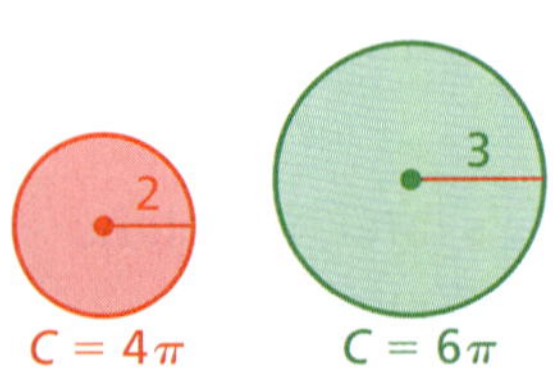

17. **CIRCUMFERENCE** Consider the circles shown.
   a. Plot the points $(2, 4\pi)$ and $(3, 6\pi)$.
   b. Write an equation of the line that passes through the two points.

18. **SOAP BOX DERBY** The table shows the changes in elevation for a Soap Box Derby track.

Track Distance	Elevation
0 ft	48 ft
100 ft	38 ft
200 ft	28 ft
350 ft	18 ft
600 ft	8 ft
989 ft	0 ft

a. Draw a Soap Box Derby track in a coordinate plane.

b. Does each section of the track have the same slope? Explain.

c. Write an equation that represents the elevation y (in feet) of the track between 100 feet and 200 feet.

19. **CAR VALUE** The value of a car decreases at a constant rate. After 3 years, the value of the car is \$15,000. After 2 more years the value of the car is \$11,000.

a. Write an equation that represents the value y (in dollars) of the car after x years.

b. Graph the equation.

c. What is the y-intercept of the line? Interpret the y-intercept.

Leaning Tower of Pisa

20. **WATERING CAN** You water the plants in your classroom at a constant rate. After 5 seconds, your watering can contains 58 ounces of water. Fifteen seconds later, the can contains 28 ounces of water.

a. Write an equation that represents the amount y (in ounces) of water in the can after x seconds.

b. How much water was in the can when you started watering the plants?

c. When is the watering can empty?

21. **Critical Thinking** The Leaning Tower of Pisa in Italy was built between 1173 and 1350.

a. Write an equation for the yellow line.

b. The tower is 56 meters tall. How far off center is the top of the tower?

Fair Game Review What you learned in previous grades & lessons

Find the percent of the number. *(Skills Review Handbook)*

22. 15% of 300

23. 140% of 125

24. 6% of −75

25. **MULTIPLE CHOICE** What is the x-intercept of the equation $3x + 5y = 30$? *(Section 2.4)*

(A) −10 (B) −6 (C) 6 (D) 10

3 Study Help

Check It Out
Graphic Organizer
BigIdeasMath.com

You can use a **notetaking organizer** to write notes, vocabulary, and questions about a topic. Here is an example of a notetaking organizer for writing equations using a slope and a point.

Write important vocabulary or formulas in this space.

Slope-intercept form:
$y = mx + b$

$m = \text{slope} = \dfrac{\text{rise}}{\text{run}}$

b = y-intercept

Writing equations using a slope and a point

Plot the point. Plot additional points using the slope. Draw a line through the points and find the y-intercept. Write the equation in slope-intercept form.

Example:

slope $= \frac{1}{2}$; $(-4, 0)$

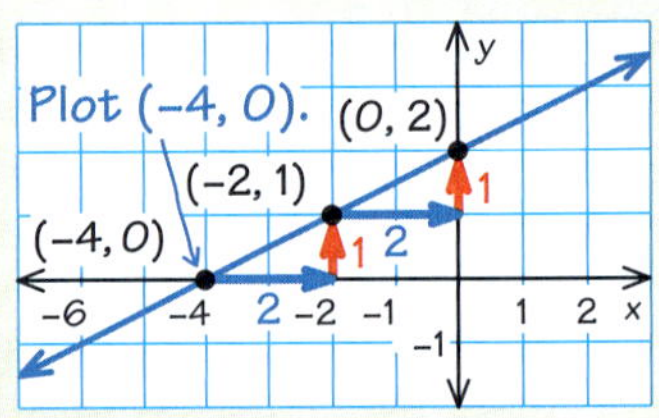

The y-intercept is 2. So, the equation is $y = \frac{1}{2}x + 2$.

How can you write an equation of a line using two points?

Write your notes about the topic in this space.

Write your questions about the topic in this space.

On Your Own

Make a notetaking organizer to help you study these topics.

1. writing equations in slope-intercept form
2. writing equations using two points

After you complete this chapter, make notetaking organizers for the following topics.

3. writing systems of linear equations
4. Think of a real-life application that can be modeled using a linear equation. Then make a notetaking organizer to help you study this application.

"My notetaking organizer has me thinking about retirement when I won't have to fetch sticks any more."

3.1–3.3 Quiz

Check It Out
Progress Check
BigIdeasMath.com

Write an equation of the line in slope-intercept form. *(Section 3.1)*

1.

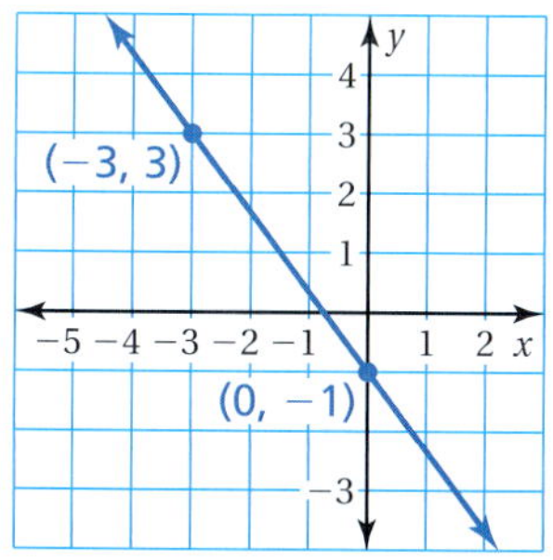

2.

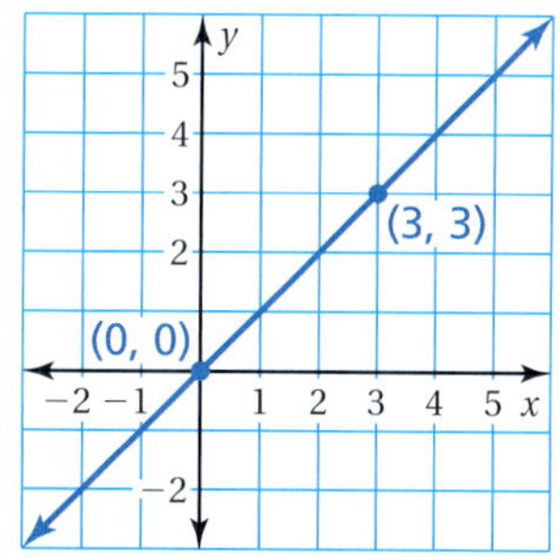

3.

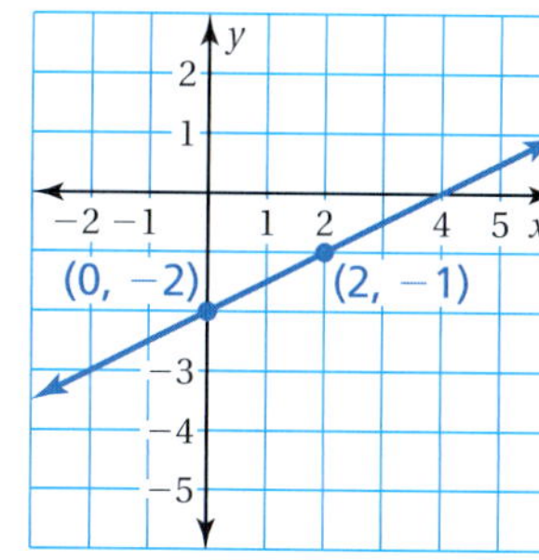

Write an equation of the line with the given slope that passes through the given point. *(Section 3.2)*

4. $m = 2;\ (1, 3)$

5. $m = \frac{1}{3};\ (-3, -2)$

6. $m = -1;\ (-1, 4)$

7. $m = -\frac{1}{8};\ (8, -5)$

Write an equation of the line that passes through the points. *(Section 3.3)*

8. $\left(0, -\frac{2}{3}\right)\left(-3, -\frac{2}{3}\right)$

9. $(4, 0), (0, 4)$

10. CONSTRUCTION A construction crew is extending a highway sound barrier that is 13 miles long. The crew builds $\frac{1}{2}$ mile per week. Write an equation for the length y (in miles) of the barrier after x weeks. *(Section 3.1)*

11. STORAGE You pay \$510 to rent a storage unit for 3 months. The total cost includes an initial deposit plus a monthly fee of \$160. *(Section 3.2)*

a. Write an equation that represents your total cost y (in dollars) after x months.

b. Interpret the y-intercept.

12. CORN After 3 weeks, a corn plant is 2 feet tall. After 9 weeks, the plant is 8 feet tall. Write an equation that represents the height y (in feet) of the corn plant after x weeks. *(Section 3.3)*

3.4 Solving Real-Life Problems

STATE STANDARDS
MA.8.A.1.1
MA.8.A.1.2

Share Your Work at... My.BigIdeasMath.com

Essential Question How can you use a linear equation in two variables to model and solve a real-life problem?

1 EXAMPLE: Writing a Story

Write a story that uses the graph at the right.

- **In your story, interpret the slope of the line, the *y*-intercept, and the *x*-intercept.**
- **Make a table that shows data from the graph.**
- **Label the axes of the graph with units.**
- **Draw pictures for your story.**

There are many possible stories. Here is one about a reef tank.

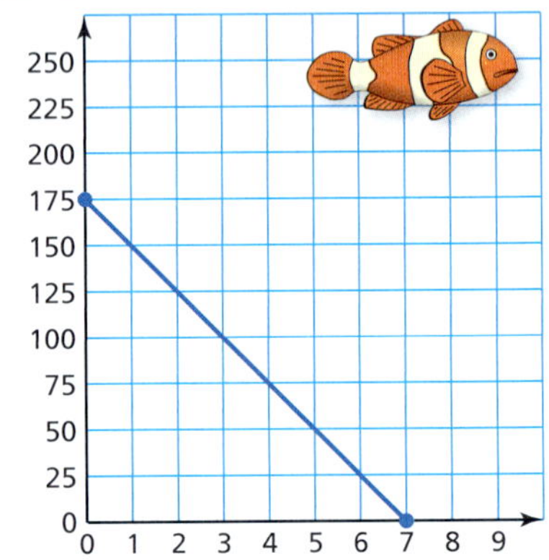

Tom works at an aquarium shop on Saturdays. One Saturday, when Tom gets to work, he is asked to clean a 175-gallon reef tank.

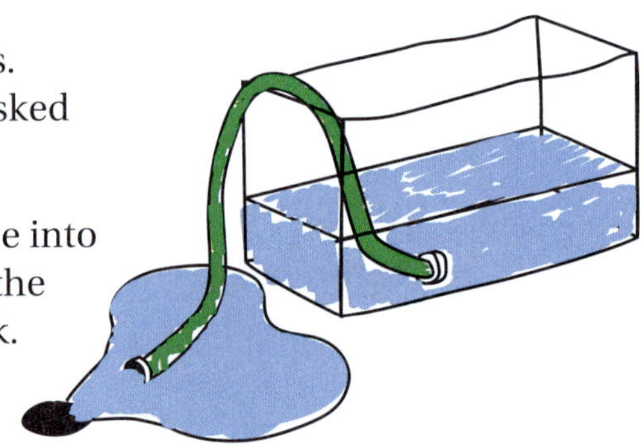

His first job is to drain the tank. He puts a hose into the tank and starts a siphon. Tom wonders if the tank will finish draining before he leaves work.

He measures the amount of water that is draining out and finds that 12.5 gallons drain out in 30 minutes. So, he figures that the rate is 25 gallons per hour. To see when the tank will be empty, Tom makes a table and draws a graph.

x-intercept: number of hours to empty the tank

x	0	1	2	3	4	5	6	7
y	175	150	125	100	75	50	25	0

y-intercept: amount of water in full tank

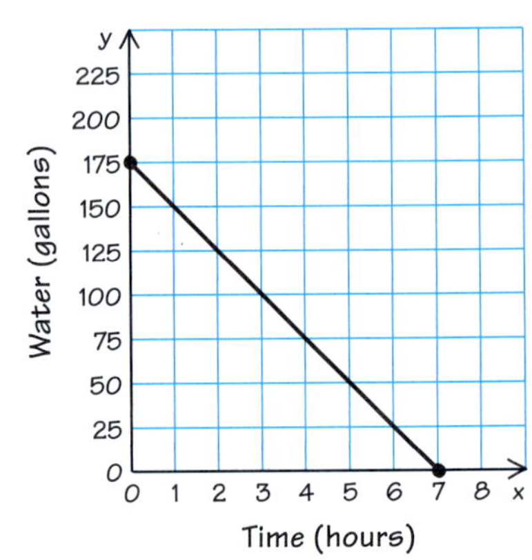

From the table and also from the graph, Tom sees that the tank will be empty after 7 hours. This will give him 1 hour to wash the tank before going home.

2 ACTIVITY: Writing a Story

Work with a partner. Write a story that uses the graph of a line.

- **In your story, interpret the slope of the line, the *y*-intercept, and the *x*-intercept.**
- **Make a table that shows data from the graph.**
- **Label the axes of the graph with units.**
- **Draw pictures for your story.**

3 ACTIVITY: Drawing Graphs

Work with a partner. Describe a real-life problem that has the given rate and intercepts. Draw a line that represents the problem.

a. Rate: -30 feet per second
y-intercept: 150 feet
x-intercept: 5 seconds

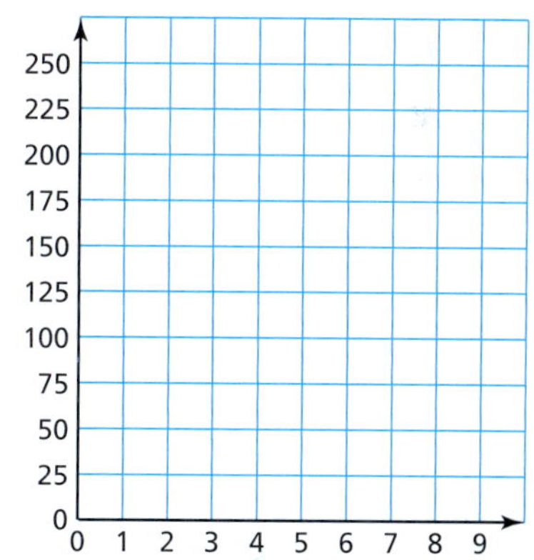

b. Rate: -25 dollars per month
y-intercept: \$200
x-intercept: 8 months

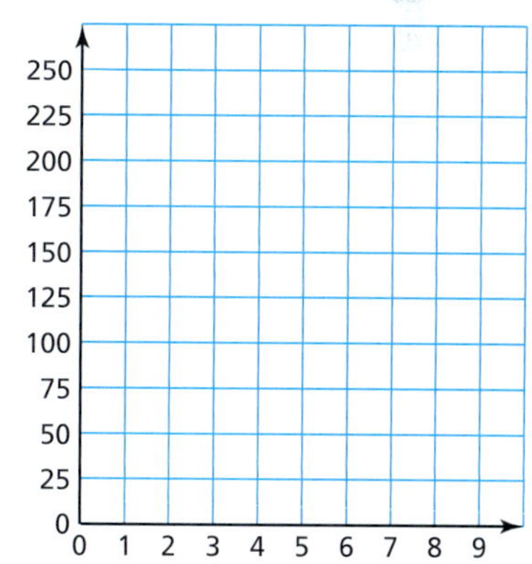

What Is Your Answer?

4. **IN YOUR OWN WORDS** How can you use a linear equation in two variables to model and solve a real-life problem? List three different rates that can be represented by slopes in real-life problems.

Practice Use what you learned about solving real-life problems to complete Exercises 4 and 5 on page 130.

3.4 Lesson

Check It Out
Lesson Tutorials
BigIdeasMath.com

EXAMPLE 1 Real-Life Application

The percent y (in decimal form) of battery power remaining x hours after you turn on a laptop computer is $y = -0.2x + 1$. (a) Graph the equation. (b) Interpret the x- and y-intercepts. (c) After how many hours is the battery power at 75%?

a. Use the slope and the y-intercept to graph the equation.

$$y = -0.2x + 1$$

slope: -0.2 ; y-intercept: 1

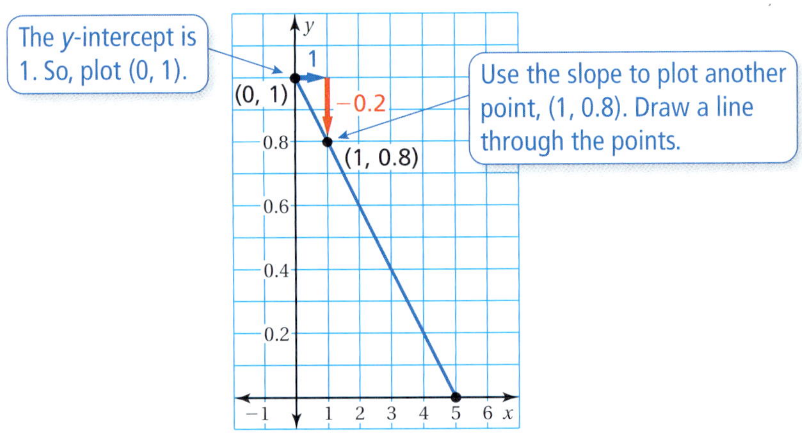

b. To find the x-intercept, substitute 0 for y in the equation.

$y = -0.2x + 1$ — Write the equation.

$0 = -0.2x + 1$ — Substitute 0 for y.

$5 = x$ — Solve for x.

The x-intercept is 5. So, the battery lasts 5 hours. The y-intercept is 1. So, the battery power is at 100% when you turn on the laptop.

75% Remaining

c. Find the value of x when $y = 0.75$.

$y = -0.2x + 1$ — Write the equation.

$0.75 = -0.2x + 1$ — Substitute 0.75 for y.

$1.25 = x$ — Solve for x.

The battery power is at 75% after 1.25 hours.

On Your Own

Now You're Ready
Exercise 6

1. The amount y (in gallons) of gasoline remaining in a gas tank after driving x hours is $y = -2x + 12$. (a) Graph the equation. (b) Interpret the x- and y-intercepts. (c) After how many hours are there 5 gallons left?

EXAMPLE 2 Real-Life Application

The graph relates temperatures y (in degrees Fahrenheit) to temperatures x (in degrees Celsius). (a) Find the slope and y-intercept. (b) Write an equation of the line. (c) What is the mean temperature of Earth in degrees Fahrenheit?

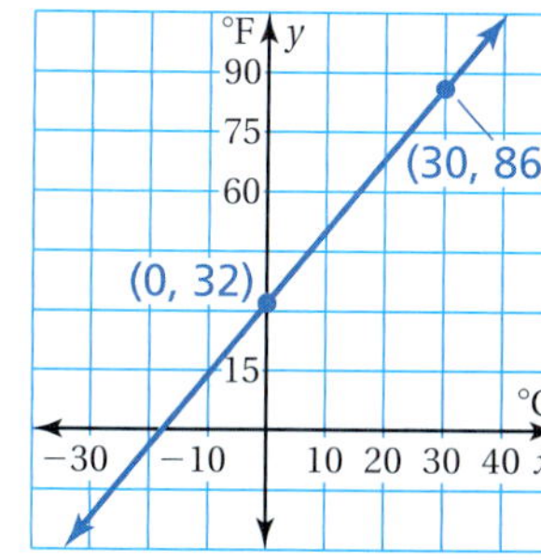

Mean Temperature: 15°C

a. slope $= \dfrac{\text{change in } y}{\text{change in } x} = \dfrac{54}{30} = \dfrac{9}{5}$

The line crosses the y-axis at (0, 32). So, the y-intercept is 32.

∴ The slope is $\frac{9}{5}$ and the y-intercept is 32.

b. Use the slope and y-intercept to write an equation.

slope → $\frac{9}{5}$; y-intercept → 32

∴ The equation is $y = \frac{9}{5}x + 32$.

c. In degrees Celsius, the mean temperature of Earth is 15°. To find the mean temperature in degrees Fahrenheit, find the value of y when $x = 15$.

$y = \frac{9}{5}x + 32$ — Write the equation.

$= \frac{9}{5}(15) + 32$ — Substitute 15 for x.

$= 59$ — Simplify.

∴ The mean temperature of Earth is 59°F.

On Your Own

Now You're Ready
Exercise 7

2. The graph shows the height y (in feet) of a flag x seconds after you start raising it up a flagpole.

   a. Find and interpret the slope.

   b. Write an equation of the line.

   c. What is the height of the flag after 9 seconds?

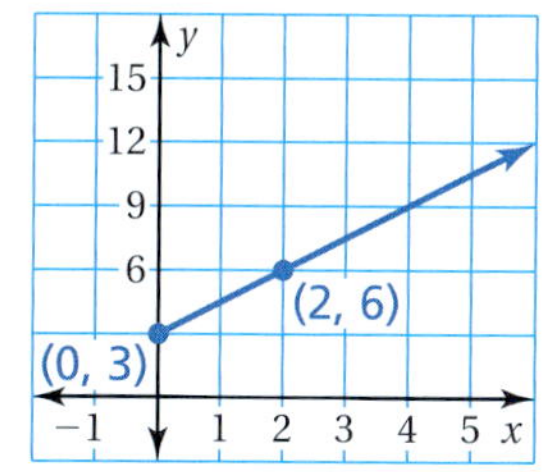

3.4 Exercises

Check It Out
Help with Homework
BigIdeasMath.com

Vocabulary and Concept Check

1. **REASONING** Explain how to find the slope, y-intercept, and x-intercept of the line shown.

2. **OPEN-ENDED** Describe a real-life situation that uses a negative slope.

3. **REASONING** In a real-life situation, what does the slope of a line represent?

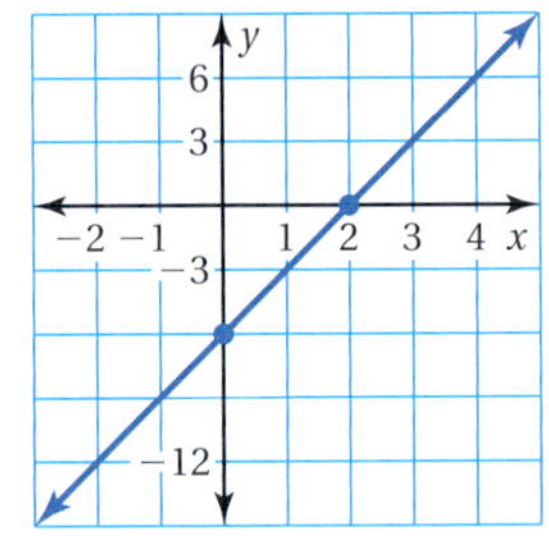

Practice and Problem Solving

Describe a real-life problem that has the given rate and intercepts. Draw a line that represents the problem.

4. Rate: -1.6 gallons per hour

   y-intercept: 16 gallons

   x-intercept: 10 hours

5. Rate: -45 pesos per week

   y-intercept: 180 pesos

   x-intercept: 4 weeks

1 6. **DOWNLOAD** You are downloading a song. The percent y (in decimal form) of megabytes remaining to download after x seconds is $y = -0.1x + 1$.

   a. Graph the equation.

   b. Interpret the x- and y-intercepts.

   c. After how many seconds is the download 50% complete?

2 7. **HIKING** The graph relates temperature y (in degrees Fahrenheit) to altitude x (in thousands of feet).

   a. Find the slope and y-intercept.

   b. Write an equation of the line.

   c. What is the temperature at sea level?

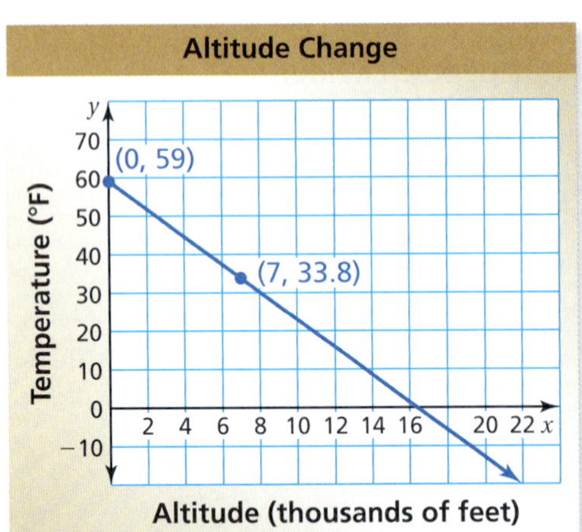

8. **TRAVEL** Your family is driving from Pensacola to Jacksonville. The graph relates your distance from Jacksonville y (in miles) and travel time x (in hours).

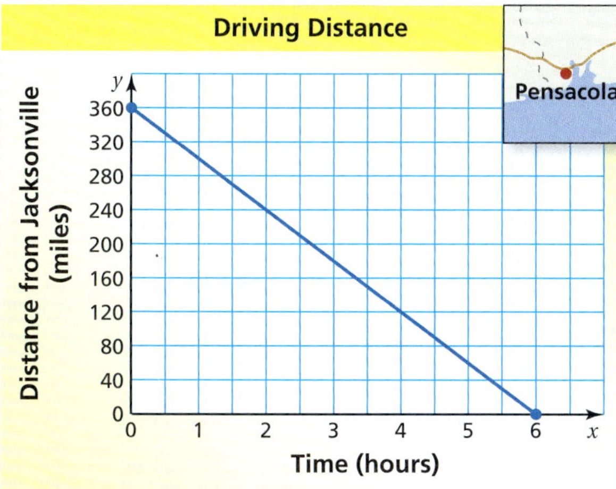

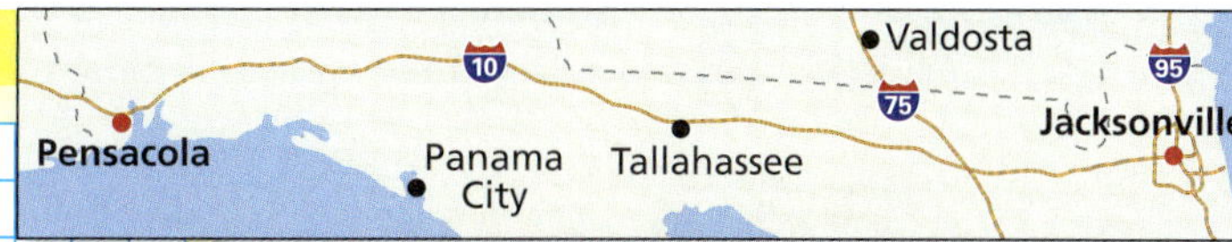

a. Interpret the x- and y-intercepts.

b. What is the slope? What does the slope represent in this situation?

c. Write an equation of the line. How would the graph and the equation change if you were able to travel in a straight line?

9. **PROJECT** Use a map or the Internet to find the latitude and longitude of your school to the nearest whole number. Then find the latitudes and longitudes of: Antananarivo, Madagascar; Denver, Colorado; Brasilia, Brazil; London, England; and Beijing, China.

a. Graph a point for each of the cities in the same coordinate plane. Let the positive y-axis represent north and the positive x-axis represent east.

b. Write an equation of the line that passes through Denver and Beijing.

c. In part (b), what geographic location does the y-intercept represent?

10. **Reasoning** A band is performing at an auditorium for a fee of \$1500. In addition to this fee, the band receives 30% of each \$20 ticket sold. The maximum capacity of the auditorium is 800 people.

a. Write an equation that represents the band's revenue R when x tickets are sold.

b. The band needs \$5000 for new equipment. How many tickets must be sold for the band to earn enough money to buy the new equipment?

Fair Game Review What you learned in previous grades & lessons

Tell whether the system has *one solution, no solution,* or *infinitely many solutions.* *(Section 2.5 and Section 2.6)*

11. $y = -x + 6$
$-4(x + y) = -24$

12. $y = 3x - 2$
$-x + 2y = 11$

13. $-9x + 3y = 12$
$y = 3x - 2$

14. **MULTIPLE CHOICE** Which equation is the slope-intercept form of $24x - 8y = 56$? *(Section 2.4)*

Ⓐ $y = -3x + 7$ Ⓑ $y = 3x - 7$ Ⓒ $y = -3x - 7$ Ⓓ $y = 3x + 7$

3.5 Writing Systems of Linear Equations

STATE STANDARDS
MA.8.A.1.3
MA.8.A.1.4

Essential Question How can you use a system of linear equations to model and solve a real-life problem?

Share Your Work at... My.BigIdeasMath.com

1 ACTIVITY: Writing a System

Work with a partner.

- **Peak Valley Middle School has 1200 students. Its enrollment is decreasing by 30 students per year.**
- **Southern Tier Middle School has 500 students. Its enrollment is increasing by 40 students per year.**
- **In how many years will the two schools have equal enrollments?**

a. **USE A TABLE** Use a table to answer the question.

Now

Year, x	0	1	2	3	4	5	6	7	8	9	10
Peak Valley MS, P	1200										
Southern Tier MS, S	500										

b. **USE A GRAPH** Write a linear equation that represents each enrollment.

$P =$

$S =$

Then graph each equation and find the point of intersection to answer the question.

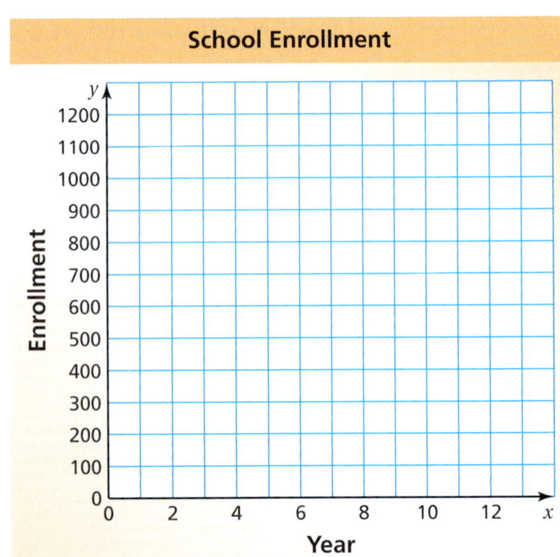

c. **USE ALGEBRA** Answer the question by setting the expressions for P and S equal to each other and solving for x.

2 ACTIVITY: Writing a System

Work with a partner. The table shows the enrollments of Sizemore Middle School and Wright Middle School for 7 years.

Year, x	0	1	2	3	4	5	6
Sizemore MS, S	1500	1438	1423	1350	1308	1247	1204
Wright MS, W	825	854	872	903	927	946	981

From the enrollment pattern, do you think the two schools will ever have the same enrollment? If so, when?

a. Plot the enrollments of each middle school.

b. Draw a line that approximately fits the points for each middle school.

c. Estimate the year in which the schools will have the same enrollment.

d. Write an equation for each line.

$S =$ ______

$W =$ ______

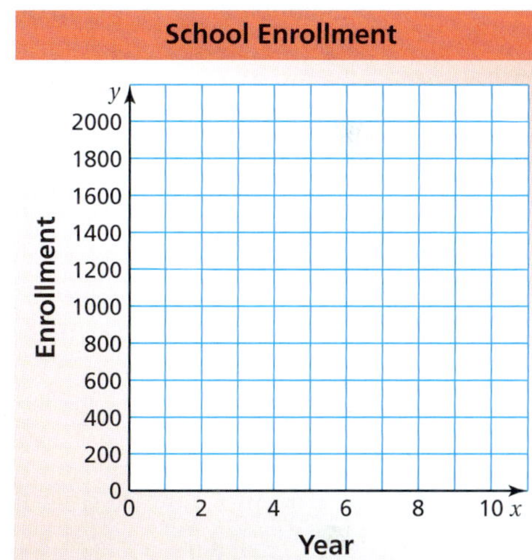

e. **USE ALGEBRA** Answer the question by setting the expressions for S and W equal to each other and solving for x.

What Is Your Answer?

3. **IN YOUR OWN WORDS** How can you use a system of linear equations to model and solve a real-life problem?

4. **PROJECT** Use the Internet, a newspaper, a magazine, or some other reference to find two sets of real-life data that can be modeled by linear equations.

   a. List the data in a table.

   b. Graph the data. Find a line to represent each data set.

   c. If possible, estimate when the two quantities will be equal.

Practice Use what you learned about writing systems of linear equations to complete Exercises 4 and 5 on page 136.

3.5 Lesson

Check It Out
Lesson Tutorials
BigIdeasMath.com

EXAMPLE 1 Writing a System of Linear Equations

A bank teller is counting \$20 bills and \$10 bills. There are 16 bills that total \$200. Write and solve a system of equations to find the number x of \$20 bills and the number y of \$10 bills.

Words	Number of \$20 bills	plus	number of \$10 bills	is	the total number of bills.	
Equation 1 → **Equation**	x	$+$	y	$=$	16	

Words	Twenty times	the number of \$20 bills	plus ten times	the number of \$10 bills	is	the total value.
Equation 2 → **Equation**	$20\,\bullet$	x	$+\,10\,\bullet$	y	$=$	200

The linear system is $x + y = 16$ and $20x + 10y = 200$.

Solve each equation for y. Then make a table of values to find the x-value that gives the same y-value for both equations.

x	0	1	2	3	4
$y = 16 - x$	16	15	14	13	12
$y = 20 - 2x$	20	18	16	14	12

The solution is (4, 12).

So, there are 4 twenty-dollar bills and 12 ten-dollar bills.

Check

Equation 1	Equation 2
$x + y = 16$	$20x + 10y = 200$
$4 + 12 \stackrel{?}{=} 16$	$20(4) + 10(12) \stackrel{?}{=} 200$
$16 = 16$ ✓	$200 = 200$ ✓

On Your Own

1. The length ℓ of the rectangle is 1 more than 3 times the width w. Write and solve a system of linear equations to find the dimensions of the rectangle.

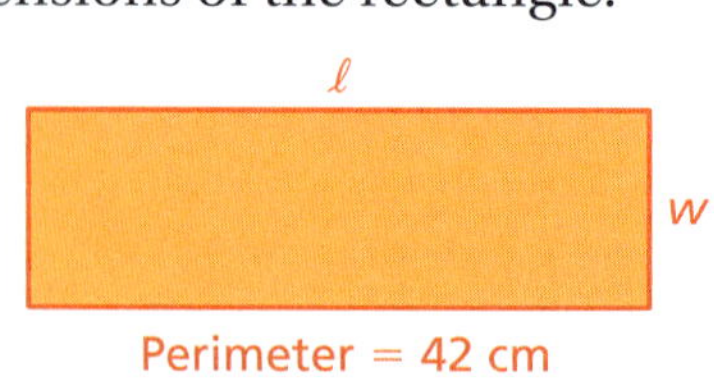

EXAMPLE 2 Standardized Test Practice

The sum of two numbers is 35. The second number y is equal to 4 times the first number x. Which system of linear equations represents the two numbers?

(A) $x + y = 35$, $x = y + 4$　(B) $x + y = 35$, $y = 4x$　(C) $x + y = 35$, $y = -4x$　(D) $x - y = 35$, $y = 4x$

Words	First number	plus	second number	is	35.
Equation 1 → **Equation**	x	$+$	y	$=$	35

Words	Second number	is equal to	4 times the	first number.
Equation 2 → **Equation**	y	$=$	$4 \cdot$	x

The system is $x + y = 35$ and $y = 4x$. The correct answer is (B).

EXAMPLE 3 Writing a System of Linear Equations

x	Airbus A320, A	Boeing 777, B
0	0	9000
1	1000	8500
2	2000	8000
3	3000	7500
4	4000	7000

The table shows the altitudes (in feet) of two jets after x minutes. After how many minutes do the jets have the same altitude?

Method 1: Plot the points and draw each line. The graphs appear to intersect at (6, 6000).

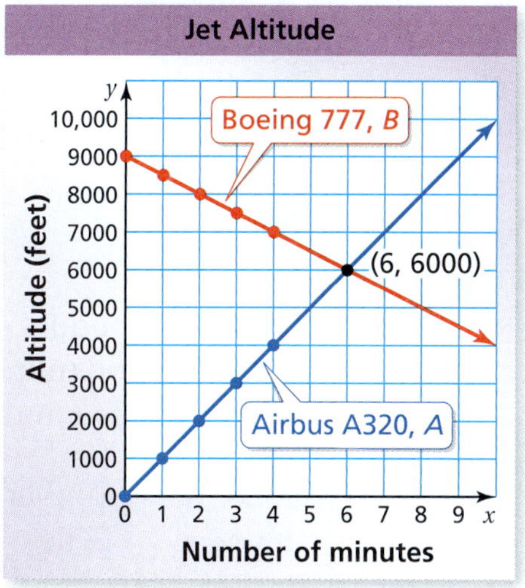

So, the jets have the same altitude after 6 minutes.

Method 2: Use the slopes and y-intercepts to write equations for A and B. Set the equations equal to each other and solve for x.

$A = 1000x$　　$B = -500x + 9000$

$$1000x = -500x + 9000$$
$$1500x = 9000$$
$$x = 6$$

The jets have the same altitude after 6 minutes.

On Your Own

Now You're Ready
Exercises 4–6

2. The sum of two numbers is 20. The second number is 3 times the first number. Write and solve a system of equations to find the two numbers.

3. **WHAT IF?** In Example 3, the altitude of the Boeing 777 decreases 800 feet each minute. After how many minutes do the jets have the same altitude? Solve using both methods.

3.5 Exercises

Check It Out
Help with Homework
BigIdeasMath.com

Vocabulary and Concept Check

1. **VOCABULARY** Why is the equation $2x - y = 4$ called a *linear* equation?
2. **VOCABULARY** What must be true for an ordered pair to be a solution of a system of two linear equations?
3. **WRITING** Describe three ways to solve a system of linear equations.

Practice and Problem Solving

In Exercises 4–6, (a) write a system of linear equations to represent the situation. Then, answer the question using (b) a table, (c) a graph, and (d) algebra.

1 2 3

4. **ATTENDANCE** The first football game has 425 adult fans and 225 student fans. The adult attendance A decreases by 15 each game. The student attendance S increases by 25 each game. After how many games x will the adult attendance equal the student attendance?

Adults:	Attendence each game	is	425	minus 15 times	number of games.
Students:	Attendence each game	is	225	plus 25 times	number of games.

5. **BOUQUET** A bouquet of lilies and tulips has 12 flowers. Lilies cost \$3 each and tulips cost \$2 each. The bouquet costs \$32. How many lilies x and tulips y are in the bouquet?

Number of flowers:		Number of lilies	plus		Number of tulips	is	12.
Cost of bouquet:	\$3 times	number of lilies	plus	\$2 times	number of tulips	is	\$32.

6. **CHORUS** There are 63 students in a middle school chorus. There are 11 more boys than girls. How many boys x and girls y are in the chorus?

Number of students:	Number of boys	plus	number of girls	is	63.
Boys and girls:	Number of boys	equals	number of girls	plus	11.

7. **WHAT IS MISSING?** You have dimes and nickels in your pocket with a total value of \$0.95. There are more dimes than nickels. How many of each coin do you have?

a. Do you have enough information to write a system of equations to answer the question? If not, what else do you need to know?

b. Find one possible solution.

x	Account *A*	Account *B*
0	420	465
1	426	468
2	432	471
3	438	474
4	444	477

8. **INTEREST** The table shows the balances (in dollars) of two accounts earning simple interest for x years. After how many years will the accounts have the same balance?

9. **CRITICAL THINKING** Is it possible for a system of two linear equations to have exactly two solutions? Explain.

10. **DINNER** How much does it cost for two specials and two glasses of milk?

11. **REASONING** A system of two linear equations has more than one solution. Describe the graph of the system.

Scottish Team

Puerto Rican Team

12. **WORLD CUP** The global competition for the World Cup is broken up into six continental zones. The number of teams in the Scottish team's zone is 17 less than twice the number of teams in the Puerto Rican team's zone. There is a total of 88 teams in both zones. How many teams are in each zone?

13. **Algebra** The graphs of the three equations form a triangle. Use algebra to find the coordinates of the vertices of the triangle.

$$x + y = 1 \qquad x + 7y = 1 \qquad x - 2y = -8$$

Fair Game Review What you learned in previous grades & lessons

Write an equation of the line that passes through the points. *(Section 3.3)*

14. $(0, -1), (1, 1)$

15. $(-4, -3), (4, -1)$

16. $(2, 1), (3, -1)$

17. **MULTIPLE CHOICE** Which function rule relates x and y for the set of ordered pairs $(2, 4), (4, 5), (6, 6)$? *(Skills Review Handbook)*

Ⓐ $y = x - 2$ Ⓑ $y = \frac{1}{2}x + 3$ Ⓒ $y = 2x + 1$ Ⓓ $y = \frac{1}{2}x - 3$

3.4–3.5 Quiz

Check It Out
Progress Check
BigIdeasMath.com

1. **FISH POND** You are draining a fish pond. The amount y (in liters) of water remaining after x hours is $y = -60x + 480$. (a) Graph the equation. (b) Interpret the x- and y-intercepts. *(Section 3.4)*

2. **CABLE CAR** The graph shows the distance y (in meters) that a cable car travels up a mountain in x minutes. *(Section 3.4)*

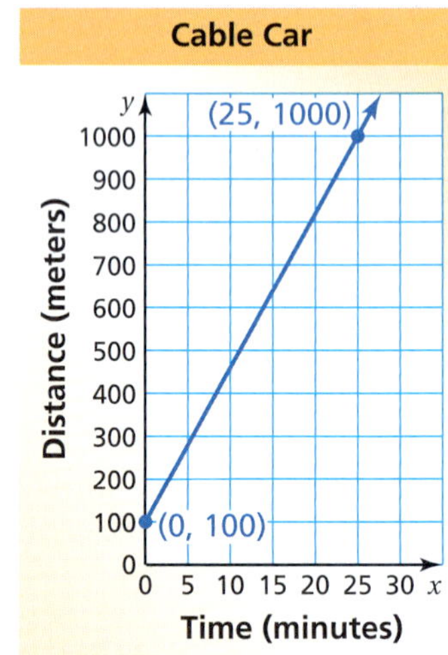

   a. Find and interpret the slope.

   b. Write an equation of the line.

   c. How far does the cable car travel in 15 minutes?

3. **BICYCLE** You need \$160 to buy a mountain bike. You earn \$40 per week for babysitting and \$20 per lawn mowed. *(Section 3.4)*

   a. Write an equation that represents your weekly income y (in dollars) for x lawns mowed.

   b. How many lawns do you need to mow to earn enough money in 1 week to buy the mountain bike?

4. **WATER** A recreation department bought bottled water to sell at a fair. The graph shows the number y of bottles remaining after each hour x. *(Section 3.4)*

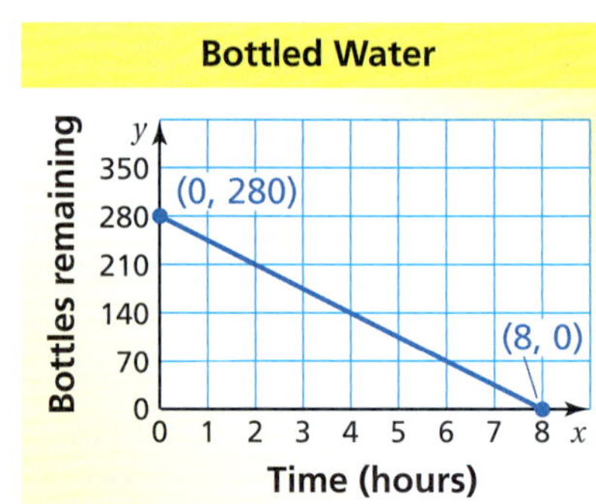

   a. Find the slope and y-intercept.

   b. Write an equation of the line.

   c. The fair started at 10 A.M. When did the recreation department run out of bottled water?

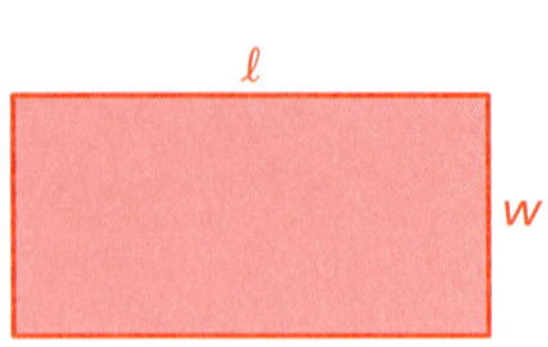

Perimeter = 36 ft

5. **RECTANGLE** The length of the rectangle is twice its width. Write and solve a system of linear equations to find the length ℓ and width w of the rectangle. *(Section 3.5)*

6. **PUZZLE** The difference of two numbers is 8. The first number x is 1 less than twice the second number y. Write and solve a system of linear equations to find the two numbers. *(Section 3.5)*

7. **SHELTER** A cat and dog shelter houses 23 animals. There are 5 more cats x than dogs y. Write and solve a system of linear equations to find the numbers of cats and dogs in the shelter. *(Section 3.5)*

3 Chapter Review

Check It Out
Vocabulary Help
BigIdeasMath.com

Review Examples and Exercises

3.1 Writing Equations in Slope-Intercept Form (pp. 106–111)

Write an equation of the line in slope-intercept form.

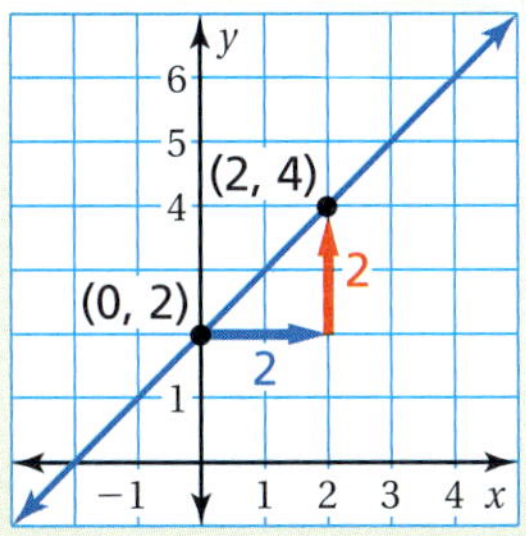

Find the slope and y-intercept.

$$\text{slope} = \frac{\text{rise}}{\text{run}} = \frac{2}{2} = 1$$

Because the line crosses the y-axis at (0, 2), the y-intercept is 2.

slope → y-intercept

So, the equation is $y = 1x + 2$, or $y = x + 2$.

Exercises

Write an equation of the line in slope-intercept form.

1.

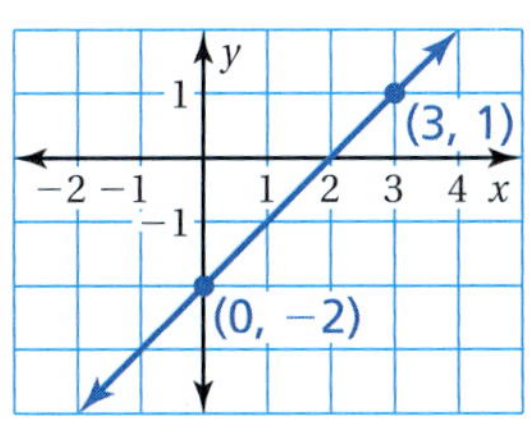

2.

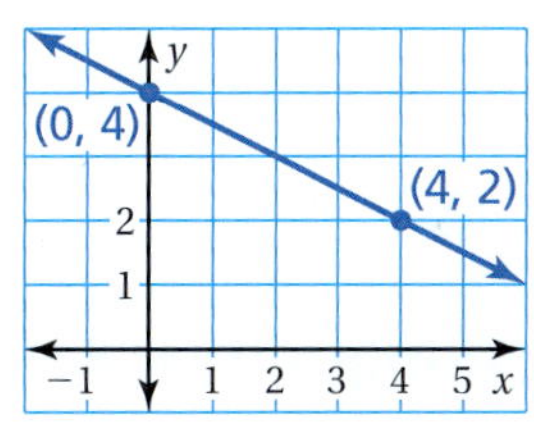

3.2 Writing Equations Using a Slope and a Point (pp. 112–117)

Write an equation of the line with a slope of $\frac{2}{3}$ that passes through the point (−3, −1).

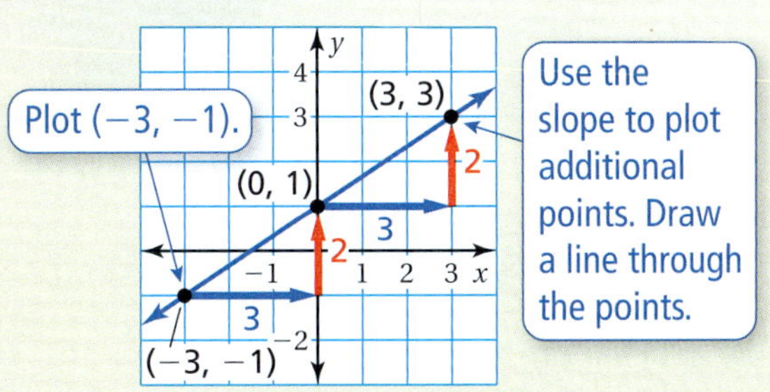

Use a graph to find the y-intercept.

Because the line crosses the y-axis at (0, 1), the y-intercept is 1.

So, the equation is $y = \frac{2}{3}x + 1$.

Exercises

Write an equation of the line with the given slope that passes through the given point.

3. $m = 3$; (4, 4) **4.** $m = 2$; (2, 6) **5.** $m = -0.5$; (−4, 2)

3.3 Writing Equations Using Two Points (pp. 118–123)

Write an equation of the line that passes through the points (2, 3) and (−2, −3).

Use a graph to find the slope and y-intercept.

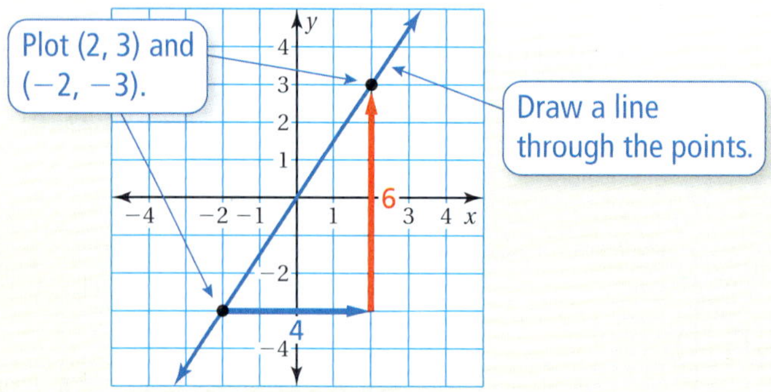

$$\text{slope} = \frac{\text{rise}}{\text{run}} = \frac{6}{4} = \frac{3}{2}$$

Because the line crosses the y-axis at (0, 0), the y-intercept is 0.

∴ So, the equation is $y = \frac{3}{2}x + 0$, or $y = \frac{3}{2}x$.

Exercises

Write an equation of the line that passes through the points.

6. (−2, 0), (2, −4)
7. (−2, −2), (4, 1)

3.4 Solving Real-Life Problems (pp. 126–131)

The amount y (in dollars) of money you have left after playing x games at a carnival is $y = -0.75x + 10$. How much money do you have after playing eight games?

$y = -0.75x + 10$	Write the equation.
$= -0.75(8) + 10$	Substitute 8 for x.
$= 4$	Simplify.

∴ You have $4 left after playing 8 games.

Exercises

8. **HAY** The amount y (in bales) of hay remaining after feeding cows for x days is $y = -3.5x + 105$. (a) Graph the equation. (b) Interpret the x- and y-intercepts. (c) How many bales are left after 10 days?

3.5 Writing Systems of Linear Equations (pp. 132–137)

You have quarters and dimes in your pocket. There are 5 coins that total \$0.80. Write and solve a system of equations to find the number x of dimes and the number y of quarters.

Words	Number of dimes	plus	number of quarters	is	five .
Equation 1 → **Equation**	x	$+$	y	$=$	5

Words	Value of a dime	times	number of dimes	plus	value of a quarter	times	number of quarters	is	the total value .
Equation 2 → **Equation**	0.10	$\cdot$	x	$+$	0.25	$\cdot$	y	$=$	0.80

The linear system is $x + y = 5$ and $0.10x + 0.25y = 0.80$.

Solve each equation for y. Then make a table of values to find the x-value that gives the same y-value for both equations.

x	0	1	2	3	4
$y = 5 - x$	5	4	3	2	1
$y = 3.2 - 0.4x$	3.2	2.8	2.4	2	1.6

The solution is (3, 2).

∴ So, you have three dimes and two quarters in your pocket.

Exercises

9. **GIFT BASKET** A gift basket that contains jars of jam and packages of muffin mix costs \$45. There are eight items in the basket. Jars of jam cost \$6 each and packages of muffin mix cost \$5 each. Write and solve a system of equations to find the number of jars of jam and the number of packages of muffin mix.

10. **MARBLES** A bag contains half as many yellow marbles as purple marbles. There are 36 marbles in the bag. Write and solve a system of equations to find the number of marbles of each color.

3 Chapter Test

Check It Out
Test Practice
BigIdeasMath.com

Write an equation of the line in slope-intercept form.

1.

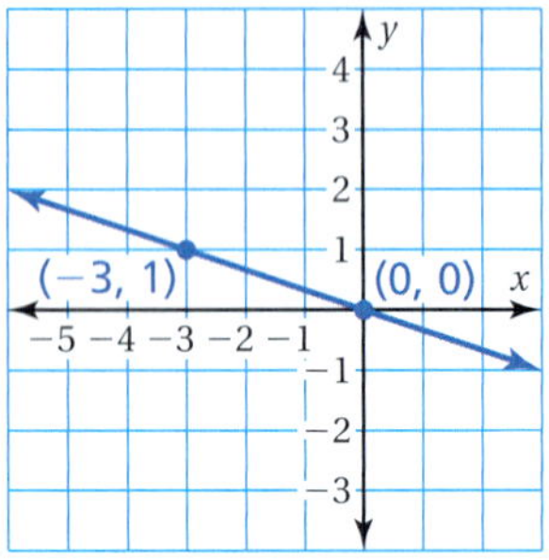

2.

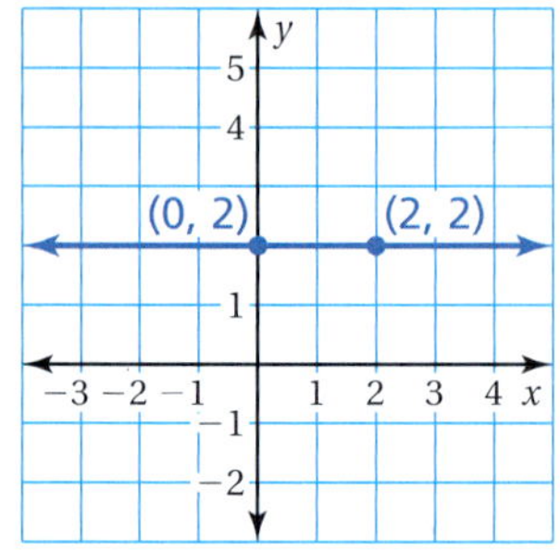

3.

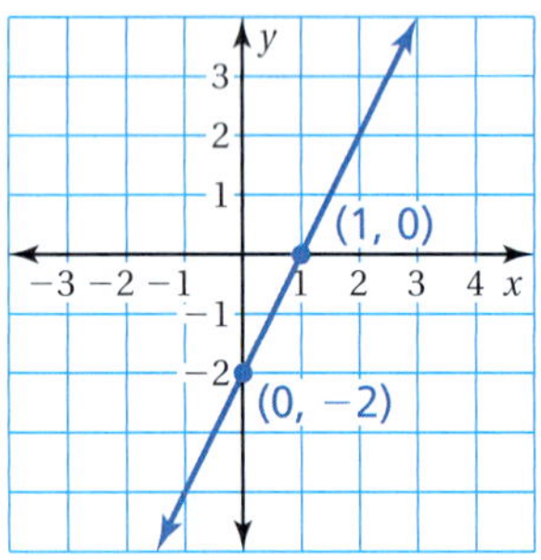

4.

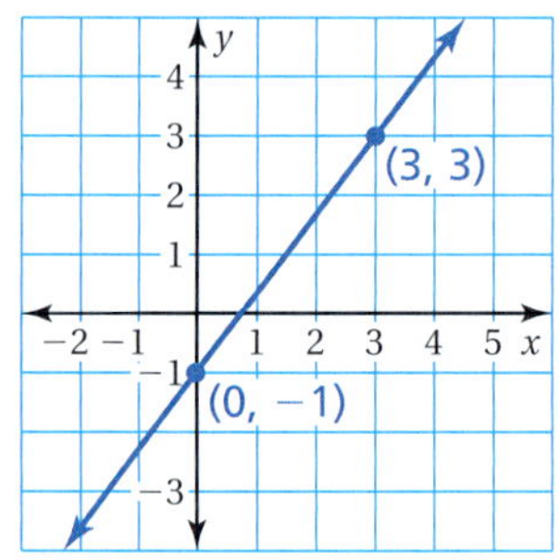

Write an equation of the line with the given slope that passes through the given point.

5. $m = -3$; $(-2, -2)$

6. $m = -\frac{1}{2}$; $(4, -1)$

7. $m = \frac{2}{3}$; $(3, 3)$

8. $m = 1$; $(4, 3)$

Write an equation of the line that passes through the points.

9. $(-1, 5)$, $(3, -3)$

10. $(-4, 1)$, $(4, 3)$

11. $(-2, 5)$, $(-1, 1)$

12. **BRAILLE** Because of its size and detail, Braille takes longer to read than text. A person reading Braille reads at 25% the rate of a person reading text. Write an equation that represents the average rate y of a Braille reader in terms of the average rate x of a text reader.

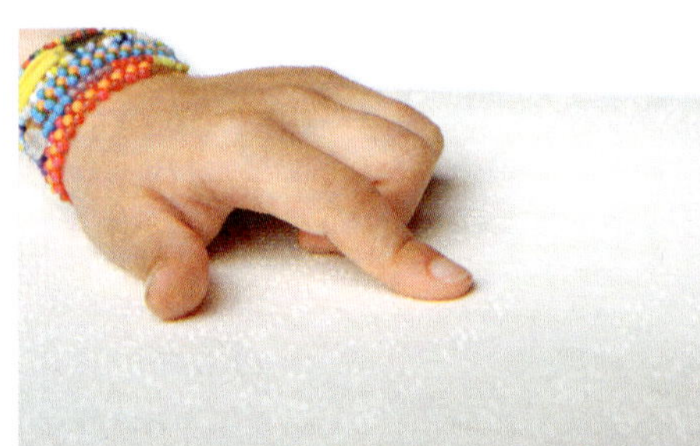

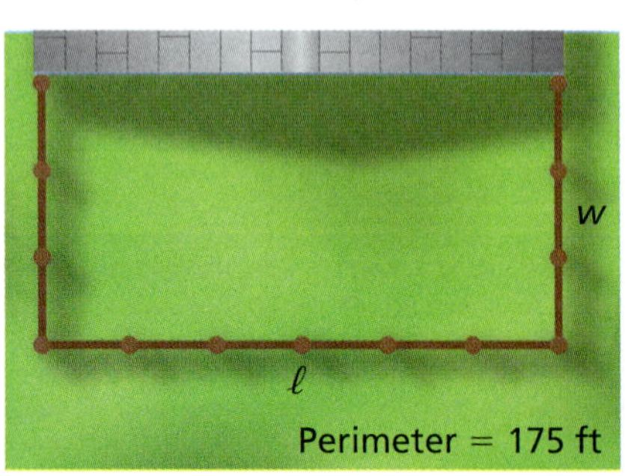

13. **PASTURE** You are building a fence for a rectangular pasture. The length of the pasture is 1 foot less than twice the width.

a. Write and solve a system of linear equations to find the length and width of the pasture.

b. Find the area of the pasture.

3 Standardized Test Practice

Test-Taking Strategy
Work Backwards

"**Work backwards** by trying 1, 2, 3, and 4. You will see that 3(4) = 12. So, D is correct."

1. A line contains the points $(-3, 5)$ and $(6, 8)$. What is the equation of the line?

   A. $y = \frac{1}{3}x$ **C.** $y = 3x - 10$

   B. $y = \frac{1}{3}x + 6$ **D.** $y = 3x + 14$

2. Two lines have the same y-intercept. The slope of one line is 1 and the slope of the other line is -1. What can you conclude?

   F. The lines are parallel.

   G. The lines meet at exactly one point.

   H. The lines meet at more than one point.

   I. The situation described is impossible.

3. What value of x makes the equation below true?

$$4x - 11 = -4$$

4. A car's value depreciates at a rate of \$2,500 per year. Three years after it was purchased, the car's value was \$21,000. Which equation can be used to find v, its value in dollars, n years after it was purchased?

   A. $v = 28{,}500 - 2{,}500n$ **C.** $v = 18{,}500 - 2{,}500n$

   B. $v = 21{,}000 - 2{,}500n$ **D.** $v = 18{,}500 - n$

5. You drink 8 fluid ounces of orange juice every morning. How many quarts of orange juice do you drink in 6 weeks?

   F. 1.5 qt **H.** 10.5 qt

   G. 7.5 qt **I.** 21 qt

6. The line $4x + 5y = 12$ is written in standard form. At what point does the graph of this line cross the x-axis?

   A. (0, 2.4) **C.** (2.4, 0)

   B. (0, 3) **D.** (3, 0)

7. Water is leaking from a jug at a constant rate. After leaking for two hours, the jug contains 48 fluid ounces of water. After leaking for five hours, the jug contains 42 fluid ounces of water.

Think Solve Explain

Part A Find the rate at which water is leaking from the jug.

Part B Find how many fluid ounces of water were in the jug before it started leaking. Show your work and explain your reasoning.

Part C Write an equation that shows how many fluid ounces y of water are left in the jug after it has been leaking for h hours.

Part D Find how many hours it will take the jug to empty entirely. Show your work and explain your reasoning.

8. What is the slope of the line given by $3x - 6y = 33$?

F. -3

G. $-\frac{1}{2}$

H. $\frac{1}{2}$

I. 3

9. You have 40 nickels and dimes. Their total value is \$3.30. Which system of equations could be used to find the number n of nickels and the number d of dimes?

A. $n + d = 40$
$n + d = 3.30$

B. $n + d = 40$
$5n + 10d = 3.30$

C. $5n + 10d = 40$
$n + d = 330$

D. $n + d = 40$
$5n + 10d = 330$

10. You bought a lead pencil for \$5 and three identical markers. You spent \$12.47 in all. Which equation could be used to find the price p of one marker?

F. $5 + p = 12.47$

G. $5 + 3p = 12.47$

H. $3(5 + p) = 12.47$

I. $3p = 12.47 + 5$

The graph below shows how many calories c are burned during m minutes of playing basketball. Use the graph for Exercises 11 and 12.

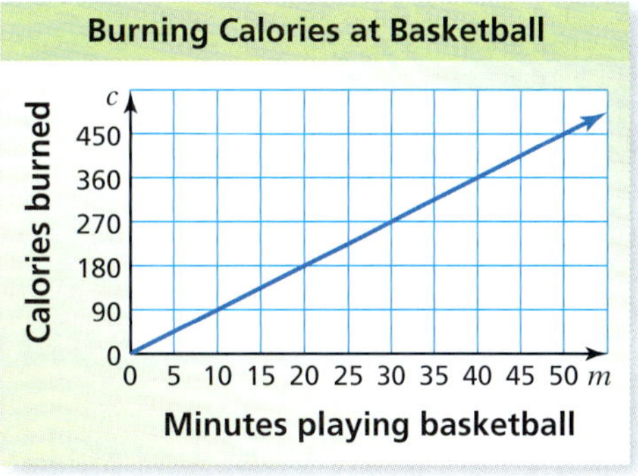

11. How many calories are burned in 25 minutes?

A. 180 cal

B. 185 cal

C. 225 cal

D. 270 cal

12. Which equation fits the data given in the graph?

F. $c = 9m$

G. $c = 90m$

H. $c = m + 80$

I. $m = 9c$

13. The line shown in the graph below has a slope of -3. What is the equation of the line?

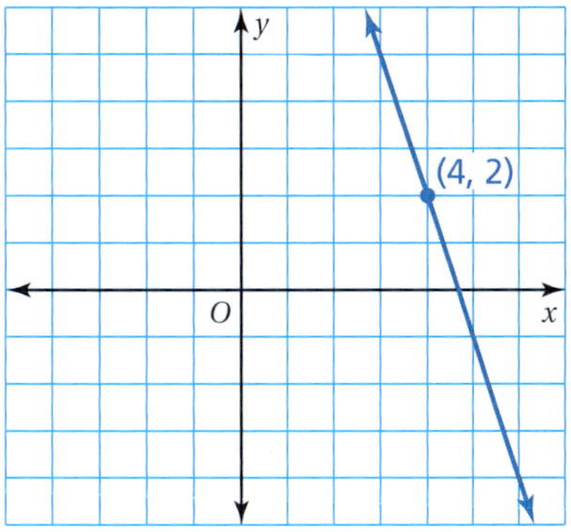

A. $y = 3x - 10$

B. $y = -3x + 10$

C. $y = -3x + 14$

D. $y = -3x - 14$

14. Solve the formula below for I.

$$A = P + PI$$

F. $I = A - 2P$

G. $I = \frac{A}{P} - P$

H. $I = A - \frac{P}{P}$

I. $I = \frac{A - P}{P}$

4 Functions

4.1 Domain and Range of a Function

4.2 Discrete and Continuous Domains

4.3 Linear Function Patterns

4.4 Comparing Linear and Nonlinear Functions

"Here's how I remember that the range is the *y*-values."

"I draw a cabin on the *y*-axis. Then, I hum 'Home, Home on the range'."

"It is my treat-converter function machine. However many cat treats I input, the machine outputs TWICE that many dog biscuits. Isn't that cool?"

What You Learned Before

"Do you think the stripes in this shirt make me look too linear?"

Recognizing Patterns (Grades 6 and 7)

Describe the pattern of inputs and outputs.

Example 1

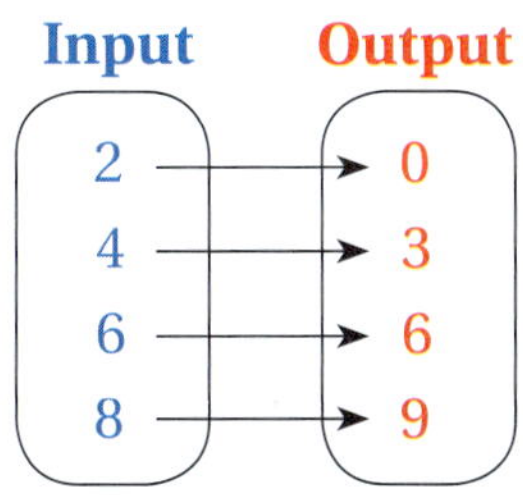

As the input increases by 2, the output increases by 3.

Example 2

Input, x	6	1	-4	-9	-14
Output, y	7	8	9	10	11

As the input x decreases by 5, the output y increases by 1.

Example 3 Draw a mapping diagram for the graph. Then describe the pattern of inputs and outputs.

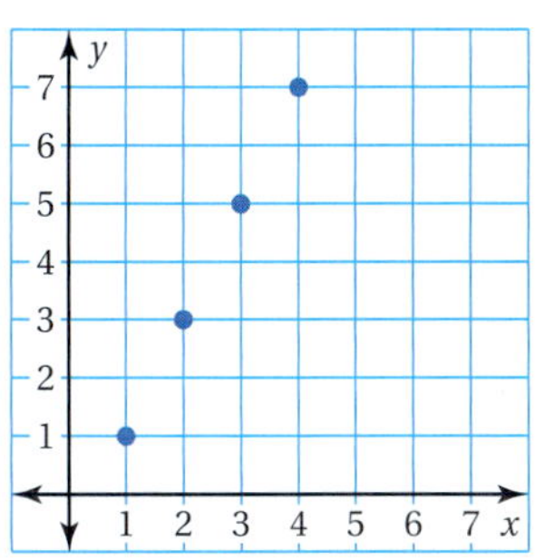

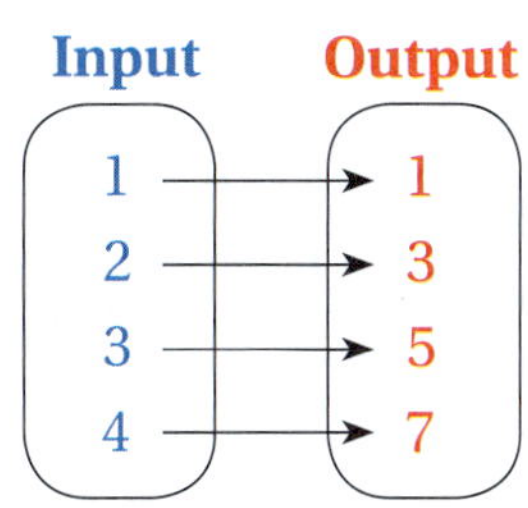

As the input increases by 1, the output increases by 2.

Try It Yourself

Describe the pattern of inputs x and outputs y.

1.

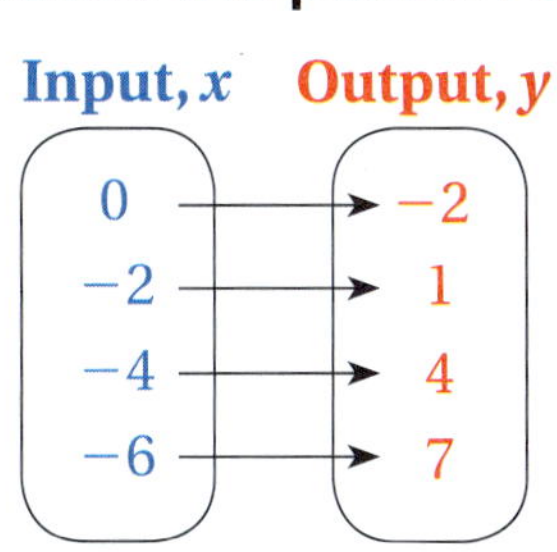

2.

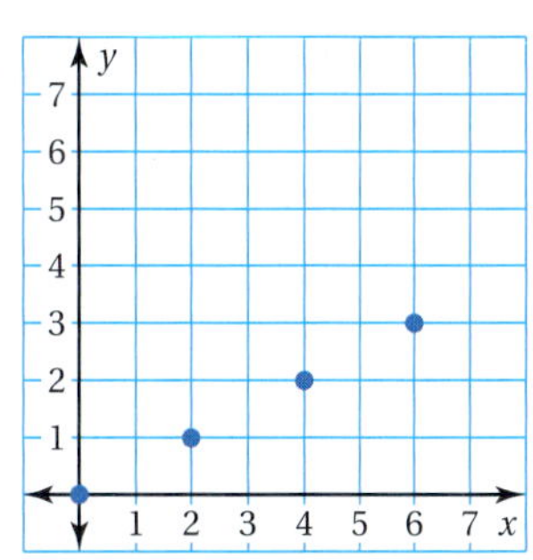

3.

Input, x	0	-1	-2	-3	-4
Output, y	7	3.5	0	-3.5	-7

4.1 Domain and Range of a Function

STATE STANDARDS
MA.8.A.1.1
MA.8.A.1.5

Essential Question How can you find the domain and range of a function?

1 ACTIVITY: The Domain and Range of a Function

Work with a partner. The table shows the number of adult and child tickets sold for a school concert.

Input →	Number of Adult Tickets, x	0	1	2	3	4
Output →	Number of Child Tickets, y	8	6	4	2	0

The variables x and y are related by the linear equation $4x + 2y = 16$.

a. Write the equation in **function form** by solving for y.

b. The **domain** of a function is the set of all input values. Find the domain of the function.

Domain =

Why is $x = 5$ not in the domain of the function?

Why is $x = \frac{1}{2}$ not in the domain of the function?

c. The **range** of a function is the set of all output values. Find the range of the function.

Range =

d. Functions can be described in many ways.

- by an equation
- by an input-output table
- in words
- by a graph
- as a set of ordered pairs

Use the graph to write the function as a set of ordered pairs.

(,), (,),
(,), (,),
(,)

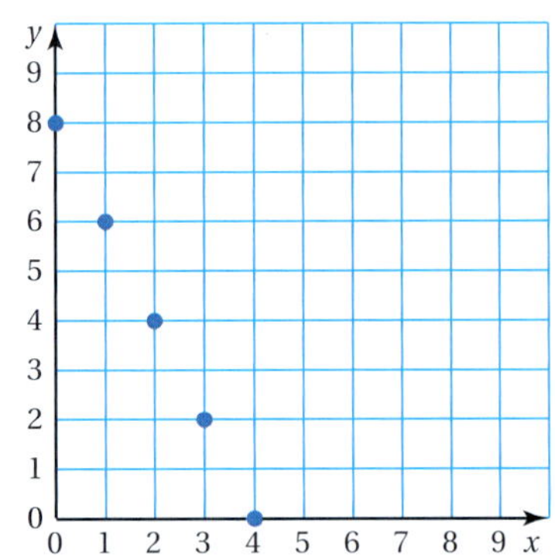

2 ACTIVITY: Finding Domains and Ranges

Work with a partner.

- **Copy and complete each input-output table.**
- **Find the domain and range of the function represented by the table.**

a. $y = -3x + 4$

x	-2	-1	0	1	2
y					

b. $y = \frac{1}{2}x - 6$

x	0	1	2	3	4
y					

c.

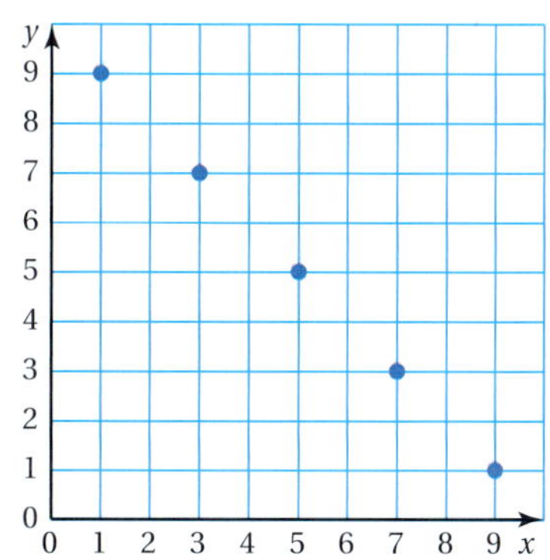

x					
y					

d.

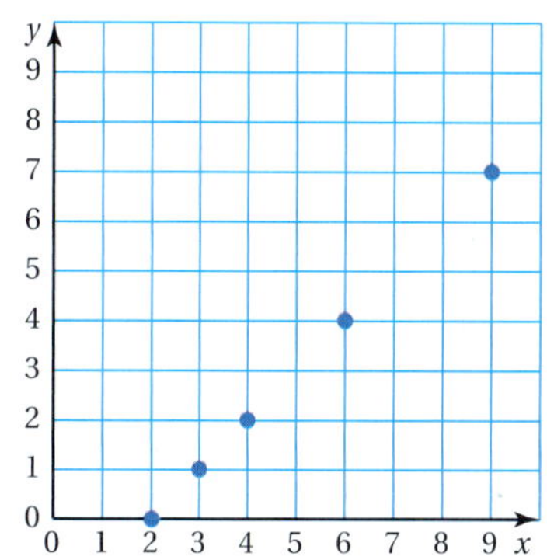

x					
y					

What Is Your Answer?

3. **IN YOUR OWN WORDS** How can you find the domain and range of a function?
4. **The following are general rules for finding a person's foot length.**

To find the length y (in inches) of a woman's foot, divide her shoe size x by 3 and add 7.

To find the length y (in inches) of a man's foot, divide his shoe size x by 3 and add 7.3.

a. Write an equation for one of the statements.

b. Make an input-output table for the function in part (a). Use shoe sizes $5\frac{1}{2}$ to 12.

c. Label the domain and range of the function on the table.

Practice Use what you learned about the domain and range of a function to complete Exercise 3 on page 152.

4.1 Lesson

Check It Out
Lesson Tutorials
BigIdeasMath.com

Key Vocabulary
function, *p. 150*
domain, *p. 150*
range, *p. 150*
function form, *p. 150*

Key Idea

Functions

A **function** is a relationship that pairs each *input* with exactly one *output*. The **domain** is the set of all possible input values. The **range** is the set of all possible output values.

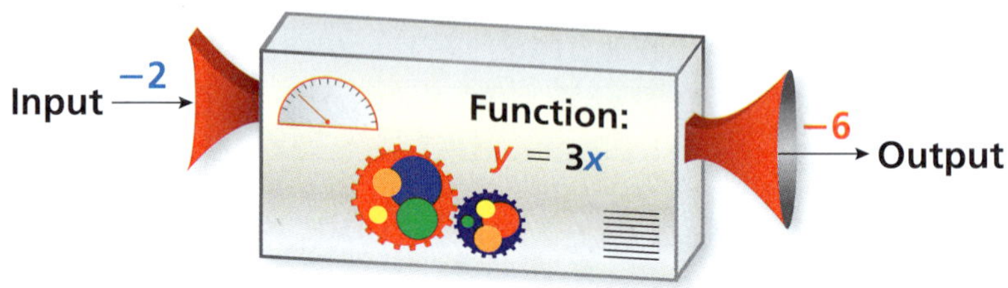

Remember

The ordered pair (x, y) shows the output y for an input x.

EXAMPLE 1 Finding Domain and Range from a Graph

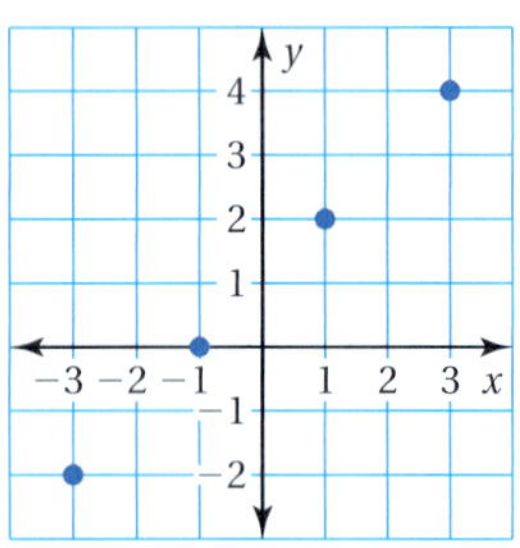

Find the domain and range of the function represented by the graph.

Write the ordered pairs. Identify the inputs and outputs.

inputs

$(-3, -2), (-1, 0), (1, 2), (3, 4)$

outputs

∴ The domain is −3, −1, 1, and 3. The range is −2, 0, 2, and 4.

On Your Own

Now You're Ready
Exercises 4–6

Find the domain and range of the function represented by the graph.

1.

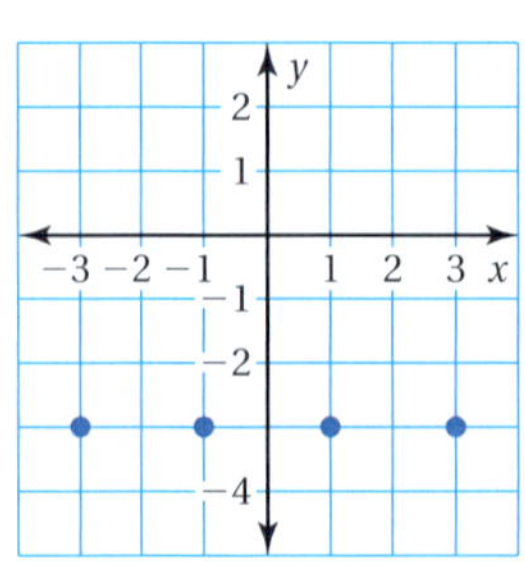

2.

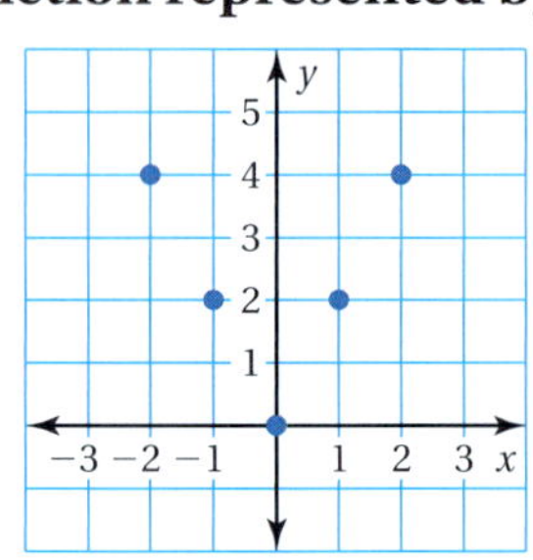

An equation is in **function form** if it is solved for y.

$x + y = 1$	$y = -x + 1$
not in function form	in function form

Multi-Language Glossary at BigIdeasMath.com.

EXAMPLE 2 Finding the Range of a Function

Input, x	$-2x + 8$	Output, y
-2	$-2(-2) + 8$	12
0	$-2(0) + 8$	8
2	$-2(2) + 8$	4
4	$-2(4) + 8$	0
6	$-2(6) + 8$	-4

The domain of the function represented by $2x + y = 8$ is -2, 0, 2, 4, and 6. What is the range of the function represented by the table?

Write the function in function form.

$$2x + y = 8$$

$$y = -2x + 8$$

Use this form to make an input-output table.

∴ The range is 12, 8, 4, 0, and -4.

EXAMPLE 3 Real-Life Application

The table shows the percent y (in decimal form) of the moon that was visible at midnight x days after January 24, 2011. (a) Interpret the domain and range. (b) What percent of the moon was visible on January 26, 2011?

x	y
0	0.76
1	0.65
2	0.54
3	0.43
4	0.32

a. Zero days after January 24 is January 24. One day after January 24 is January 25. So, the domain of 0, 1, 2, 3, and 4 represents January 24, 25, 26, 27, and 28.

The range is 0.76, 0.65, 0.54, 0.43, and 0.32. These amounts are decreasing, so the moon was less visible each day.

b. January 26, 2011 corresponds to the input $x = 2$. When $x = 2$, $y = 0.54$. So, 0.54, or 54% of the moon was visible on January 26, 2011.

On Your Own

Now You're Ready
Exercises 9–11

Copy and complete the input-output table for the function. Then find the domain and range of the function represented by the table.

3. $y = 2x - 3$

x	-1	0	1	2
y				

4. $x + y = -3$

x	0	1	2	3
y				

5. The table shows the percent y (in decimal form) of the moon that was visible at midnight x days after December 17, 2012.
(a) Interpret the domain and range.
(b) What percent of the moon was visible on December 21, 2012?

x	0	1	2	3	4
y	0.2	0.3	0.4	0.5	0.6

4.1 Exercises

Check It Out
Help with Homework
BigIdeasMath.com

Vocabulary and Concept Check

1. **VOCABULARY** Is the equation $2x - 3y = 4$ in function form? Explain.

2. **DIFFERENT WORDS, SAME QUESTION** Which is different? Find "both" answers.

Find the range of the function represented by the table.

Find the inputs of the function represented by the table.

Find the x-values of the function represented by $(2, 7)$, $(4, 5)$, and $(6, -1)$.

Find the domain of the function represented by $(2, 7)$, $(4, 5)$, and $(6, -1)$.

x	2	4	6
y	7	5	-1

Practice and Problem Solving

3. The number of earrings and headbands you can buy with \$24 is represented by the equation $8x + 4y = 24$. The table shows the number of earrings and headbands.

   a. Write the equation in function form.
   b. Find the domain and range.
   c. Why is $x = 6$ not in the domain of the function?

Earrings, x	0	1	2	3
Headbands, y	6	4	2	0

Find the domain and range of the function represented by the graph.

1 4.

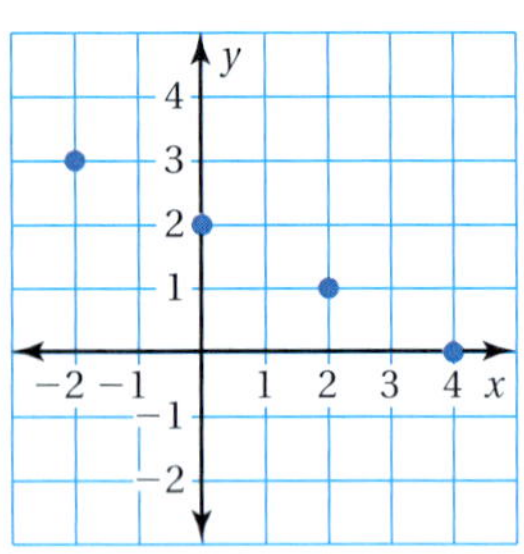

5.

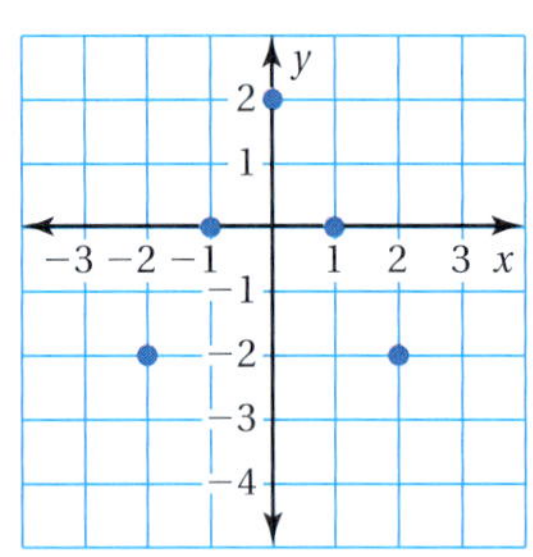

6.

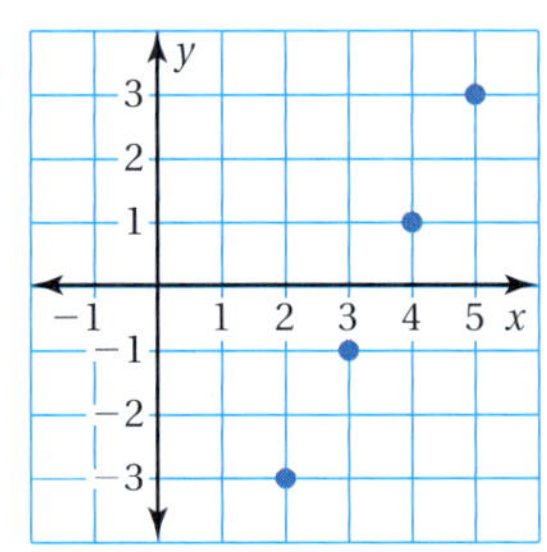

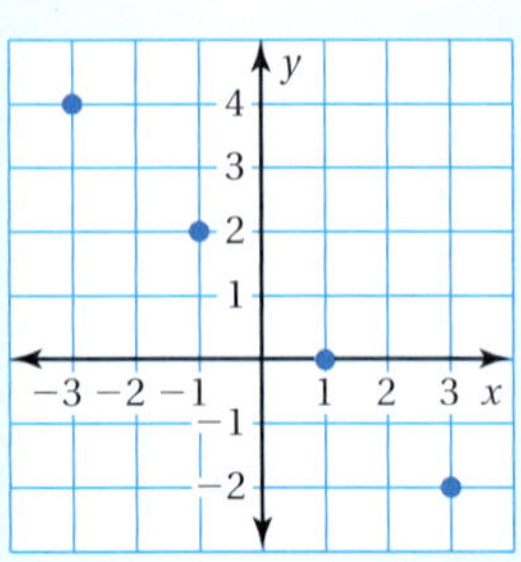

The domain is −2, 0, 2, and 4.

The range is −3, −1, 1, 3.

7. **ERROR ANALYSIS** Describe and correct the error in finding the domain and range of the function represented by the graph.

8. **REASONING** Find the domain and range of the function represented by the table.

Tickets, x	2	3	5	8
Cost, y	\$14	\$21	\$35	\$56

Copy and complete the input-output table for the function. Then find the domain and range of the function represented by the table.

9. $y = 6x + 2$

x	−1	0	1	2
y				

10. $y = -\frac{1}{4}x - 2$

x	0	4	8	12
y				

11. $y = 1.5x + 3$

x	−1	0	1	2
y				

12. VAULTING In the sport of vaulting, a vaulter performs a routine while on a moving horse. For each round x of competition, the vaulter receives a score y from 1 to 10.

x	y
1	6.856
2	7.923
3	8.135

a. Find the domain and range of the function represented by the table.

b. Interpret the domain and range.

c. What is the mean score of the vaulter?

13. MANATEE Florida's state marine mammal is the manatee. A manatee eats about 12% of its body weight each day.

a. Write an equation in function form that represents the amount y (in pounds) of food a manatee eats each day for its weight x.

b. Create an input-output table for the equation in part (a). Use the inputs 150, 300, 450, 600, 750, and 900.

c. Find the domain and range of the function represented by the table.

d. An aquatic center has manatees that weigh 300 pounds, 750 pounds, and 1050 pounds. How many pounds of food do all three manatees eat in a day? in a week?

14. Critical Thinking Describe the domain and range of the function.

a. $y = |x|$ **b.** $y = -|x|$ **c.** $y = |x| - 6$ **d.** $y = -|x| + 4$

Fair Game Review What you learned in previous grades & lessons

Graph the linear equation. *(Section 2.1)*

15. $y = 2x + 8$ **16.** $5x + 6y = 12$ **17.** $-x - 3y = 2$ **18.** $y = 7x - 5$

19. MULTIPLE CHOICE The minimum number of people needed for a group rate at an amusement park is 8. Which inequality represents the number of people needed to get the group rate? *(Skills Review Handbook)*

Ⓐ $x \leq 8$ Ⓑ $x > 8$ Ⓒ $x < 8$ Ⓓ $x \geq 8$

4.2 Discrete and Continuous Domains

STATE STANDARDS
MA.8.A.1.1
MA.8.A.1.5

Essential Question How can you decide whether the domain of a function is discrete or continuous?

1 EXAMPLE: Discrete and Continuous Domains

In Activities 1 and 2 in Section 2.4, you studied two real-life problems represented by the same equation.

$$4x + 2y = 16 \quad \text{or} \quad y = -2x + 8$$

a.

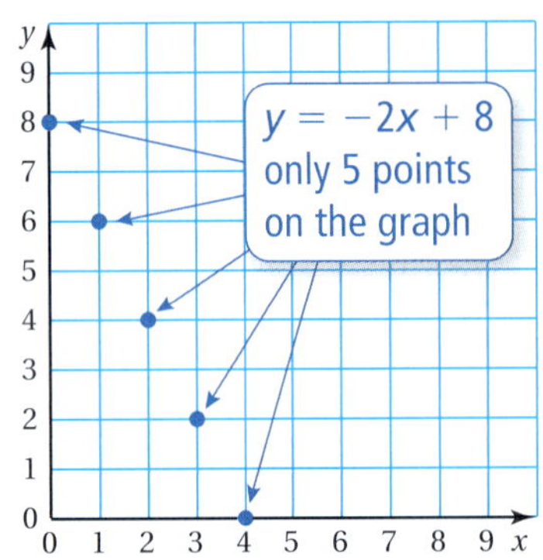

Domain (x-values): 0, 1, 2, 3, 4

Range (y-values): 8, 6, 4, 2, 0

The domain is **discrete** because it consists of only the numbers 0, 1, 2, 3, and 4.

b.

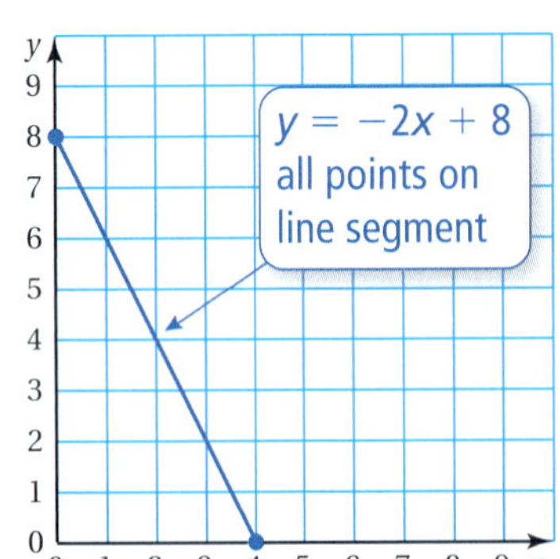

Domain (x-values): $x \geq 0$ and $x \leq 4$
(All numbers from 0 to 4)

Range (y-values): $y \geq 0$ and $y \leq 8$
(All numbers from 0 to 8)

The domain is **continuous** because it consists of all numbers from 0 to 4 on the number line.

2 ACTIVITY: Discrete and Continuous Domains

Work with a partner.

- **Write a function to represent each problem.**
- **Graph each function.**
- **Describe the domain and range of each function. Is the domain discrete or continuous?**

a. You are in charge of reserving hotel rooms for a youth soccer team. Each room costs \$69, plus \$6 tax, per night. You need each room for two nights. You need 10 to 16 rooms. Write a function for the total hotel cost.

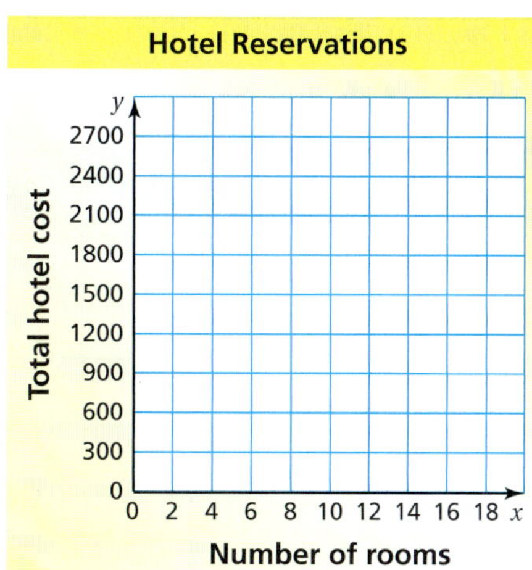

b. The airline you are using for the soccer trip needs an estimate of the total weight of the team's luggage. You determine that there will be 36 pieces of luggage and each piece will weigh from 25 to 45 pounds. Write a function for the total weight of the luggage.

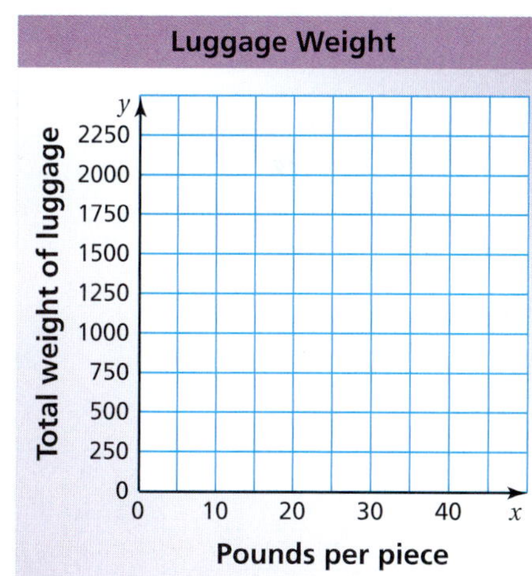

What Is Your Answer?

3. **IN YOUR OWN WORDS** How can you decide whether the domain of a function is discrete or continuous? Describe two real-life examples of functions: one with a discrete domain and one with a continuous domain.

Practice Use what you learned about discrete and continuous domains to complete Exercises 3 and 4 on page 158.

4.2 Lesson

Check It Out
Lesson Tutorials
BigIdeasMath.com

Key Vocabulary
discrete domain, *p. 156*
continuous domain, *p. 156*

Key Idea

Discrete and Continuous Domains

A **discrete domain** is a set of input values that consists of only certain numbers in an interval.

Example: Integers from 1 to 5

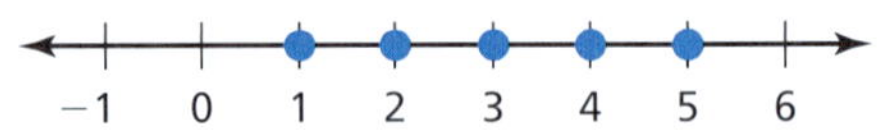

A **continuous domain** is a set of input values that consists of all numbers in an interval.

Example: All numbers from 1 to 5.

EXAMPLE 1 Graphing Discrete Data

The function $y = 15.95x$ represents the cost y (in dollars) of x tickets for the South Florida Museum. Graph the function using a domain of 0, 1, 2, 3, and 4. Is the domain of the graph discrete or continuous? Explain.

Make an input-output table.

Input, x	$15.95x$	Output, y	Ordered Pair, (x, y)
0	15.95(0)	0	(0, 0)
1	15.95(1)	15.95	(1, 15.95)
2	15.95(2)	31.9	(2, 31.9)
3	15.95(3)	47.85	(3, 47.85)
4	15.95(4)	63.8	(4, 63.8)

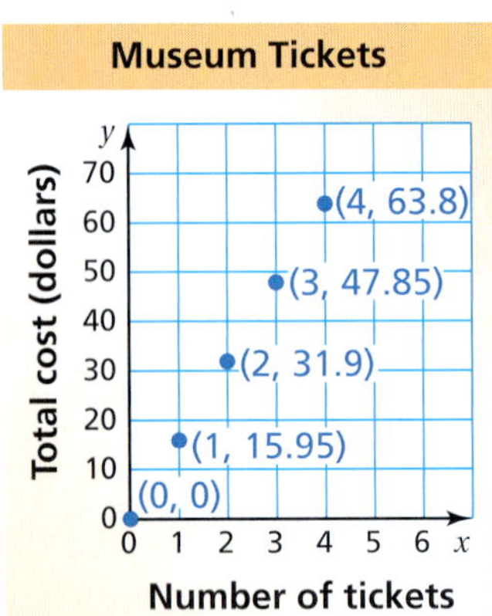

Plot the ordered pairs. Because you cannot buy part of a ticket, the graph consists of individual points.

∴ So, the domain is discrete.

On Your Own

1. The function $m = 50 - 9d$ represents the amount of money m (in dollars) you have after buying d DVDs. Graph the function. Is the domain discrete or continuous? Explain.

Multi-Language Glossary at BigIdeasMath.com.

EXAMPLE 2 Graphing Continuous Data

A cereal bar contains 130 calories. The number c of calories consumed is a function of the number b of bars eaten. Graph the function. Is the domain of the graph discrete or continuous?

Make an input-output table.

Input, b	Output, c	Ordered Pair, (b, c)
0	0	(0, 0)
1	130	(1, 130)
2	260	(2, 260)
3	390	(3, 390)
4	520	(4, 520)

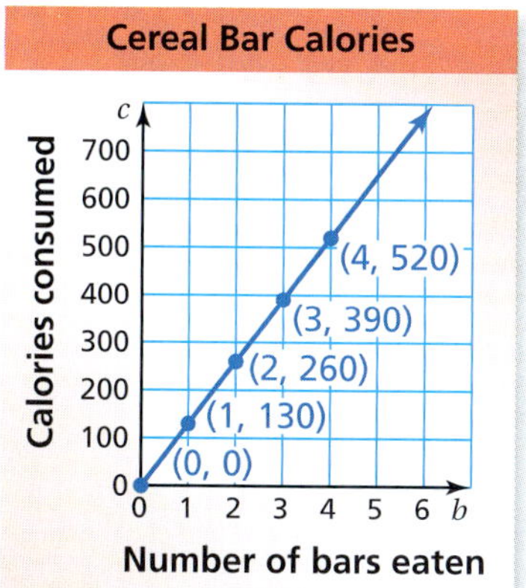

Plot the ordered pairs. Because you can eat part of a cereal bar, b can be any value greater than or equal to 0. Draw a line through the points.

So, the domain is continuous.

EXAMPLE 3 Standardized Test Practice

You conduct an experiment on the speed of sound waves in dry air at 86 °F. You record your data in a table. Which of the following is true?

Input Time, t (seconds)	Output Distance, d (miles)
2	0.434
4	0.868
6	1.302
8	1.736
10	2.170

(A) The domain is $t \geq 2$ and $t \leq 10$ and it is discrete.

(B) The domain is $t \geq 2$ and $t \leq 10$ and it is continuous.

(C) The domain is $d \geq 0.434$ and $d \leq 2.17$ and it is discrete.

(D) The domain is $d \geq 0.434$ and $d \leq 2.17$ and it is continuous.

The domain is the set of possible input values, or the time t. The time t can be any value from 2 to 10. So, the domain is continuous.

The correct answer is (B).

On Your Own

Now You're Ready
Exercises 5–8

2. A 20-gallon bathtub is draining at a rate of 2.5 gallons per minute. The number g of gallons remaining is a function of the number m of minutes. Graph the function. Is the domain discrete or continuous?

3. Are the data shown in the table discrete or continuous? Explain.

Number of Stories	1	2	3	4	5
Height of Building (feet)	12	24	36	48	60

4.2 Exercises

Check It Out
Help with Homework
BigIdeasMath.com

Vocabulary and Concept Check

1. **VOCABULARY** Describe the difference between a discrete domain and a continuous domain.

2. **WRITING** Describe how you can use a graph to determine whether a domain is discrete or continuous.

Practice and Problem Solving

Describe the domain and range of the function. Is the domain discrete or continuous?

3. 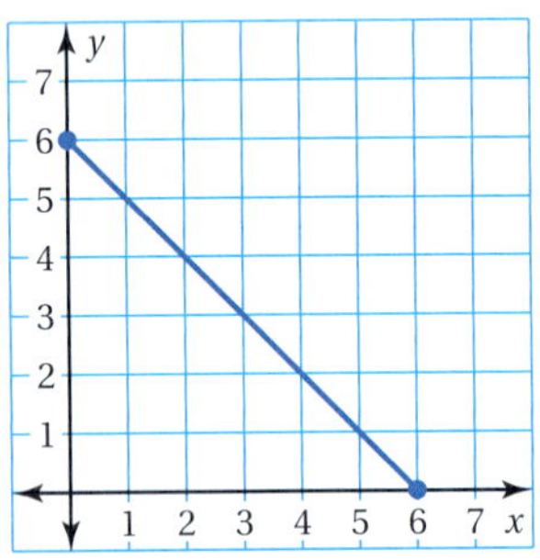

4. 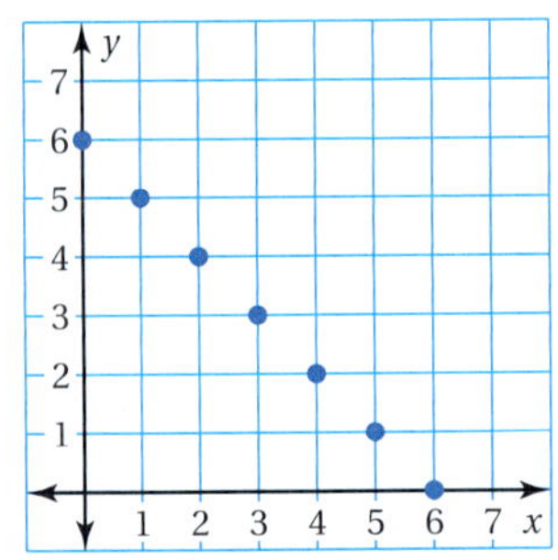

Graph the function. Is the domain of the graph discrete or continuous?

1 2 5.

Input Bags, x	Output Marbles, y
2	20
4	40
6	60

6.

Input Years, x	Output Height of a Tree, y (feet)
0	3
1	6
2	9

7.

Input Width, x (inches)	Output Volume, y (cubic inches)
5	50
10	100
15	150

8.

Input Hats, x	Output Cost, y (dollars)
0	0
1	8.45
2	16.9

9. **ERROR ANALYSIS** Describe and correct the error in classifying the domain.

The domain is continuous.

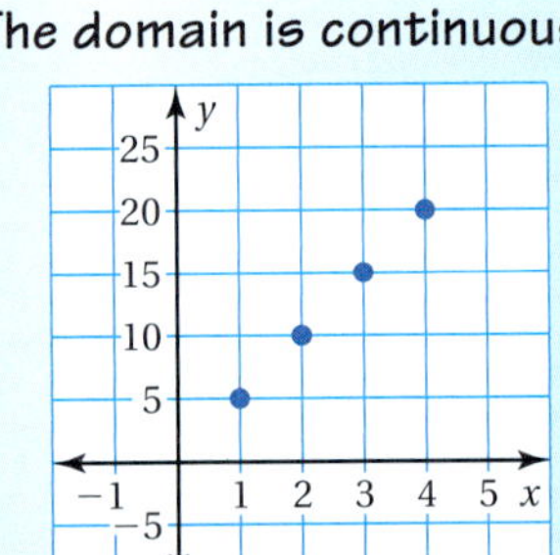

10. **YARN** The function $m = 40 - 8.5b$ represents the amount m of money (in dollars) that you have after buying b balls of yarn. Graph the function using a domain of 0, 1, 2, and 3. Is the domain discrete or continuous?

11. **REASONING** The input of one function is *length*. The input of another function is *number of shirts*. Which function has a continuous domain? Explain.

12. **DISTANCE** The function $y = 3.28x$ converts length from x meters to y feet. Graph the function. Is the domain discrete or continuous?

13. **AREA** The area A of the triangle is a function of the height h. Graph the function. Is the domain discrete or continuous?

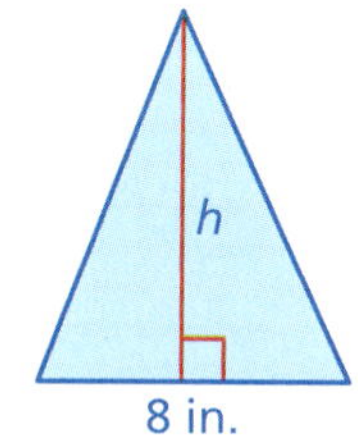

14. **PACKING** You are packing books into boxes. The function $y = 20x$ represents the number y of books that will fit into x boxes.
    a. Is 4 in the domain? Explain.
    b. Is 60 in the range? Explain.

15. **Reasoning** You want to fill a 2-foot shelf with framed pictures. There are x pictures in 4-inch frames and y pictures in 8-inch frames.
    a. Write a function for this situation.
    b. Graph the function.
    c. Is the domain discrete or continuous?

Fair Game Review What you learned in previous grades & lessons

Find the slope of the line. *(Section 2.2)*

16. 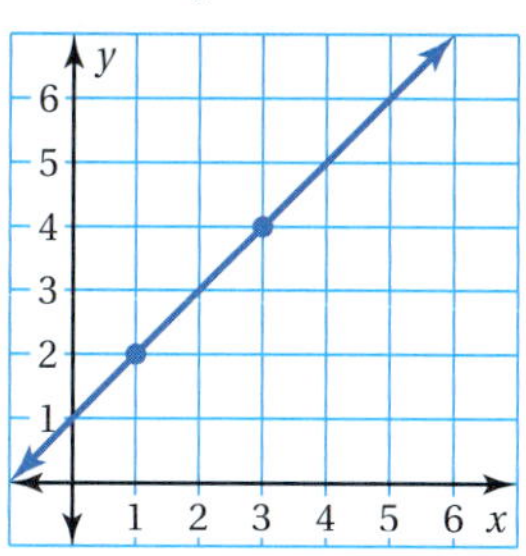

17. 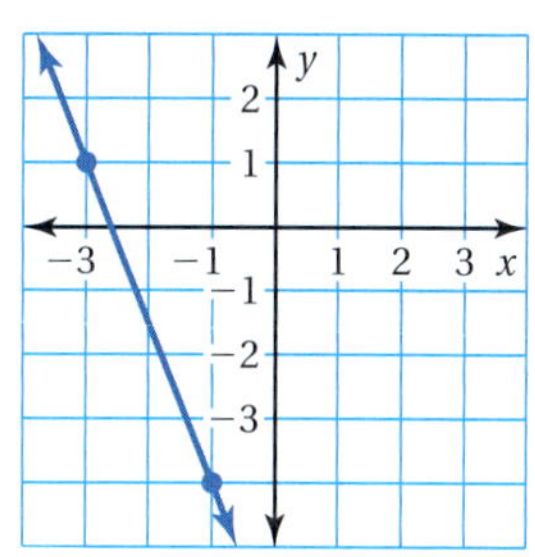

18. 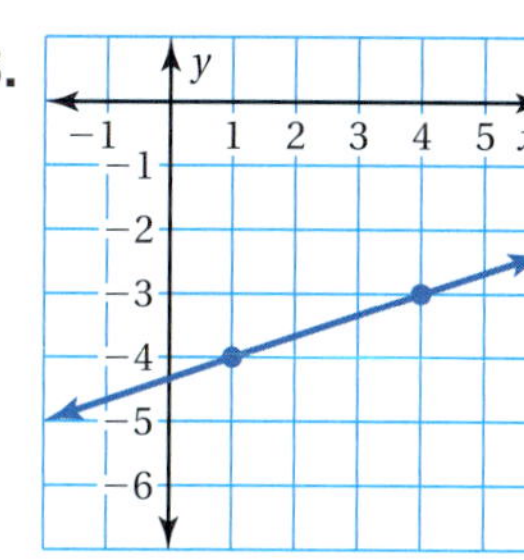

19. **MULTIPLE CHOICE** What is the y-intercept of the graph of the linear equation? *(Section 2.3)*

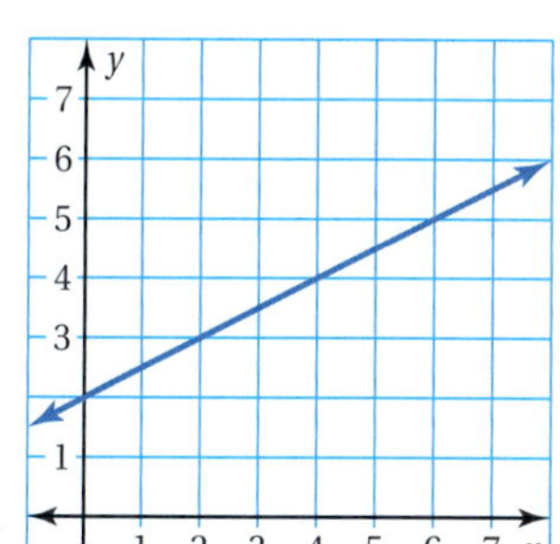

Ⓐ -4 Ⓑ -2 Ⓒ 2 Ⓓ 4

4 Study Help

Check It Out
Graphic Organizer
BigIdeasMath.com

You can use a **comparison chart** to compare two topics. Here is an example of a comparison chart for domain and range.

<table>
<tr><th></th><th>Domain</th><th>Range</th></tr>
<tr><td>Definition</td><td>the set of all possible input values</td><td>the set of all possible output values</td></tr>
<tr><td>Algebra
Example: $y = mx + b$</td><td>x-values</td><td>corresponding y-values</td></tr>
<tr><td>Ordered pairs
Example: (−4, 0), (−3, 1), (−2, 2), (−1, 3)</td><td>−4, −3, −2, −1</td><td>0, 1, 2, 3</td></tr>
<tr><td>Table
Example:<table><tr><td>x</td><td>−1</td><td>0</td><td>2</td><td>3</td></tr><tr><td>y</td><td>1</td><td>0</td><td>4</td><td>9</td></tr></table></td><td>−1, 0, 2, 3</td><td>0, 1, 4, 9</td></tr>
<tr><td>Graph
Example:
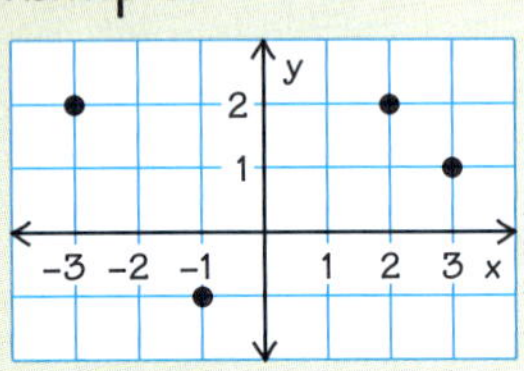</td><td>−3, −1, 2, 3</td><td>−1, 1, 2</td></tr>
</table>

On Your Own

Make a comparison chart to help you study and compare these topics.

1. discrete data and continuous data

After you complete this chapter, make comparison charts for the following topics.

2. linear functions with positive slopes and linear functions with negative slopes
3. linear functions and nonlinear functions

"Creating a comparison chart causes canines to crystalize concepts."

4.1–4.2 Quiz

Check It Out
Progress Check
BigIdeasMath.com

Find the domain and range of the function represented by the graph. *(Section 4.1)*

1. 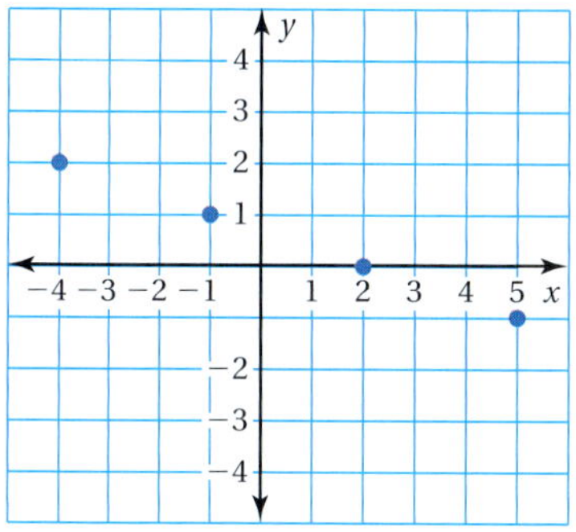

2. 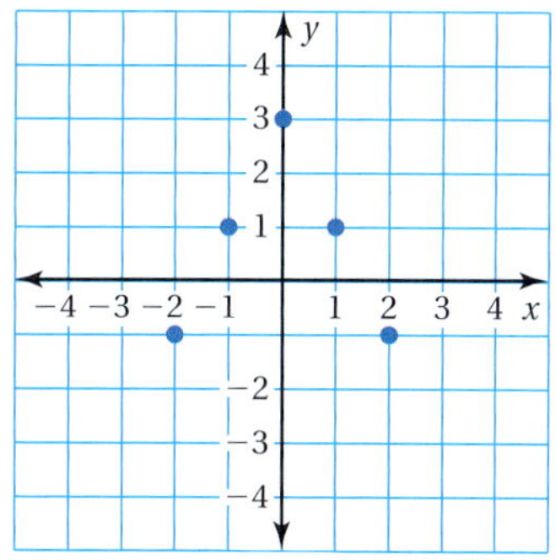

3. 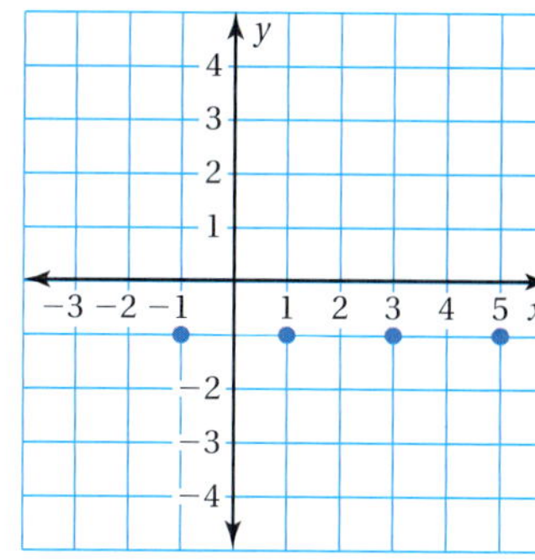

Copy and complete the input-output table for the function. Then find the domain and range of the function represented by the table. *(Section 4.1)*

4. $y = 5x - 6$

x	0	1	2	3
y				

5. $2x + y = 2$

x	−1	0	1	2
y				

Graph the function. Is the domain of the graph discrete or continuous? *(Section 4.2)*

6.

Rulers, x	Cost, y
0	0
1	1.5
2	3
3	4.5

7.

Gallons, x	Miles Remaining, y
0	300
1	265
2	230
3	195

8.

Minutes, x	0	10	20	30
Height, y	40	35	30	25

9.

Relay Teams, x	2	4	6	8
Athletes, y	8	16	24	32

10. VIDEO GAME The function $m = 30 - 3r$ represents the amount m (in dollars) of money you have after renting r video games. Graph the function using a domain of 0, 1, 2, 3, and 4. Is the domain of the graph discrete or continuous? *(Section 4.2)*

11. WATER Water accounts for about 60% of a person's body weight. *(Section 4.1)*

a. Write an equation in function form that represents the water weight y of a person that weighs x pounds.

b. Make an input-output table for the function in part (a). Use the inputs 100, 120, 140, and 160.

4.3 Linear Function Patterns

STATE STANDARDS
MA.8.A.1.1
MA.8.A.1.5

Essential Question How can you use a linear function to describe a linear pattern?

1 ACTIVITY: Finding Linear Patterns

Work with a partner.

- **Plot the points from the table in a coordinate plane.**
- **Write a linear equation for the function represented by the graph.**

a.

x	0	2	4	6	8
y	150	125	100	75	50

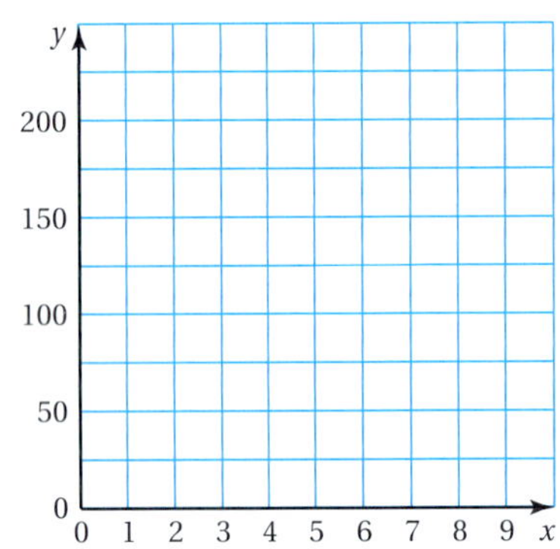

b.

x	4	6	8	10	12
y	15	20	25	30	35

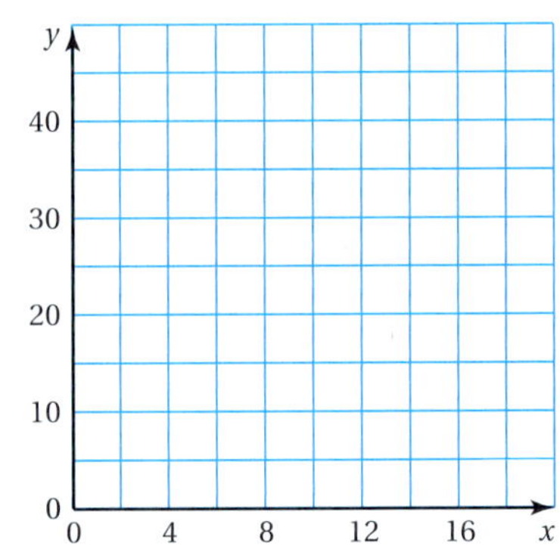

c.

x	−4	−2	0	2	4
y	4	6	8	10	12

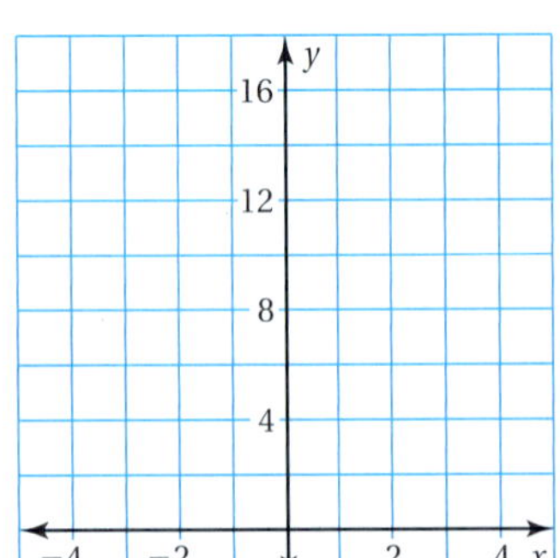

d.

x	−4	−2	0	2	4
y	1	0	−1	−2	−3

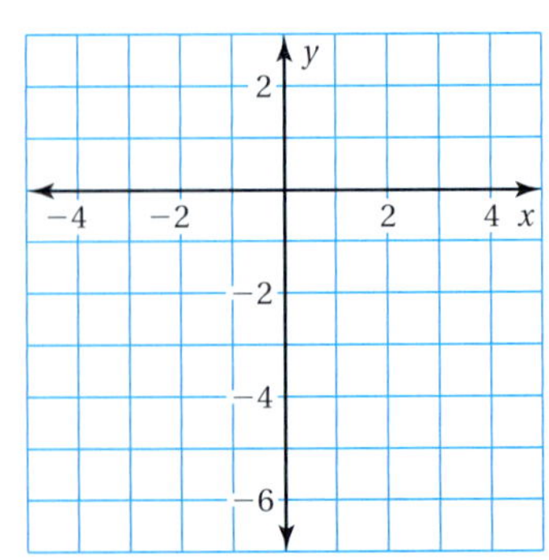

2 ACTIVITY: Finding Linear Patterns

Work with a partner. The table shows a familiar linear pattern from geometry.

- **Write a linear function that relates *y* to *x*.**
- **What do the variables *x* and *y* represent?**
- **Graph the linear function.**

a.

x	1	2	3	4	5
y	2π	4π	6π	8π	10π

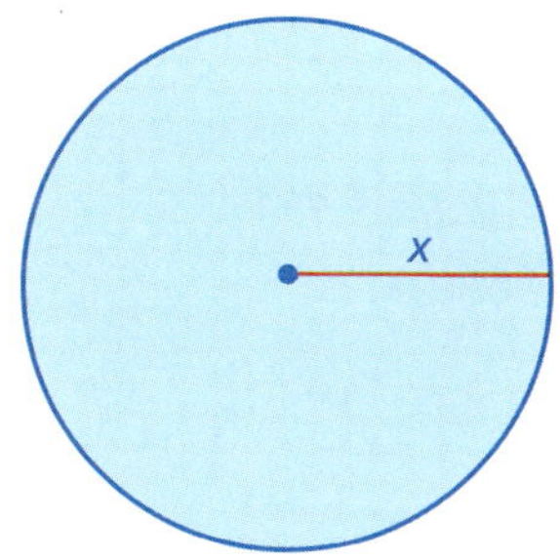

b.

x	1	2	3	4	5
y	10	12	14	16	18

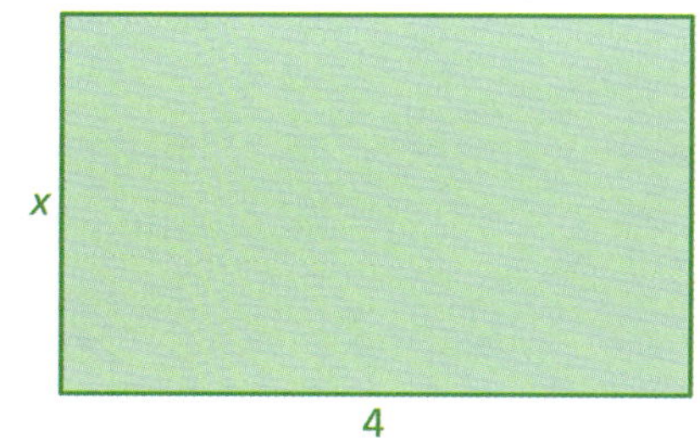

c.

x	1	2	3	4	5
y	5	6	7	8	9

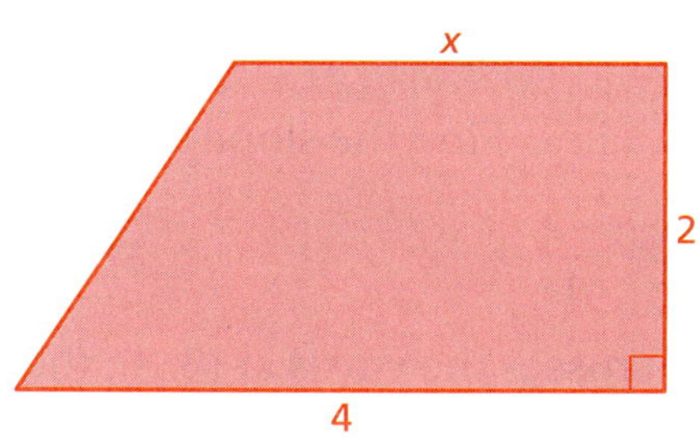

d.

x	1	2	3	4	5
y	28	40	52	64	76

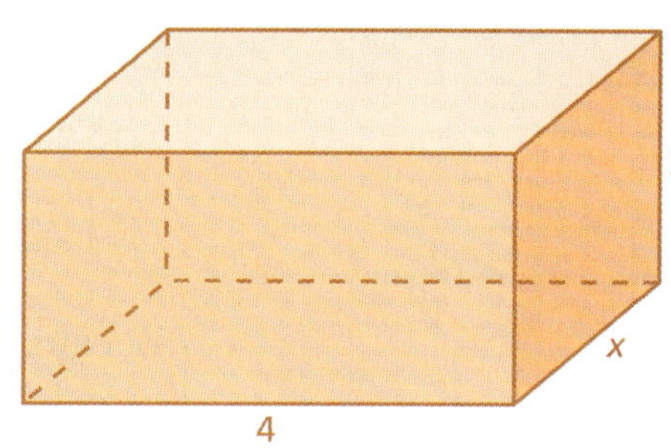

What Is Your Answer?

3. **IN YOUR OWN WORDS** How can you use a linear function to describe a linear pattern?

4. Describe the strategy you used to find the linear functions in Activities 1 and 2.

Practice Use what you learned about linear function patterns to complete Exercises 3 and 4 on page 166.

Key Vocabulary
linear function, p. 164

A **linear function** is a function whose graph is a line.

EXAMPLE 1 Finding a Linear Function Using a Graph

Use the graph to write a linear function that relates y to x.

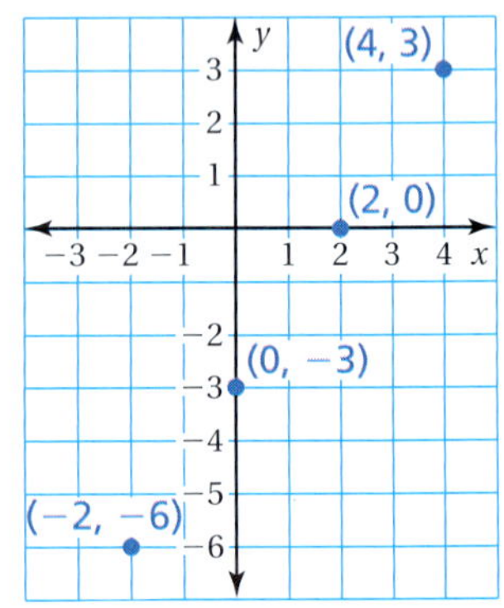

The points lie on a line. Find the slope and y-intercept of the line.

$$\text{slope} = \frac{\text{rise}}{\text{run}} = \frac{3}{2}$$

Because the line crosses the y-axis at $(0, -3)$, the y-intercept is -3.

So, the linear function is $y = \frac{3}{2}x - 3$.

EXAMPLE 2 Finding a Linear Function Using a Table

Use the table to write a linear function that relates y to x.

x	−3	−2	−1	0
y	9	7	5	3

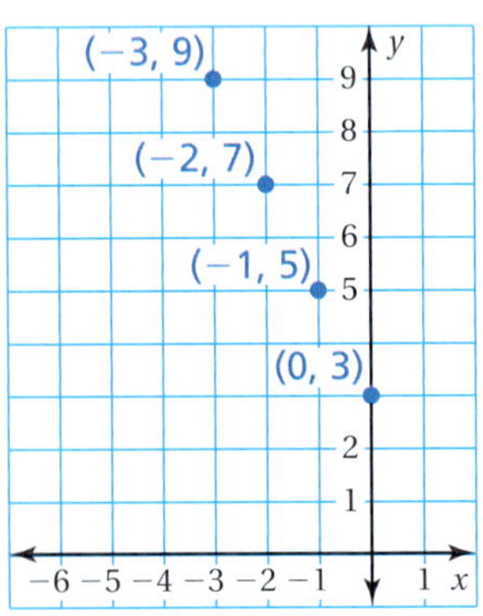

Plot the points in the table.

The points lie on a line. Find the slope and y-intercept of the line.

$$\text{slope} = \frac{\text{rise}}{\text{run}} = \frac{-2}{1} = -2$$

Because the line crosses the y-axis at $(0, 3)$, the y-intercept is 3.

So, the linear function is $y = -2x + 3$.

On Your Own

Now You're Ready
Exercises 5–10

Use the graph or table to write a linear function that relates y to x.

1. 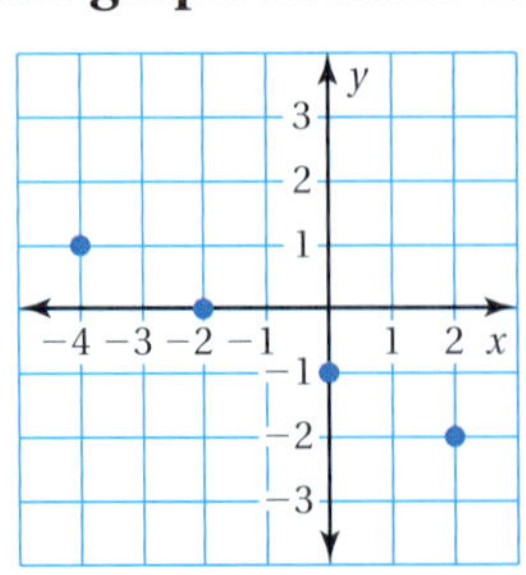

2.

x	−2	−1	0	1
y	2	2	2	2

EXAMPLE 3 Real-Life Application

Hours Kayaking, x	Calories Burned, y
2	600
4	1200
6	1800
8	2400

Graph the data in the table. (a) Is the domain discrete or continuous? (b) Write a linear function that relates y to x. (c) How many calories do you burn in 4.5 hours?

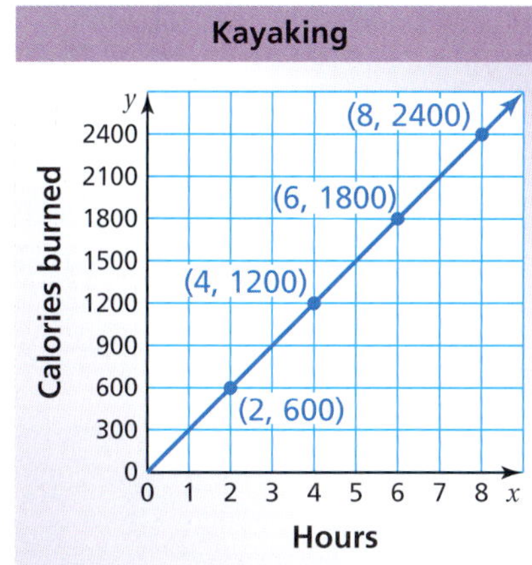

a. Plot the points. Time can represent any value greater than or equal to 0, so the domain is continuous. Draw a line through the points.

b. The slope is $\frac{600}{2} = 300$ and the y-intercept is 0.

So, the linear function is $y = 300x$.

c. Find the value of y when $x = 4.5$.

$y = 300x$ — Write the equation.

$= 300(4.5)$ — Substitute 4.5 for x.

$= 1350$ — Multiply.

You burn 1350 calories in 4.5 hours of kayaking.

On Your Own

Hours Rock Climbing, x	Calories Burned, y
3	1950
6	3900
9	5850
12	7800

3. Graph the data in the table.

a. Is the domain discrete or continuous?

b. Write a linear function that relates y to x.

c. How many calories do you burn in 5.5 hours?

Summary

Representing a Function

Words An output is 2 more than the input.

Equation $y = x + 2$

Input-Output Table

Input, x	−1	0	1	2
Output, y	1	2	3	4

Graph

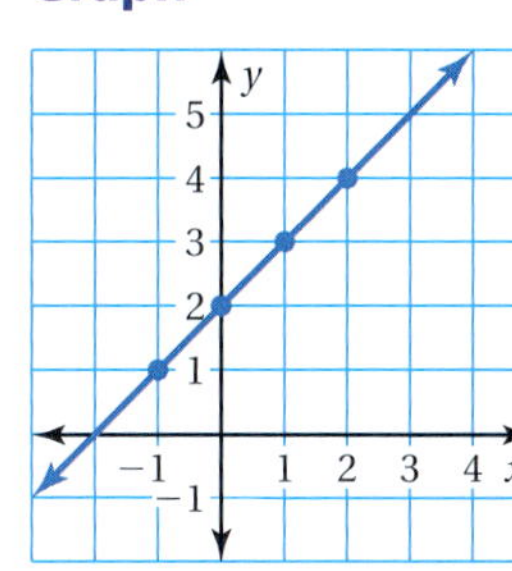

4.3 Exercises

Check It Out
Help with Homework
BigIdeasMath.com

Vocabulary and Concept Check

1. **VOCABULARY** Describe four ways to represent a function.

2. **VOCABULARY** Is the function represented by the graph a linear function? Explain.

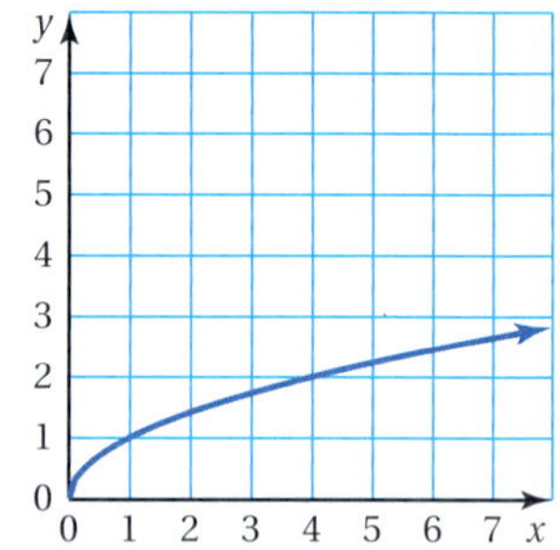

Practice and Problem Solving

The table shows a familiar linear pattern from geometry. Write a linear function that relates *y* to *x*. What do the variables *x* and *y* represent? Graph the linear function.

3.

x	1	2	3	4	5
y	π	2π	3π	4π	5π

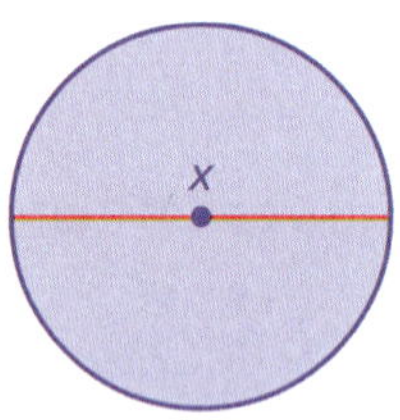

4.

x	1	2	3	4	5
y	2	4	6	8	10

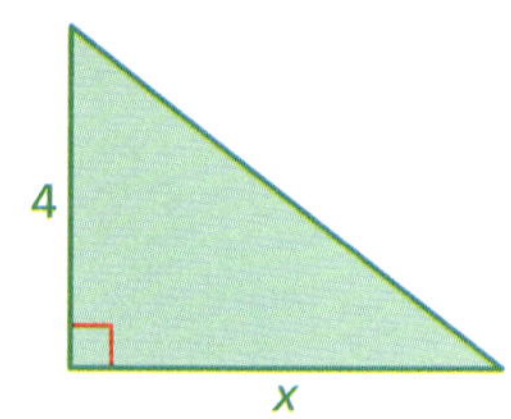

Use the graph or table to write a linear function that relates *y* to *x*.

1 2 5.

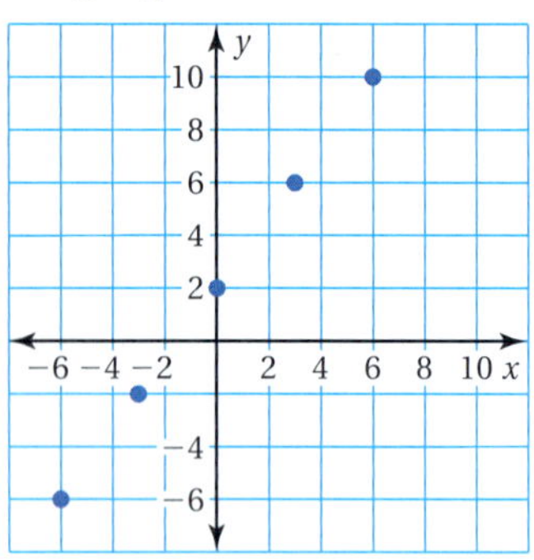

6.

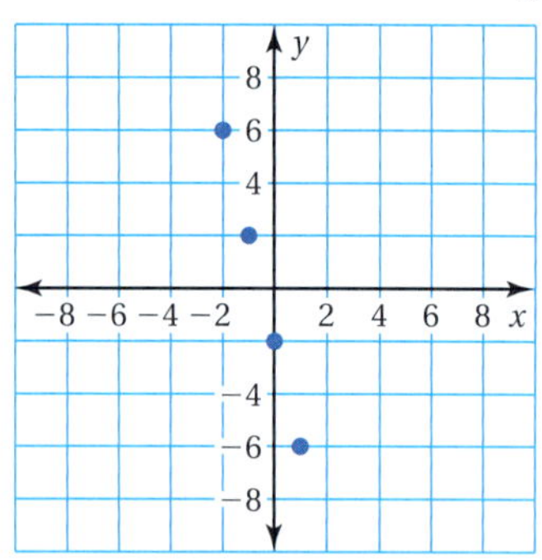

7.

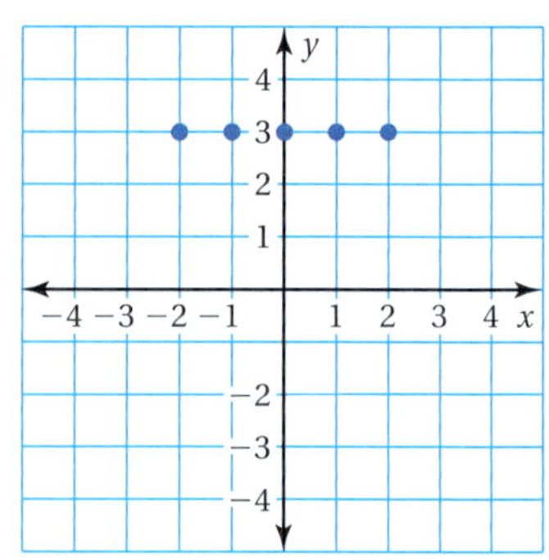

8.

x	−2	−1	0	1
y	−4	−2	0	2

9.

x	−8	−4	0	4
y	2	1	0	−1

10.

x	−3	0	3	6
y	3	5	7	9

11. **MOVIES** The table shows the cost *y* (in dollars) of renting *x* movies.

   a. Graph the data. Is the domain of the graph discrete or continuous?

   b. Write a linear function that relates *y* to *x*.

   c. How much does it cost to rent three movies?

Number of Movies, *x*	0	1	2	4
Cost, *y*	0	3	6	12

12. **BIKE JUMPS** A bunny hop is a bike trick in which the rider brings both tires off the ground without using a ramp. The table shows the height y (in inches) of a bunny hop on a bike that weighs x pounds.

Weight, x	19	21	23
Height, y	10.2	9.8	9.4

a. Graph the data. Then describe the pattern.

b. Write a linear function that relates the height of a bunny hop to the weight of the bike.

c. What is the height of a bunny hop on a bike that weighs 21.5 pounds?

Years of Education, x	Annual Salary, y
0	28
2	40
4	52
6	64
10	88

13. **SALARY** The table shows a person's annual salary y (in thousands of dollars) after x years of education beyond high school.

a. Graph the data.

b. Write a linear function that relates the person's annual salary to the number of years of education beyond high school.

c. What is the annual salary of the person after 8 years of education beyond high school?

14. **Critical Thinking** The Heat Index is calculated using the relative humidity and the temperature. For every 1 degree increase in the temperature from 94°F to 98°F at 75% relative humidity, the Heat Index rises 4°F.

a. On a summer day in Pensacola, the relative humidity is 75%, the temperature is 94°F, and the Heat Index is 122°F. Construct a table that relates the temperature t to the Heat Index H. Start the table at 94°F and end it at 98°F.

b. Write a linear function that represents this situation.

c. Estimate the Heat Index when the temperature is 100°F.

Fair Game Review What you learned in previous grades & lessons

Find the annual simple interest rate. *(Skills Review Handbook)*

15. $I = \$60$, $P = \$400$, $t = 3$ years

16. $I = \$45$, $P = \$1000$, $t = 18$ months

17. **MULTIPLE CHOICE** You buy a pair of gardening gloves for \$2.25 and x packets of seeds for \$0.88 each. Which equation represents the total cost y? *(Skills Review Handbook)*

Ⓐ $y = 0.88x - 2.25$

Ⓑ $y = 0.88x + 2.25$

Ⓒ $y = 2.25x - 0.88$

Ⓓ $y = 2.25x + 0.88$

4.4 Comparing Linear and Nonlinear Functions

STATE STANDARDS
MA.8.A.1.6

Essential Question How can you recognize when a pattern in real life is linear or nonlinear?

1 ACTIVITY: Finding Patterns for Similar Figures

Work with a partner. Copy and complete each table for the sequence of similar rectangles. Graph the data in each table. Decide whether each pattern is linear or nonlinear.

a. Perimeters of Similar Rectangles

x	1	2	3	4	5
P					

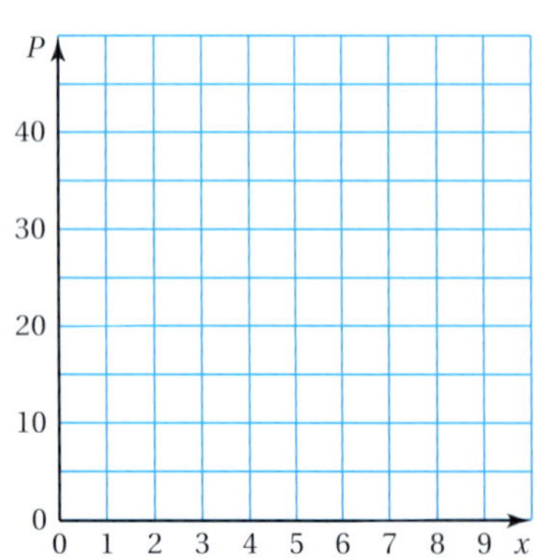

b. Areas of Similar Rectangles

x	1	2	3	4	5
A					

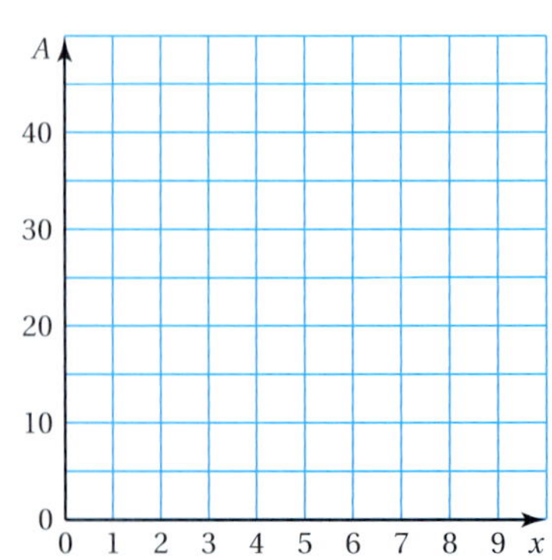

2 ACTIVITY: Comparing Linear and Nonlinear Functions

Work with a partner. The table shows the height *h* (in feet) of a falling object at *t* seconds.

- **Graph the data in the table.**
- **Decide whether the graph is linear or nonlinear.**
- **Compare the two falling objects. Which one has an increasing speed?**

a. Falling parachute jumper

t	0	1	2	3	4
h	300	285	270	255	240

b. Falling bowling ball

t	0	1	2	3	4
h	300	284	236	156	44

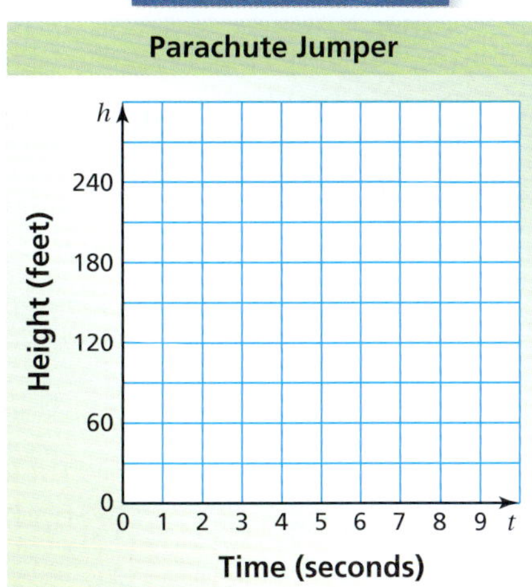

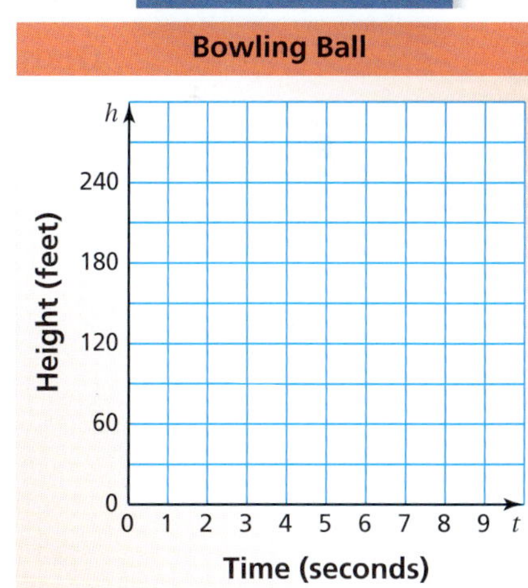

What Is Your Answer?

3. **IN YOUR OWN WORDS** How can you recognize when a pattern in real life is linear or nonlinear? Describe two real-life patterns: one that is linear and one that is nonlinear. Use patterns that are different from those described in Activities 1 and 2.

Practice Use what you learned about comparing linear and nonlinear functions to complete Exercises 3–6 on page 172.

Check It Out
Lesson Tutorials
BigIdeasMath.com

Key Vocabulary
nonlinear function, *p. 170*

The graph of a linear function shows a constant rate of change. A **nonlinear function** does not have a constant rate of change. So, its graph is *not* a line.

EXAMPLE 1 Identifying Functions from Tables

Does the table represent a *linear* or *nonlinear* function? Explain.

a.

+3 +3 +3

x	3	6	9	12
y	40	32	24	16

−8 −8 −8

As x increases by 3, y decreases by 8. The rate of change is constant. So, the function is linear.

b.

+2 +2 +2

x	1	3	5	7
y	2	11	33	88

+9 +22 +55

As x increases by 2, y increases by different amounts. The rate of change is *not* constant. So, the function is nonlinear.

EXAMPLE 2 Identifying Functions from Graphs

Does the graph represent a *linear* or *nonlinear* function? Explain.

a.

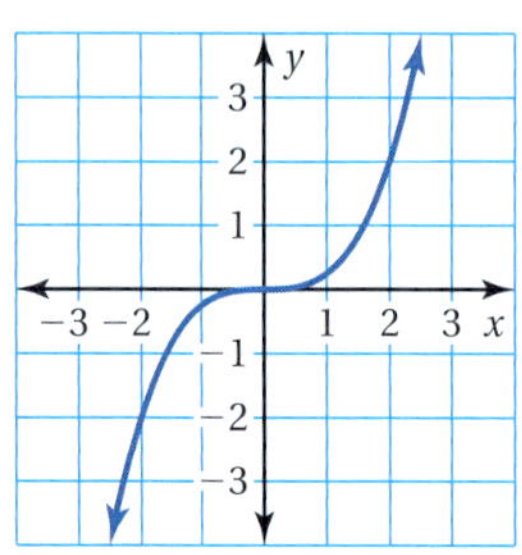

The graph is *not* a line.
So, the function is nonlinear.

b.

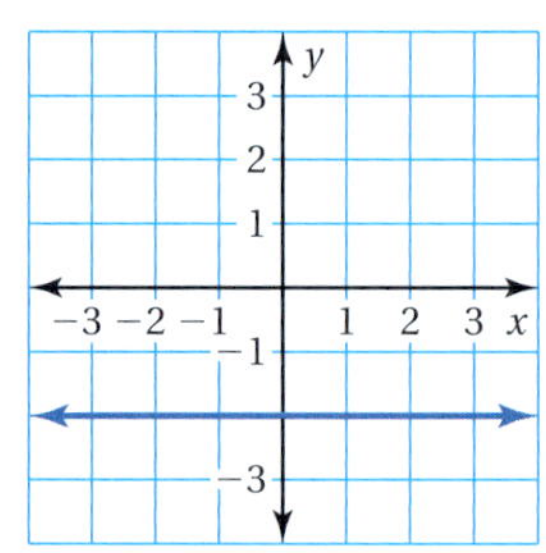

The graph is a line.
So, the function is linear.

On Your Own

Now You're Ready
Exercises 3–11

Does the table or graph represent a *linear* or *nonlinear* function? Explain.

1.

x	*y*
0	25
7	20
14	15
21	10

2.

x	*y*
2	8
4	4
6	0
8	−4

3.

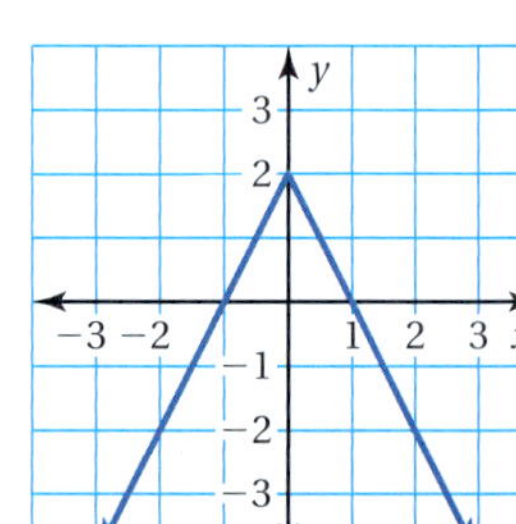

Multi-Language Glossary at BigIdeasMath.com.

EXAMPLE 3 Standardized Test Practice

Which equation represents a *nonlinear* function?

Ⓐ $y = 4.7$ Ⓑ $y = \pi x$

Ⓒ $y = \frac{4}{x}$ Ⓓ $y = 4(x - 1)$

The equations $y = 4.7$, $y = \pi x$, and $y = 4(x - 1)$ can be rewritten in slope-intercept form. So, they are linear functions.

The equation $y = \frac{4}{x}$ cannot be rewritten in slope-intercept form. So, it is a nonlinear function.

∴ The correct answer is Ⓒ.

EXAMPLE 4 Real-Life Application

Account A earns simple interest. Account B earns compound interest. The table shows the balances for 5 years. Graph the data and compare the graphs.

Remember

The simple interest formula is given by $I = Prt$.

- I is the simple interest
- P is the principal
- r is the annual interest rate
- t is the time in years

Year, t	Account A Balance	Account B Balance
0	\$100	\$100
1	\$110	\$110
2	\$120	\$121
3	\$130	\$133.10
4	\$140	\$146.41
5	\$150	\$161.05

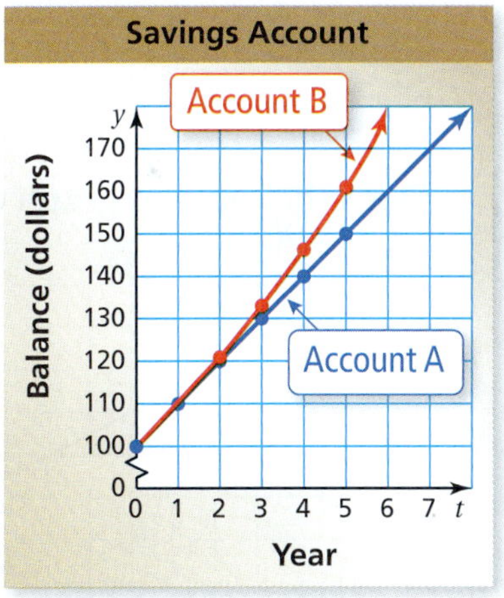

The balance of Account A has a constant rate of change of \$10. So, the function representing the balance of Account A is linear.

The balance of Account B increases by different amounts each year. Because the rate of change is not constant, the function representing the balance of Account B is nonlinear.

On Your Own

Now You're Ready
Exercises 12–14

Does the equation represent a *linear* or *nonlinear* function? Explain.

4. $y = x + 5$ **5.** $y = \frac{4x}{3}$ **6.** $y = 1 - x^2$

4.4 Exercises

Check It Out
Help with Homework
BigIdeasMath.com

Vocabulary and Concept Check

1. **VOCABULARY** Describe the difference between a linear function and a nonlinear function.

2. **WHICH ONE DOESN'T BELONG?** Which equation does *not* belong with the other three? Explain your reasoning.

$5y = 2x$ $\quad$ $y = \frac{2}{5}x$ $\quad$ $10y = 4x$ $\quad$ $5xy = 2$

Practice and Problem Solving

Graph the data in the table. Decide whether the function is *linear* or *nonlinear*.

1 3.

x	0	1	2	3
y	4	8	12	16

4.

x	1	2	3	4
y	1	2	6	24

5.

x	6	5	4	3
y	21	15	10	6

6.

x	−1	0	1	2
y	−7	−3	1	5

Does the table or graph represent a *linear* or *nonlinear* function? Explain.

2 7.

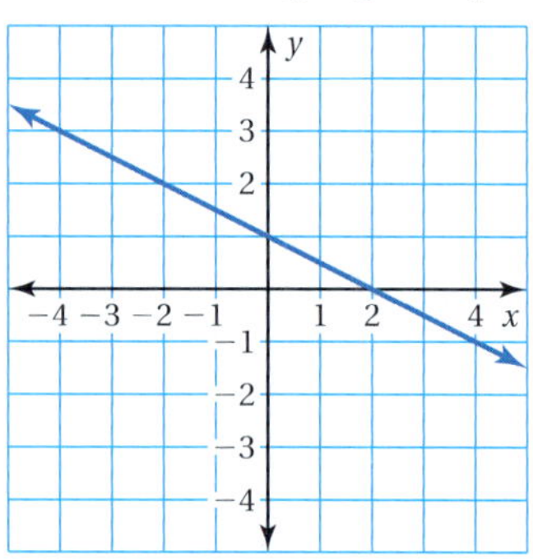

8.

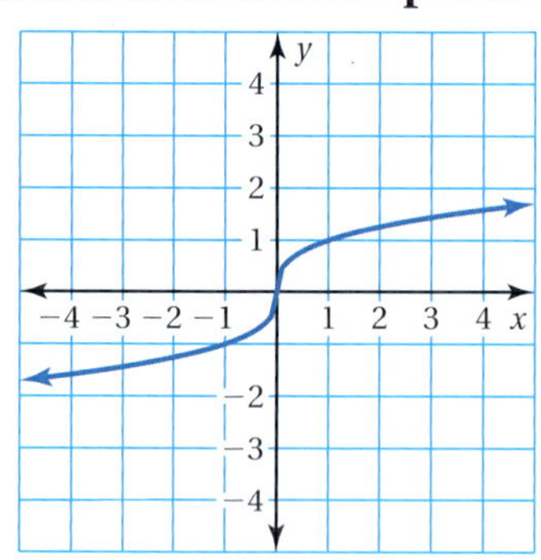

9.

x	5	11	17	23
y	7	11	15	19

10.

x	−3	−1	1	3
y	9	1	1	9

11. **VOLUME** The table shows the volume V (in cubic feet) of a cube with a side length of x feet. Does the table represent a linear or nonlinear function? Explain.

Side Length, *x*	1	2	3	4	5	6	7	8
Volume, *V*	1	8	27	64	125	216	343	512

Does the equation represent a *linear* or *nonlinear* function? Explain.

3 **12.** $2x + 3y = 7$ **13.** $y + x = 4x + 5$ **14.** $y = \frac{8}{x^2}$

15. SUNFLOWER SEEDS The table shows the cost y (in dollars) of x pounds of sunflower seeds.

Pounds, x	Cost, y
2	2.80
3	?
4	5.60

a. What is the missing y-value that makes the table represent a linear function?

b. Write a linear function that represents the cost y of x pounds of seeds.

16. LIGHT The frequency y (in terahertz) of a light wave is a function of its wavelength x (in nanometers). Does the table represent a linear or nonlinear function? Explain.

Color	Red	Yellow	Green	Blue	Violet
Wavelength, x	660	595	530	465	400
Frequency, y	454	504	566	645	749

17. LIGHTHOUSES The table shows the heights x (in feet) of four Florida lighthouses and the number y of steps in each. Does the table represent a linear or nonlinear function? Explain.

Lighthouse	Height, x	Steps, y
Ponce de Leon Inlet	175	213
St. Augustine	167	219
Cape Canaveral	145	179
Key West	86	98

18. PROJECT The wooden bars of a xylophone produce different musical notes when struck. The pitch of a note is determined by the length of the bar. Use the Internet or some other reference to decide whether the pitch of a note is a linear function of the length of the bar.

19. Geometry The radius of the base of a cylinder is 3 feet. Is the volume of the cylinder a linear or nonlinear function of the height of the cylinder?

Fair Game Review What you learned in previous grades & lessons

Classify the angle as *acute, obtuse, right,* or *straight.* *(Skills Review Handbook)*

20. 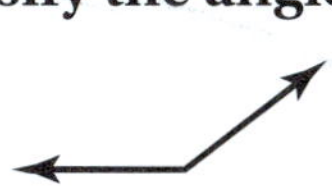

21.

22. 

23. 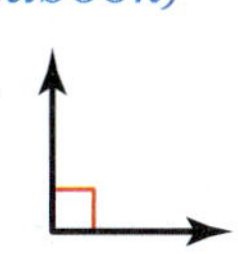

24. MULTIPLE CHOICE What is the value of x? *(Section 1.1)*

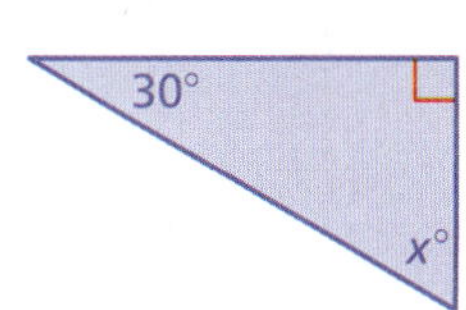

Ⓐ 30 Ⓑ 60 Ⓒ 90 Ⓓ 180

4.3–4.4 Quiz

Check It Out
Progress Check
BigIdeasMath.com

Use the graph or table to write a linear function that relates *y* to *x*. *(Section 4.3)*

1.

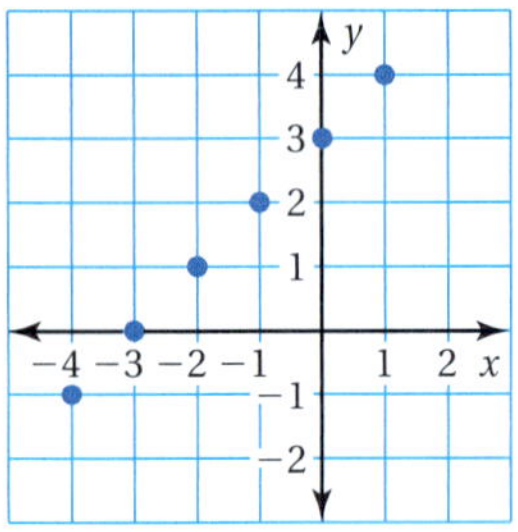

2.

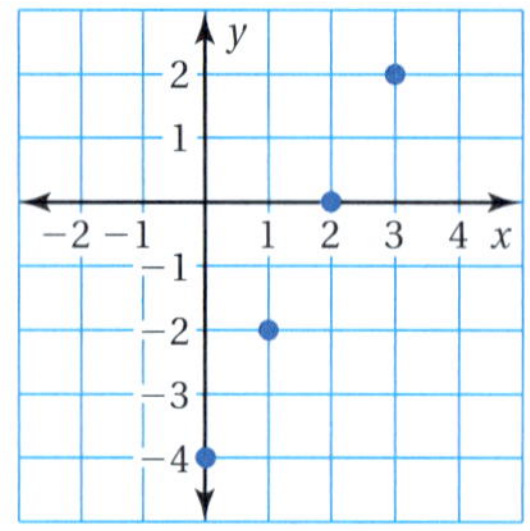

3.

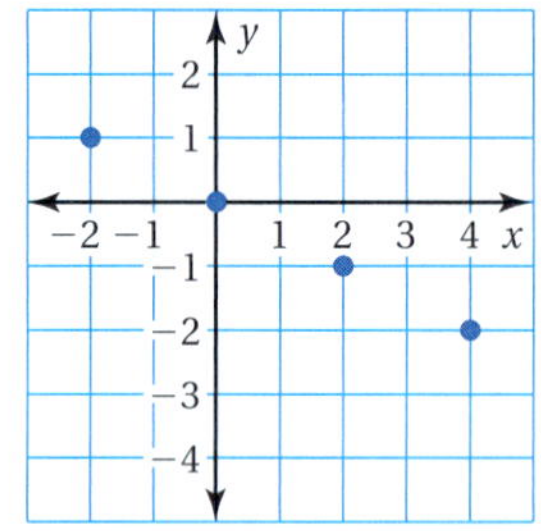

4.

x	0	1	2	3
y	2	1	0	−1

5.

x	−3	0	3	6
y	−3	−1	1	3

Does the table or graph represent a *linear* or *nonlinear* function? Explain. *(Section 4.4)*

6.

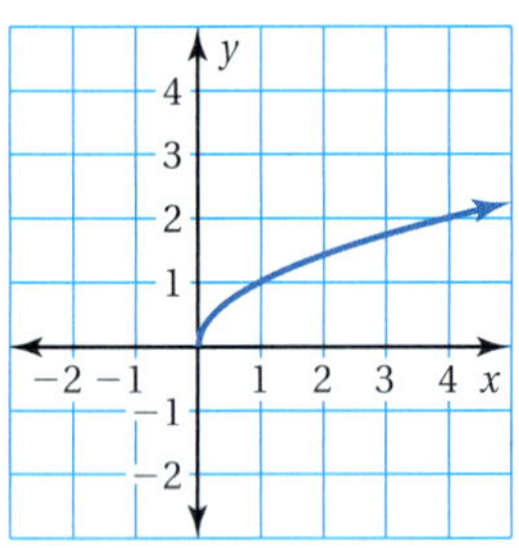

7.

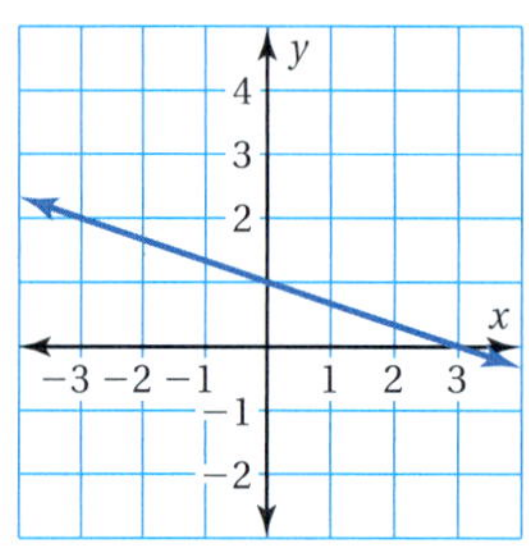

8.

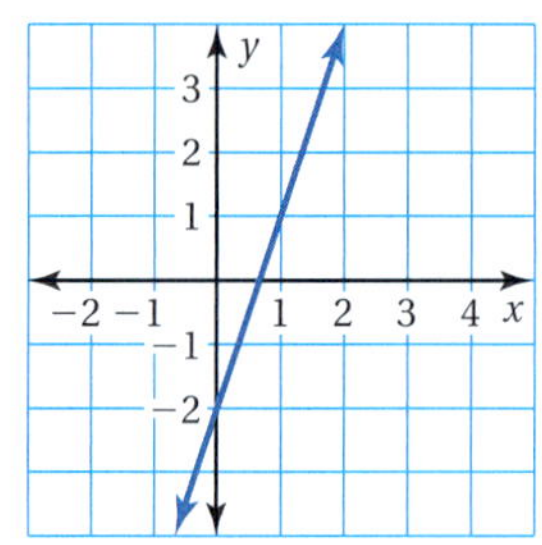

9.

x	*y*
0	0
2	−2
4	−4
6	−6

10.

x	*y*
1	−2
3	7
5	23
7	47

11.

x	*y*
0	3
3	0
6	3
9	6

12. **ADVERTISING** The table shows the revenue *R* (in millions of dollars) of a company when it spends *A* (in millions of dollars) on advertising. *(Section 4.3)*

Advertising, *A*	Revenue, *R*
0	2
2	6
4	10
6	14
8	18

a. Write a linear function that relates the revenue to the advertising cost.

b. What is the revenue of the company when it spends \$10 million on advertising?

13. **CHICKEN SALAD** The equation $y = 7.9x$ represents the cost *y* (in dollars) of buying *x* pounds of chicken salad. Does this equation represent a linear or nonlinear function? Explain. *(Section 4.4)*

4 Chapter Review

Check It Out
Vocabulary Help
BigIdeasMath.com

Review Key Vocabulary

function, *p. 150*	function form, *p. 150*	linear function, *p. 164*
domain, *p. 150*	discrete domain, *p. 156*	nonlinear function, *p. 170*
range, *p. 150*	continuous domain, *p. 156*	

Review Examples and Exercises

4.1 Domain and Range of a Function *(pp. 148–153)*

Find the domain and range of the function represented by the graph.

Write the ordered pairs. Identify the inputs and outputs.

inputs

$(-2, -3), (0, -1), (2, 1), (4, 3)$

outputs

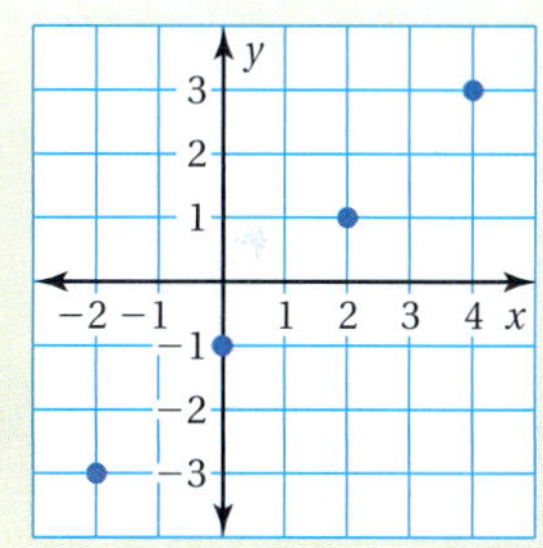

∴ The domain is −2, 0, 2, and 4. The range is −3, −1, 1, and 3.

Exercises

Find the domain and range of the function represented by the graph.

1.

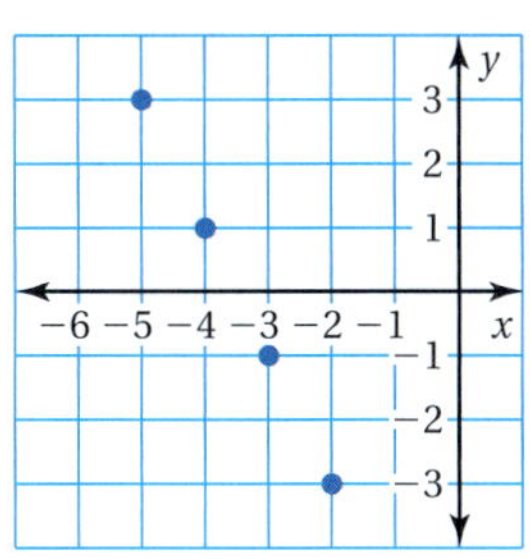

2.

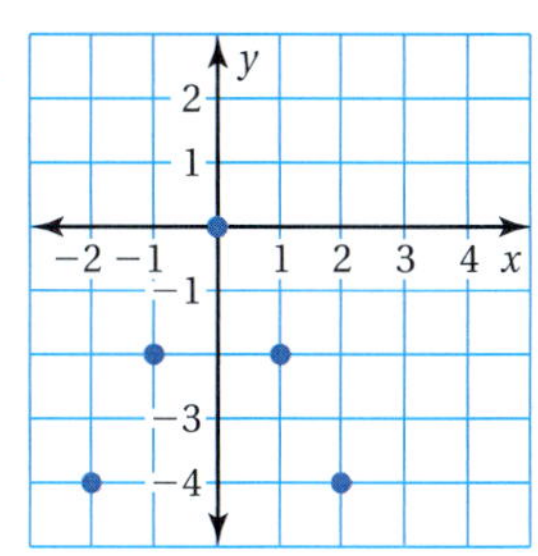

Copy and complete the input-output table for the function. Then find the domain and range of the function represented by the table.

3. $y = 3x - 1$

x	*y*
−1	
0	
1	
2	

4. $4x + y = 2$

x	*y*
0	
1	
2	
3	

4.2 Discrete and Continuous Domains (pp. 154–159)

The function $y = 19.5x$ represents the cost y (in dollars) of x yearbooks. Graph the function. Is the domain of the graph discrete or continuous?

Make an input-output table.

Input, x	$19.5x$	Output, y	Ordered Pair, (x, y)
0	19.5(0)	0	(0, 0)
1	19.5(1)	19.5	(1, 19.5)
2	19.5(2)	39	(2, 39)
3	19.5(3)	58.5	(3, 58.5)
4	19.5(4)	78	(4, 78)

Plot the ordered pairs.

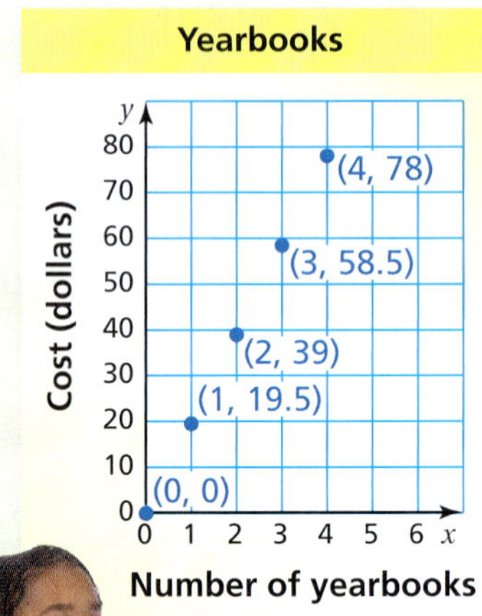

Because you cannot buy part of a yearbook, the graph consists of individual points.

∴ So, the domain is discrete.

Exercises

Graph the function. Is the domain of the graph discrete or continuous?

5.

Hours, x	Miles, y
0	0
1	4
2	8
3	12
4	16

6.

Stamps, x	Cost, y
20	8.4
40	16.8
60	25.2
80	33.6
100	42

4.3 Linear Function Patterns (pp. 162–167)

Use the graph to write a linear function that relates y to x.

The points lie on a line. Find the slope and y-intercept of the line.

$$\text{slope} = \frac{\text{rise}}{\text{run}} = \frac{2}{1} = 2$$

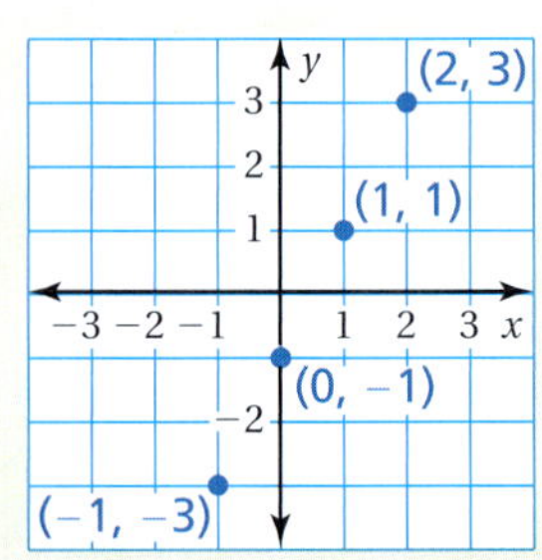

Because the line crosses the y-axis at $(0, -1)$, the y-intercept is -1.

∴ So, the linear function is $y = 2x - 1$.

Exercises

Use the graph or table to write a linear function that relates *y* to *x*.

7.

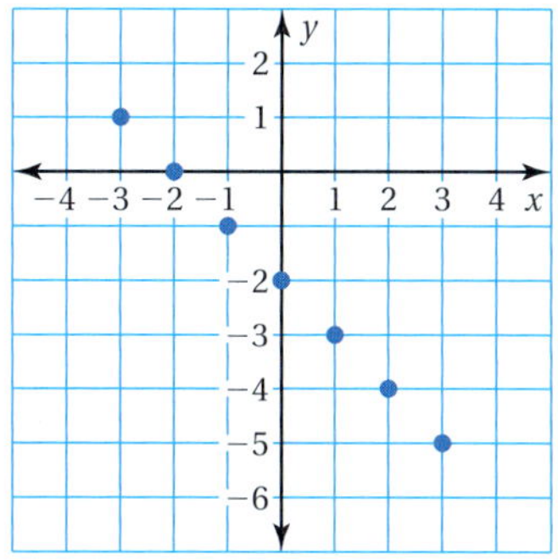

8.

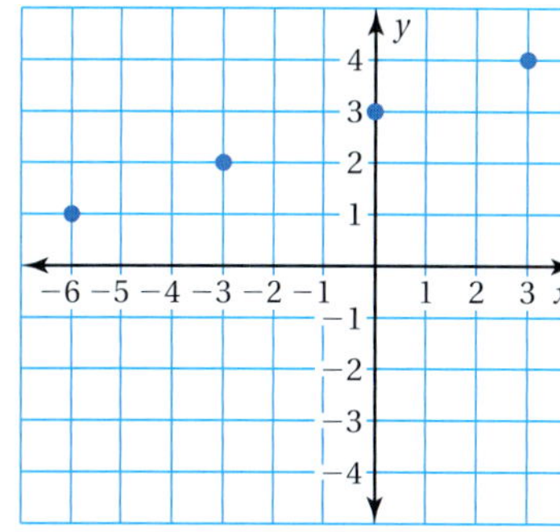

9.

x	−2	−1	0	1
y	−5	−2	1	4

10.

x	−2	0	2	4
y	−7	−7	−7	−7

4.4 Comparing Linear and Nonlinear Functions *(pp. 168–173)*

Does the table represent a *linear* or *nonlinear* function? Explain.

a.

+2 +2 +2

x	0	2	4	6
y	0	1	4	9

+1 +3 +5

As *x* increases by 2, *y* increases by different amounts. The rate of change is *not* constant. So, the function is nonlinear.

b.

x	*y*
0	50
5	40
10	30
15	20

+5 +5 +5 (x); −10 −10 −10 (y)

As *x* increases by 5, *y* decreases by 10. The rate of change is constant. So, the function is linear.

Exercises

Does the table or graph represent a *linear* or *nonlinear* function? Explain.

11.

x	*y*
3	1
6	10
9	19
12	28

12.

x	*y*
1	3
3	1
5	1
7	3

13.

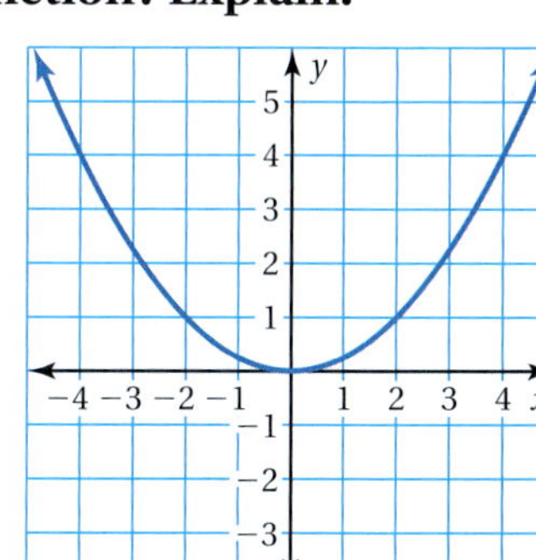

4 Chapter Test

Check It Out
Test Practice
BigIdeasMath.com

1. Find the domain and range of the function represented by the graph.

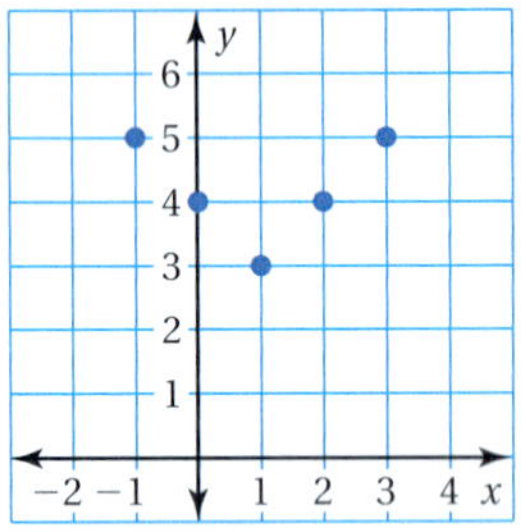

2. Copy and complete the input-output table for the function $y = 7x - 3$. Then find the domain and range of the function represented by the table.

x	−1	0	1	2
y				

Graph the function. Is the domain of the graph discrete or continuous?

3.

Hair Clips, x	Cost, y
0	0
1	1.5
2	3
3	4.5

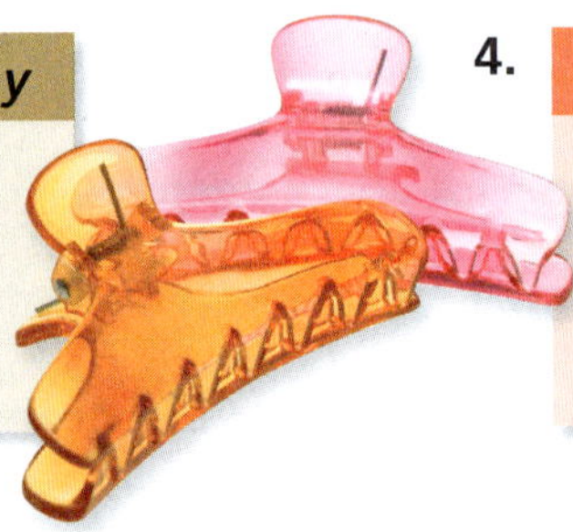

4.

Minutes, x	Gallons, y
0	60
5	45
10	30
15	15

5. Write a linear function that relates y to x.

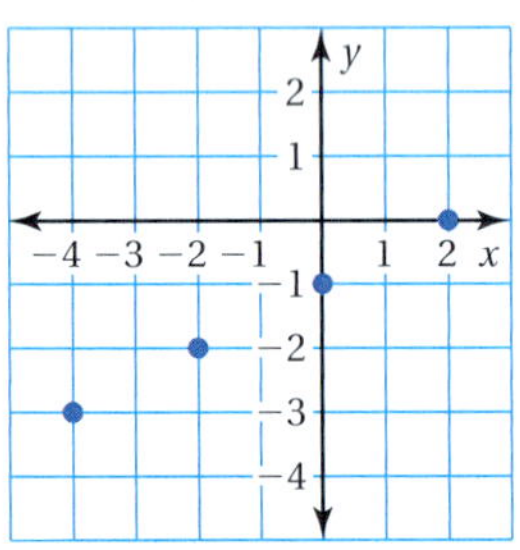

6. Does the table represent a *linear* or *nonlinear* function? Explain.

x	0	2	4	6
y	8	0	−8	−16

7. SAVINGS You save 15% of your monthly earnings x (in dollars).

a. Write an equation in function form that represents the amount y (in dollars) you save each month.

b. Create an input-output table for the equation in part (a). Use the inputs 25, 30, 35, and 40.

c. What is the total amount saved during those 4 months?

8. FOOD DRIVE You are putting cans of food into boxes for a food drive. One box holds 30 cans of food. Write a linear function that represents the number y of cans of food that will fit in x boxes.

9. SURFACE AREA The table shows the surface area S (in square inches) of a cube with a side length of x feet. Does the table represent a linear or nonlinear function? Explain.

Side Length, x	1	2	3	4
Surface Area, S	6	24	54	96

4 Standardized Test Practice

Test-Taking Strategy
Read Question Before Answering

"Take your time and read the question carefully before choosing your answer."

1. The domain of the function $y = 0.2x - 5$ is 5, 10, 15, 20. What is the range of this function?

A. 20, 15, 10, 5 **C.** 4, 3, 2, 1

B. 0, 5, 10, 15 **D.** $-4, -3, -2, -1$

2. A toy runs on a rechargeable battery. During use, the battery loses power at a constant rate. The percent P of total power left in the battery x hours after being fully charged, can be found using the equation shown below. When will the battery be fully discharged?

$$P = -0.25x + 1$$

F. After 4 hours of use

G. After 1 hour of use

H. After 0.75 hour of use

I. After 0.25 hour of use

3. A limousine company charges a fixed cost for a limousine and an hourly rate for its driver. It costs \$500 to rent the limousine for 5 hours and \$800 to rent the limousine for 10 hours. What is the fixed cost, in dollars, to rent the limousine?

4. Which graph shows a nonlinear function?

A.
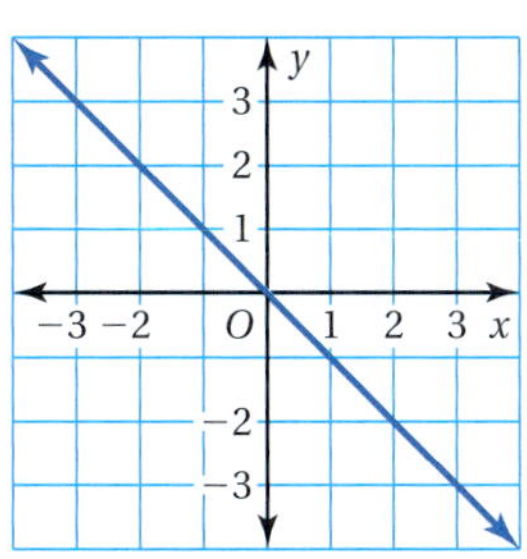

C.
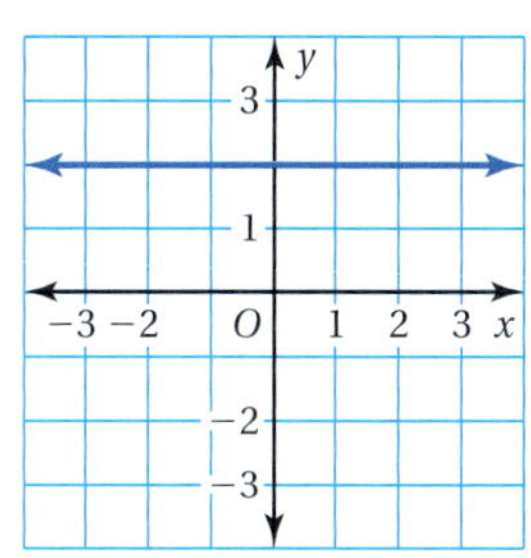

B.
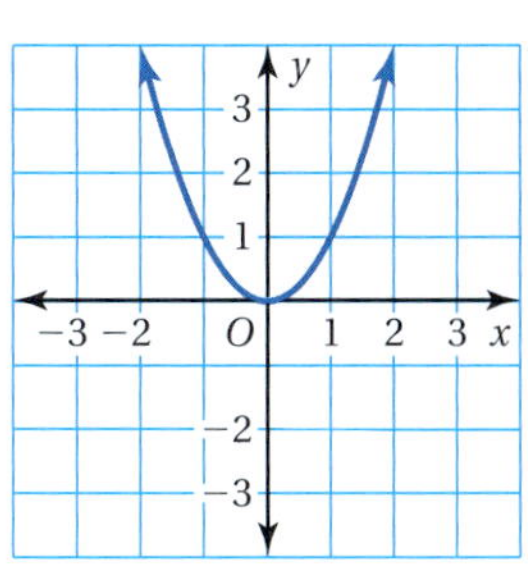

D.
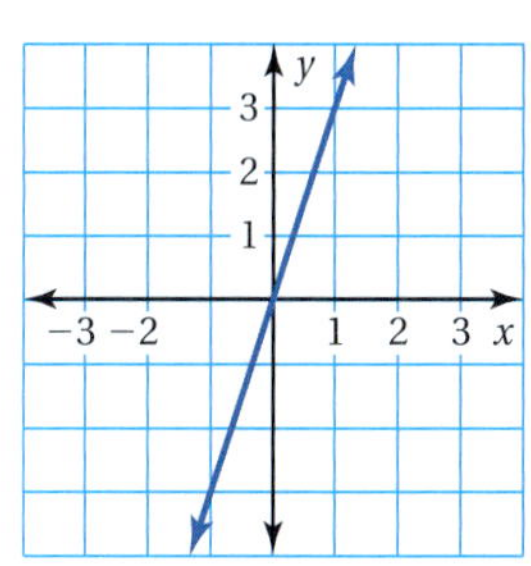

5. The equations $y = -x + 4$ and $y = \frac{1}{2}x - 8$ form a system of linear equations. The table below shows the (x, y) values for these equations at six different values of x.

x	0	2	4	6	8	10
$y = -x + 4$	4	2	0	−2	−4	−6
$y = \frac{1}{2}x - 8$	−8	−7	−6	−5	−4	−3

What can you conclude from the table?

F. The system has one solution, when $x = 0$.

G. The system has one solution, when $x = 4$.

H. The system has one solution, when $x = 8$.

I. The system has no solution.

6. The temperature fell from 54 degrees Fahrenheit to 36 degrees Fahrenheit over a six-hour period. The temperature fell by the same number of degrees each hour. How many degrees Fahrenheit did the temperature fall each hour?

7. What is the domain of the function graphed in the coordinate plane below?

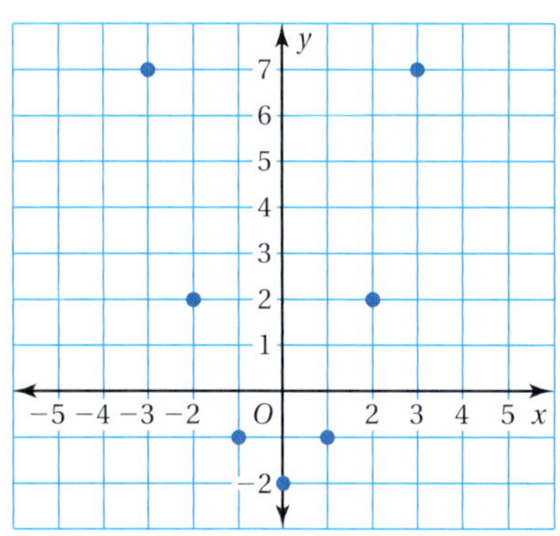

A. 0, 1, 2, 3

B. −2, −1, 2, 7

C. −3, −2, −1, 0, 1, 2, 3

D. −2, −1, 0, 1, 2, 3, 7

8. What value of w makes the equation below true?

$$\frac{w}{3} = 3(w - 1) - 1$$

F. $\frac{3}{2}$

G. $\frac{5}{4}$

H. $\frac{3}{4}$

I. $\frac{1}{2}$

9. What is the slope of the line shown in the graph below?

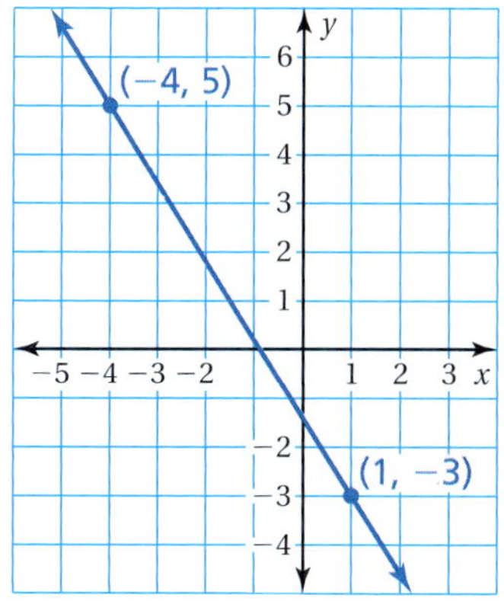

A. $-\frac{2}{5}$

B. $-\frac{2}{3}$

C. $-\frac{8}{5}$

D. $-\frac{8}{3}$

10. A line with slope of $\frac{1}{3}$ contains the point (6, 1). What is the equation of the line?

F. $y = \frac{1}{3}x$

G. $y = \frac{1}{3}x + 1$

H. $x - 3y = 3$

I. $x + 3y = 3$

11. Think Solve Explain The tables show how the perimeter and area of a square are related to its side length. Examine the data in the table.

Side Length	1	2	3	4	5	6
Perimeter	4	8	12	16	20	24

Side Length	1	2	3	4	5	6
Area	1	4	9	16	25	36

Part A Does the first table show a linear function? Explain your reasoning.

Part B Does the second table show a linear function? Explain your reasoning.

12. A bottle of orange extract marked 25 mL costs \$2.49. What is the cost per liter?

A. \$2490.00 per L

B. \$99.60 per L

C. \$9.96 per L

D. \$0.00249 per L

5 Angles and Similarity

What You Learned Before

Finding Unknown Measures in Similar Triangles (Grade 7)

Example 1 **The two triangles are similar. Find the value of *x*.**

$\frac{16}{18} = \frac{12}{x}$ Write a proportion.

$16x = 216$ Use Cross Products Property.

$x = 13.5$ Divide each side by 16.

So, x is 13.5 yards.

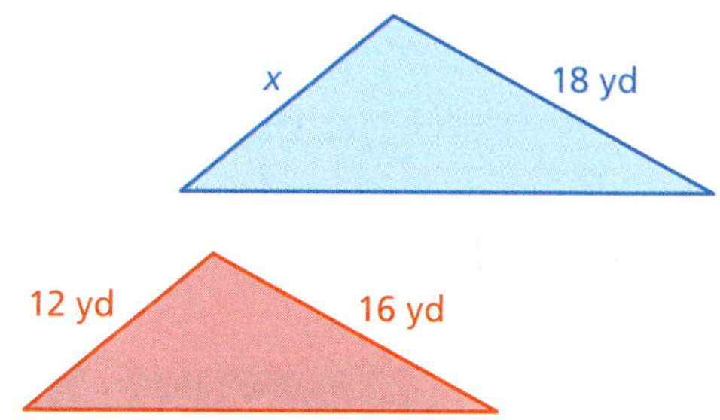

Example 2 **The two quadrilaterals are similar. The ratio of their perimeters is 4 : 5. Find the value of *x*.**

$\frac{4}{5} = \frac{x}{25}$ Write a proportion.

$100 = 5x$ Use Cross Products Property.

$20 = x$ Divide each side by 5.

So, x is 20 centimeters.

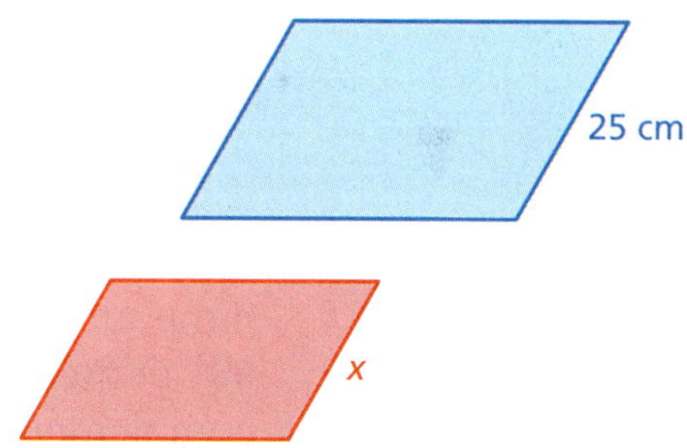

Try It Yourself

The polygons are similar. Find the value of *x*.

1.

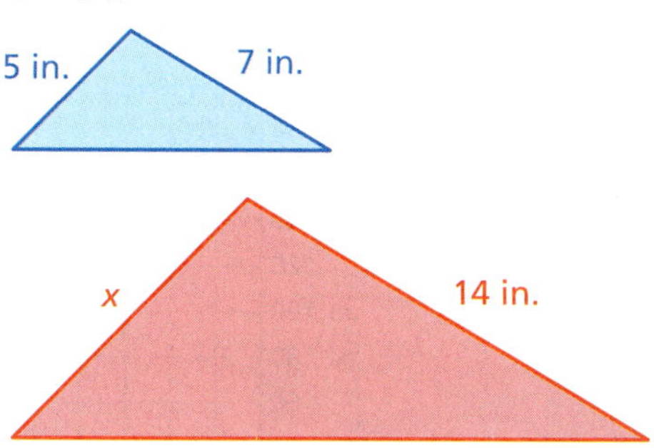

2. The ratio of the perimeters is 2 : 1.

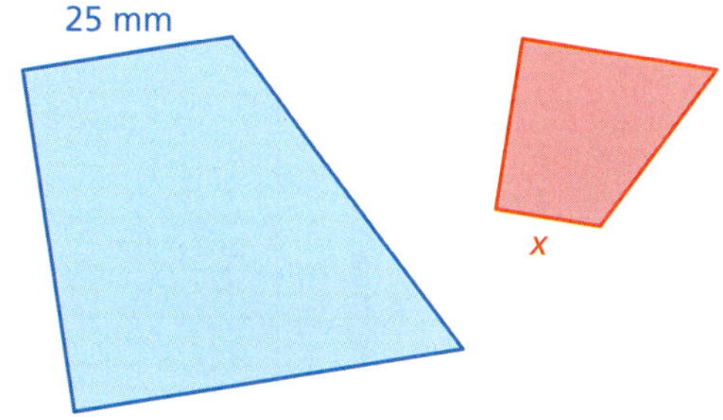

5.1 Classifying Angles

STATE STANDARDS
MA.8.G.2.2

Essential Question How can you classify two angles as complementary or supplementary?

Classification of Angles

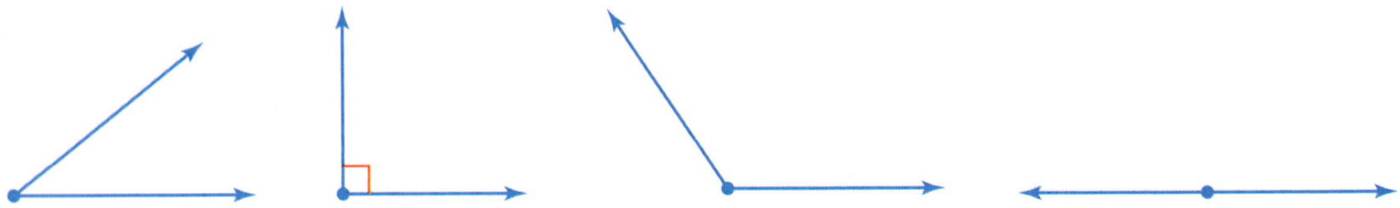

Acute:	***Right:***	***Obtuse:***	***Straight:***
Less than 90°	Equal to 90°	Greater than 90° and less than 180°	Equal to 180°

1 ACTIVITY: Complementary and Supplementary Angles

Work with a partner.

- **Copy and complete each table.**
- **Graph each function. Is the function linear?**
- **Write an equation for *y* as a function of *x*.**
- **Describe the domain of each function.**

a. Two angles are **complementary** if the sum of their measures is 90°. In the table, *x* and *y* are complementary.

x	15°	30°	45°	60°	75°
y					

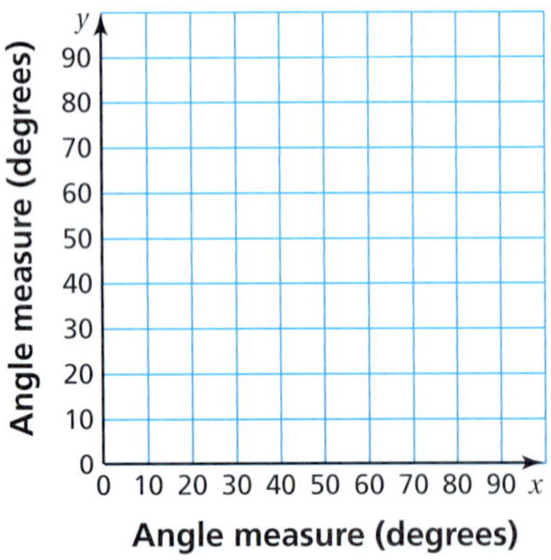

b. Two angles are **supplementary** if the sum of their measures is 180°. In the table, *x* and *y* are supplementary.

x	30°	60°	90°	120°	150°
y					

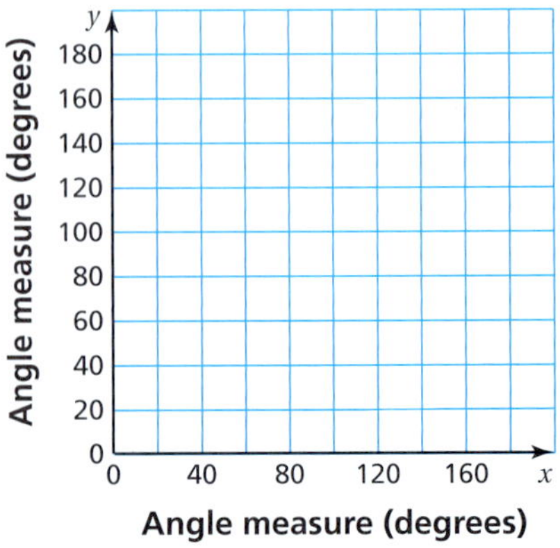

2 ACTIVITY: Exploring Rules About Angles

Work with a partner. Copy and complete each sentence with *always*, *sometimes*, or *never*.

a. If x and y are complementary angles, then both x and y are ________ acute.

b. If x and y are supplementary angles, then x is ________ acute.

c. If x is a right angle, then x is ________ acute.

3 ACTIVITY: Naming Angles

Some angles, such as $\angle A$, can be named by a single letter. When this does not clearly identify an angle, you should use three letters, as follows.

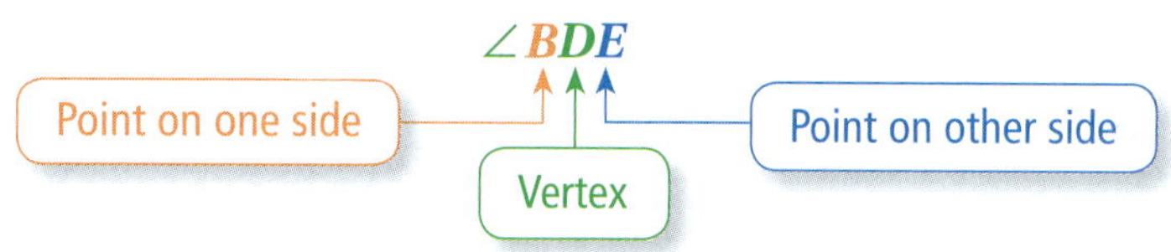

ABEF and *BCDE* are squares.

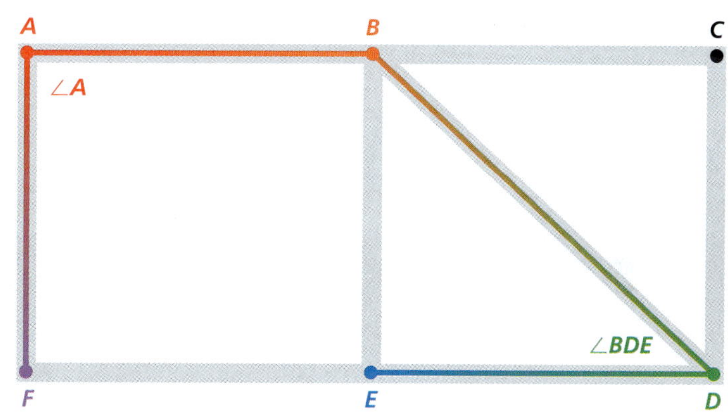

Work with a partner.

a. Name all pairs of complementary angles in the diagram above.

b. Name all pairs of supplementary angles in the diagram above.

What Is Your Answer?

4. **IN YOUR OWN WORDS** How can you classify two angles as complementary or supplementary? Give examples of each type.

5. Find examples of real-life objects that use complementary and supplementary angles. Make a drawing of each object and approximate the degree measure of each angle.

Practice Use what you learned about classifying angles to complete Exercises 3–5 on page 188.

5.1 Lesson

Check It Out
Lesson Tutorials
BigIdeasMath.com

Key Vocabulary

complementary angles, *p. 186*
supplementary angles, *p. 186*
congruent angles, *p. 187*
vertical angles, *p. 187*

Key Ideas

Complementary Angles

Words Two angles are **complementary angles** if the sum of their measures is 90°.

Examples

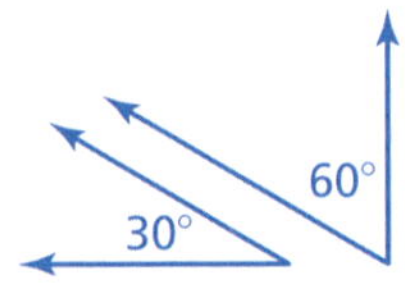

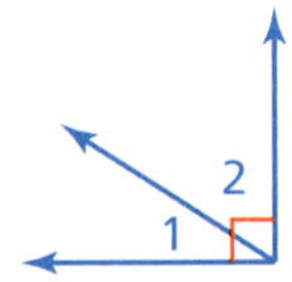

$\angle 1$ and $\angle 2$ are complementary angles.

Supplementary Angles

Words Two angles are **supplementary angles** if the sum of their measures is 180°.

Examples

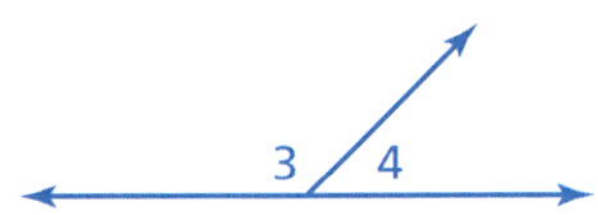

$\angle 3$ and $\angle 4$ are supplementary angles.

EXAMPLE 1 Classifying Pairs of Angles

Tell whether the angles are *complementary*, *supplementary*, or *neither*.

a.

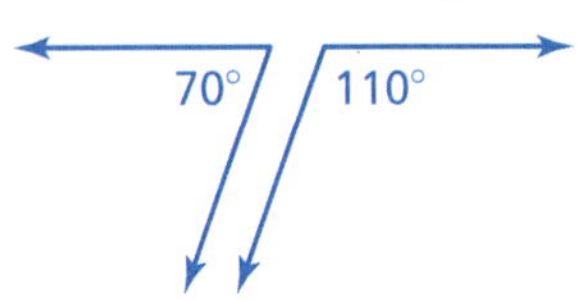

$70° + 110° = 180°$

So, the angles are supplementary.

b.

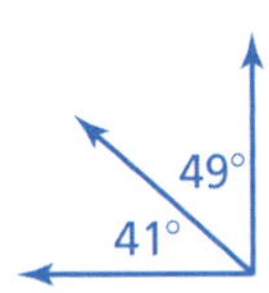

$41° + 49° = 90°$

So, the angles are complementary.

c.

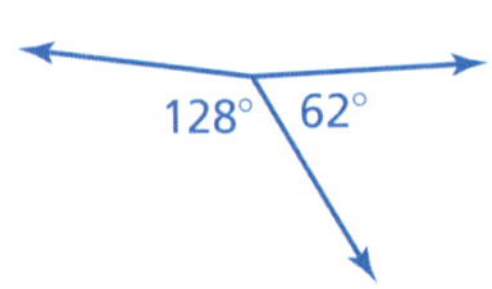

$128° + 62° = 190°$

So, the angles are *neither* complementary nor supplementary.

On Your Own

Now You're Ready
Exercises 6–11

Tell whether the angles are *complementary*, *supplementary*, or *neither*.

1.

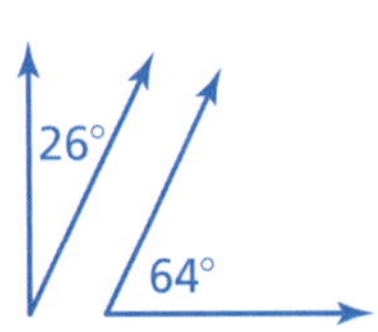

2.

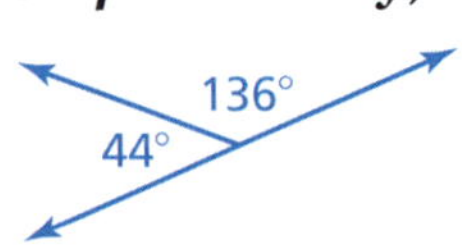

3.

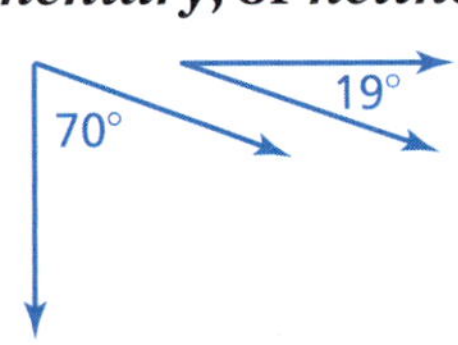

Multi-Language Glossary at BigIdeasMath.com.

Key Ideas

Congruent Angles

Words Two angles are **congruent** if they have the same measure.

Examples

> **Reading**
> Arcs are used to indicate congruent angles.

Vertical Angles

Words Two angles are **vertical angles** if they are opposite angles formed by the intersection of two lines. Vertical angles are congruent.

Examples

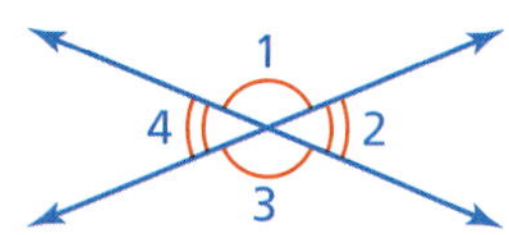

$\angle 1$ and $\angle 3$ are vertical angles.
$\angle 2$ and $\angle 4$ are vertical angles.

EXAMPLE 2 Finding Angle Measures

Find the value of x.

a.

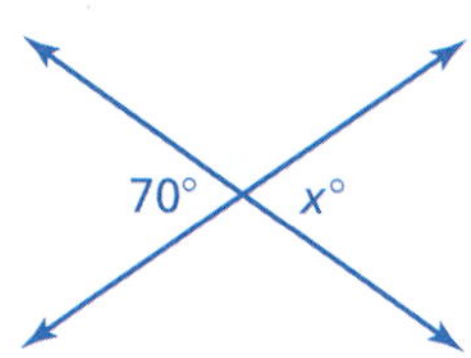

The angles are vertical angles. Because vertical angles are congruent, the angles have the same measure.

So, x is 70.

b.

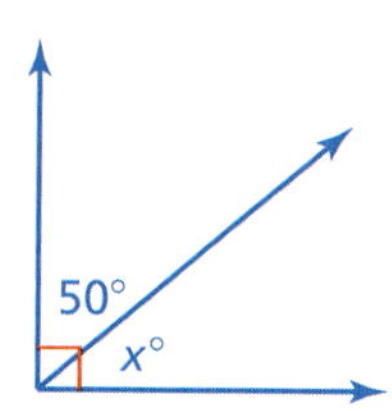

The angles are complementary. So, the sum of their measures is 90°.

$$x + 50 = 90$$

$$x = 40$$

So, x is 40.

On Your Own

> **Now You're Ready**
> Exercises 12–14

Find the value of x.

4.

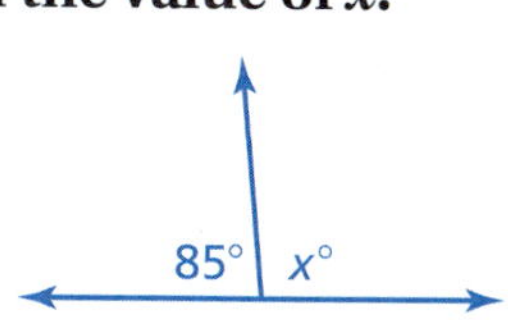

5.

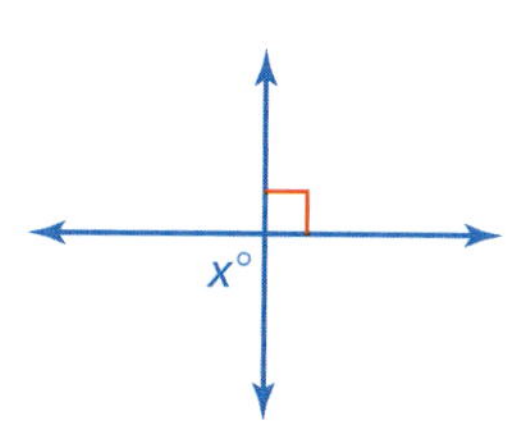

6.

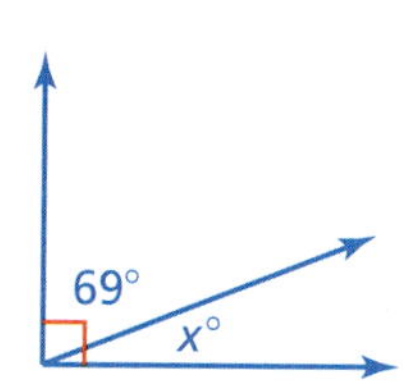

5.1 Exercises

Check It Out
Help with Homework
BigIdeasMath.com

Vocabulary and Concept Check

1. **VOCABULARY** Explain the difference between complementary angles and supplementary angles.
2. **WRITING** When two lines intersect, how many pairs of vertical angles are formed? Explain.

Practice and Problem Solving

Tell whether the statement is *always*, *sometimes*, or *never* true. Explain.

3. If x and y are supplementary angles, then x is obtuse.
4. If x and y are right angles, then x and y are supplementary angles.
5. If x and y are complementary angles, then y is a right angle.

Tell whether the angles are *complementary*, *supplementary*, or *neither*.

1 6.

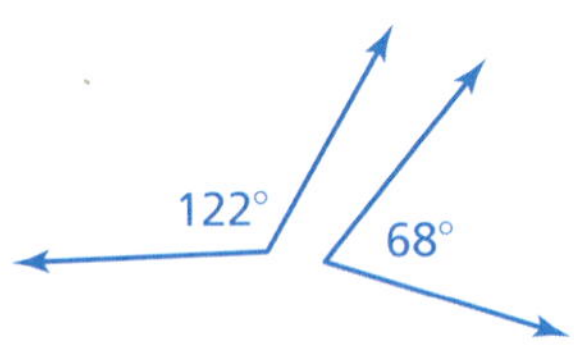

7.

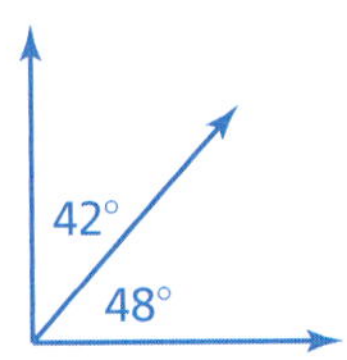

8.

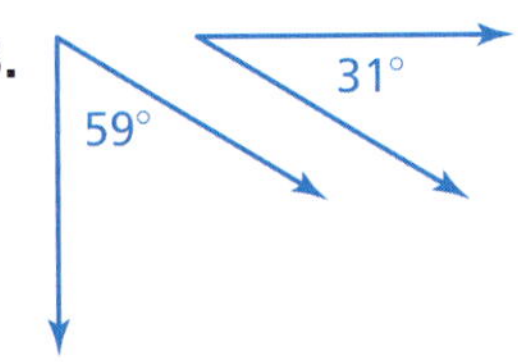

9.

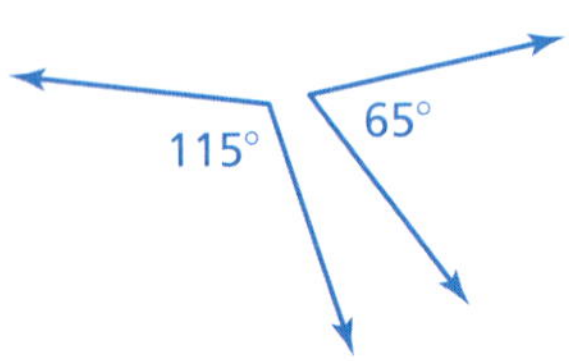

10.

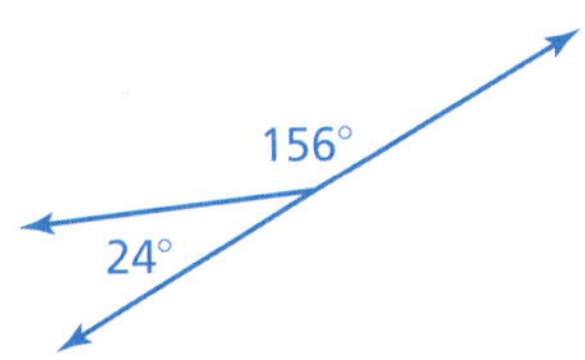

11.

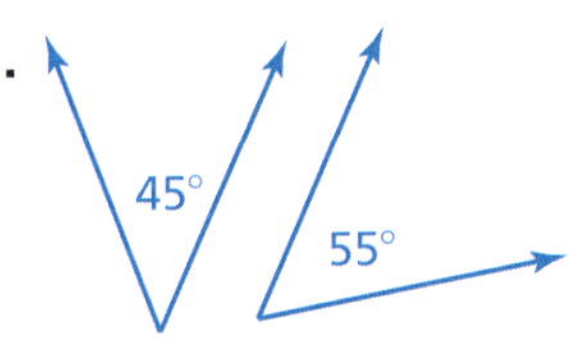

Find the value of x.

2 12.

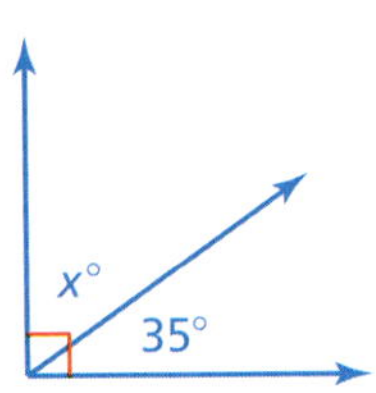

13.

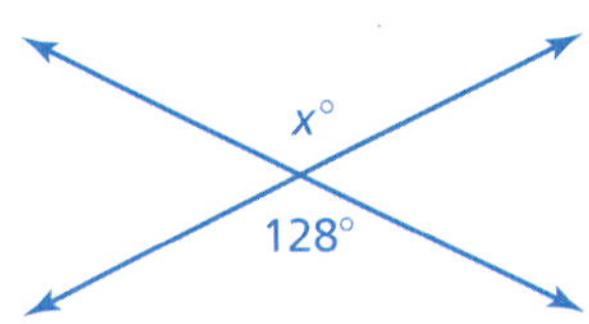

14.

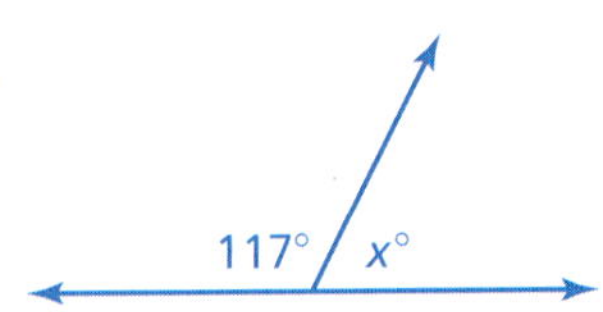

15. **ERROR ANALYSIS** Describe and correct the error in finding the value of x.

> ✗ The value of x is 55 because vertical angles are complementary.

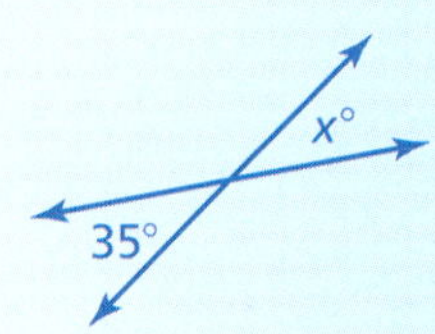

16. **TRIBUTARY** A tributary joins a river at an angle. Find the value of x.

Find the value of *x*.

17.

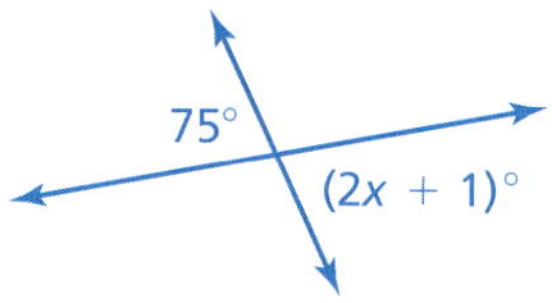

18.

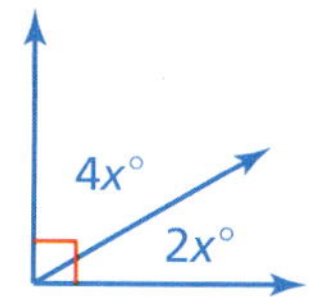

19.

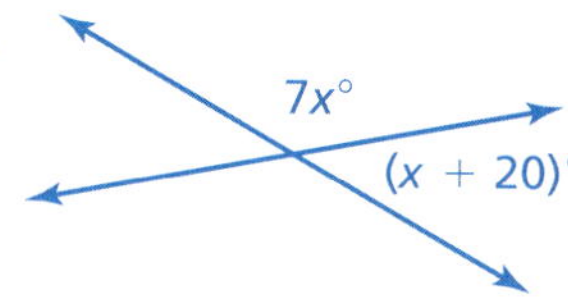

20. OPEN-ENDED Give an example of an angle that can be a supplementary angle but cannot be a complementary angle. Explain.

21. VANISHING POINT The vanishing point of the picture is represented by point B.

a. Name two pairs of complementary angles.

b. Name three pairs of supplementary angles.

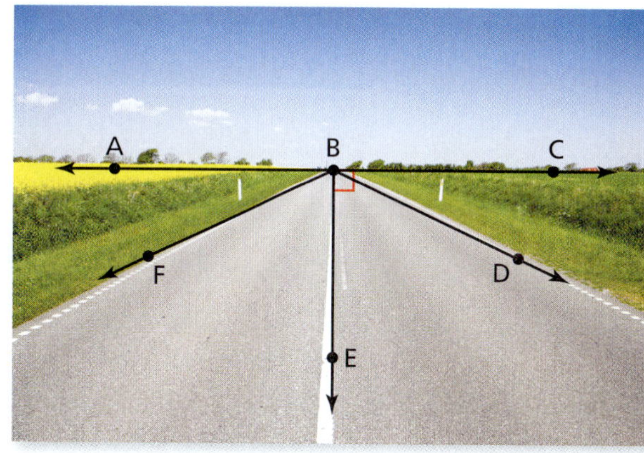

22. INTERSECTION What are the measures of the other three angles formed by the intersection of I-75 and I-10 near Lake City, Florida?

23. RATIO The measures of two complementary angles have a ratio of 3 : 2. What is the measure of the larger angle?

24. REASONING Two angles are vertical angles. What are their measures if they are also complementary angles? supplementary angles?

25. Critical Thinking Write and solve a system of equations to find the values of x and y.

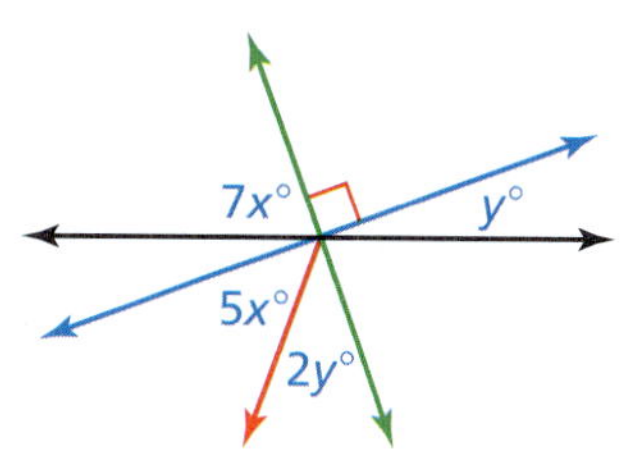

Fair Game Review What you learned in previous grades & lessons

Solve the equation. Check your solution. *(Section 1.1 and Section 1.2)*

26. $x + 60 + 45 = 180$ **27.** $x + 58.5 + 92.2 = 180$ **28.** $x + x + 110 = 180$

29. MULTIPLE CHOICE The graph of which equation has a slope of $-\frac{1}{2}$ and passes through the point (6, 4)? *(Section 3.2)*

Ⓐ $y = x + 3$ Ⓑ $y = -\frac{1}{2}x + 7$ Ⓒ $y = -\frac{1}{2}x + 1$ Ⓓ $y = \frac{1}{2}x - 3$

5.2 Angles and Sides of Triangles

STATE STANDARDS
MA.8.G.2.3

Essential Question How can you classify triangles by their angles?

1 ACTIVITY: Exploring the Angles of a Triangle

Work with a partner.

a. Draw a triangle that has an obtuse angle. Label the angles *A*, *B*, and *C*.

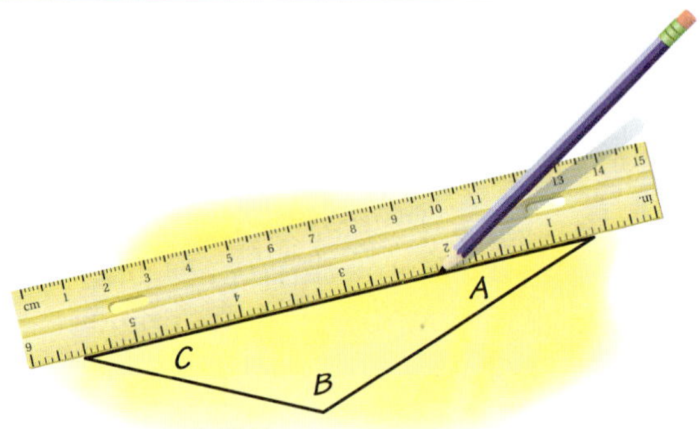

b. Carefully cut out the triangle. Tear off the three corners of the triangle.

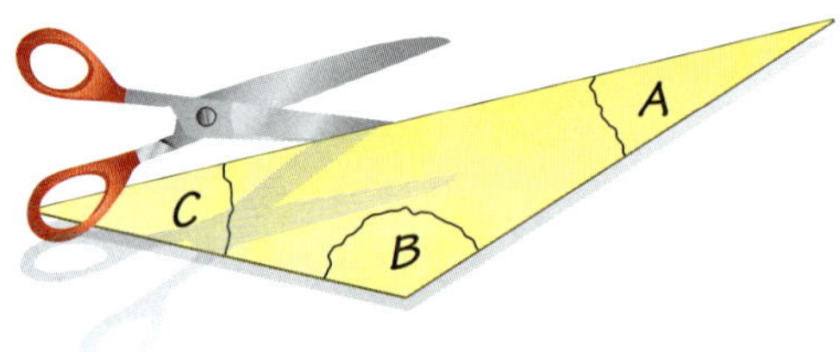

c. Draw a straight line on a piece of paper. Arrange angles *A* and *B* as shown.

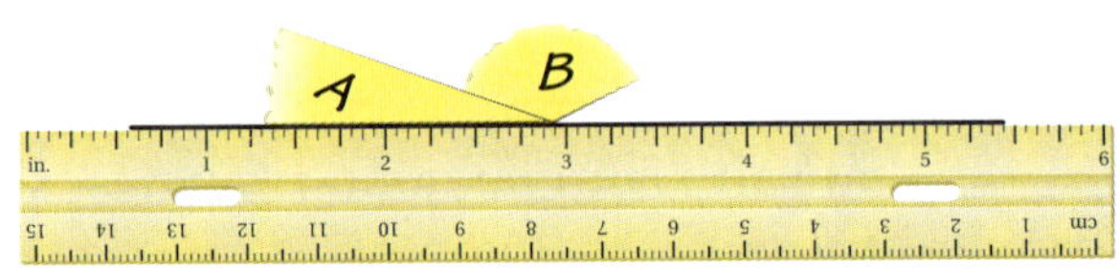

d. Place the third angle as shown. What does this tell you about the sum of the measures of the angles?

e. Draw three other triangles that have different shapes. Repeat parts (b)–(d) for each one. Do you get the same result as in part (d)? Explain.

f. Write a rule about the sum of the measures of the angles of a triangle. Compare your rule with the rule you wrote in Activity 2 in Section 1.1. Did you get the same result? Explain.

2 ACTIVITY: Thinking About Vocabulary

Work with a partner. Talk about the meaning of each name. Use reasoning to define each name. Then match each name with a triangle.

Note: Each triangle has at least one name, but some have more than one name.

a. Right triangle

b. Acute triangle

c. Obtuse triangle

d. Equiangular triangle

e. Equilateral triangle

f. Isosceles triangle

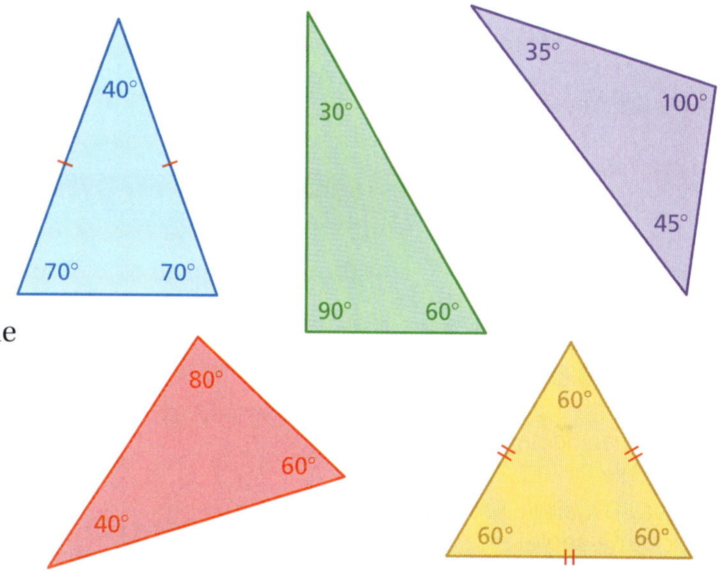

3 ACTIVITY: Triangles in Art

Work with a partner.

a. Trace four triangles in the painting. Classify each triangle using the names in Activity 2.

b. Design your own abstract art painting. How many different types of triangles did you use in your painting?

Abstract II by Linda Bahner

What Is Your Answer?

4. **IN YOUR OWN WORDS** How can you classify triangles by their angles?

5. Find examples of real-life triangles in architecture. Name each type of triangle that you find.

Practice Use what you learned about angles of triangles to complete Exercises 3–5 on page 194.

5.2 Lesson

Check It Out
Lesson Tutorials
BigIdeasMath.com

Key Vocabulary

isosceles triangle, *p. 192*
congruent sides, *p. 192*
equilateral triangle, *p. 192*
equiangular triangle, *p. 192*

Key Idea

Angle Measures of a Triangle

Words The sum of the angle measures of a triangle is 180°.

Algebra $x + y + z = 180$

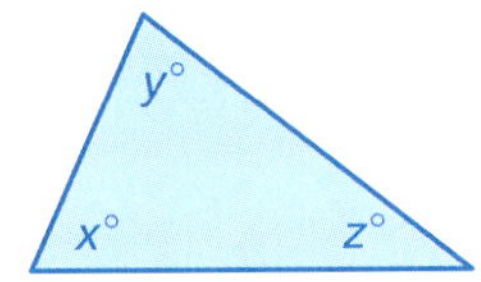

EXAMPLE 1 Finding Angle Measures

Find each value of x. Then classify each triangle.

a.

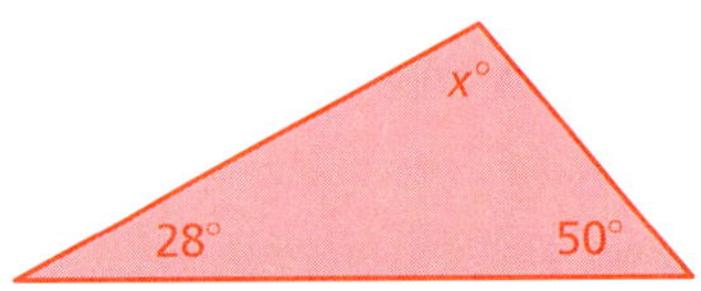

$$x + 28 + 50 = 180$$
$$x + 78 = 180$$
$$x = 102$$

The value of x is 102. The triangle has an obtuse angle. So, it is an obtuse triangle.

b.

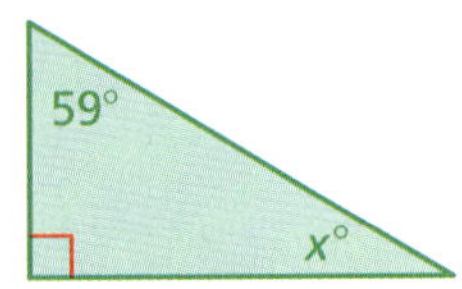

$$x + 59 + 90 = 180$$
$$x + 149 = 180$$
$$x = 31$$

The value of x is 31. The triangle has a right angle. So, it is a right triangle.

Remember

An *acute triangle* has all acute angles.

A *right triangle* has one right angle.

An *obtuse triangle* has one obtuse angle.

On Your Own

Find the value of x. Then classify the triangle.

Now You're Ready
Exercises 6–8

1.

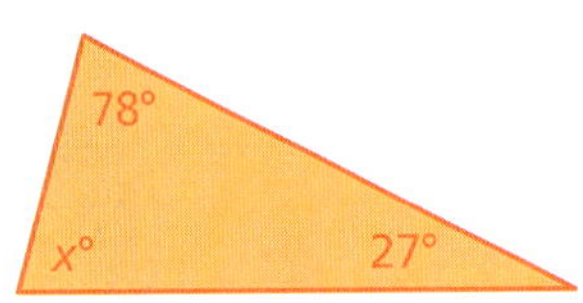

2.

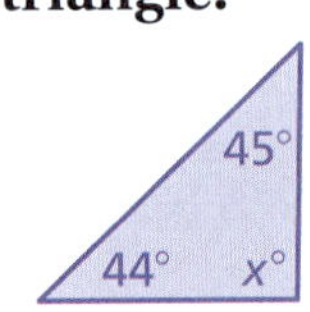

Key Ideas

Isosceles Triangle

An **isosceles triangle** has at least two sides that are **congruent** (have the same length).

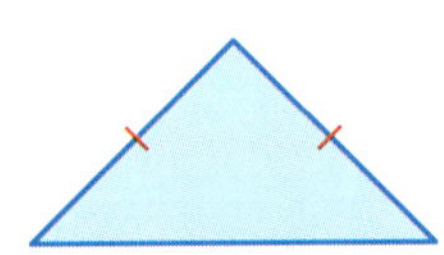

Equilateral Triangle

An **equilateral triangle** has three congruent sides.

An equilateral triangle is also **equiangular** (three congruent angles).

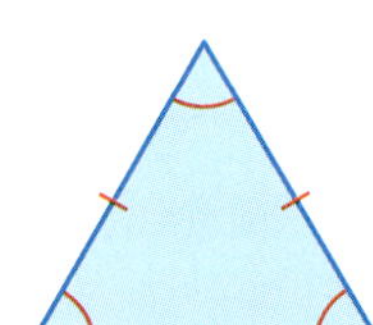

Reading

Small line segments are used to indicate congruent sides.

Multi-Language Glossary at BigIdeasMath.com.

EXAMPLE 2 Finding Angle Measures

Find the value of *x*. Then classify each triangle.

a. Flag of Jamaica

$$x + x + 128 = 180$$
$$2x + 128 = 180$$
$$2x = 52$$
$$x = 26$$

The value of x is 26. Two of the sides are congruent. So, it is an isosceles triangle.

b. Flag of Cuba

$$x + x + 60 = 180$$
$$2x + 60 = 180$$
$$2x = 120$$
$$x = 60$$

The value of x is 60. All three angles are congruent. So, it is an equilateral and equiangular triangle.

EXAMPLE 3 Standardized Test Practice

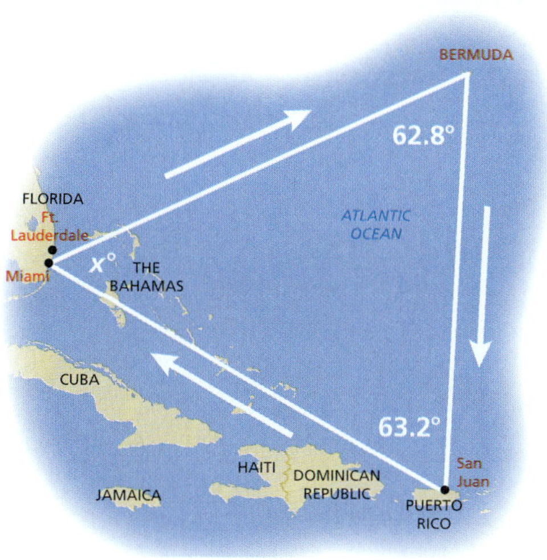

An airplane leaves from Miami and travels around the Bermuda Triangle. What is the value of *x*?

Ⓐ 26.8 Ⓑ 27.2 Ⓒ 54 Ⓓ 64

Use what you know about the angle measures of a triangle to write an equation.

$x + 62.8 + 63.2 = 180$	Write equation.
$x + 126 = 180$	Add.
$x = 54$	Subtract 126 from each side.

The value of x is 54. The correct answer is Ⓒ.

On Your Own

Now You're Ready
Exercises 9–11

Find the value of *x*. Then classify the triangle in as many ways as possible.

3.

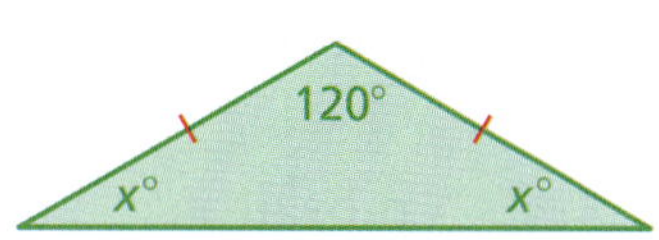

4.

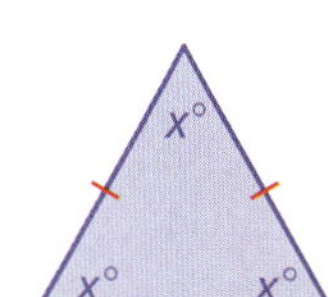

5. In Example 3, the airplane leaves from Fort Lauderdale. The angle measure at Bermuda is 63.9° and the angle measure at San Juan is 61.8°. Find the value of x.

5.2 Exercises

Check It Out
Help with Homework
BigIdeasMath.com

Vocabulary and Concept Check

1. **VOCABULARY** Compare equilateral and isosceles triangles.
2. **REASONING** Describe how to find the missing angle of the triangle.

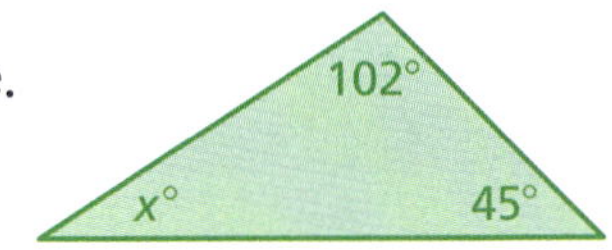

Practice and Problem Solving

Classify the triangle in as many ways as possible.

3. 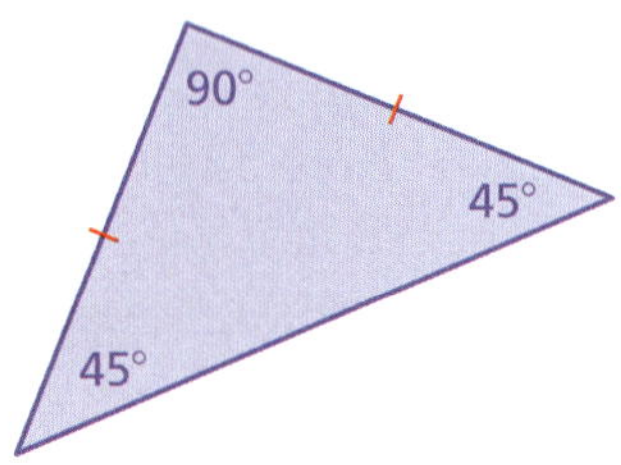

4. 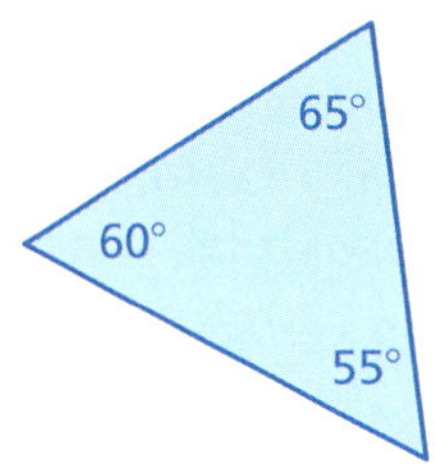

5. 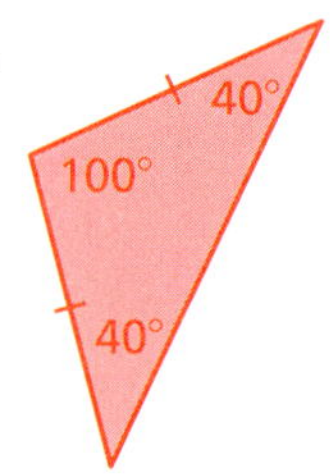

Find the value of x. Then classify the triangle in as many ways as possible.

1 6. 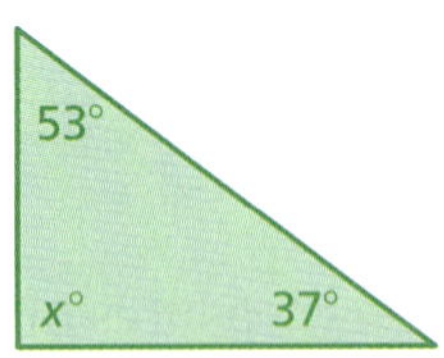

7. 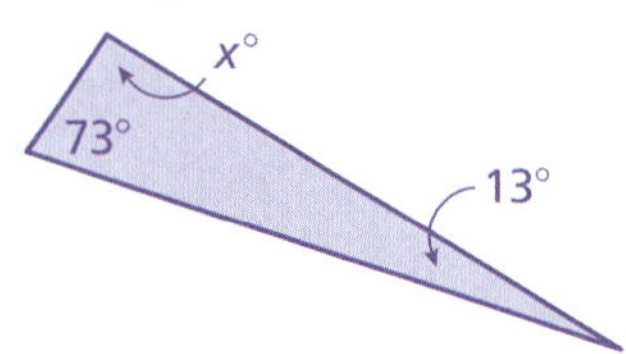

8. 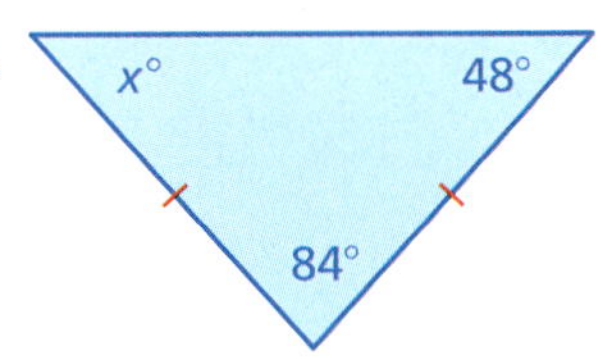

2 9. 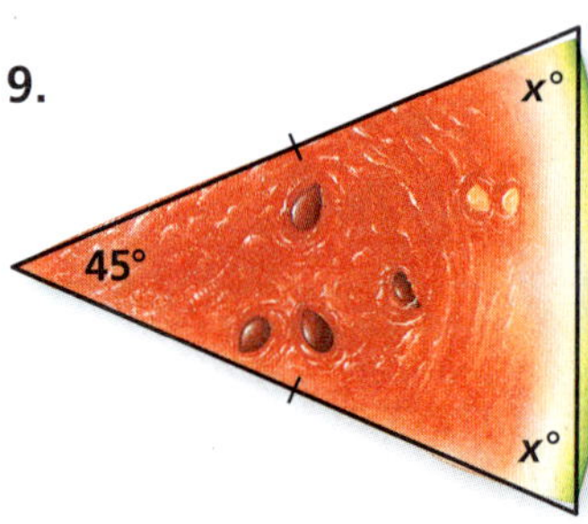

10. 

11. 

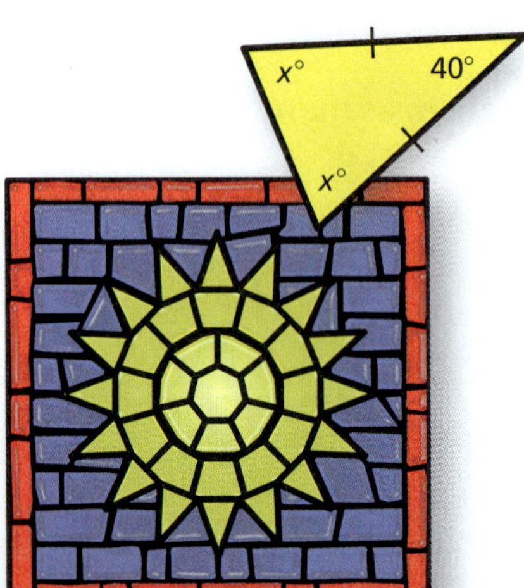

12. **ERROR ANALYSIS** Describe and correct the error in classifying the triangle.

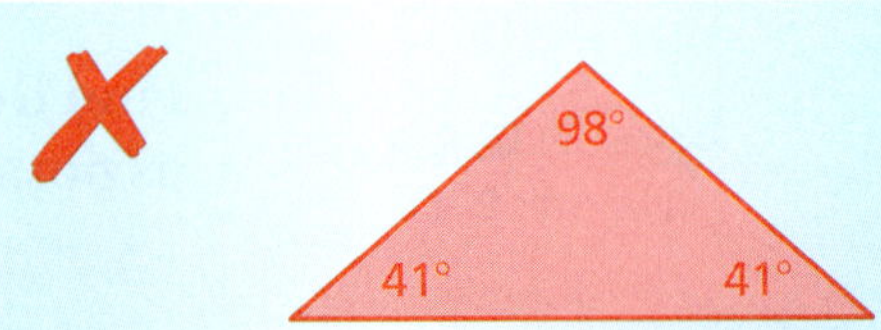

The triangle is an acute triangle, because it has acute angles.

13. **MOSAIC TILE** A mosaic is a pattern or picture made of small pieces of colored material.
    a. Find the value of x.
    b. Classify the triangle used in the mosaic in two ways.

Tell whether a triangle can have the given angle measures. If not, change the first angle measure so that the angle measures form a triangle.

14. $76.2°, 81.7°, 22.1°$

15. $115.1°, 47.5°, 93°$

16. $5\frac{2}{3}°, 64\frac{1}{3}°, 87°$

17. $31\frac{3}{4}°, 53\frac{1}{2}°, 94\frac{3}{4}°$

18. CRITICAL THINKING Consider the three isosceles triangles.

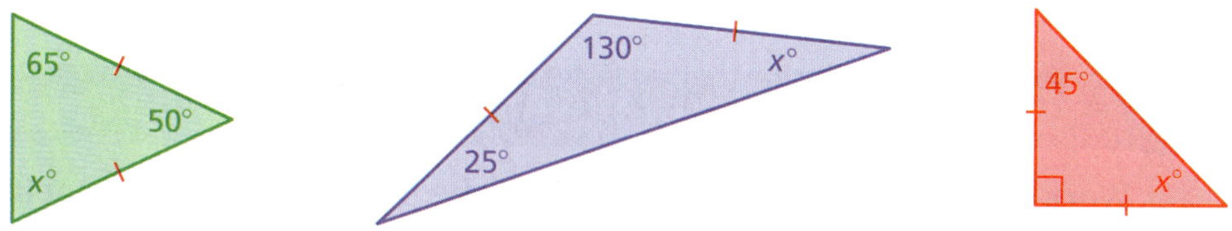

a. Find the value of x for each triangle.

b. What do you notice about the angle measures of each triangle?

c. Write a rule about the angle measures of an isosceles triangle.

19. REASONING Explain why all triangles have at least two acute angles.

20. CARDS One method of stacking cards is shown.

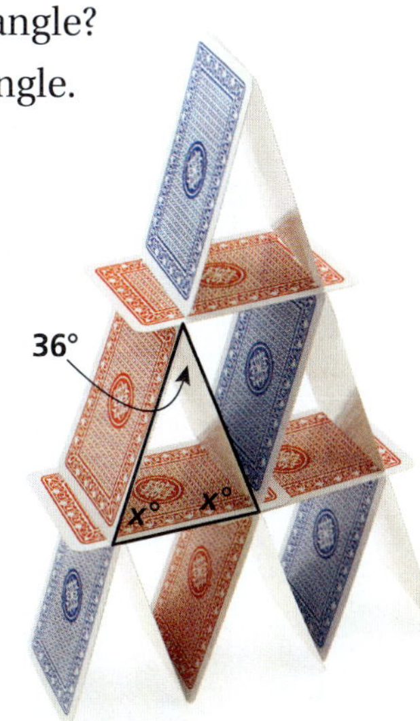

a. Find the value of x.

b. **Critical Thinking** Describe how to stack the cards with different angles. Is the value of x limited? If so, what are the limitations? Explain your reasoning.

Fair Game Review What you learned in previous grades & lessons

Write and solve an equation to find x. Use 3.14 for π. *(Skills Review Handbook)*

21. $P = 48$ cm

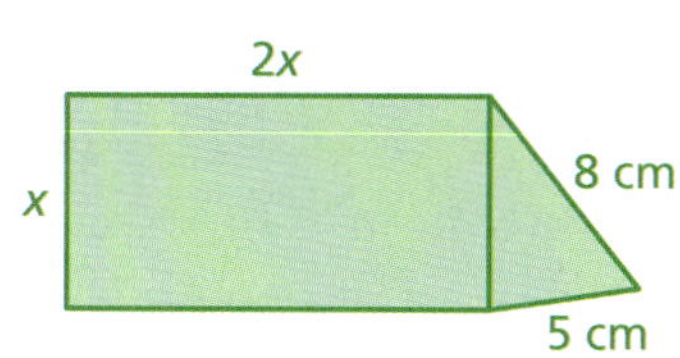

22. $P = 28$ in.

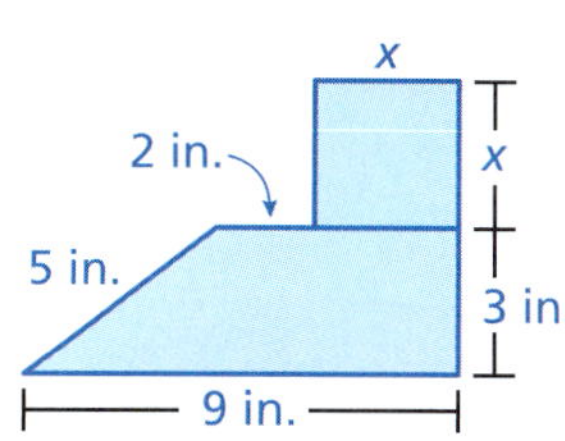

23. $P = 25.42$ m

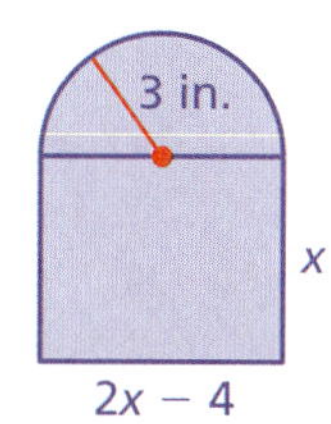

24. MULTIPLE CHOICE You have \$10 for text messages. Each message costs \$0.25. Which equation represents the amount of money you have after x messages? *(Section 3.1)*

Ⓐ $y = -0.25x + 10$

Ⓑ $y = 0.25x - 10$

Ⓒ $y = -0.25x - 10$

Ⓓ $y = 0.25x + 10$

5.3 Angles of Polygons

STATE STANDARDS
MA.8.G.2.3

Essential Question How can you find a formula for the sum of the angle measures of any polygon?

1 ACTIVITY: The Sum of the Angle Measures of a Polygon

Work with a partner. Find the sum of the angle measures of each polygon with n sides.

a. Sample: Quadrilateral: $n = 4$

Draw a line that divides the quadrilateral into two triangles.

Because the sum of the angle measures of each triangle is 180°, the sum of the angle measures of the quadrilateral is 360°.

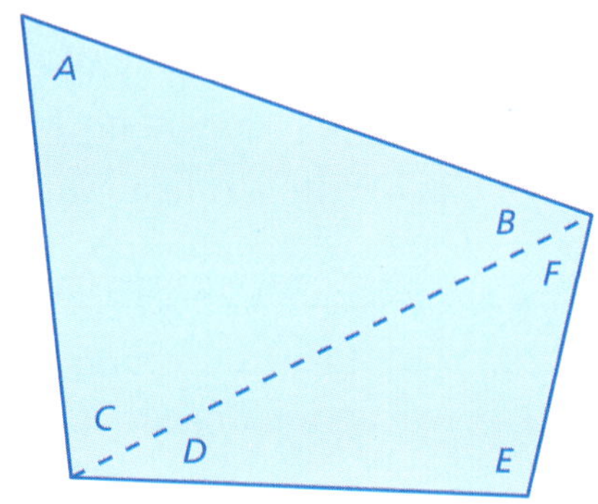

$$(A + B + C) + (D + E + F) = 180° + 180°$$
$$= 360°$$

b. Pentagon: $n = 5$

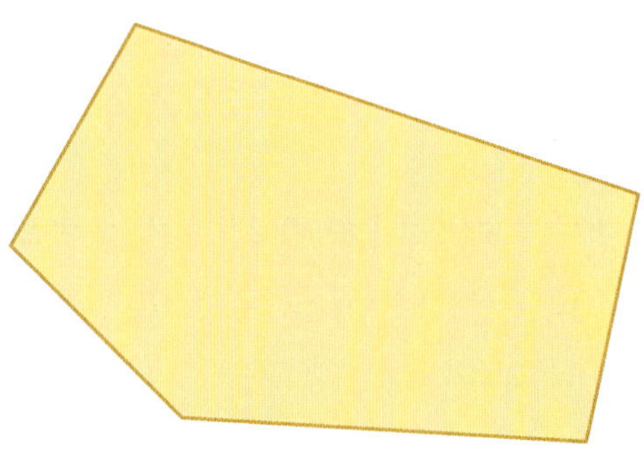

c. Hexagon: $n = 6$

d. Heptagon: $n = 7$

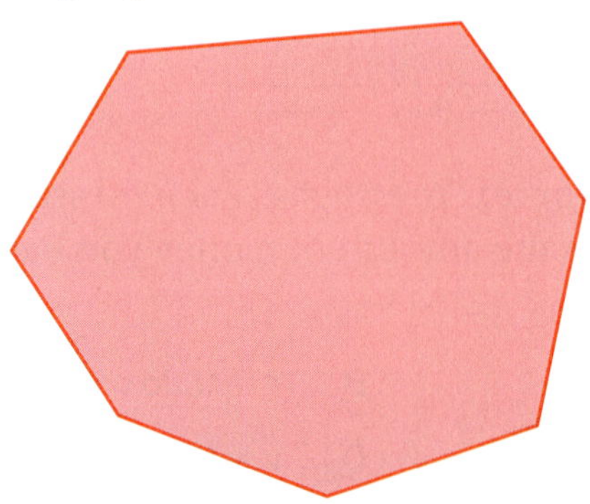

e. Octagon: $n = 8$

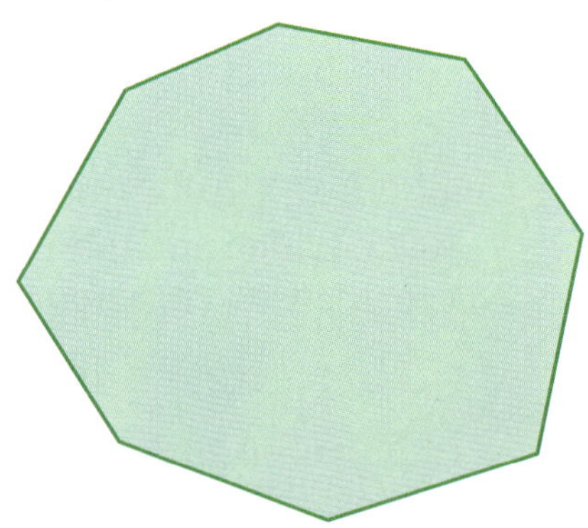

2 ACTIVITY: The Sum of the Angle Measures of a Polygon

Work with a partner.

a. Use the table to organize your results from Activity 1.

Sides, n	3	4	5	6	7	8
Angle Sum, S						

b. Plot the points in the table in a coordinate plane.

c. Write a linear equation that relates S to n.

d. What is the domain of the function? Explain your reasoning.

e. Use the function to find the sum of the angle measures of a polygon with 10 sides.

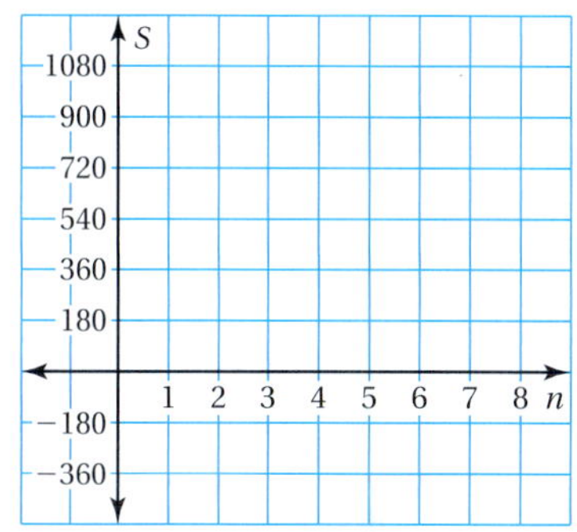

3 ACTIVITY: The Sum of the Angle Measures of a Polygon

Work with a partner.

A polygon is convex if the line segment connecting any two vertices lies entirely inside the polygon. A polygon that is not convex is called concave.

Does the equation you found in Activity 2 apply to concave polygons? Explain.

How can you define the measure of an angle so that your equation applies to *any* polygon?

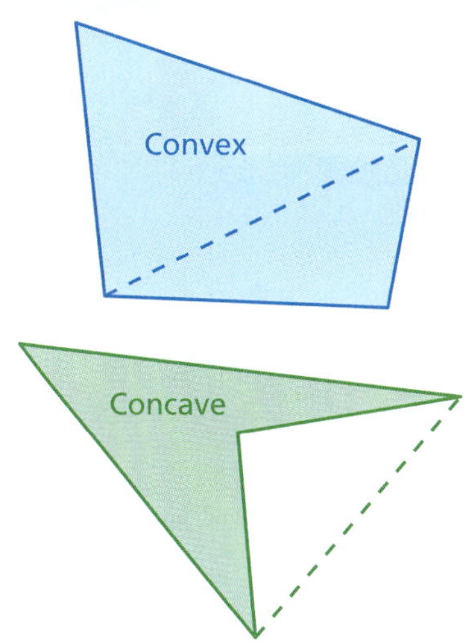

What Is Your Answer?

4. **IN YOUR OWN WORDS** How can you find a formula for the sum of the angle measures of any polygon?

Practice Use what you learned about angles of polygons to complete Exercises 4–6 on page 201.

5.3 Lesson

Check It Out
Lesson Tutorials
BigIdeasMath.com

Key Vocabulary
polygon, *p. 198*
regular polygon, *p. 199*
convex polygon, *p. 200*
concave polygon, *p. 200*

A **polygon** is a closed plane figure made up of three or more line segments that intersect only at their endpoints.

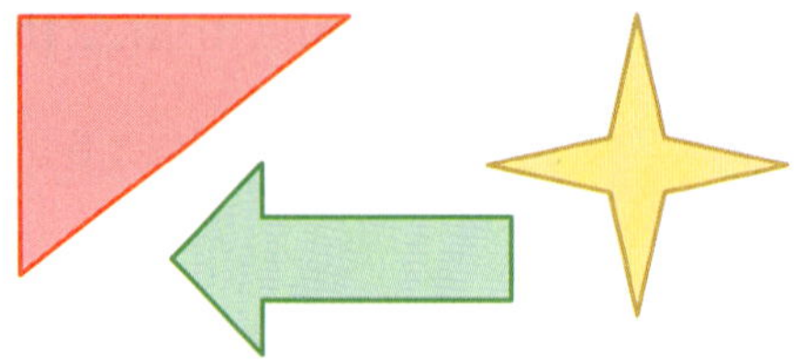

Polygons

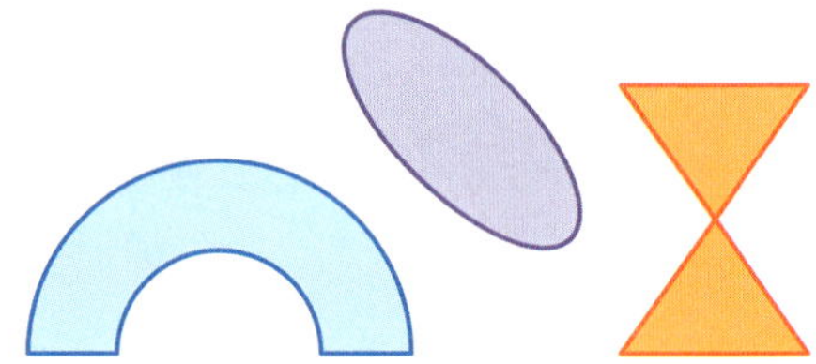

Not polygons

Key Idea

Angle Measures of a Polygon

The sum S of the angle measures of a polygon with n sides is

$$S = (n - 2) \cdot 180^\circ.$$

EXAMPLE 1 Finding the Sum of the Angle Measures of a Polygon

An entrance to Booker Middle School in Sarasota, Florida is outlined in red. Find the sum of the angle measures of the entrance.

The entrance is in the shape of a pentagon. It has 5 sides.

$S = (n - 2) \cdot 180^\circ$	Write the formula.
$= (5 - 2) \cdot 180^\circ$	Substitute 5 for n.
$= 3 \cdot 180^\circ$	Subtract.
$= 540^\circ$	Multiply.

∴ The sum of the angle measures is 540°.

Reading

For polygons whose names you have not learned, you can use the phrase "n-gon," where n is the number of sides. For example, a 15-gon is a polygon with 15 sides.

On Your Own

Now You're Ready
Exercises 7–9

Find the sum of the angle measures of the green polygon.

1.

2.

Multi-Language Glossary at BigIdeasMath.com.

EXAMPLE 2 Finding an Angle Measure of a Polygon

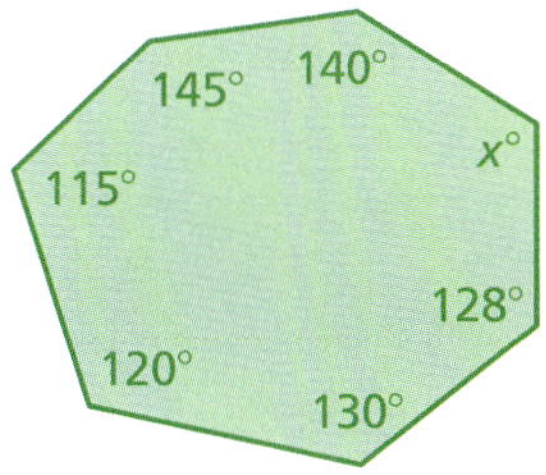

Find the value of x.

Step 1: The polygon has 7 sides. Find the sum of the angle measures.

$S = (n - 2) \cdot 180°$ Write the formula.

$= (7 - 2) \cdot 180°$ Substitute 7 for n.

$= 900°$ Simplify. The sum of the angle measures is 900°.

Step 2: Write and solve an equation.

$$140 + 145 + 115 + 120 + 130 + 128 + x = 900$$

$$778 + x = 900$$

$$x = 122$$

The value of x is 122.

On Your Own

Now You're Ready
Exercises 12–14

Find the value of x.

3. 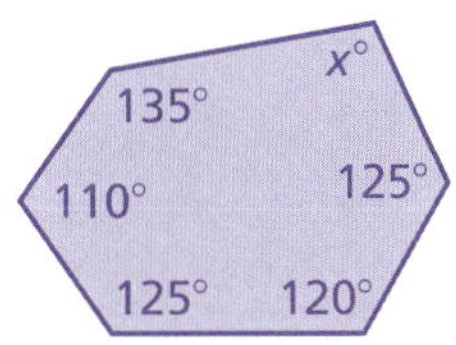

4. 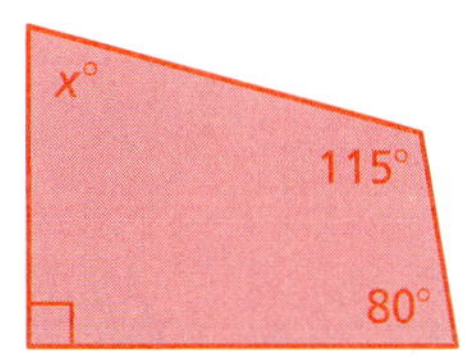

5. 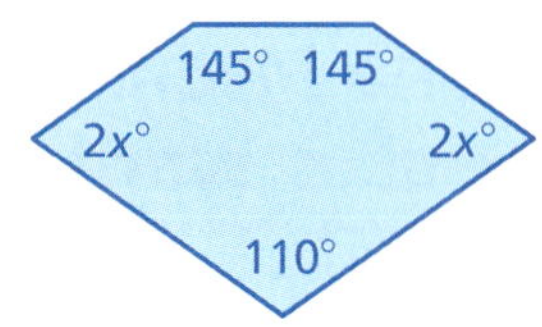

In a **regular polygon**, all of the sides are congruent and all of the angles are congruent.

EXAMPLE 3 Real-Life Application

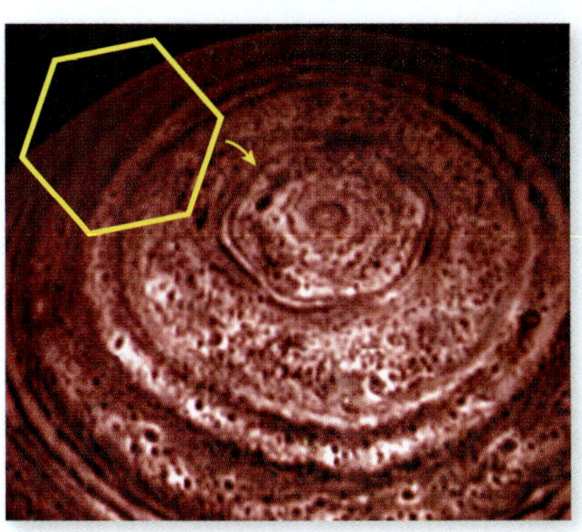

The hexagon is about 15,000 miles across. Approximately four Earths could fit inside it.

A cloud system discovered on Saturn is in the approximate shape of a regular hexagon. Find the measure of each angle of the hexagon.

Step 1: A hexagon has 6 sides. Find the sum of the angle measures.

$S = (n - 2) \cdot 180°$ Write the formula.

$= (6 - 2) \cdot 180°$ Substitute 6 for n.

$= 720°$ Simplify. The sum of the angle measures is 720°.

Step 2: Divide the sum by the number of angles, 6.

$$720° \div 6 = 120°$$

The measure of each angle is 120°.

On Your Own

Now You're Ready
Exercises 16–18

Find the measure of each angle of the regular polygon.

6. octagon
7. decagon
8. 18-gon

Key Idea

Convex and Concave Polygons

A polygon is **convex** if every line segment connecting any two vertices lies entirely inside the polygon.

A polygon is **concave** if at least one line segment connecting any two vertices lies outside the polygon.

EXAMPLE 4 Identifying Convex and Concave Polygons

The Meaning of a Word

Concave

To remember the term con**cave**, think of a polygon that is "**cave**d in."

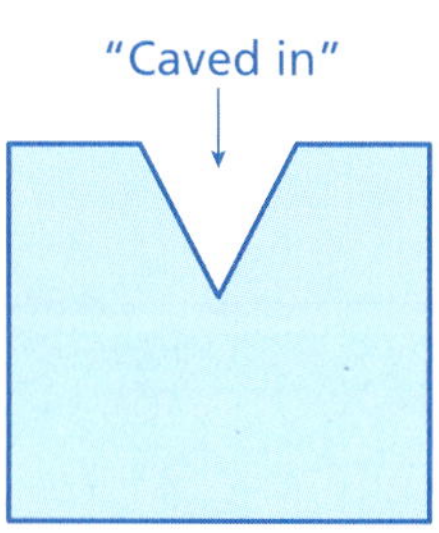

Tell whether the polygon is *convex* or *concave*. Explain.

a.

A line segment connecting two vertices lies outside the polygon. So, the polygon is concave.

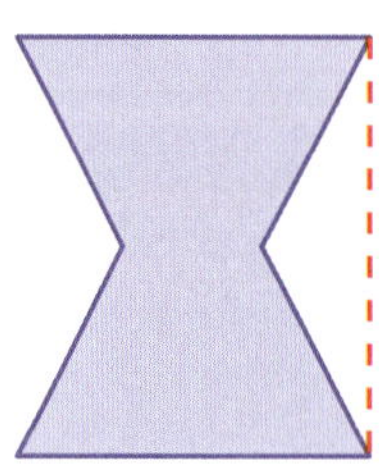

b.

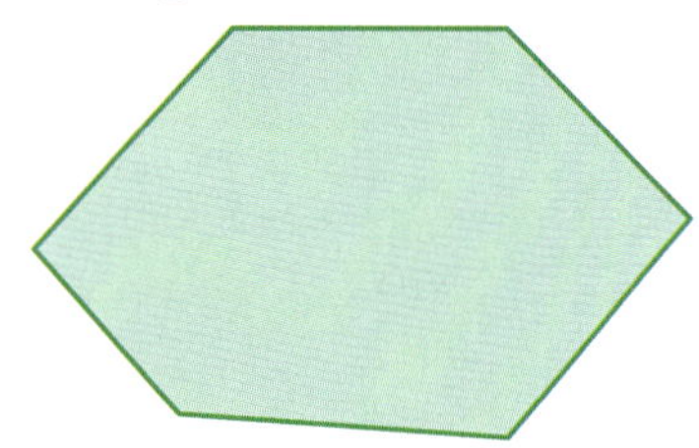

No line segment connecting two vertices lies outside the polygon. So, the polygon is convex.

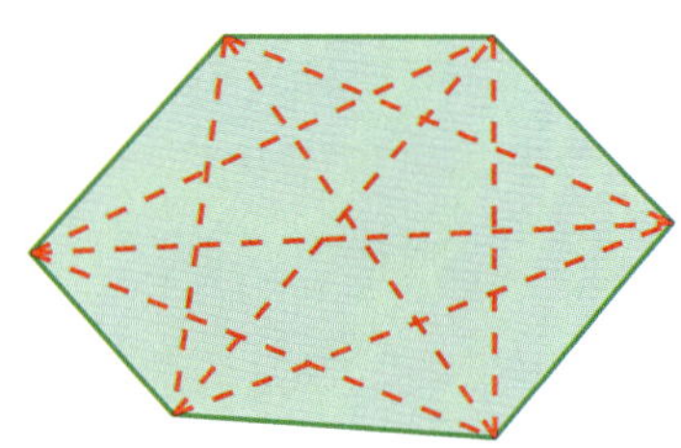

On Your Own

Now You're Ready
Exercises 22–24

Tell whether the polygon is *convex* or *concave*. Explain.

9. 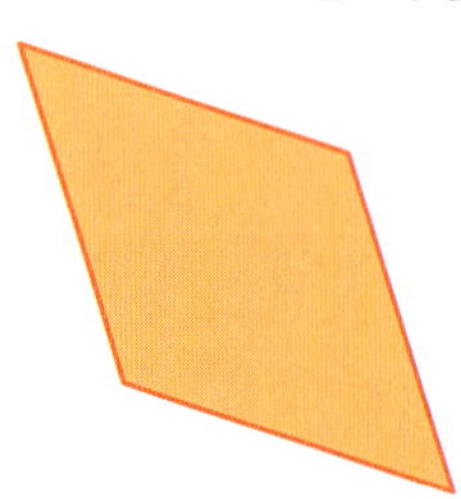

10. 

11. 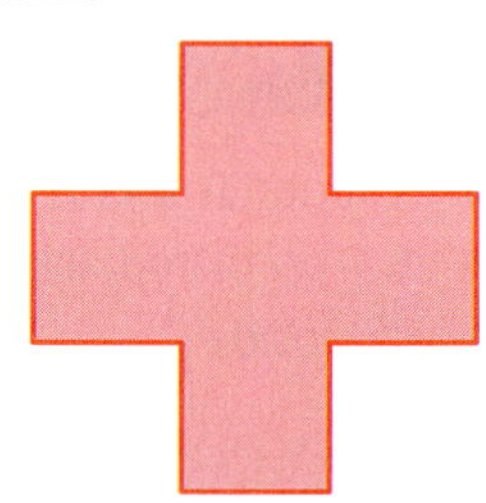

5.3 Exercises

Vocabulary and Concept Check

1. **VOCABULARY** Draw a regular polygon that has three sides.

2. **WHICH ONE DOESN'T BELONG?** Which figure does *not* belong with the other three? Explain your reasoning.

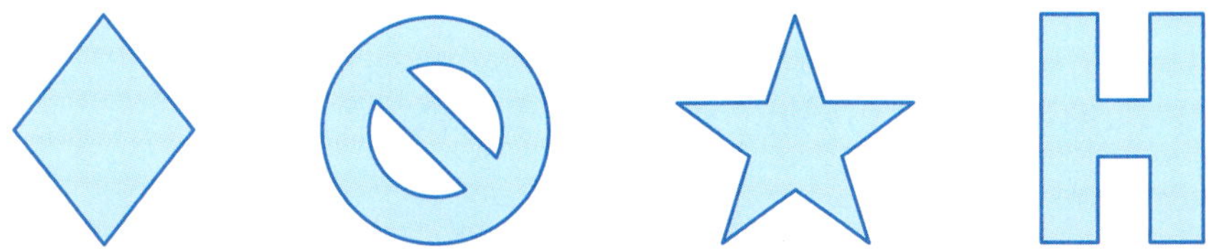

3. **DIFFERENT WORDS, SAME QUESTION** Which is different? Find "both" answers.

What is the measure of an angle of a regular pentagon?	What is the sum of the angle measures of a convex pentagon?
What is the sum of the angle measures of a regular pentagon?	What is the sum of the angle measures of a concave pentagon?

Practice and Problem Solving

Use triangles to find the sum of the angle measures of the polygon.

4. 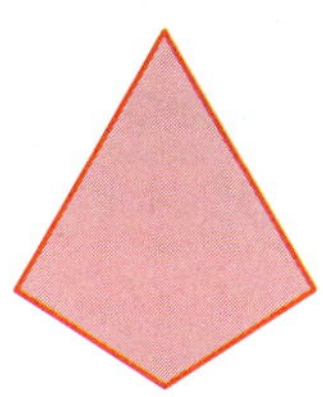

5. 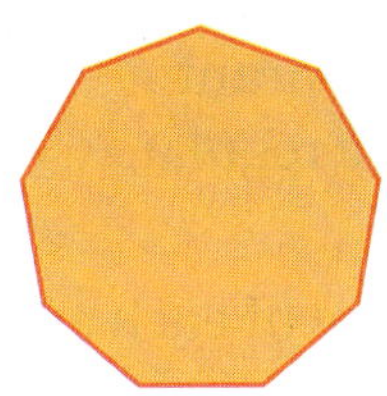

6. 

Find the sum of the angle measures of the polygon.

1 7. 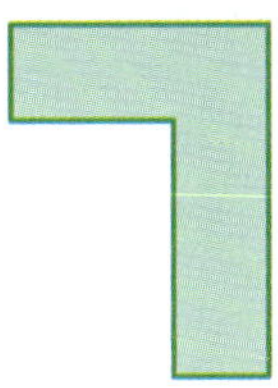

8. 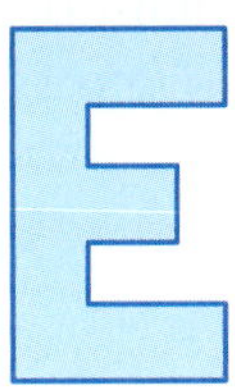

9. 

10. **ERROR ANALYSIS** Describe and correct the error in finding the sum of the angle measures of a 13-gon.

✗
$$S = n \cdot 180^\circ$$
$$= 13 \cdot 180^\circ$$
$$= 2340^\circ$$

11. **NUMBER SENSE** Can a pentagon have angles that measure 120°, 105°, 65°, 150°, and 95°? Explain.

Find the value of x.

2 **12.**

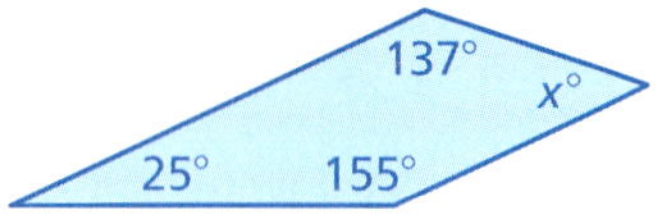

13.

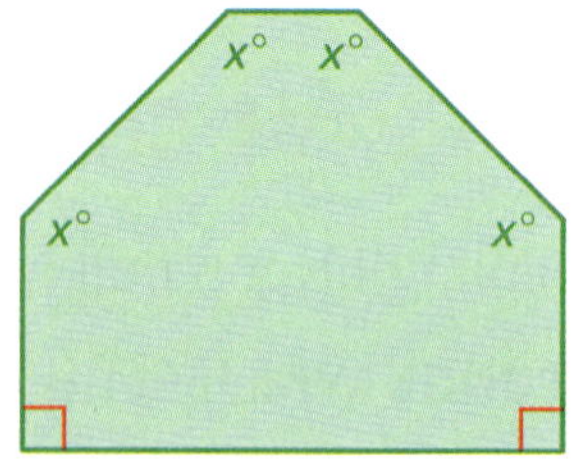

14.

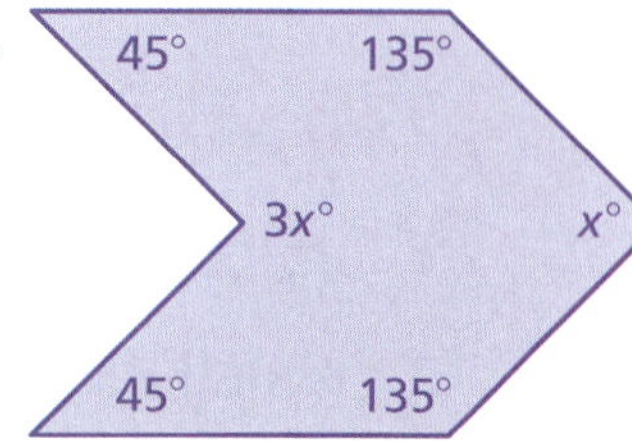

15. REASONING The sum of the angle measures in a regular polygon is 1260°. What is the measure of one of the angles of the polygon?

Find the measure of each angle of the regular polygon.

3 **16.**

17.

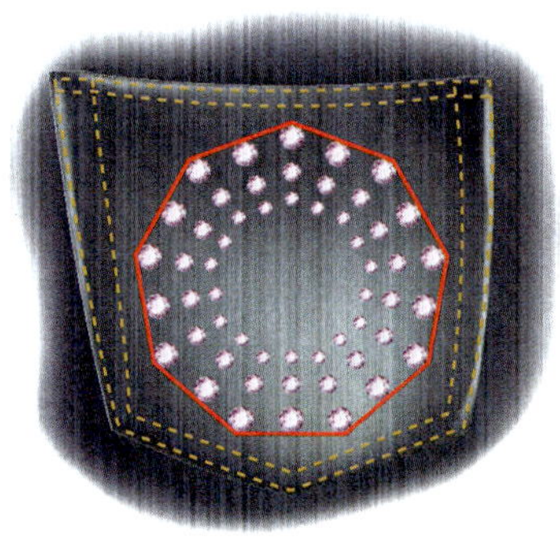

18.

19. ERROR ANALYSIS Describe and correct the error in finding the measure of each angle of a regular 20-gon.

$S = (n - 2) \cdot 180°$

$= (20 - 2) \cdot 180°$

$= 18 \cdot 180°$

$= 3240°$

$3240° \div 18 = 180$

The measure of each angle is 180°.

20. FIRE HYDRANT A fire hydrant bolt is in the shape of a regular pentagon.

a. What is the measure of each angle?

b. Why are fire hydrants made this way?

21. PUZZLE The angles of a regular polygon each measure 165°. How many sides does the polygon have?

Tell whether the polygon is *convex* or *concave*. Explain.

4 **22.**

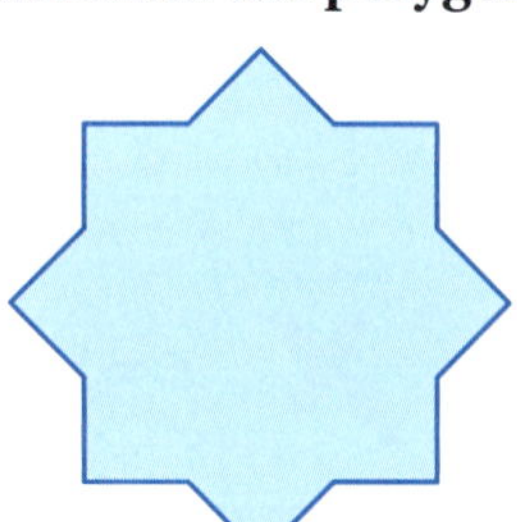

23.

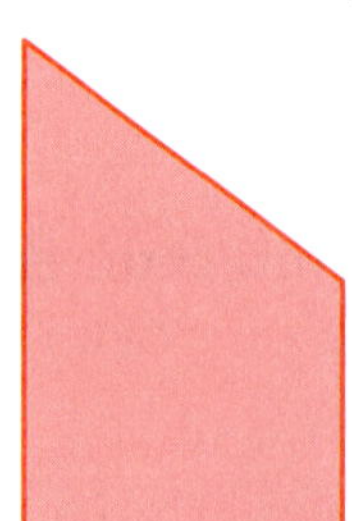

24.

25. CRITICAL THINKING Can a concave polygon be regular? Explain.

26. OPEN-ENDED Draw a polygon that has congruent sides but is not regular.

27. **STAINED GLASS** The center of the stained glass window is in the shape of a regular polygon. What is the measure of each angle of the polygon?

28. **PENTAGON** Draw a pentagon that has two right angles, two 45° angles, and one 270° angle.

29. **GAZEBO** The floor of a gazebo is in the shape of a heptagon. Four of the angles measure 135°. The other angles have equal measures. Find the measure of each of the remaining angles.

30. **MONEY** The border of a Susan B. Anthony dollar is in the shape of a regular polygon.

  a. How many sides does the polygon have?

  b. What is the measure of each angle of the border? Round your answer to the nearest degree.

31. **REASONING** Copy and complete the table. Does the table represent a linear function? Explain.

Sides of a Regular Polygon, n	3	4	5	6	7	8	9	10
Measure of One Angle, a								

32. **Geometry** When tiles can be used to cover a floor with no empty spaces, the collection of tiles is called a *tessellation*.

  a. Create a tessellation using equilateral triangles.

  b. Find two more regular polygons that form tessellations.

  c. Create a tessellation that uses two different regular polygons.

Fair Game Review What you learned in previous grades & lessons

Solve the proportion. *(Skills Review Handbook)*

33. $\frac{x}{12} = \frac{3}{4}$

34. $\frac{14}{21} = \frac{x}{3}$

35. $\frac{x}{9} = \frac{2}{6}$

36. $\frac{4}{10} = \frac{x}{15}$

37. **MULTIPLE CHOICE** The ratio of tulips to daisies is 3 : 5. Which of the following could be the total number of tulips and daisies? *(Skills Review Handbook)*

Ⓐ 6 Ⓑ 10 Ⓒ 15 Ⓓ 16

5 Study Help

Check It Out
Graphic Organizer
BigIdeasMath.com

You can use an **example and non-example chart** to list examples and non-examples of a vocabulary word or item. Here is an example and non-example chart for complementary angles.

Complementary Angles

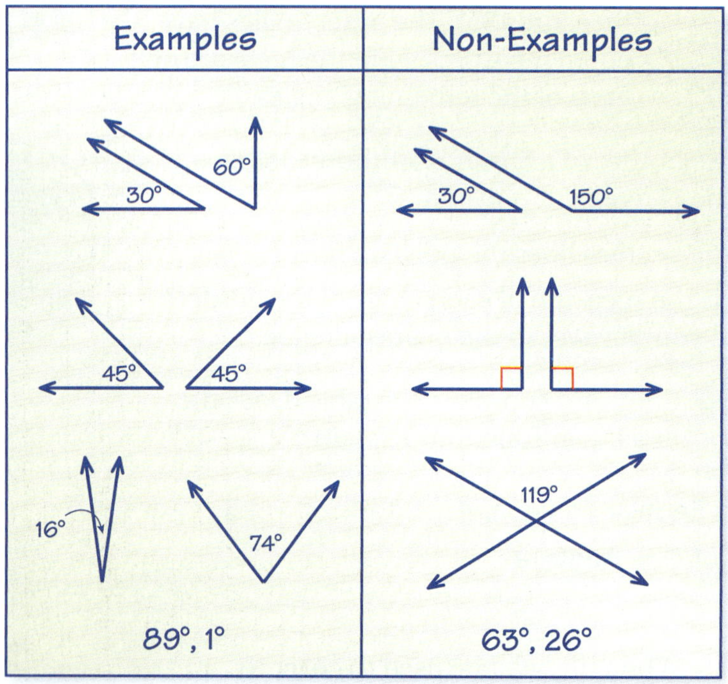

On Your Own

Make an example and non-example chart to help you study these topics.

1. isosceles triangles
2. equilateral triangles
3. regular polygons
4. convex polygons
5. concave polygons

After you complete this chapter, make example and non-example charts for the following topics.

6. similar triangles
7. transversals
8. interior angles

"What do you think of my **example & non-example chart** for popular cat toys?"

5.1–5.3 Quiz

Check It Out
Progress Check
BigIdeasMath.com

Tell whether the angles are *complementary*, *supplementary*, or *neither*. *(Section 5.1)*

1.

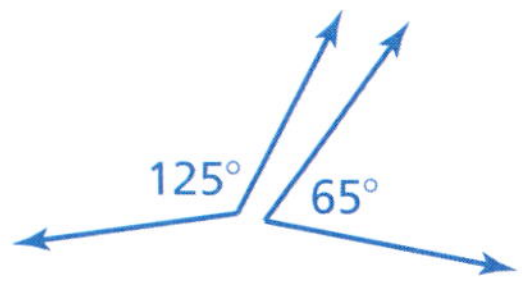

2.

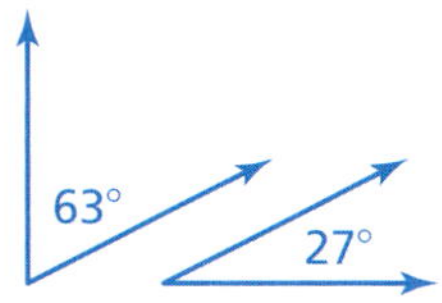

3.

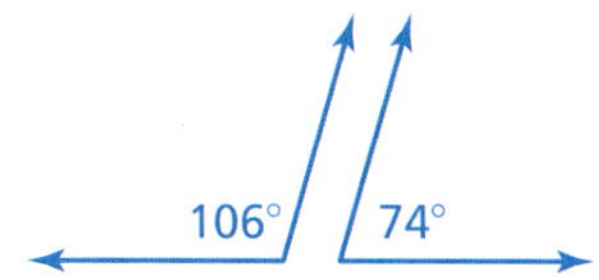

Find the value of *x*. *(Section 5.1)*

4.

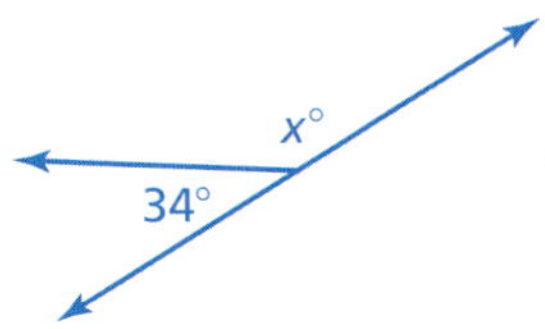

5.

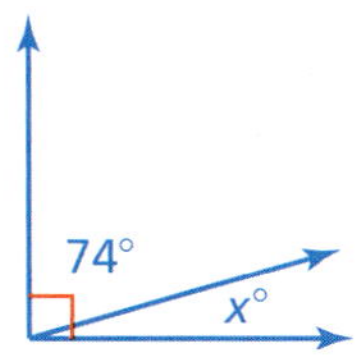

6.

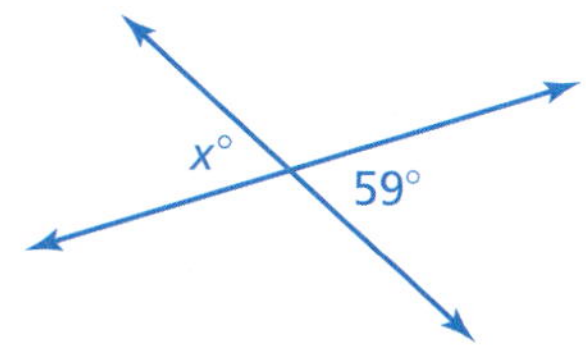

Find the value of *x*. Then classify the triangle in as many ways as possible. *(Section 5.2)*

7.

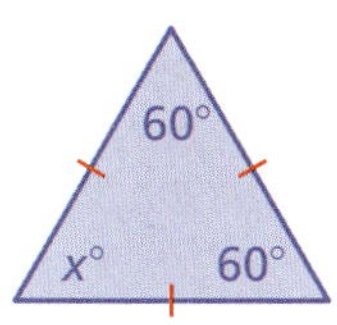

8.

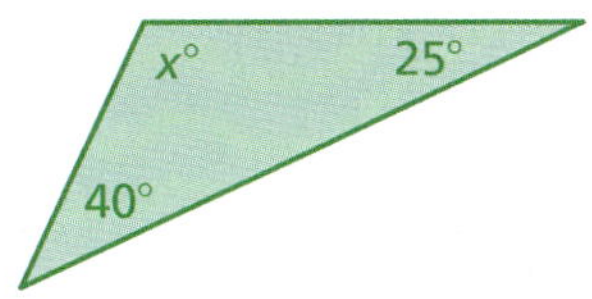

9.

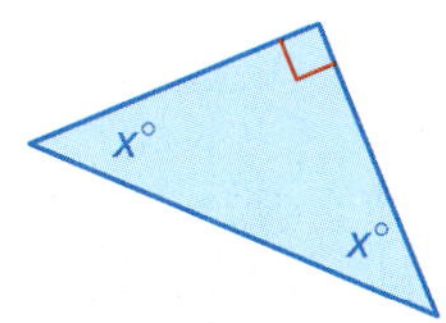

10. Find the sum of the angle measures of the polygon. *(Section 5.3)*

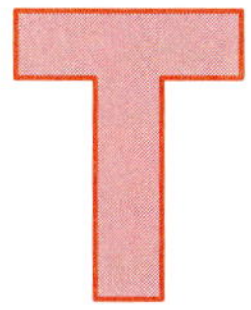

11. Tell whether the polygon is concave or convex. *(Section 5.3)*

Find the value of *x*. *(Section 5.3)*

12.

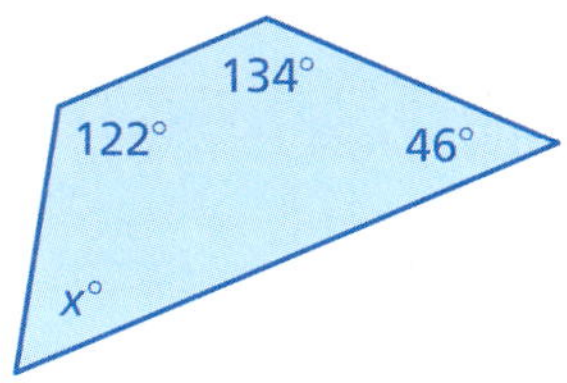

13.

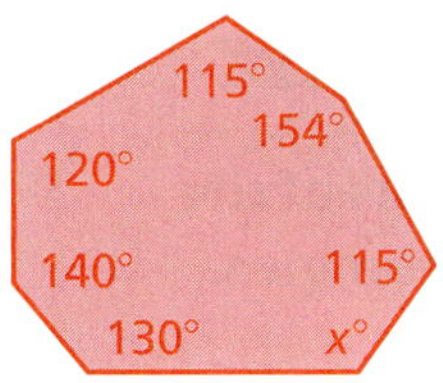

14.

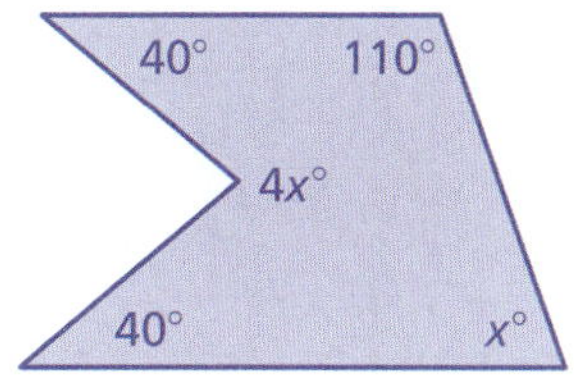

15. RAILROAD CROSSING What are the measures of the other three angles formed by the intersection of the road and the railroad tracks? *(Section 5.1)*

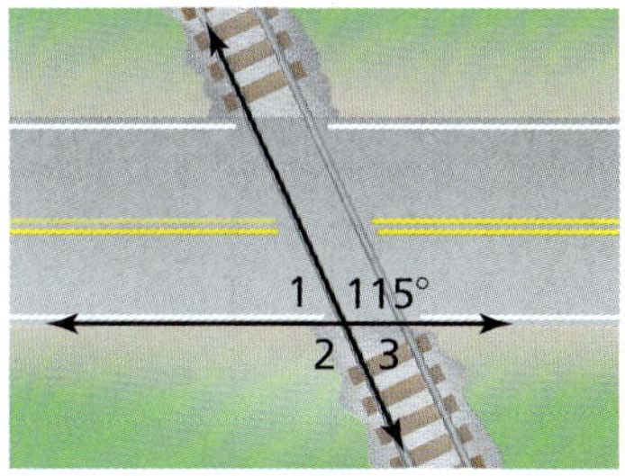

16. REASONING The sum of the angle measures of a polygon is 4140°. How many sides does the polygon have? *(Section 5.3)*

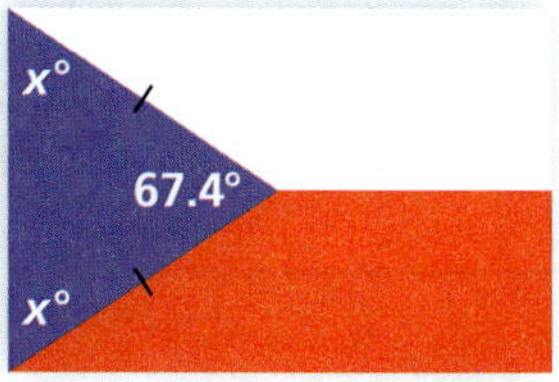

17. FLAG Classify the triangle on the flag of the Czech Republic in as many ways as possible. *(Section 5.2)*

5.4 Using Similar Triangles

STATE STANDARDS
MA.8.G.2.1

Essential Question Which properties of triangles make them special among all other types of polygons?

You already know that two triangles are **similar** if and only if the ratios of their corresponding side lengths are equal.

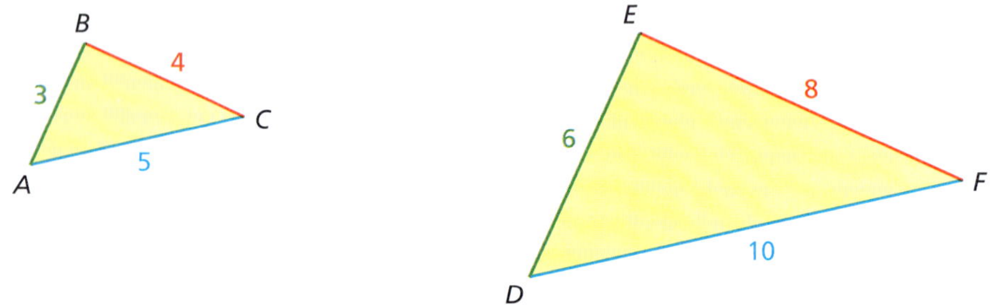

For example, $\triangle ABC$ is similar to $\triangle DEF$ because the ratios of their corresponding side lengths are equal.

$$\frac{6}{3} = \frac{10}{5} = \frac{8}{4}$$

1 ACTIVITY: Angles of Similar Triangles

Work with a partner.

- **Discuss how to make a triangle that is larger than $\triangle XYZ$ and has the *same* angle measures as $\triangle XYZ$.**
- **Measure the lengths of the sides of the two triangles.**
- **Find the ratios of the corresponding side lengths. Are they all the same? What can you conclude?**

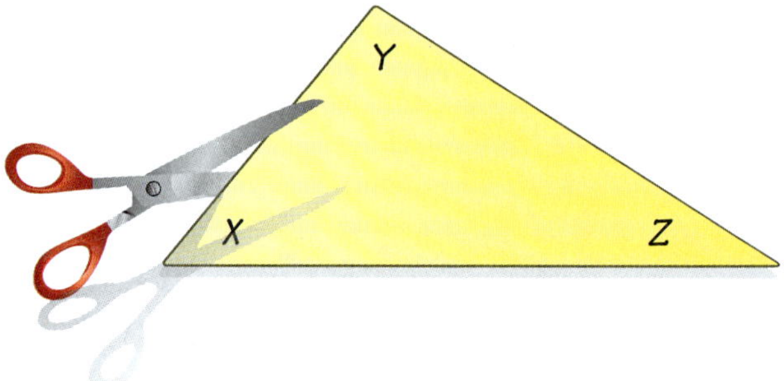

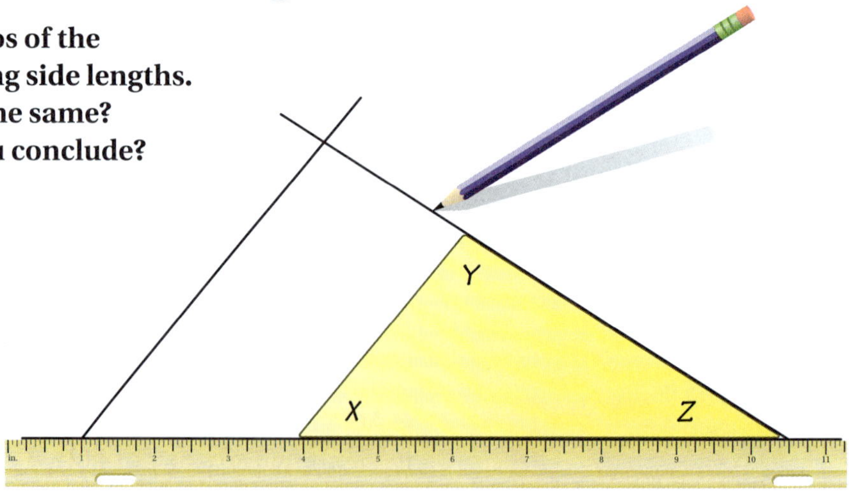

2 ACTIVITY: Amazing Triangles

Work with a partner. Use what you know about polygons to decide whether each statement is true. In each case, explain your reasoning.

a. If two triangles are similar, then the ratios of their corresponding side lengths are equal.

If two quadrilaterals are similar, then the ratios of their corresponding side lengths are equal.

b. If the ratios of the corresponding sides of two triangles are equal, then the triangles are similar.

If the ratios of the corresponding sides of two quadrilaterals are equal, then the quadrilaterals are similar.

c. If two triangles are similar, then their corresponding angles are congruent.

If two quadrilaterals are similar, then their corresponding angles are congruent.

d. If the corresponding angles in two triangles are congruent, then the triangles are similar.

If the corresponding angles in two quadrilaterals are congruent, then the quadrilaterals are similar.

e. If the corresponding sides of two triangles are congruent, then the two triangles have identical shapes.

If the corresponding sides of two quadrilaterals are congruent, then the two quadrilaterals have identical shapes.

What Is Your Answer?

3. **IN YOUR OWN WORDS** Which properties of triangles make them special among all other types of polygons? Describe two careers in which the special properties of triangles are used.

Practice Use what you learned about similar triangles to complete Exercises 3 and 4 on page 210.

5.4 Lesson

Check It Out
Lesson Tutorials
BigIdeasMath.com

Key Vocabulary
similar triangles, *p. 208*
indirect measurement, *p. 209*

Triangles that have the same shape but not necessarily the same size are **similar triangles**.

Key Idea

Angles of Similar Triangles

Words Two triangles have the same angle measures if and only if they are similar.

Example

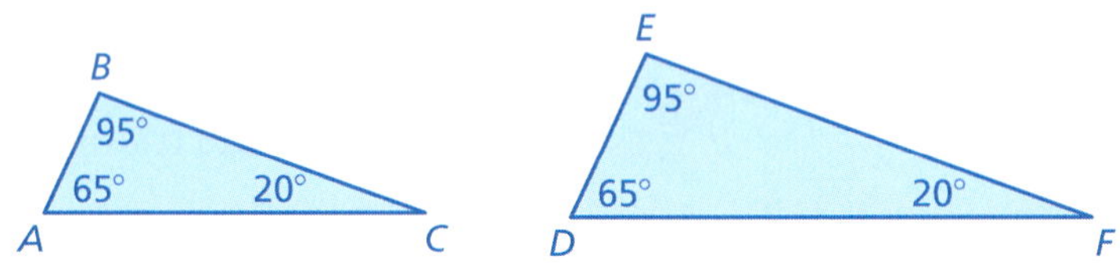

Triangle ABC is similar to triangle DEF: $\triangle ABC \sim \triangle DEF$.

Study Tip

If two angles in one triangle are congruent to two angles in another triangle, then the third angles are also congruent.

EXAMPLE 1 Identifying Similar Triangles

Tell whether the triangles are similar. Explain.

a.

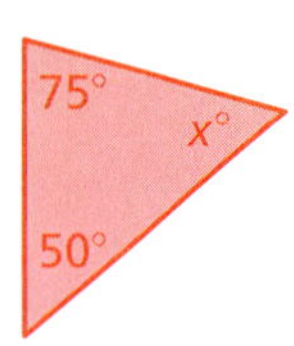

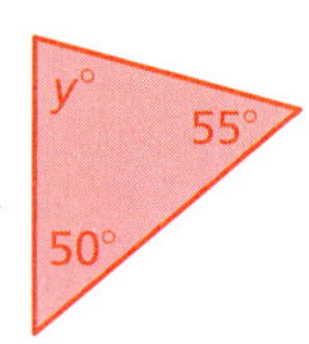

$$75 + 50 + x = 180$$
$$125 + x = 180$$
$$x = 55$$

$$y + 50 + 55 = 180$$
$$y + 105 = 180$$
$$y = 75$$

The triangles have the same angle measures, 75°, 50°, and 55°. So, they are similar.

b.

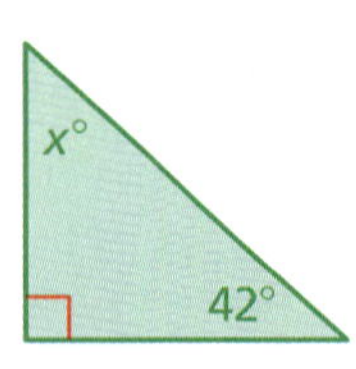

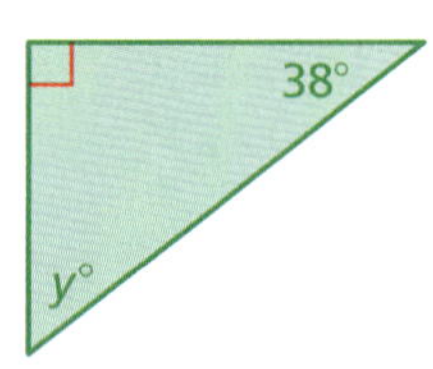

$$x + 90 + 42 = 180$$
$$x + 132 = 180$$
$$x = 48$$

$$90 + 38 + y = 180$$
$$128 + y = 180$$
$$y = 52$$

The triangles do not have the same angle measures. So, they are not similar.

Multi-Language Glossary at BigIdeasMath.com.

On Your Own

Now You're Ready
Exercises 5–8

Tell whether the triangles are similar. Explain.

1. 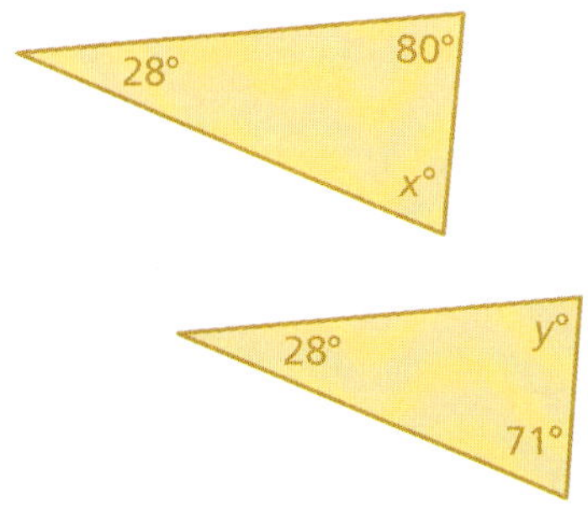

2. 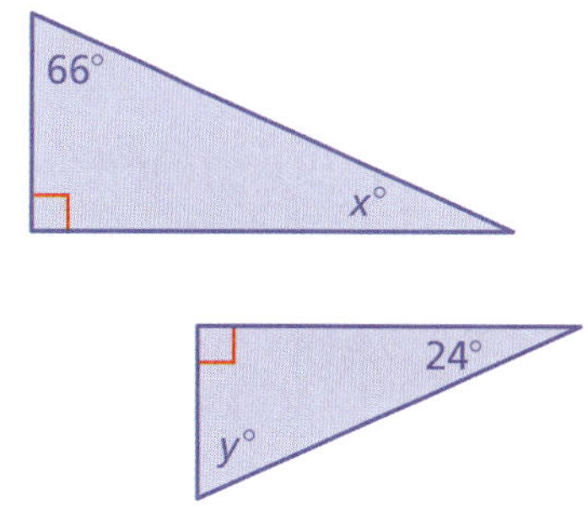

Indirect measurement uses similar figures to find a missing measure when it is difficult to find directly.

EXAMPLE 2 Using Indirect Measurement

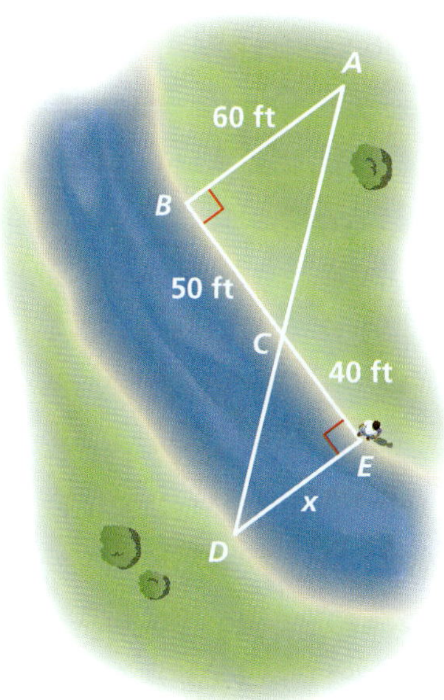

You plan to cross a river and want to know how far it is to the other side. You take measurements on your side of the river and make the drawing shown. (a) Explain why $\triangle ABC$ and $\triangle DEC$ are similar. (b) What is the distance x across the river?

a. $\angle B$ and $\angle E$ are right angles, so they are congruent. $\angle ACB$ and $\angle DCE$ are vertical angles, so they are congruent.

Because two angles in $\triangle ABC$ are congruent to two angles in $\triangle DEC$, the third angles are also congruent. The triangles have the same angle measures, so they are similar.

b. The ratios of the corresponding side lengths in similar triangles are equal. Write and solve a proportion to find x.

$$\frac{x}{60} = \frac{40}{50} \qquad \text{Write a proportion.}$$

$$60 \cdot \frac{x}{60} = 60 \cdot \frac{40}{50} \qquad \text{Multiply each side by 60.}$$

$$x = 48 \qquad \text{Simplify.}$$

The distance across the river is 48 feet.

On Your Own

Now You're Ready
Exercises 10 and 11

3. **WHAT IF?** In Example 2, the distance from vertex A to vertex B is 55 feet. What is the distance across the river?

5.4 Exercises

Check It Out
Help with Homework
BigIdeasMath.com

Vocabulary and Concept Check

1. **REASONING** How can you use similar triangles to find a missing measurement?

2. **WHICH ONE DOESN'T BELONG?** Which triangle does *not* belong with the other three? Explain your reasoning.

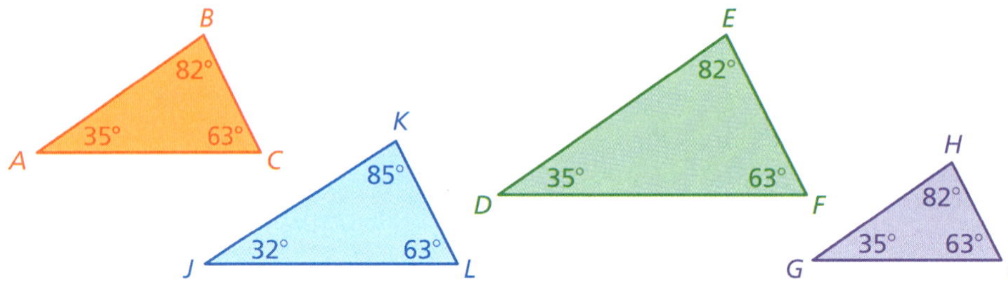

Practice and Problem Solving

Make a triangle that is larger than the one given and has the same angle measures. Find the ratios of the corresponding side lengths.

3. 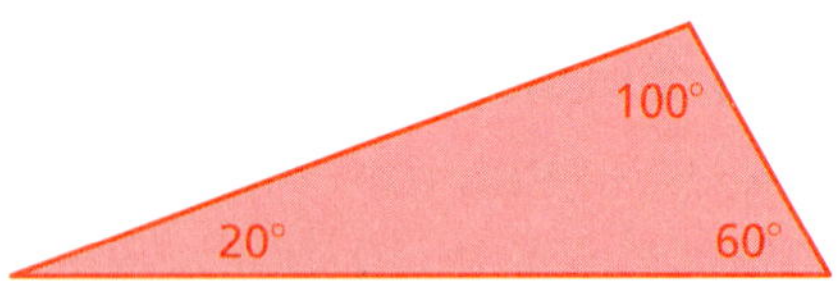

4. 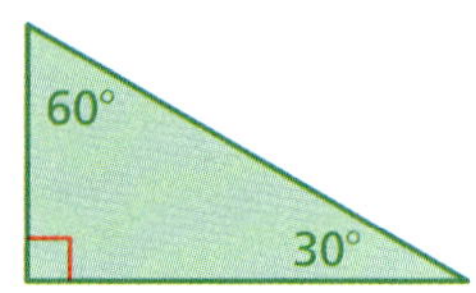

Tell whether the triangles are similar. Explain.

1 5. 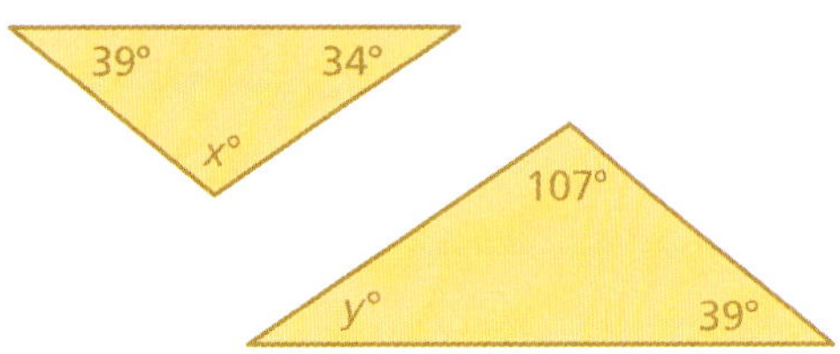

6. 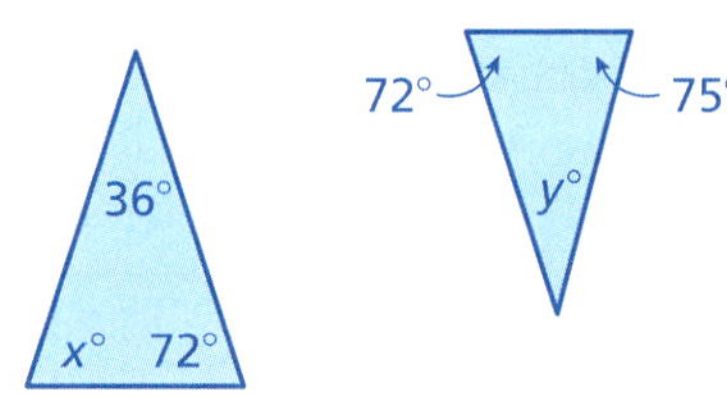

7. 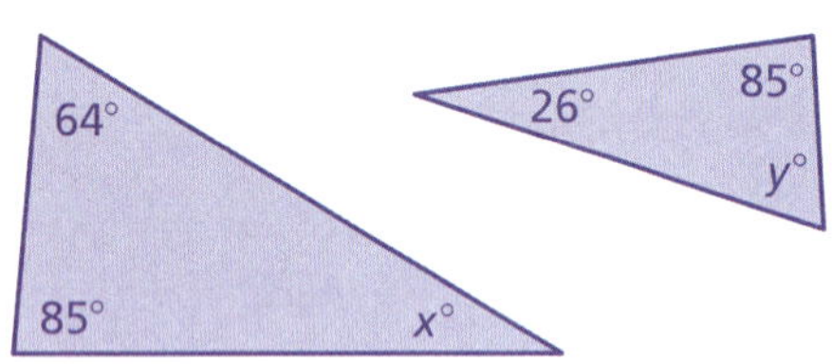

8. 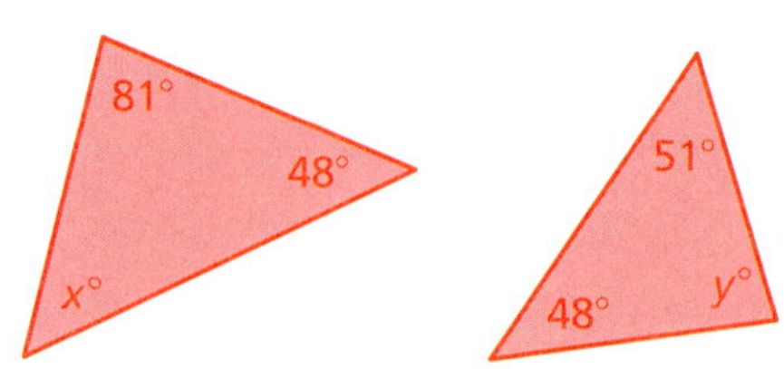

9. **ERROR ANALYSIS** Describe and correct the error in using indirect measurement.

$$\frac{16}{18} = \frac{x}{8}$$

$$18x = 128$$

$$x \approx 7$$

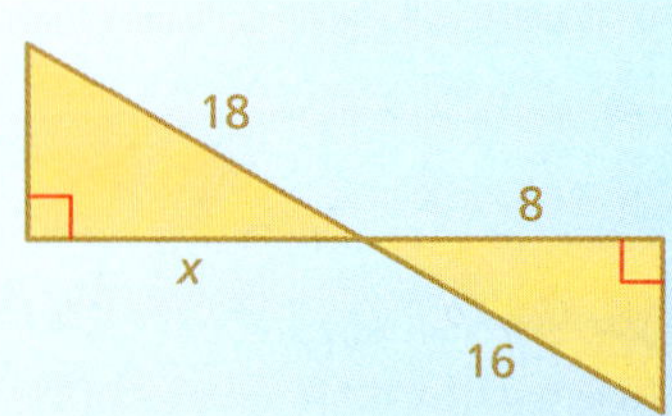

The triangles are similar. Find the value of x.

10.

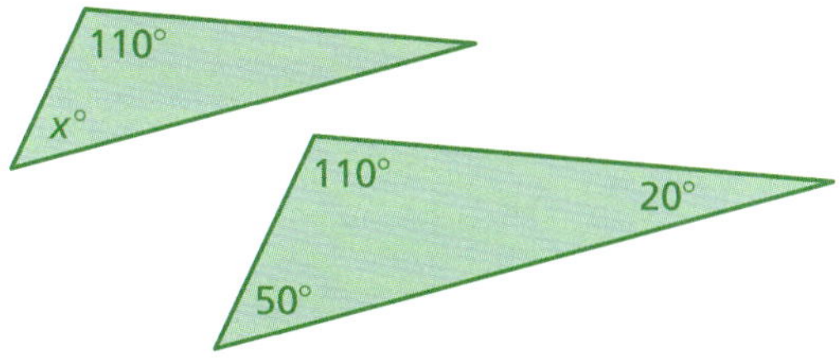

11.

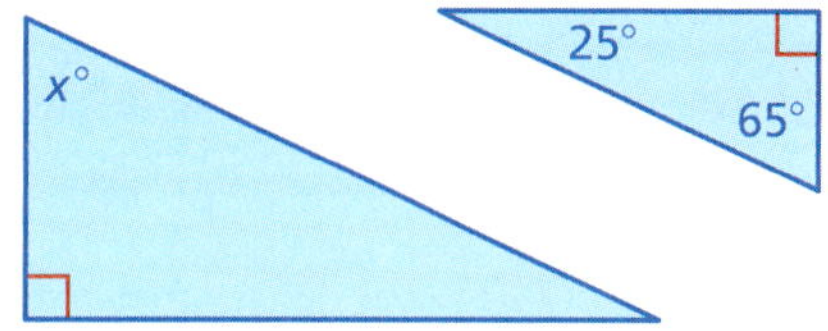

② **12.** **TREASURE** The map shows the number of steps you must take to get to the treasure. However, the map is old and the last dimension is unreadable. How many steps do you take from the pyramids to the treasure?

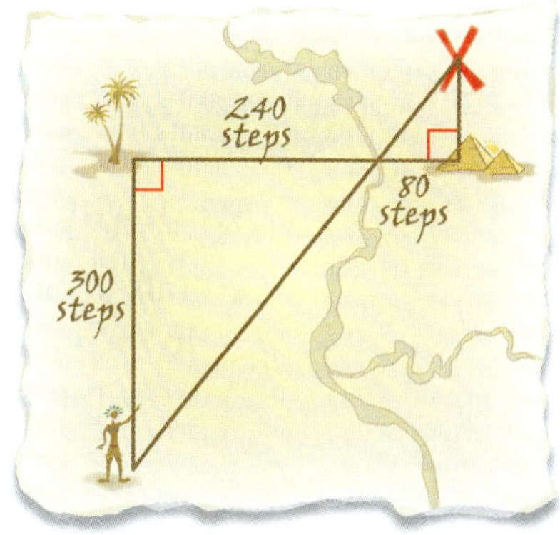

13. **CRITICAL THINKING** The side lengths of a triangle are increased by 50% to make a similar triangle. Does the area increase by 50% as well? Explain.

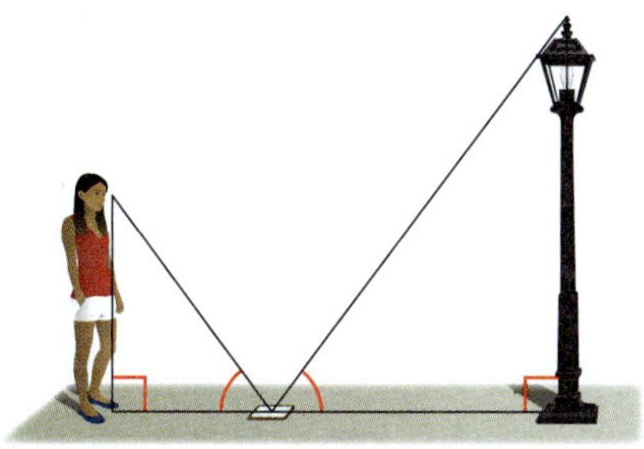

14. **PROJECT** Using a mirror, a tape measure, and indirect measurement, you can find the height of a lamppost. Place the mirror flat on the ground 6 feet from the lamppost. Move away from the mirror and the lamppost until you can see the top of the lamppost in the mirror. Measure the distance between yourself and the mirror. Then use similar triangles to find the height of the lamppost.

15. Geometry The Florida Shuffleboard Association uses a standard shuffleboard court. $\triangle DAE \sim \triangle BAG \sim \triangle CAF$. The lengths of segments AG, GF, and FE are equal.

a. Find x. **b.** Find CF.

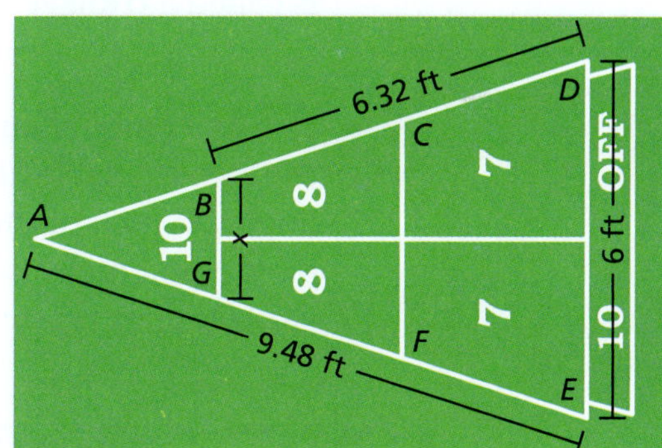

Fair Game Review What you learned in previous grades & lessons

Does the equation represent a *linear* or *nonlinear* function? Explain. *(Section 4.4)*

16. $y = \dfrac{5}{x}$ **17.** $y = -5.4x + \pi$ **18.** $y = 2x - 8$ **19.** $y = 6x^2 + x - 1$

20. **MULTIPLE CHOICE** Which two lines are parallel? *(Section 2.2)*

Ⓐ blue and red Ⓑ red and green

Ⓒ green and blue Ⓓ all three are parallel

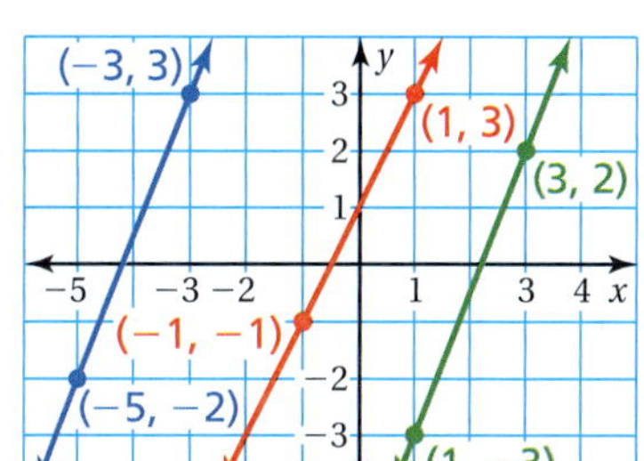

5.5 Parallel Lines and Transversals

STATE STANDARDS
MA.8.G.2.2

Essential Question How can you use properties of parallel lines to solve real-life problems?

Share Your Work at... My.BigIdeasMath.com

1 ACTIVITY: A Property of Parallel Lines

Work with a partner.

- Talk about what it means for two lines to be parallel. Decide on a strategy for drawing two parallel lines.

- Use your strategy to carefully draw two lines that are parallel.

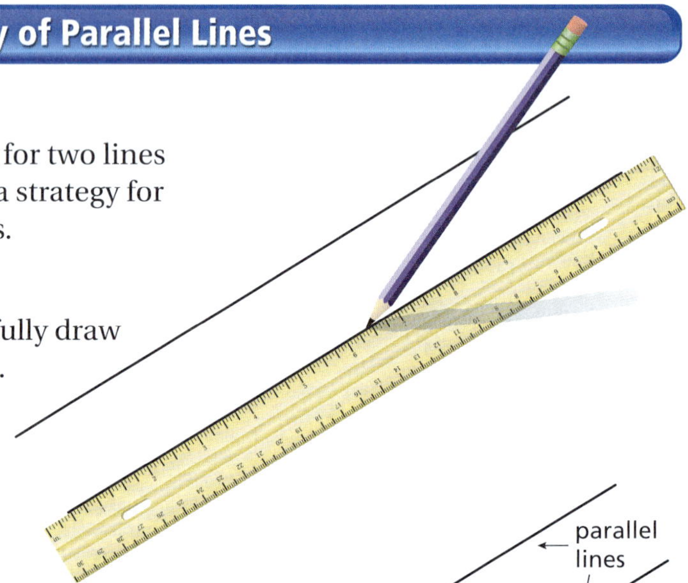

- Now, draw a third line that intersects the two parallel lines. This line is called a **transversal**.

- The two parallel lines and the transversal form eight angles. Which of these angles have equal measures? Explain your reasoning.

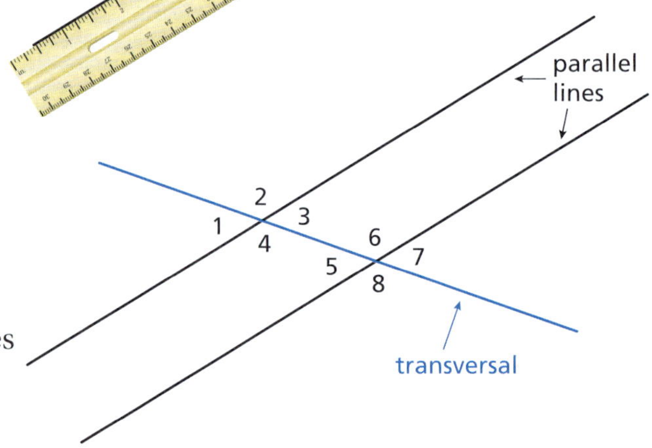

2 ACTIVITY: Creating Parallel Lines

Work with a partner.

a. If you were building the house in the photograph, how could you make sure that the studs are parallel to each other?

b. Identify sets of parallel lines and transversals in the photograph.

3 ACTIVITY: Indirect Measurement

Work with a partner.

a. Use the fact that two rays from the Sun are parallel to explain why $\triangle ABC$ and $\triangle DEF$ are similar.

b. Explain how to use similar triangles to find the height of the flagpole.

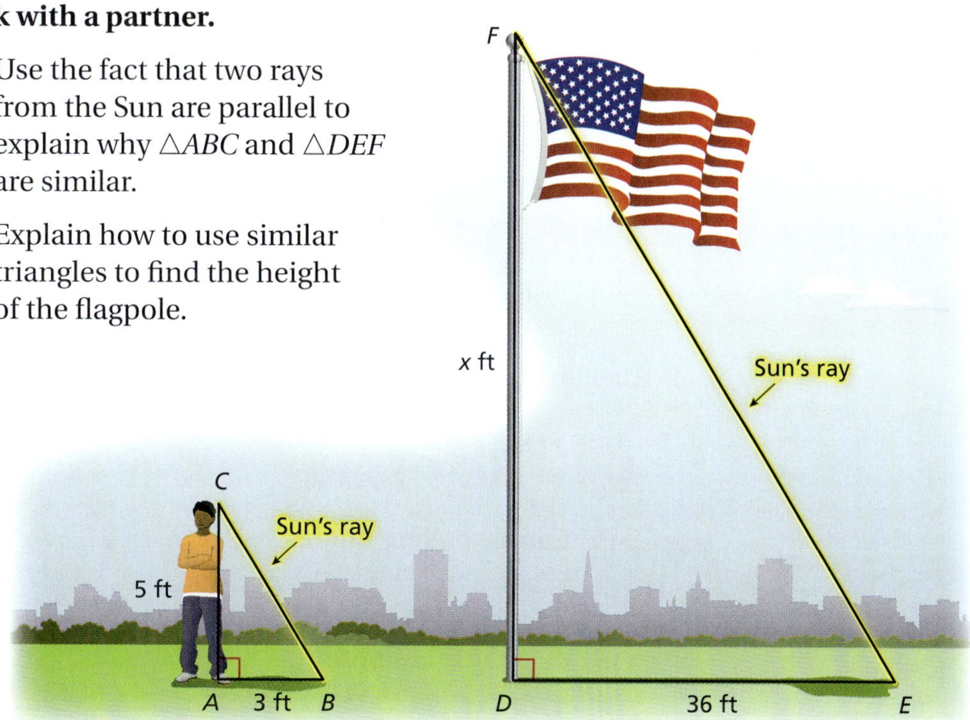

What Is Your Answer?

4. **IN YOUR OWN WORDS** How can you use properties of parallel lines to solve real-life problems? Describe some examples.

5. **INDIRECT MEASUREMENT PROJECT** Work with a partner or in a small group.

   a. Explain why the process in Activity 3 is called "indirect" measurement.

   b. Use indirect measurement to measure the height of something outside your school (a tree, a building, a flagpole). Before going outside, decide what you need to take with you to do the measurement.

   c. Draw a diagram of the indirect measurement process you used. In the diagram, label the lengths that you actually measured and also the lengths that you calculated.

Practice Use what you learned about parallel lines and transversals to complete Exercises 3–6 on page 217.

5.5 Lesson

Check It Out
Lesson Tutorials
BigIdeasMath.com

Key Vocabulary
perpendicular lines, *p. 214*
transversal, *p. 214*
interior angles, *p. 215*
exterior angles, *p. 215*

Lines in the same plane that do not intersect are called parallel lines. Lines that intersect at right angles are called **perpendicular lines**.

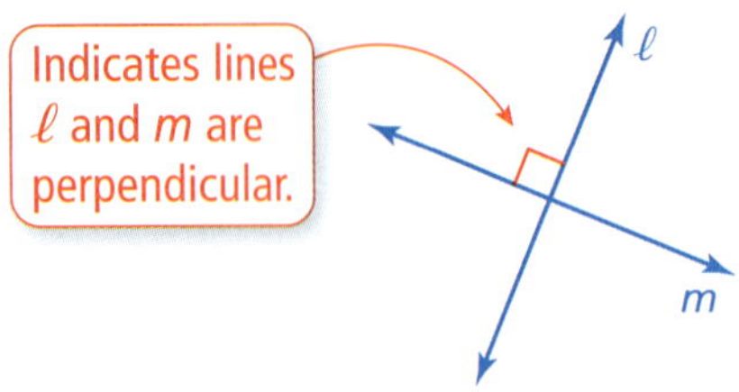

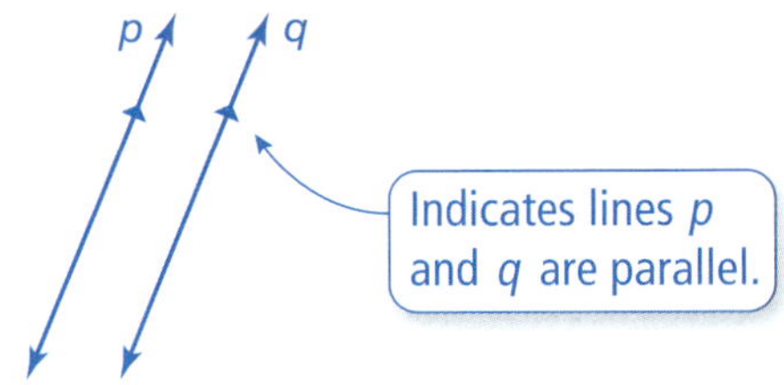

A line that intersects two or more lines is called a **transversal**. When parallel lines are cut by a transversal, several pairs of congruent angles are formed.

Key Idea

Corresponding Angles

When a transversal intersects parallel lines, corresponding angles are congruent.

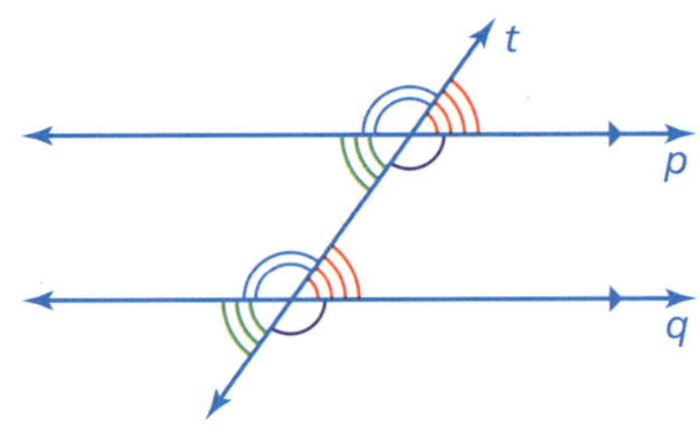

Corresponding angles

Study Tip

Corresponding angles lie on the same side of the transversal in corresponding positions.

EXAMPLE 1 Finding Angle Measures

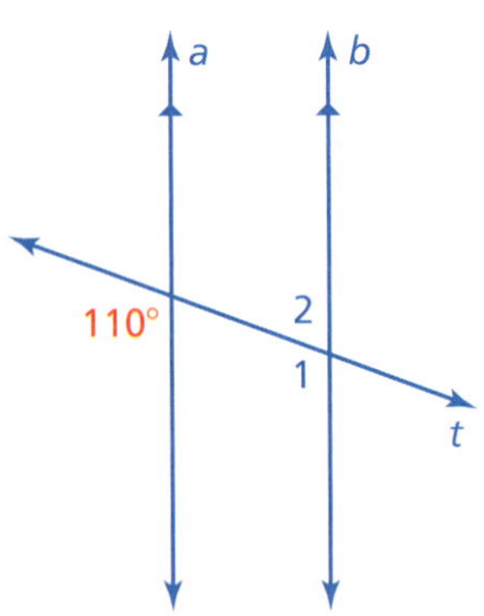

Use the figure to find the measures of (a) $\angle 1$ and (b) $\angle 2$.

a. $\angle 1$ and the 110° angle are corresponding angles. They are congruent.

So, the measure of $\angle 1$ is 110°.

b. $\angle 1$ and $\angle 2$ are supplementary.

$\angle 1 + \angle 2 = 180^\circ$	Definition of supplementary angles
$110^\circ + \angle 2 = 180^\circ$	Substitute 110° for $\angle 1$.
$\angle 2 = 70^\circ$	Subtract 110° from each side.

So, the measure of $\angle 2$ is 70°.

On Your Own

Now You're Ready
Exercises 7–9

Use the figure to find the measure of the angle. Explain your reasoning.

1. $\angle 1$ **2.** $\angle 2$

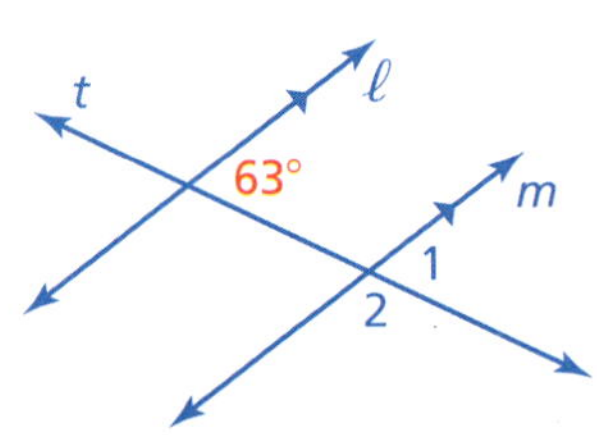

Multi-Language Glossary at BigIdeasMath.com.

EXAMPLE 2 Using Corresponding Angles

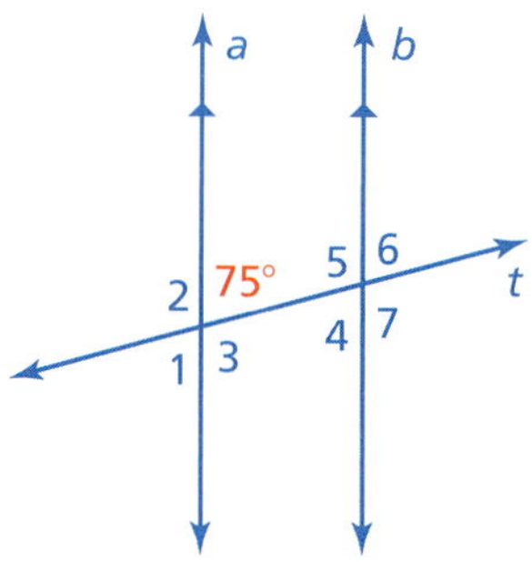

Use the figure to find the measures of the numbered angles.

∠1: ∠1 and the 75° angle are vertical angles. They are congruent.

So, the measure of ∠1 is 75°.

∠2 and ∠3: The 75° angle is supplementary to both ∠2 and ∠3.

$75^\circ + \angle 2 = 180^\circ$ Definition of supplementary angles

$\angle 2 = 105^\circ$ Subtract 75° from each side.

So, the measures of ∠2 and ∠3 are 105°.

∠4, ∠5, ∠6, and ∠7: Using corresponding angles, the measures of ∠4 and ∠6 are 75°, and the measures of ∠5 and ∠7 are 105°.

On Your Own

Now You're Ready
Exercises 15–17

3. Use the figure to find the measures of the numbered angles.

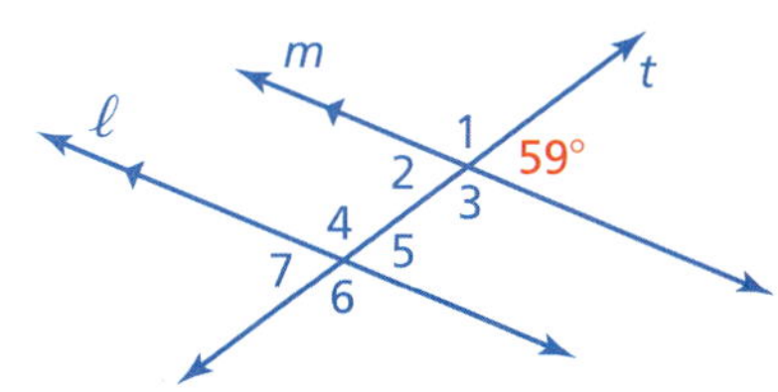

When two parallel lines are cut by a transversal, four **interior angles** are formed on the inside of the parallel lines and four **exterior angles** are formed on the outside of the parallel lines.

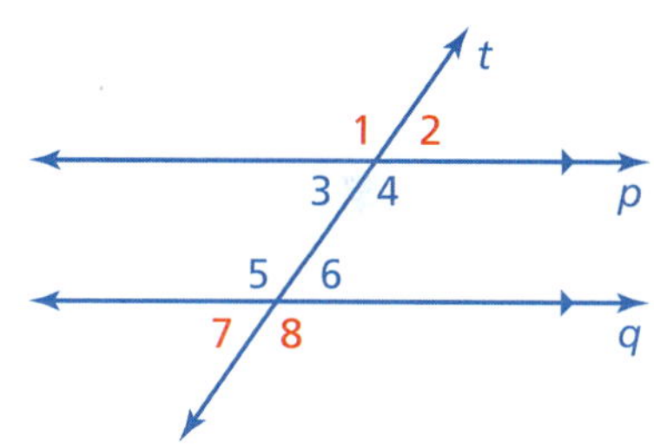

∠3, ∠4, ∠5, and ∠6 are interior angles.
∠1, ∠2, ∠7, and ∠8 are exterior angles.

EXAMPLE 3 Standardized Test Practice

A store owner uses pieces of tape to paint a window advertisement. The letters are slanted at an 80° angle. What is the measure of ∠1?

Ⓐ 80° Ⓑ 100° Ⓒ 110° Ⓓ 120°

Because all of the letters are slanted at an 80° angle, the dashed lines are parallel. The piece of tape is the transversal.

Using the corresponding angles, the 80° angle is congruent to the angle that is supplementary to ∠1, as shown.

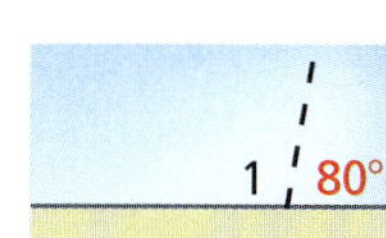

The measure of ∠1 is 180° − 80° = 100°. The correct answer is Ⓑ.

On Your Own

Now You're Ready
Exercises 18 and 19

4. **WHAT IF?** In Example 3, the letters are slanted at a 65° angle. What is the measure of ∠1?

Key Idea

Study Tip

Alternate interior angles and alternate exterior angles lie on opposite sides of the transversal.

Alternate Interior Angles and Alternate Exterior Angles

When a transversal intersects parallel lines, alternate interior angles are congruent and alternate exterior angles are congruent.

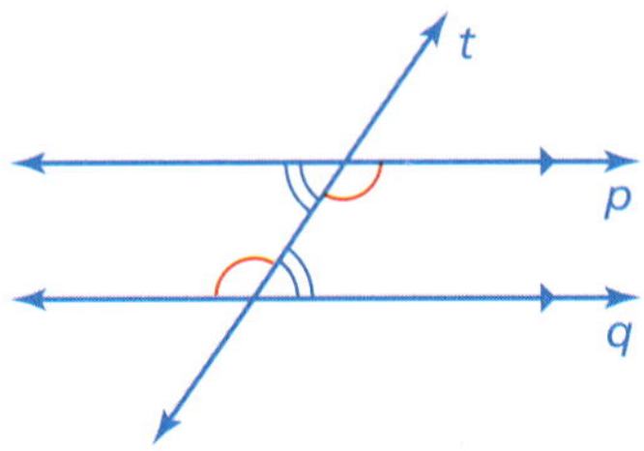

Alternate interior angles

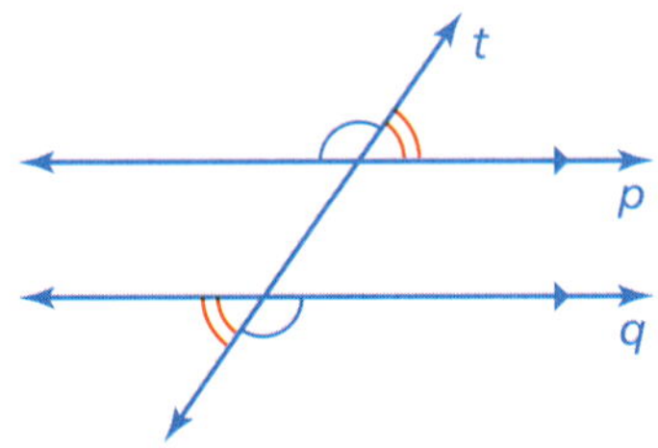

Alternate exterior angles

EXAMPLE 4 Identifying Alternate Interior and Alternate Exterior Angles

The photo shows a portion of the St. Petersburg-Clearwater International Airport. Describe the relationship between each pair of angles.

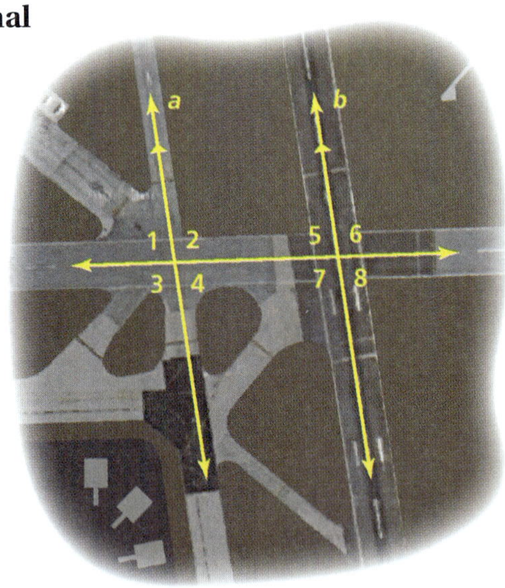

a. **∠3 and ∠6**

∠3 and ∠6 are alternate exterior angles.

So, ∠3 is congruent to ∠6.

b. **∠2 and ∠7**

∠2 and ∠7 are alternate interior angles.

So, ∠2 is congruent to ∠7.

On Your Own

Now You're Ready
Exercises 20 and 21

In Example 4, the measure of ∠4 is 84°. Find the measure of the angle. Explain your reasoning.

5. ∠3
6. ∠5
7. ∠6

5.5 Exercises

Vocabulary and Concept Check

1. **VOCABULARY** Draw two parallel lines and a transversal. Label a pair of corresponding angles.

2. **WHICH ONE DOES NOT BELONG?** Which statement does *not* belong with the other three? Explain your reasoning. Refer to the figure for Exercises 3–6.

The measure of $\angle 2$	The measure of $\angle 5$
The measure of $\angle 6$	The measure of $\angle 8$

Practice and Problem Solving

In Exercises 3–6, use the figure.

3. Identify the parallel lines.
4. Identify the transversal.
5. How many angles are formed by the transversal?
6. Which of the angles are congruent?

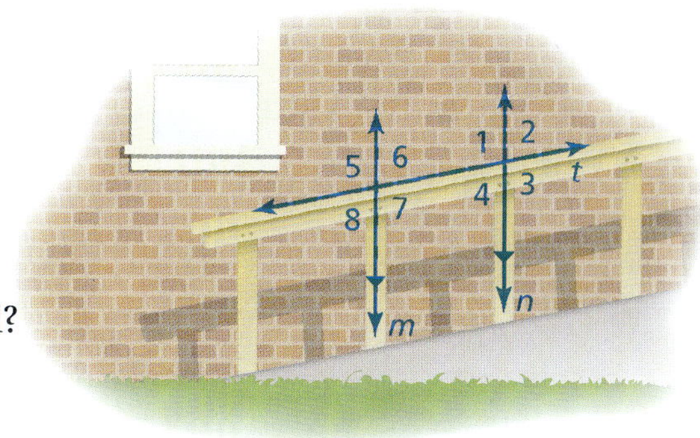

Use the figure to find the measures of the numbered angles.

1 7.

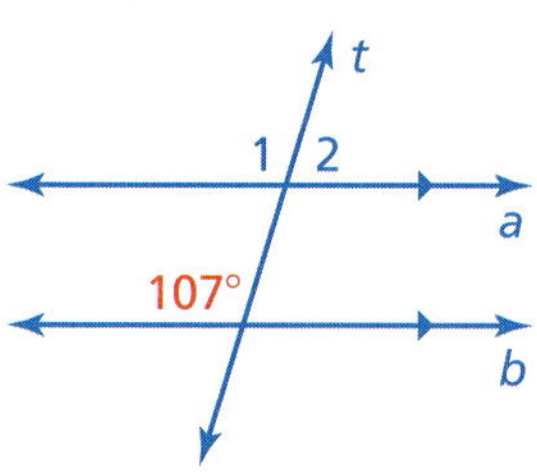

8.

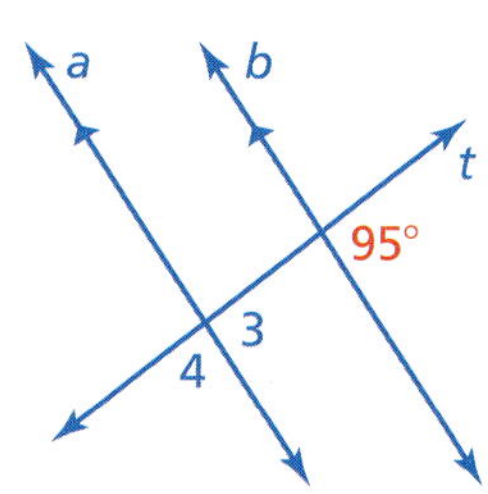

9.

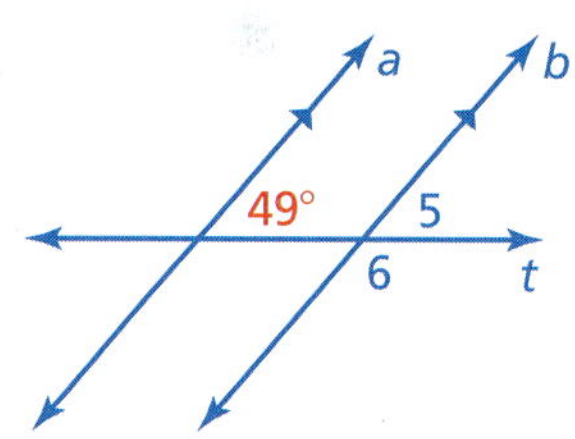

10. **ERROR ANALYSIS** Describe and correct the error in describing the relationship between the angles.

$\angle 5$ is congruent to $\angle 6$.

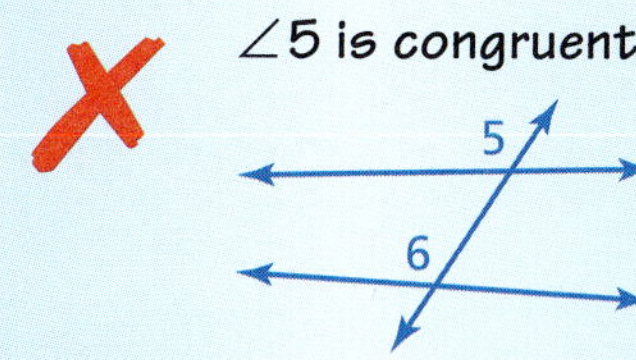

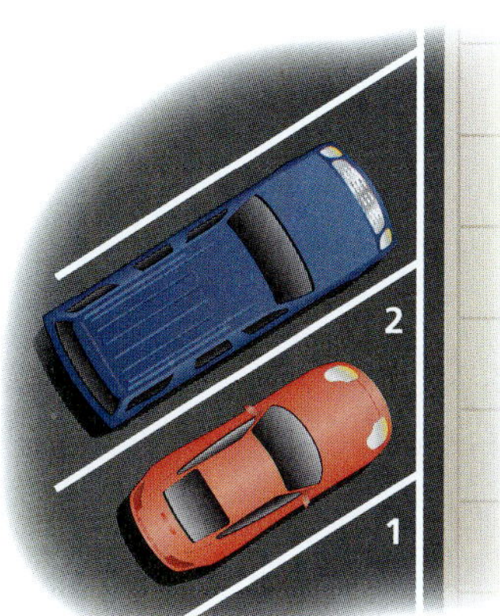

11. **PARKING** The painted lines that separate parking spaces are parallel. The measure of $\angle 1$ is 60°. What is the measure of $\angle 2$? Explain.

12. **OPEN-ENDED** Describe two real-life situations that use parallel lines.

13. **PROJECT** Draw two horizontal lines and a transversal on a piece of notebook paper. Label the angles as shown. Use a pair of scissors to cut out the angles. Compare the angles to determine which angles are congruent.

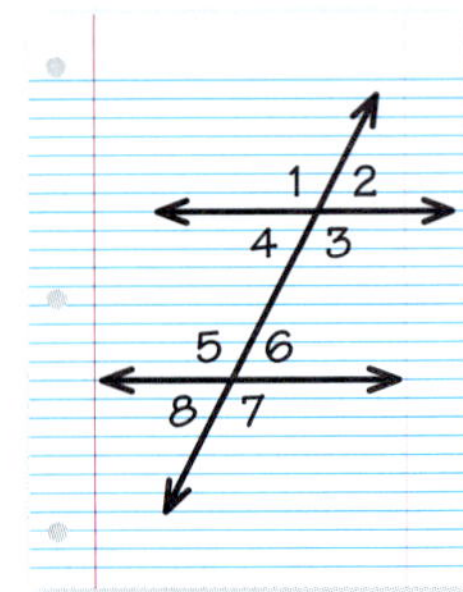

14. **REASONING** Refer to the figure for Exercise 13. What is the least number of angle measures you need to know in order to find the measure of every angle? Explain your reasoning.

Use the figure to find the measures of the numbered angles. Explain your reasoning.

(2) 15.

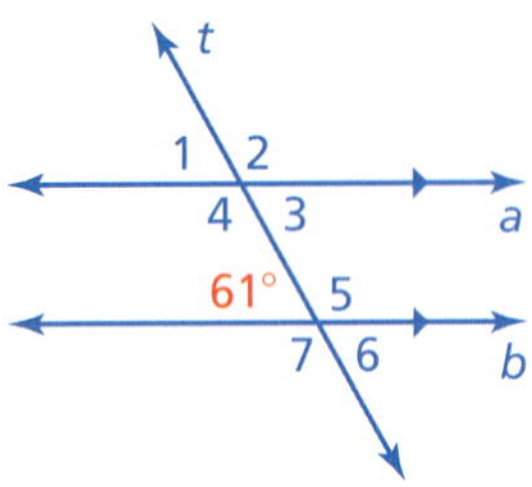

16.

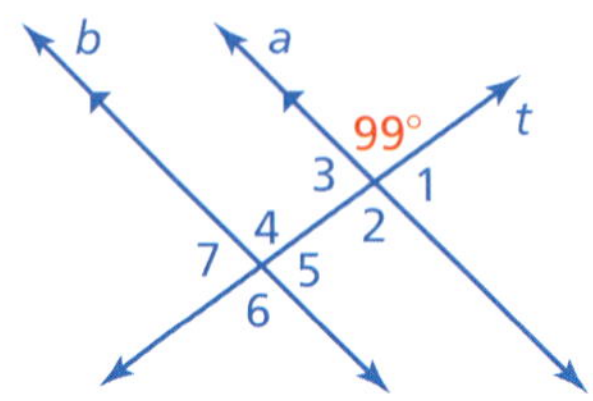

17.

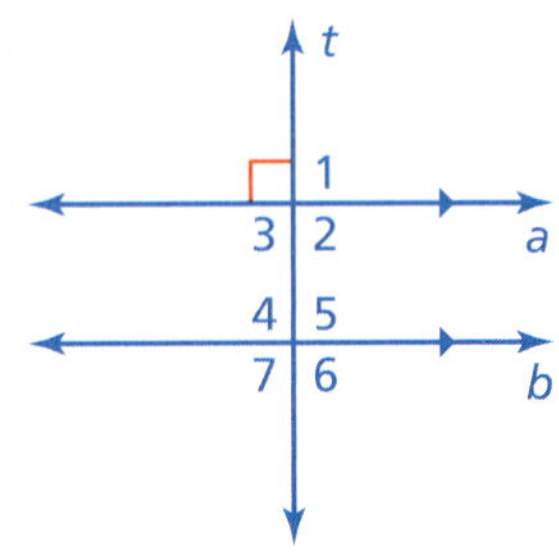

Complete the statement. Explain your reasoning.

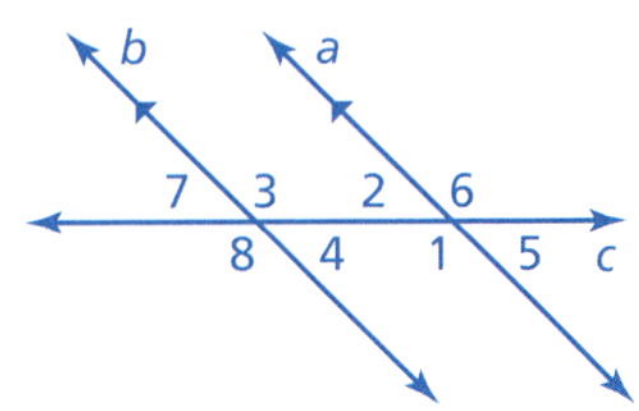

(3) 18. If the measure of $\angle 1 = 124°$, then the measure of $\angle 4 =$ ____.

19. If the measure of $\angle 2 = 48°$, then the measure of $\angle 3 =$ ____.

(4) 20. If the measure of $\angle 4 = 55°$, then the measure of $\angle 2 =$ ____.

21. If the measure of $\angle 6 = 120°$, then the measure of $\angle 8 =$ ____.

22. If the measure of $\angle 7 = 50.5°$, then the measure of $\angle 6 =$ ____.

23. If the measure of $\angle 3 = 118.7°$, then the measure of $\angle 2 =$ ____.

24. **RAINBOW** A rainbow is formed when sunlight reflects off raindrops at different angles. For blue light, the measure of $\angle 2$ is 40°. What is the measure of $\angle 1$?

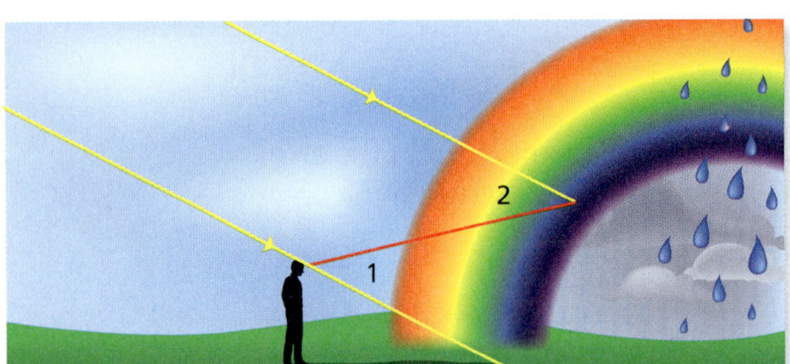

25. **REASONING** If a transversal is perpendicular to two parallel lines, what can you conclude about the angles formed? Explain.

26. **WRITING** Describe two ways you can show that $\angle 1$ is congruent to $\angle 7$.

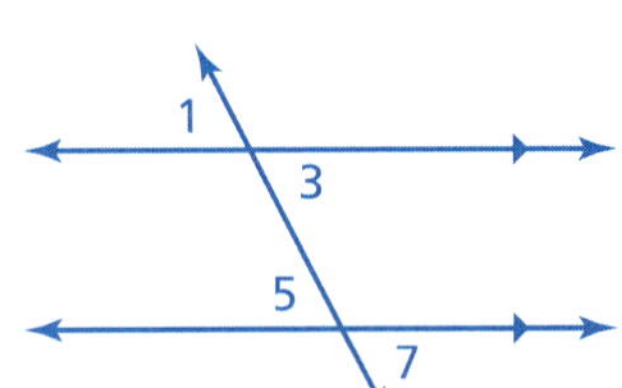

CRITICAL THINKING **Find the value of x.**

27. 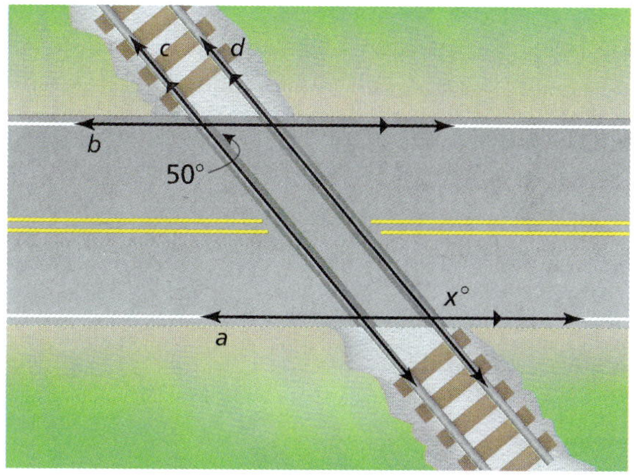

28. 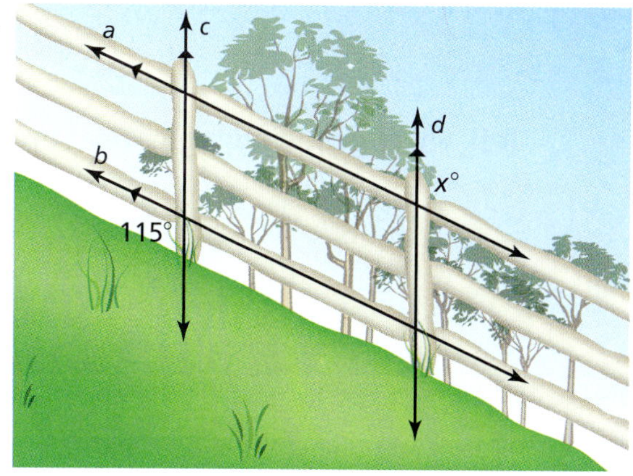

29. **OPTICAL ILLUSION** Refer to the figure.

   a. Do the horizontal lines appear to be parallel? Explain.
   b. Draw your own optical illusion using parallel lines.

30. **Geometry** The figure shows the angles used to make a double bank shot in an air hockey game.
   a. Find the value of x.
   b. Can you still get the red puck in the goal if x is increased by a little? by a lot? Explain.

Fair Game Review What you learned in previous grades & lessons

Evaluate the expression. *(Skills Review Handbook)*

31. $4 + 3^2$ 32. $5(2)^2 - 6$ 33. $11 + (-7)^2 - 9$ 34. $8 \div 2^2 + 1$

35. **MULTIPLE CHOICE** The volume of the cylinder is 20π cubic inches. What is the radius of the base? *(Skills Review Handbook)*

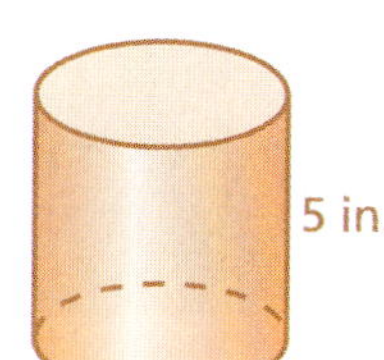

Ⓐ 1 inch Ⓑ 2 inches
Ⓒ 3 inches Ⓓ 4 inches

5.4–5.5 Quiz

Check It Out
Progress Check
BigIdeasMath.com

Tell whether the triangles are similar. Explain. *(Section 5.4)*

1.

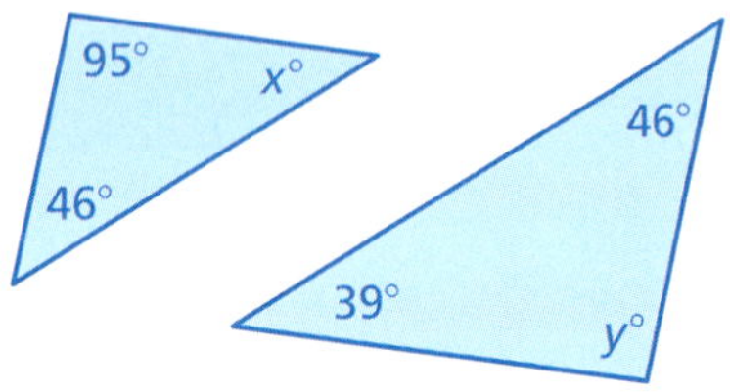

2.

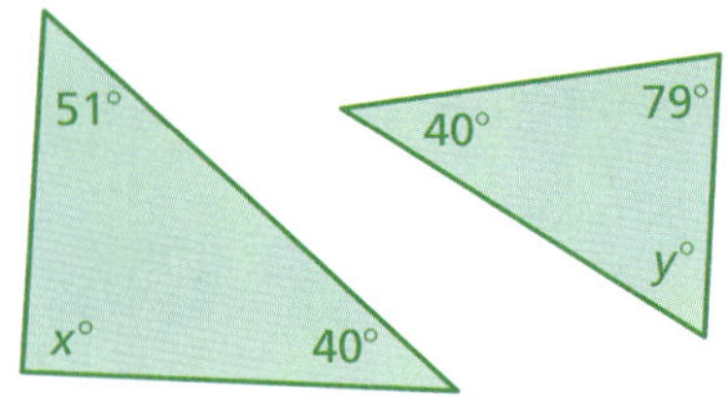

The triangles are similar. Find the value of *x*. *(Section 5.4)*

3.

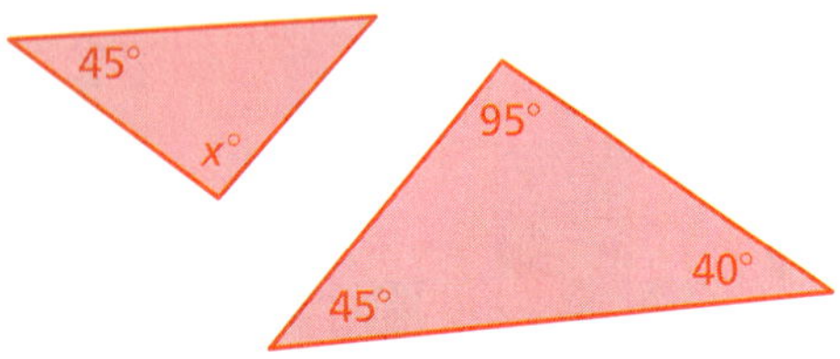

4.

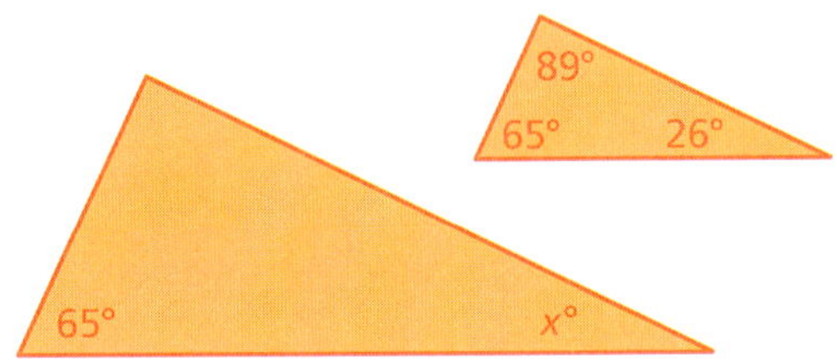

Use the figure to find the measure of the angle. Explain your reasoning. *(Section 5.5)*

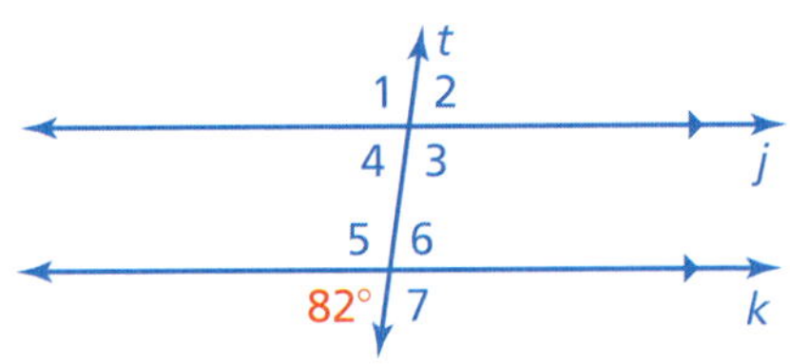

5. $\angle 2$

6. $\angle 6$

7. $\angle 4$

8. $\angle 1$

Complete the statement. Explain your reasoning. *(Section 5.5)*

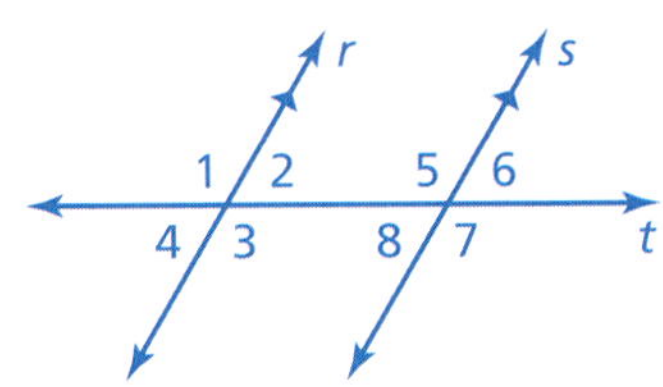

9. If the measure of $\angle 1 = 123°$, then the measure of $\angle 7 =$ ____.

10. If the measure of $\angle 2 = 58°$, then the measure of $\angle 5 =$ ____.

11. If the measure of $\angle 5 = 119°$, then the measure of $\angle 3 =$ ____.

12. If the measure of $\angle 4 = 60°$, then the measure of $\angle 6 =$ ____.

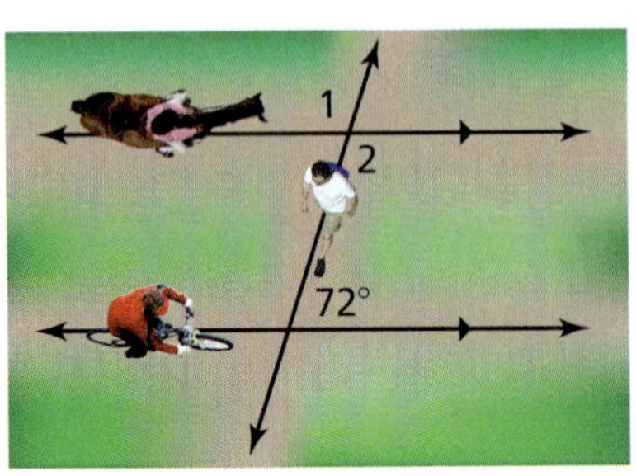

13. **PARK** In a park, a bike path and a horse riding path are parallel. In one part of the park, a hiking trail intersects the two paths. Find the measures of $\angle 1$ and $\angle 2$. Explain your reasoning. *(Section 5.5)*

14. **PERIMETER** The side lengths of a right triangle are doubled to make a similar triangle. Does the perimeter double as well? Explain. *(Section 5.4)*

5 Chapter Review

Check It Out
Vocabulary Help
BigIdeasMath.com

Review Key Vocabulary

complementary angles, *p. 186*
supplementary angles, *p. 186*
congruent angles, *p. 187*
vertical angles, *p. 187*
isosceles triangle, *p. 192*
congruent sides, *p. 192*
equilateral triangle, *p. 192*
equiangular triangle, *p. 192*
polygon, *p. 198*
regular polygon, *p. 199*
convex polygon, *p. 200*
concave polygon, *p. 200*
similar triangles, *p. 208*
indirect measurement, *p. 209*
perpendicular lines, *p. 214*
transversal, *p. 214*
interior angles, *p. 215*
exterior angles, *p. 215*

Review Examples and Exercises

5.1 Classifying Angles *(pp. 184–189)*

Find the value of x.

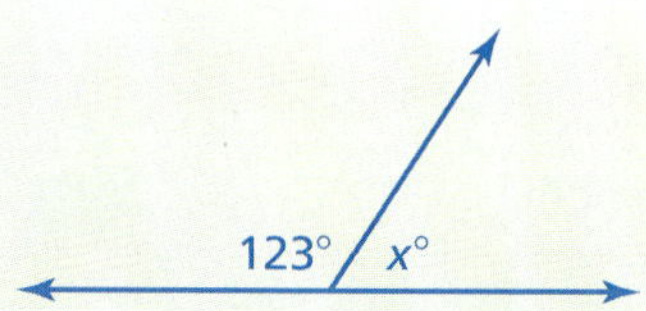

The angles are supplementary angles. So, the sum of their measures is 180°.

$$x + 123 = 180$$

$$x = 57$$

So, x is 57.

Exercises

Find the value of x.

1.

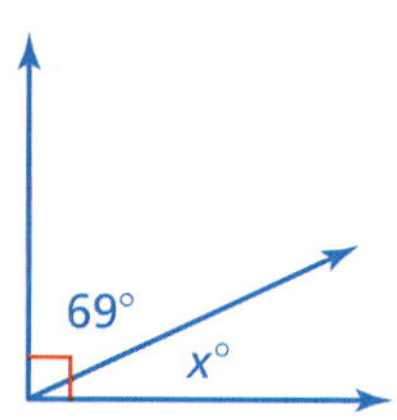

2.

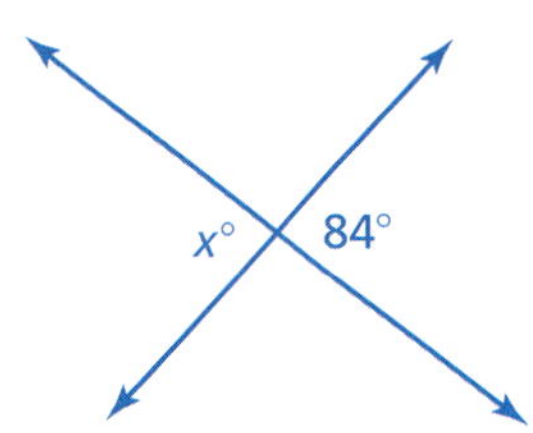

5.2 Angles and Sides of Triangles *(pp. 190–195)*

Find the value of x. Then classify the triangle.

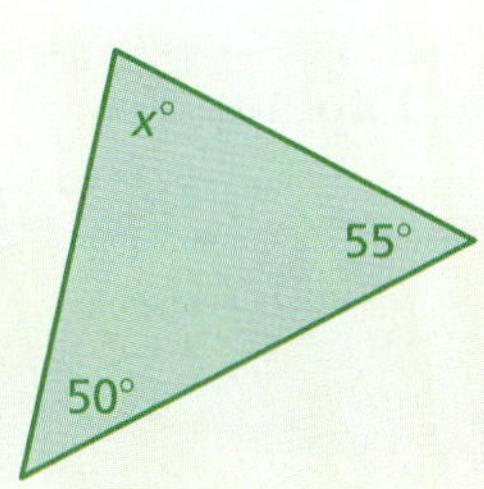

$$x + 50 + 55 = 180$$

$$x + 105 = 180$$

$$x = 75$$

The value of x is 75. The triangle has three acute angle measures, 50°, 55°, and 75°. So, it is an acute triangle.

Exercises

Find the value of x. Then classify the triangle in as many ways as possible.

3.

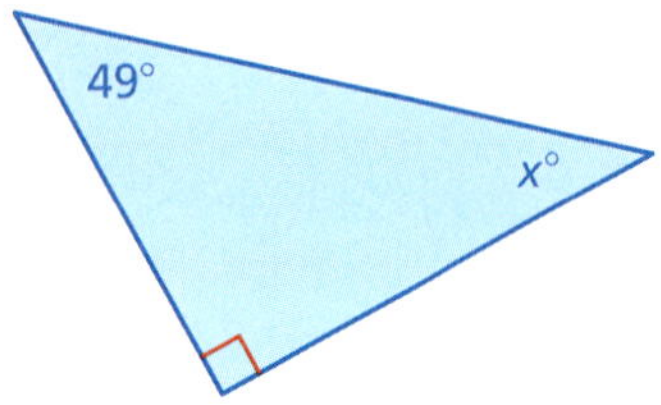

4.

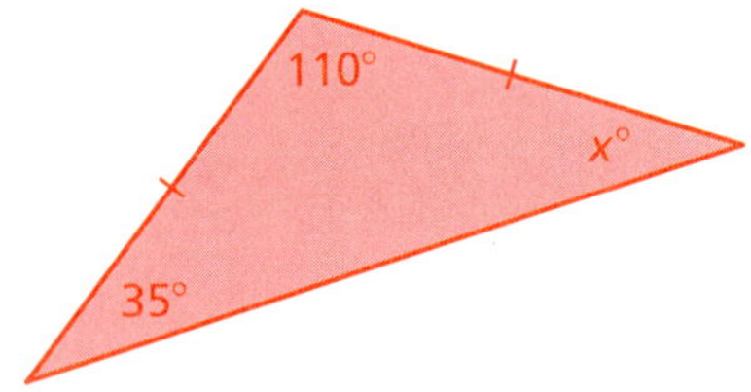

5.3 Angles of Polygons *(pp. 196–203)*

Find the value of x.

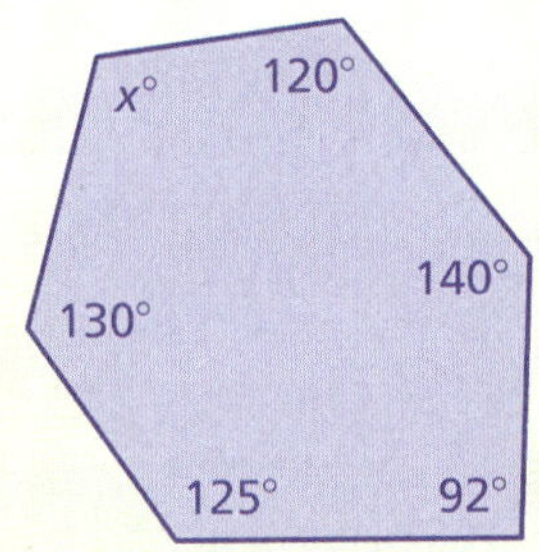

Step 1: The polygon has 6 sides. Find the sum of the angle measures.

$S = (n - 2) \cdot 180°$ Write the formula.

$= (6 - 2) \cdot 180°$ Substitute 6 for n.

$= 720$ Simplify. The sum of the angle measures is 720°.

Step 2: Write and solve an equation.

$$130 + 125 + 92 + 140 + 120 + x = 720$$

$$607 + x = 720$$

$$x = 113$$

The value of x is 113.

Exercises

Find the value of x.

5.

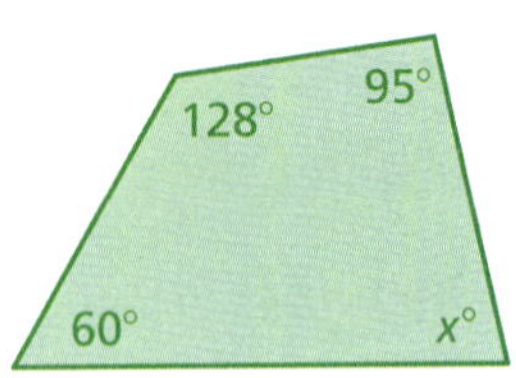

6.

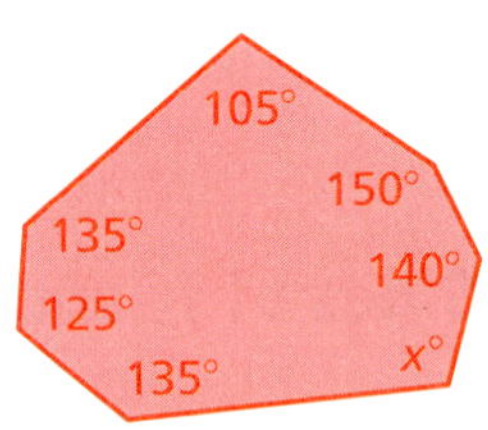

7.

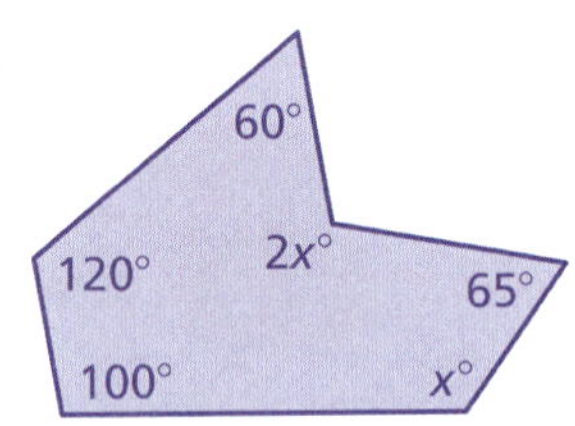

Tell whether the polygon is *convex* or *concave*. Explain.

8.

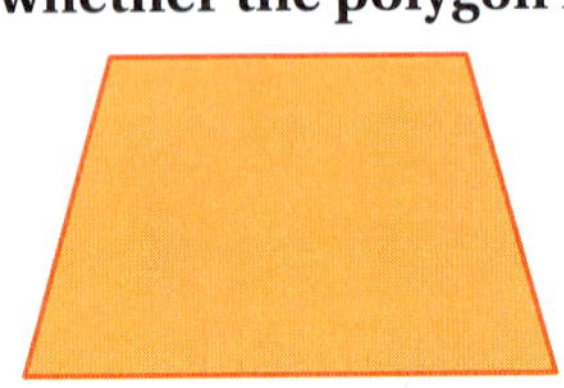

9.

10.

5.4 Using Similar Triangles (pp. 206–211)

Tell whether the triangles are similar. Explain.

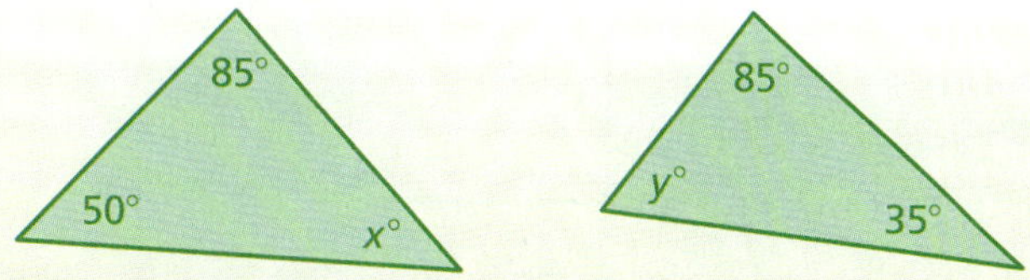

$$50 + 85 + x = 180$$
$$135 + x = 180$$
$$x = 45$$

$$y + 85 + 35° = 180$$
$$y + 120 = 180$$
$$y = 60$$

∴ The triangles do not have the same angle measures. So, they are not similar.

Exercises

11. Tell whether the triangles are similar. Explain.

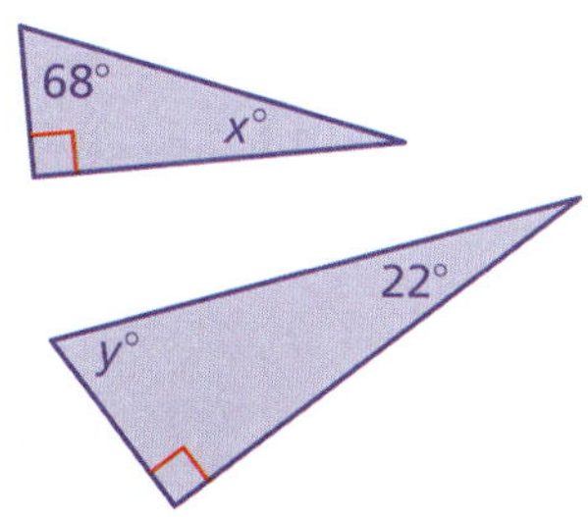

12. The triangles are similar Find the value of x.

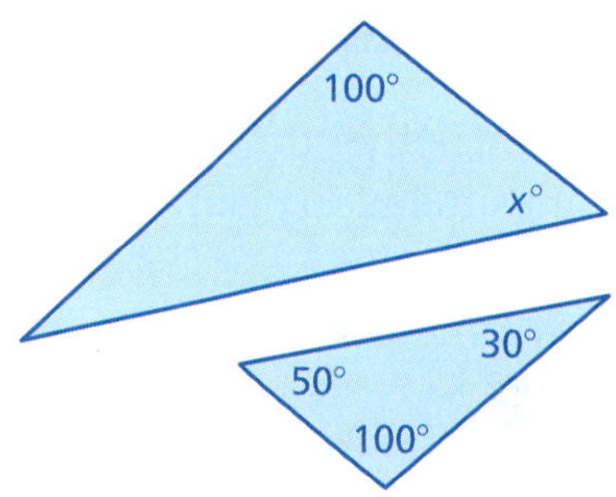

5.5 Parallel Lines and Transversals (pp. 212–219)

Use the figure to find the measure of ∠6.

∠2 and the 55° angle are supplementary.
So, the measure of ∠2 is 180° − 55° = 125°.

∠2 and ∠6 are corresponding angles.
They are congruent.

∴ So, the measure of ∠6 is 125°.

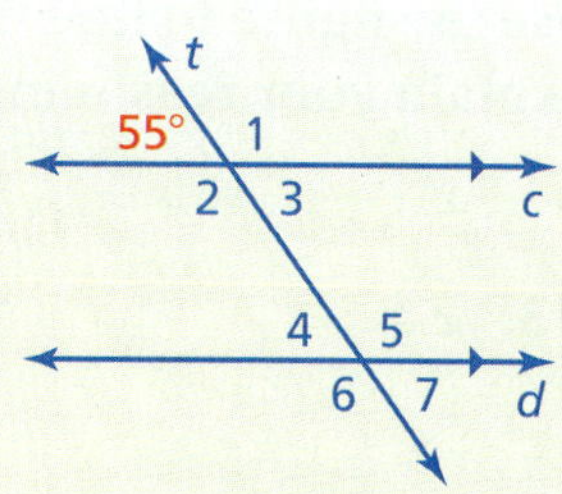

Exercises

Use the figure to find the measure of the angle. Explain your reasoning.

13. ∠8 **14.** ∠5

15. ∠7 **16.** ∠2

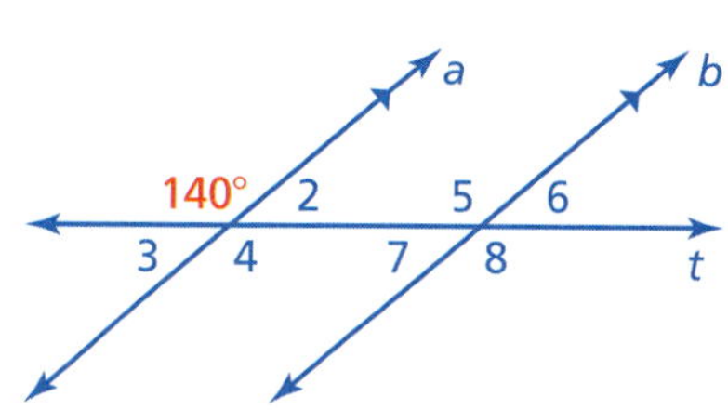

5 Chapter Test

Check It Out
Test Practice
BigIdeasMath.com

Find the value of x.

1. 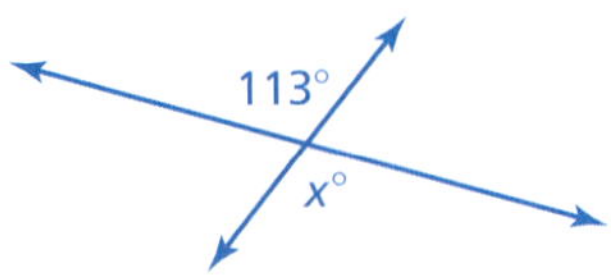

2. 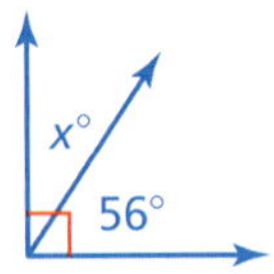

3. 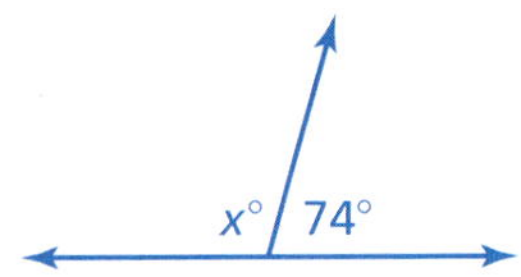

Find the value of x. Then classify the triangle in as many ways as possible.

4. 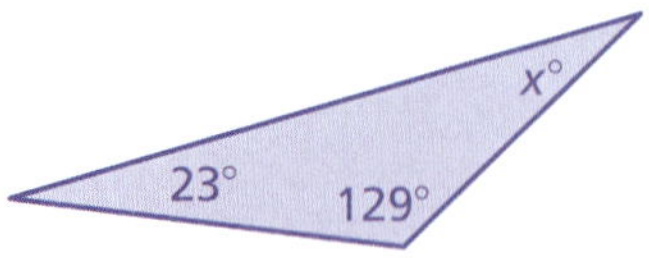

5. 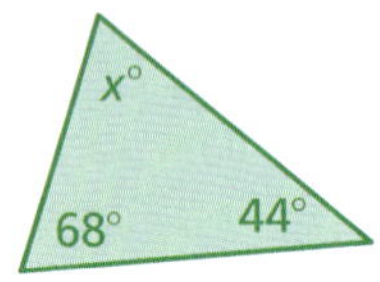

6. 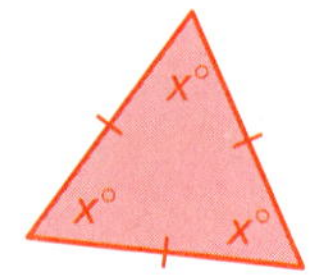

7. Tell whether the polygon is *convex* or *concave*. Explain.

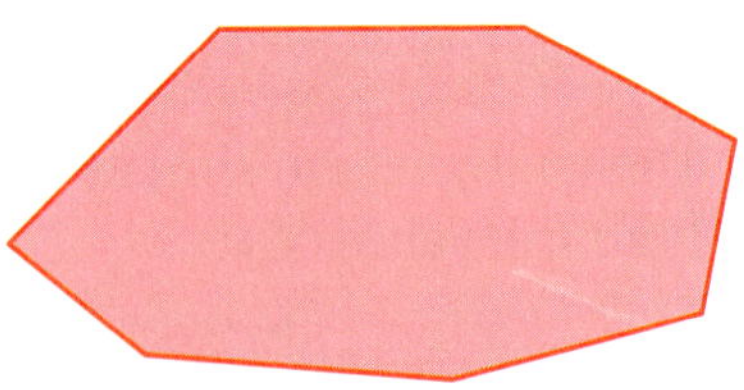

8. Find the value of x.

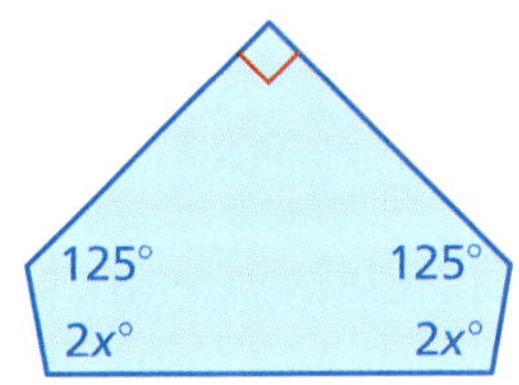

9. Tell whether the triangles are similar. Explain.

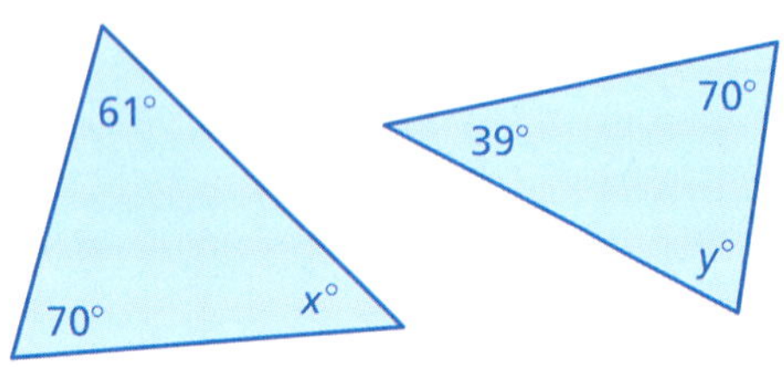

10. The triangles are similar. Find the value of x.

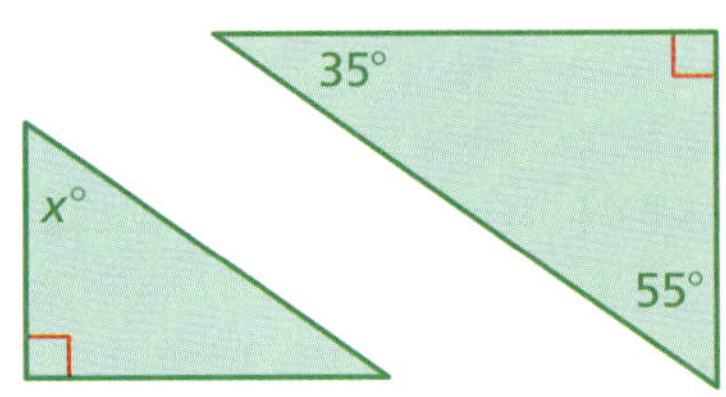

Use the figure to find the measure of the angle. Explain your reasoning.

11. $\angle 1$

12. $\angle 8$

13. $\angle 4$

14. $\angle 5$

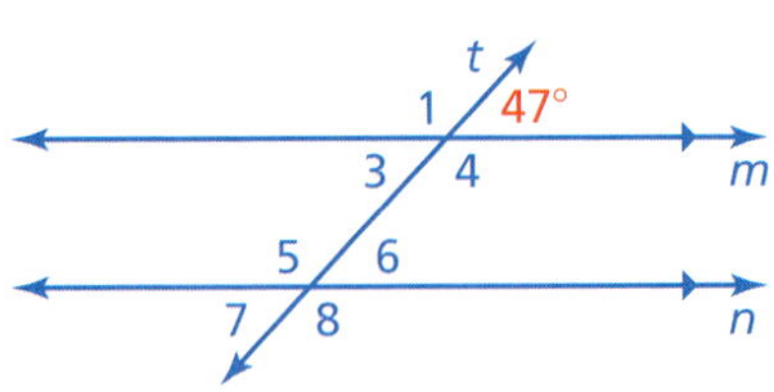

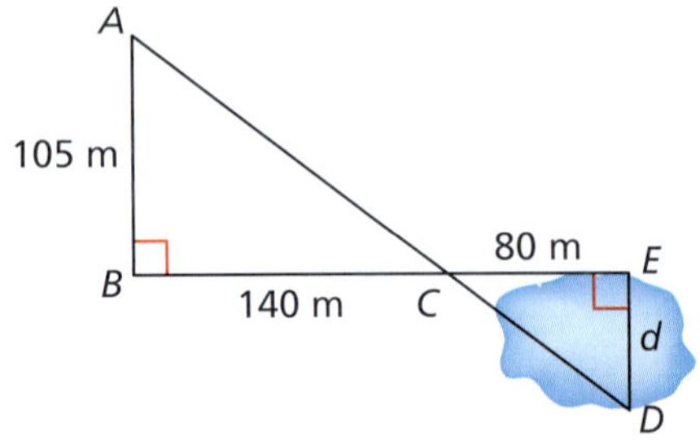

15. **POND** Use the given measurements to find the distance d across the pond.

5 Standardized Test Practice

Test-Taking Strategy
Solve Problem Before Looking at Choices

"**Solve the problem before looking at the choices.** You know 180 − 2(70) = 40. So the answer is C."

1. The border of a Canadian one-dollar coin is shaped like an 11-sided regular polygon. The shape was chosen to help visually-impaired people identify the coin. How many degrees are in each angle along the border? Round your answer to the nearest degree.

2. A public utility charges its residential customers for natural gas based on the number of therms used each month. The formula below shows how the monthly cost C in dollars is related to the number t of therms used.

$$C = 11 + 1.6t$$

Solve this formula for t.

A. $t = \dfrac{C}{12.6}$

B. $t = \dfrac{C - 11}{1.6}$

C. $t = \dfrac{C}{1.6} - 11$

D. $t = C - 12.6$

3. Which equation matches the line shown in the graph?

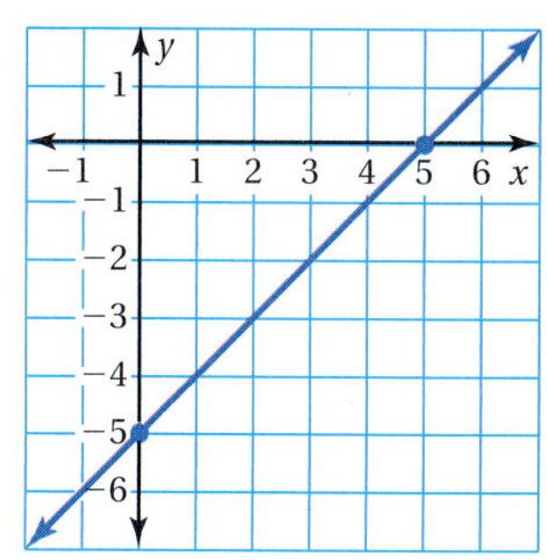

F. $y = x - 5$

G. $y = x + 5$

H. $y = -x - 5$

I. $y = -x + 5$

4. $\angle 1$ and $\angle 2$ form a straight angle. $\angle 1$ has a measure of 28°. Find the measure of $\angle 2$, in degrees.

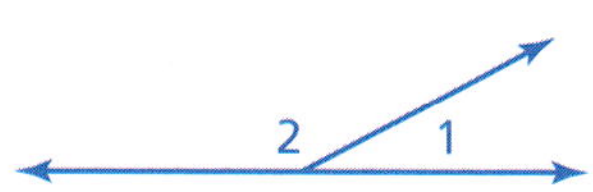

5. Which equation represents a linear function?

A. $y = x^2$

B. $y = \dfrac{2}{x}$

C. $xy = 1$

D. $x + y = 1$

6. A shipment of 2,000 laptop and desktop computers weighs 34,000 pounds. Each laptop computer weighs 8 pounds and each desktop computer weighs 20 pounds. Let ℓ represent the number of laptop computers and d represent the number of desktop computers. Which system of equations could be used to find how many laptop computers are in the shipment?

F. $\ell + d = 2{,}000$
$20\ell + 8d = 34{,}000$

H. $\ell + d = 2{,}000$
$8\ell + 20d = 34{,}000$

G. $\ell + d = 34{,}000$
$20\ell + 8d = 2{,}000$

I. $\ell + d = 34{,}000$
$8\ell + 20d = 2{,}000$

7. What is the domain of the function graphed in the coordinate plane?

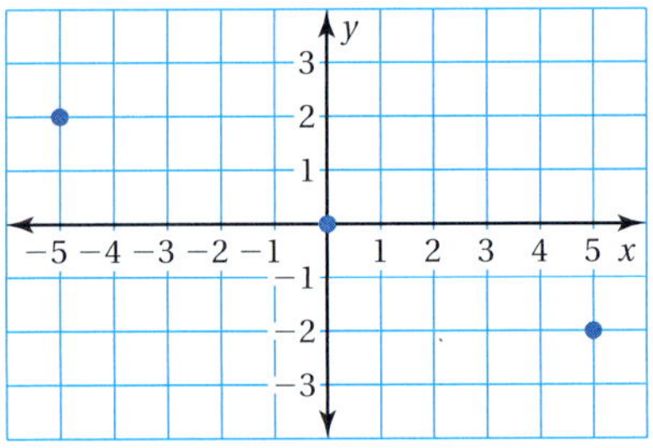

A. $-5, 0, 5$

C. $-5, -2, 0, 2, 5$

B. $-2, 0, 2$

D. $-5, 2$

8. The sum S of the angle measures of a polygon with n sides can be found using a formula.

Think Solve Explain

Part A Write the formula.

Part B A quadrilateral has angles measuring 100, 90, and 90 degrees. Find the measure of its fourth angle. Show your work and explain your reasoning.

Part C The sum of the measures of the angles of the pentagon shown is 540 degrees. Divide the pentagon into triangles to show why this must be true. Show your work and explain your reasoning.

9. The line shown in the graph has a slope of $\frac{2}{5}$.

What is the equation of the line?

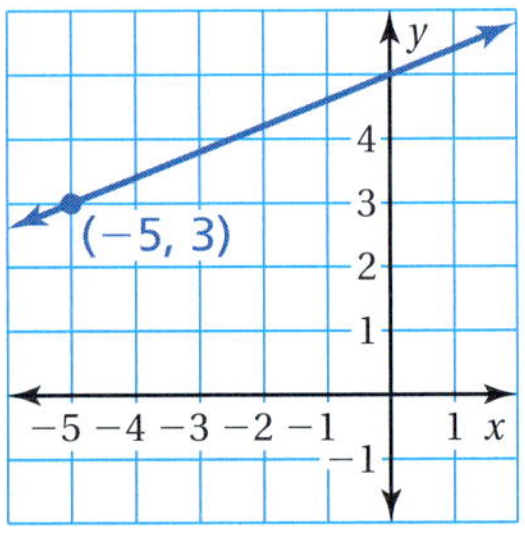

F. $x = \frac{2}{5}y + 5$

H. $x = \frac{2}{5}y + 1$

G. $y = \frac{2}{5}x + 5$

I. $y = \frac{2}{5}x + 1$

10. On a hot summer day, the temperature was 95°F, the relative humidity was 75%, and the Heat Index was 122°F. For every degree that the temperature rises, the Heat Index increases by 4 degrees. The temperature rises to 98°F. What is the Heat Index?

A. 99°F

B. 107°F

C. 126°F

D. 134°F

11. Which value of x makes the equation below true?

$$5x - 3 = 11$$

F. 1.6

G. 2.8

H. 40

I. 70

12. In the diagram below, $\triangle ABC \sim \triangle DEF$. What is the value of x?

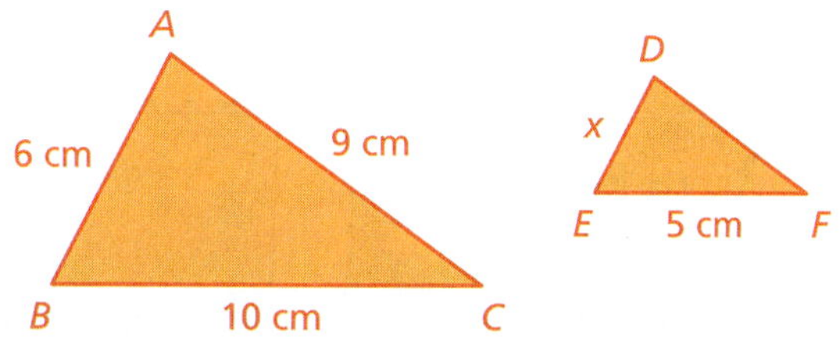

A. 1 cm

B. 3 cm

C. 4.5 cm

D. 6 cm

13. A system of linear equations is shown in the coordinate plane below. What is the solution for this system?

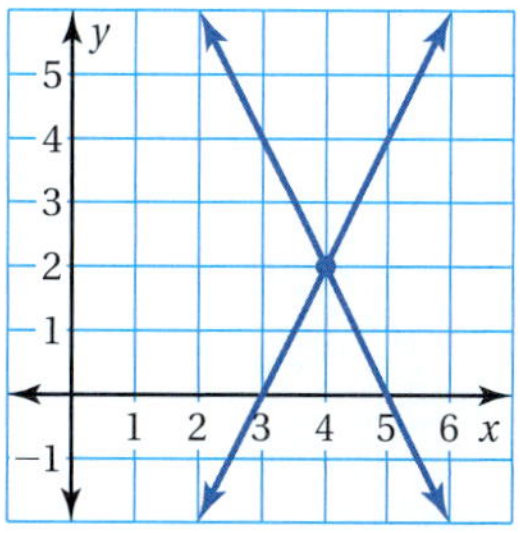

F. (0, 10)

G. (3, 0)

H. (4, 2)

I. (5, 0)

6 Square Roots and the Pythagorean Theorem

6.1 Finding Square Roots

6.2 The Pythagorean Theorem

6.3 Approximating Square Roots

6.4 Simplifying Square Roots

6.5 Using the Pythagorean Theorem

What You Learned Before

"Here's how I remember the square root of 2. February is the 2nd month. It has 28 days. Split 28 into 14 and 14. Move the decimal to get 1.414."

Comparing Decimals (Grades 5 and 7)

Complete the number sentence with <, >, or =.

Example 1 **1.1 ____ 1.01**

Because $\frac{110}{100}$ is greater than $\frac{101}{100}$, 1.1 is greater than 1.01.

∴ So, $1.1 > 1.01$.

Example 2 **−0.3 ____ −0.003**

Because $-\frac{300}{1000}$ is less than $-\frac{3}{1000}$, -0.3 is less than -0.003.

∴ So, $-0.3 < -0.003$.

Example 3 **Find three decimals that make the number sentence −5.12 > ____ true.**

Any decimal less than -5.12 will make the sentence true.

∴ *Sample answer:* -10.1, -9.05, -8.25

Try It Yourself

Complete the number sentence with <, >, or =.

1. 2.10 ____ 2.1 **2.** −4.5 ____ −4.25 **3.** π ____ 3.2

Find three decimals that make the number sentence true.

4. $-0.01 \leq$ ____ **5.** $1.75 >$ ____ **6.** $0.75 \geq$ ____

Using Order of Operations (Grades 5 and 7)

Example 4 **Evaluate $8^2 \div (32 \div 2) - 2(3 - 5)$.**

First: Parentheses $8^2 \div (32 \div 2) - 2(3 - 5) = 8^2 \div 16 - 2(-2)$

Second: Exponents $= 64 \div 16 - 2(-2)$

Third: Multiplication and Division (from left to right) $= 4 + 4$

Fourth: Addition and Subtraction (from left to right) $= 8$

Try It Yourself

Evaluate the expression.

7. $15\left(\frac{12}{3}\right) - 7^2 - 2 \cdot 7$ **8.** $3^2 \cdot 4 \div 18 + 30 \cdot 6 - 1$ **9.** $-1 + \left(\frac{4}{2}(6 - 1)\right)^2$

6.1 Finding Square Roots

STATE STANDARDS
MA.8.A.6.4

Essential Question How can you find the side length of a square when you are given the area of the square?

When you multiply a number by itself, you square the number.

Symbol for squaring is 2nd power. → $4^2 = 4 \cdot 4$

$= 16$ 4 squared is 16.

To "undo" this, take the **square root** of the number.

Symbol for square root is a radical sign. → $\sqrt{16} = \sqrt{4^2} = 4$ The square root of 16 is 4.

1 ACTIVITY: Finding Square Roots

Work with a partner. Use a square root symbol to write the side length of the square. Then find the square root. Check your answer by multiplying.

a. Sample: $s = \sqrt{121} = 11$ ft

Area = 121 ft^2

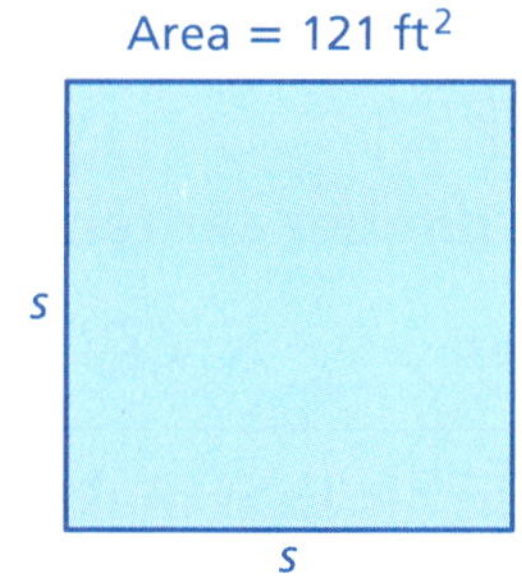

Check

$$\begin{array}{r} 11 \\ \times 11 \\ \hline 11 \\ 110 \\ \hline 121 \end{array}$$ ✓

The side length of the square is 11 feet.

b. Area = 81 yd^2

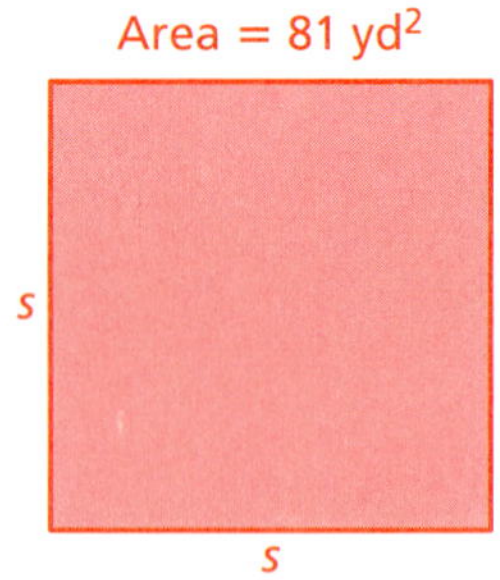

c. Area = 324 cm^2

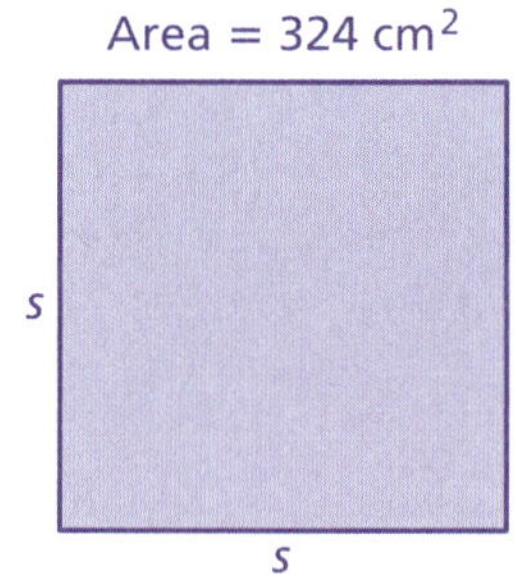

d. Area = 361 mi^2

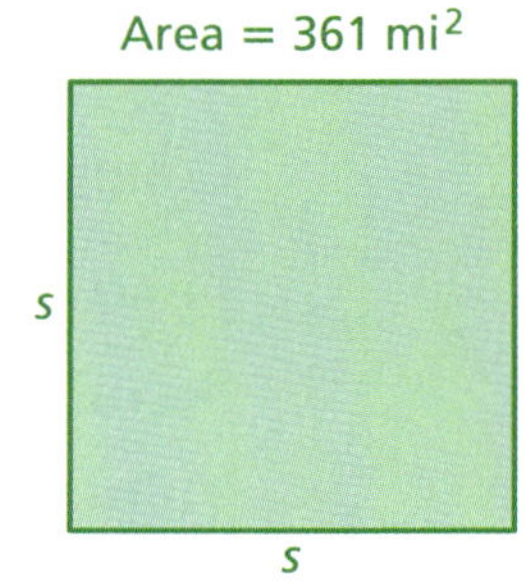

e. Area = 2.89 in.^{2}

f. Area = 4.41 m^2

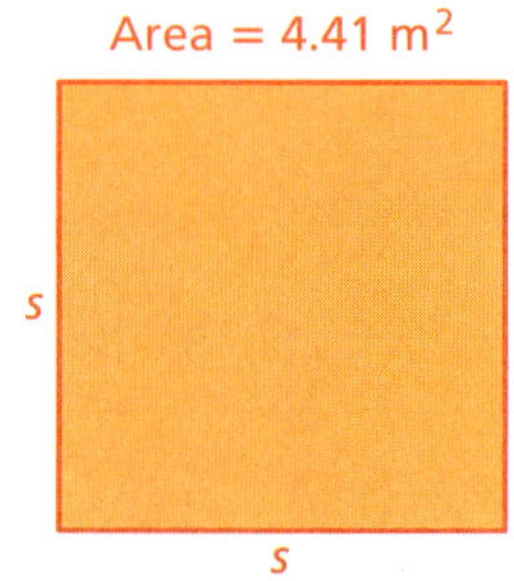

g. Area = $\frac{4}{9}$ ft^2

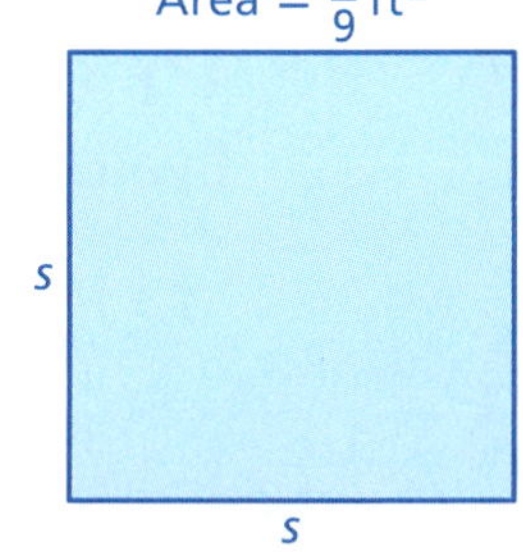

2 ACTIVITY: The Period of a Pendulum

Work with a partner.

The period of a pendulum is the time (in seconds) it takes the pendulum to swing back *and* forth.

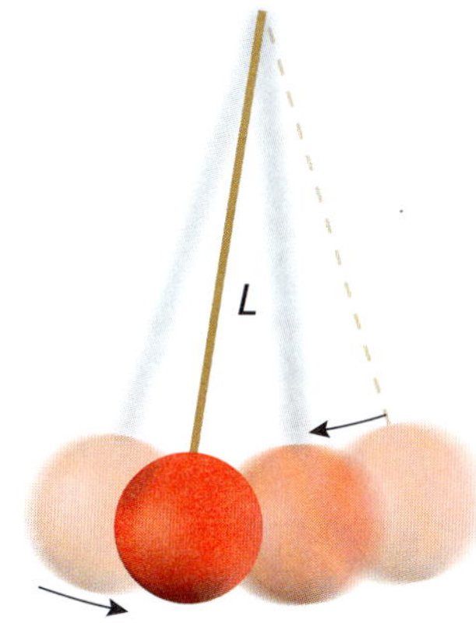

The period T is represented by $T = 1.1\sqrt{L}$, where L is the length of the pendulum (in feet).

Copy and complete the table. Then graph the function. Is the function linear?

L	1.00	1.96	3.24	4.00	4.84	6.25	7.29	7.84	9.00
T									

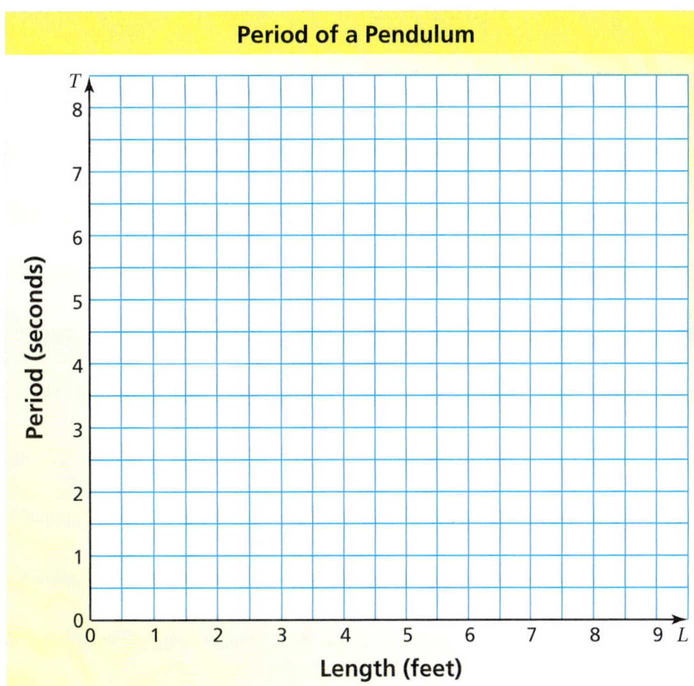

What Is Your Answer?

3. **IN YOUR OWN WORDS** How can you find the side length of a square when you are given the area of the square? Give an example. How can you check your answer?

Practice Use what you learned about finding square roots to complete Exercises 4–6 on page 234.

6.1 Lesson

Check It Out
Lesson Tutorials
BigIdeasMath.com

Key Vocabulary
square root, *p. 232*
perfect square, *p. 232*
radical sign, *p. 232*
radicand, *p. 232*

A **square root** of a number is a number that when multiplied by itself, equals the given number. Every positive number has a positive *and* a negative square root. A **perfect square** is a number with integers as its square roots.

EXAMPLE 1 Finding Square Roots of a Perfect Square

Study Tip
Zero has one square root, which is 0.

Find the two square roots of 49.

$7 \cdot 7 = 49$ and $(-7) \cdot (-7) = 49$

So, the square roots of 49 are 7 and -7.

The symbol $\sqrt{}$ is called a **radical sign**. It is used to represent a square root. The number under the radical sign is called the **radicand**.

Positive Square Root $\sqrt{}$	Negative Square Root $-\sqrt{}$	Both Square Roots $\pm\sqrt{}$
$\sqrt{16} = 4$	$-\sqrt{16} = -4$	$\pm\sqrt{16} = \pm 4$

EXAMPLE 2 Finding Square Roots

Find the square root(s).

a. $\sqrt{25}$

Because $5^2 = 25$, $\sqrt{25} = \sqrt{5^2} = 5$.

$\sqrt{25}$ represents the *positive* square root.

b. $-\sqrt{\frac{9}{16}}$

Because $\left(\frac{3}{4}\right)^2 = \frac{9}{16}$, $-\sqrt{\frac{9}{16}} = -\sqrt{\left(\frac{3}{4}\right)^2} = -\frac{3}{4}$.

$-\sqrt{\frac{9}{16}}$ represents the *negative* square root.

c. $\pm\sqrt{2.25}$

Because $1.5^2 = 2.25$, $\pm\sqrt{2.25} = \pm\sqrt{1.5^2} = 1.5$ and -1.5.

$\pm\sqrt{2.25}$ represents both the *positive and negative* square roots.

On Your Own

Now You're Ready
Exercises 7–16

Find the two square roots of the number.

1. 36 **2.** 100 **3.** 121

Find the square root(s).

4. $-\sqrt{1}$ **5.** $\pm\sqrt{\frac{4}{25}}$ **6.** $\sqrt{12.25}$

EXAMPLE 3 Evaluating Expressions Involving Square Roots

Evaluate the expression.

a. $5\sqrt{36} + 7$

$$\begin{aligned} 5\sqrt{36} + 7 &= 5(6) + 7 && \text{Evaluate the square root.} \\ &= 30 + 7 && \text{Multiply.} \\ &= 37 && \text{Add.} \end{aligned}$$

b. $\frac{1}{4} + \sqrt{\frac{18}{2}}$

$$\begin{aligned} \frac{1}{4} + \sqrt{\frac{18}{2}} &= \frac{1}{4} + \sqrt{9} && \text{Simplify.} \\ &= \frac{1}{4} + 3 && \text{Evaluate the square root.} \\ &= 3\frac{1}{4} && \text{Add.} \end{aligned}$$

Squaring a positive number and finding a square root are inverse operations. Use this relationship to solve equations involving squares.

EXAMPLE 4 Real-Life Application

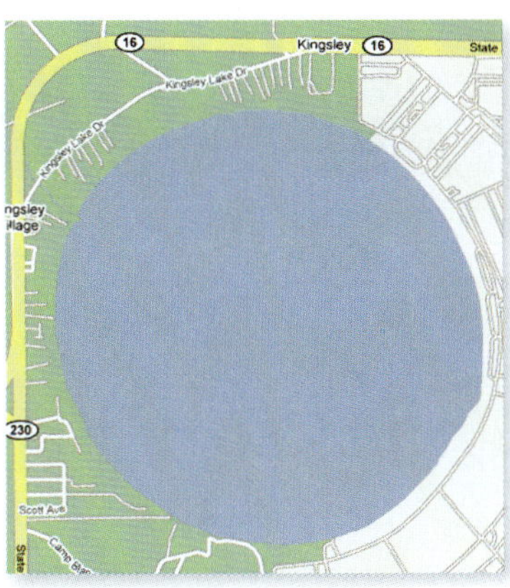

Kingsley Lake in Clay County, Florida is a circular lake that covers an area of about 8,038,400 square meters.

What is the radius of Kingsley Lake? Use 3.14 for π.

$$\begin{aligned} A &= \pi r^2 && \text{Write the formula for the area of a circle.} \\ 8{,}038{,}400 &\approx 3.14r^2 && \text{Substitute 8,038,400 for } A \text{ and 3.14 for } \pi. \\ 2{,}560{,}000 &= r^2 && \text{Divide each side by 3.14.} \\ \sqrt{2{,}560{,}000} &= \sqrt{r^2} && \text{Take positive square root of each side.} \\ 1600 &= r && \text{Simplify.} \end{aligned}$$

The radius of the lake is about 1600 meters.

On Your Own

Now You're Ready
Exercises 18–23

Evaluate the expression.

7. $12 - 3\sqrt{25}$

8. $\sqrt{\frac{28}{7}} + 2.4$

9. $5(\sqrt{49} - 10)$

10. The area of a circle is 2826 square feet. Write and solve an equation to find the radius of the circle. Use 3.14 for π.

6.1 Exercises

Check It Out
Help with Homework
BigIdeasMath.com

Vocabulary and Concept Check

1. **VOCABULARY** Is 26 a perfect square? Explain.
2. **REASONING** Can the square of an integer be a negative number? Explain.
3. **NUMBER SENSE** Does $\sqrt{256}$ represent the positive square root of 256, the negative square root of 256, or both? Explain.

Practice and Problem Solving

Find the side length of the square. Check your answer by multiplying.

4. Area = 441 cm^2

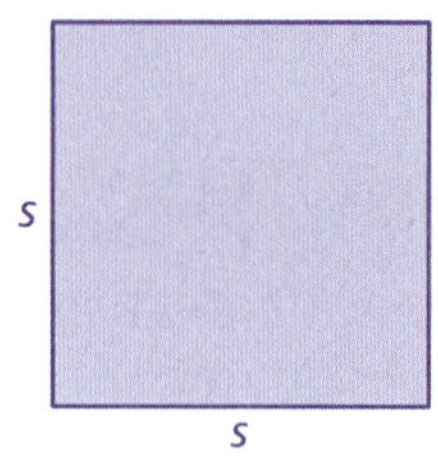

5. Area = 1.69 km^2

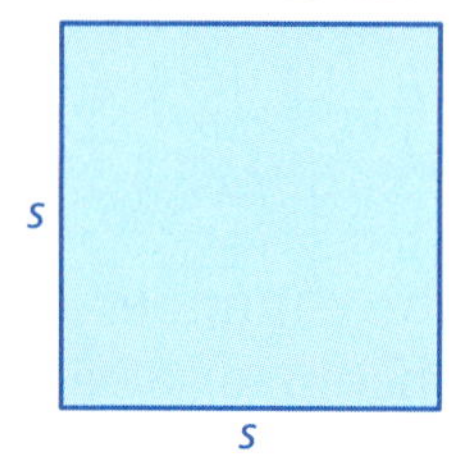

6. Area = $\frac{25}{36}$ yd^2

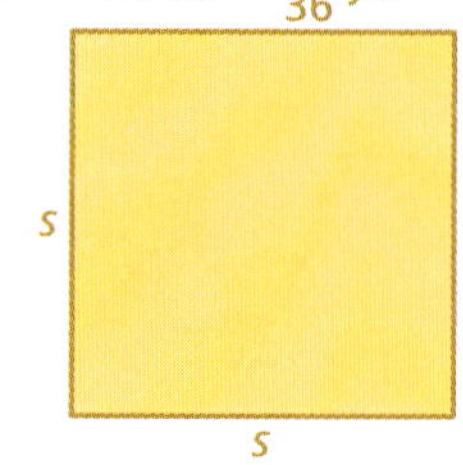

Find the two square roots of the number.

(1) 7. 9   8. 64   9. 4   10. 144

Find the square root(s).

(2) 11. $\sqrt{625}$ 12. $-\sqrt{\frac{9}{100}}$ 13. $\pm\sqrt{\frac{1}{961}}$

14. $\sqrt{7.29}$ 15. $\pm\sqrt{4.84}$ 16. $-\sqrt{361}$

17. **ERROR ANALYSIS** Describe and correct the error in finding the square roots.

✗ $\pm\sqrt{\frac{1}{4}} = \frac{1}{2}$

Evaluate the expression.

(3) 18. $3\sqrt{16} - 5$ 19. $10 - 4\sqrt{\frac{1}{16}}$ 20. $\sqrt{6.76} + 5.4$

21. $8\sqrt{8.41} + 1.8$ 22. $2\left(\sqrt{\frac{80}{5}} - 5\right)$ 23. $4\left(\sqrt{\frac{147}{3}} + 3\right)$

24. **NOTEPAD** The area of the base of a square notepad is 9 square inches. What is the length of one side of the base of the notepad?

25. **CRITICAL THINKING** There are two square roots of 25. Why is there only one answer for the radius of the button?

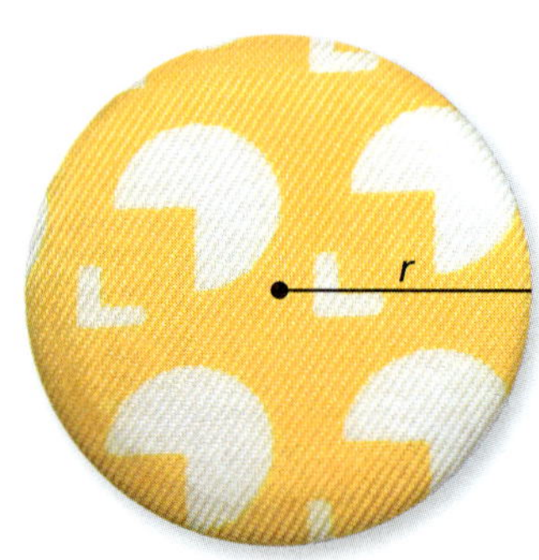

$A = 25\pi\ mm^2$

Copy and complete the statement with <, >, or =.

26. $\sqrt{81}$ ▭ 8 **27.** 0.5 ▭ $\sqrt{0.25}$ **28.** $\frac{3}{2}$ ▭ $\sqrt{\frac{25}{4}}$

29. SAILBOAT The area of a sail is $40\frac{1}{2}$ square feet. The base and the height of the sail are equal. What is the height of the sail (in feet)?

30. REASONING Is the product of two perfect squares always a perfect square? Explain your reasoning.

31. ENERGY The kinetic energy K (in joules) of a falling apple is represented by $K = \frac{v^2}{2}$, where v is the speed of the apple (in meters per second). How fast is the apple traveling when the kinetic energy is 32 joules?

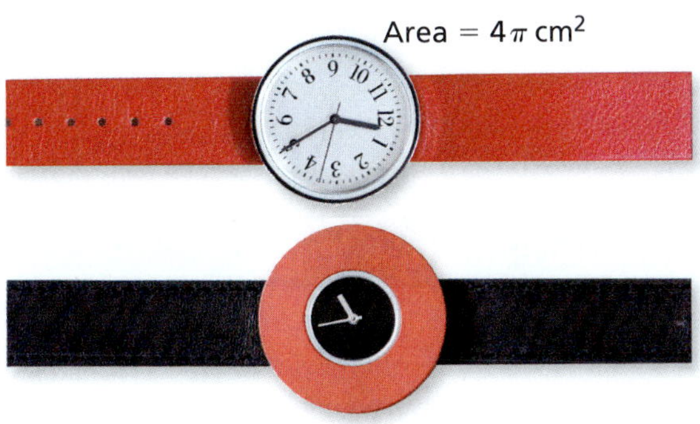

32. WATCHES The areas of the two watch faces have a ratio of 16 : 25.

a. What is the ratio of the radius of the smaller watch face to the radius of the larger watch face?

b. What is the radius of the larger watch face?

33. WINDOW The cost C (in dollars) of making a square window with a side length of n inches is represented by $C = \frac{n^2}{5} + 175$. A window costs $355. What is the length (in feet) of the window?

34. Geometry The area of the triangle is represented by the formula $A = \sqrt{s(s-21)(s-17)(s-10)}$, where s is equal to half the perimeter. What is the height of the triangle?

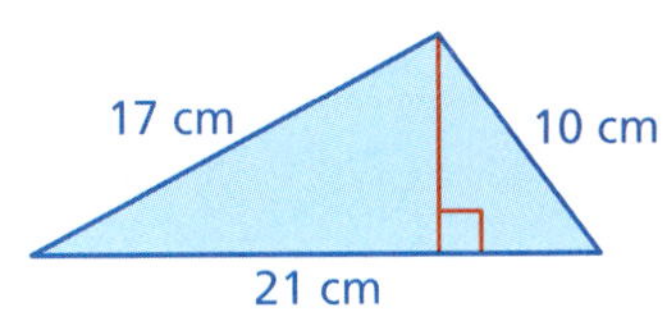

Fair Game Review What you learned in previous grades & lessons

Evaluate the expression. *(Skills Review Handbook)*

35. $3^2 + 4^2$ **36.** $8^2 + 15^2$ **37.** $13^2 - 5^2$ **38.** $25^2 - 24^2$

39. MULTIPLE CHOICE Which of the following describes the triangle? *(Section 5.2)*

Ⓐ Acute Ⓑ Right

Ⓒ Obtuse Ⓓ Equiangular

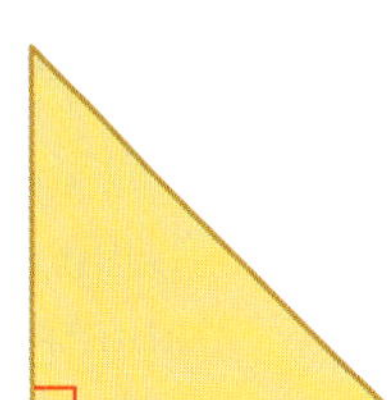

6.2 The Pythagorean Theorem

STATE STANDARDS
MA.8.G.2.4
MA.8.A.6.4

Essential Question How are the lengths of the sides of a right triangle related?

Pythagoras was a Greek mathematician and philosopher who discovered one of the most famous rules in mathematics. In mathematics, a rule is called a **theorem**. So, the rule that Pythagoras discovered is called the Pythagorean Theorem.

Pythagoras
(c. 570 B.C.–c. 490 B.C.)

1 ACTIVITY: Discovering the Pythagorean Theorem

Work with a partner.

a. On grid paper, draw any right triangle. Label the lengths of the two shorter sides (the **legs**) a and b.

b. Label the length of the longest side (the **hypotenuse**) c.

c. Draw squares along each of the three sides. Label the areas of the three squares a^2, b^2, and c^2.

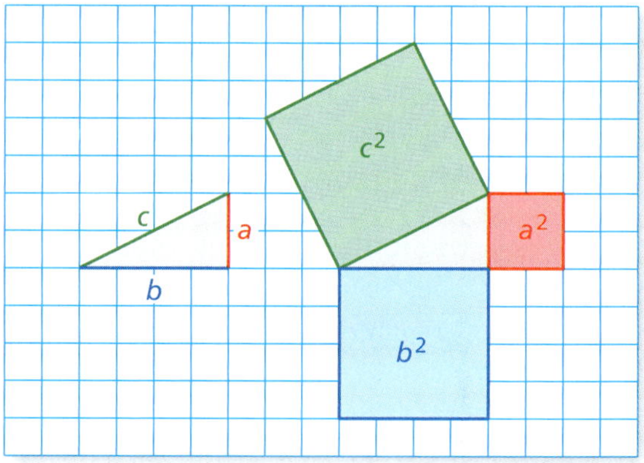

d. Cut out the three squares. Make eight copies of the right triangle and cut them out. Arrange the figures to form two identical larger squares.

e. What does this tell you about the relationship among a^2, b^2, and c^2?

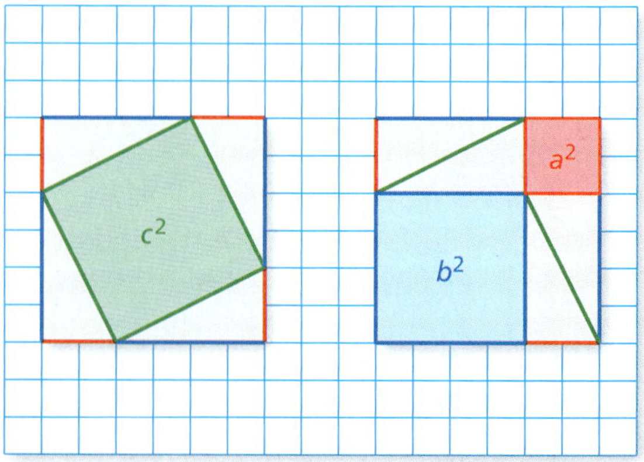

2 ACTIVITY: Finding the Length of the Hypotenuse

Work with a partner. Use the result of Activity 1 to find the length of the hypotenuse of each right triangle.

a.

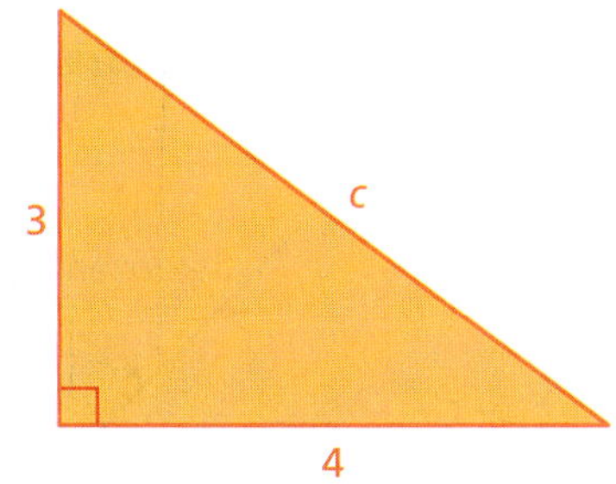

b.

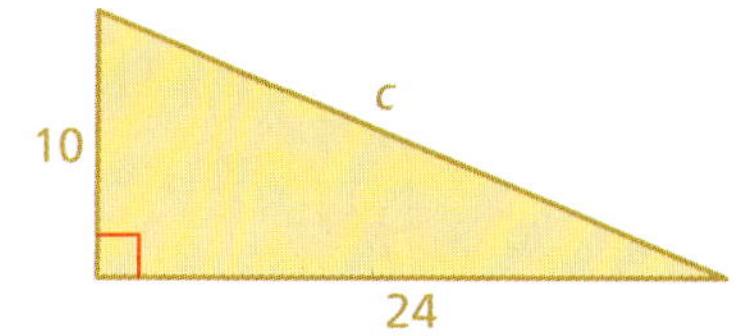

c.

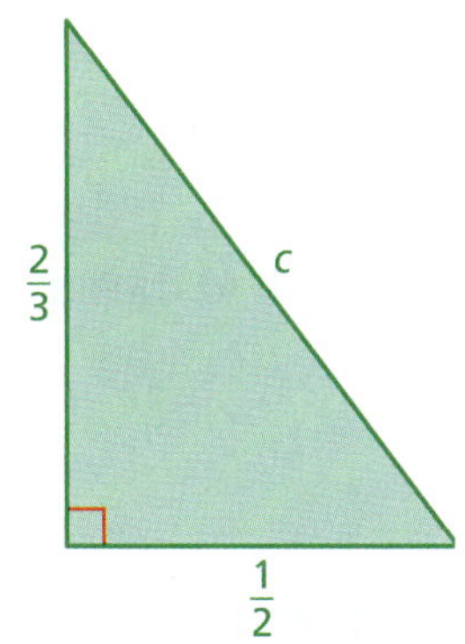

d.

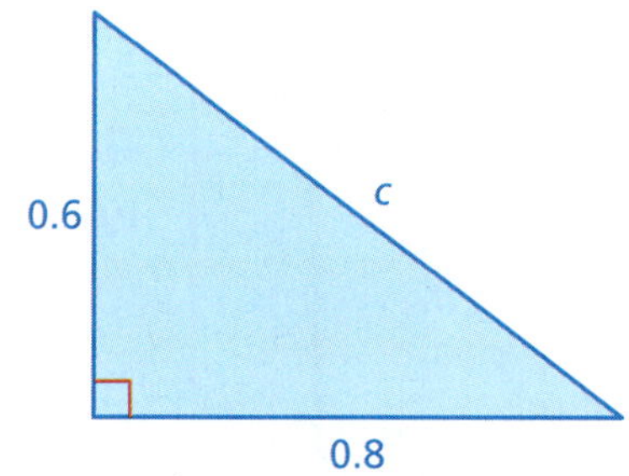

3 ACTIVITY: Finding the Length of a Leg

Work with a partner. Use the result of Activity 1 to find the length of the leg of each right triangle.

a.

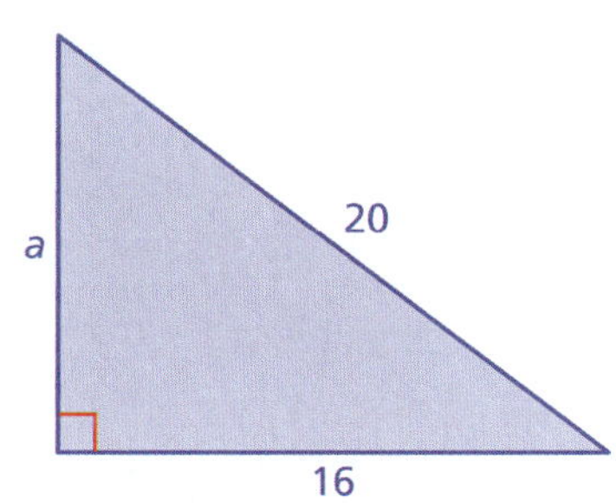

b.

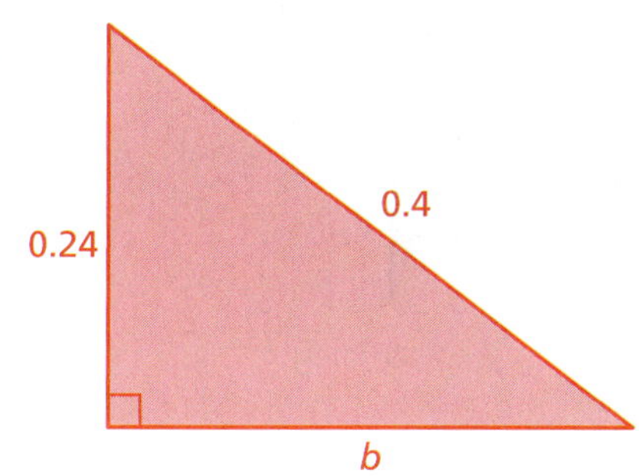

What Is Your Answer?

4. **IN YOUR OWN WORDS** How are the lengths of the sides of a right triangle related? Give an example using whole numbers.

Practice Use what you learned about the Pythagorean Theorem to complete Exercises 3–5 on page 240.

6.2 Lesson

Check It Out
Lesson Tutorials
BigIdeasMath.com

Key Vocabulary

theorem, *p. 236*
legs, *p. 238*
hypotenuse, *p. 238*
Pythagorean Theorem, *p. 238*

Key Ideas

Sides of a Right Triangle

The sides of a right triangle have special names.

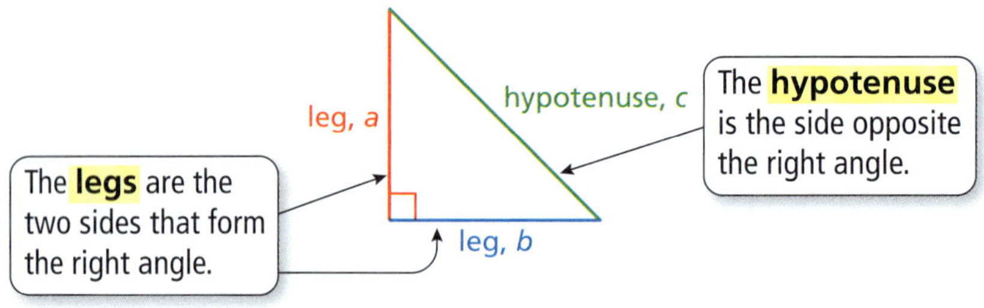

Study Tip

In a right triangle, the legs are the shorter sides and the hypotenuse is always the longest side.

The Pythagorean Theorem

Words In any right triangle, the sum of the squares of the lengths of the legs is equal to the square of the length of the hypotenuse.

Algebra $a^2 + b^2 = c^2$

EXAMPLE 1 Finding the Length of a Hypotenuse

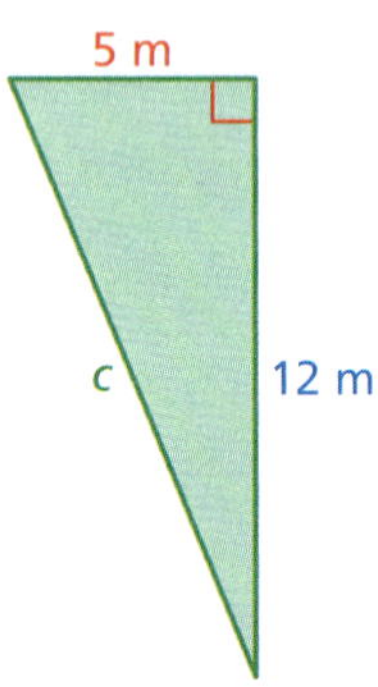

Find the length of the hypotenuse of the triangle.

$a^2 + b^2 = c^2$	Write the Pythagorean Theorem.
$5^2 + 12^2 = c^2$	Substitute 5 for *a* and 12 for *b*.
$25 + 144 = c^2$	Evaluate powers.
$169 = c^2$	Add.
$\sqrt{169} = \sqrt{c^2}$	Take positive square root of each side.
$13 = c$	Simplify.

The length of the hypotenuse is 13 meters.

On Your Own

Find the length of the hypotenuse of the triangle.

1.

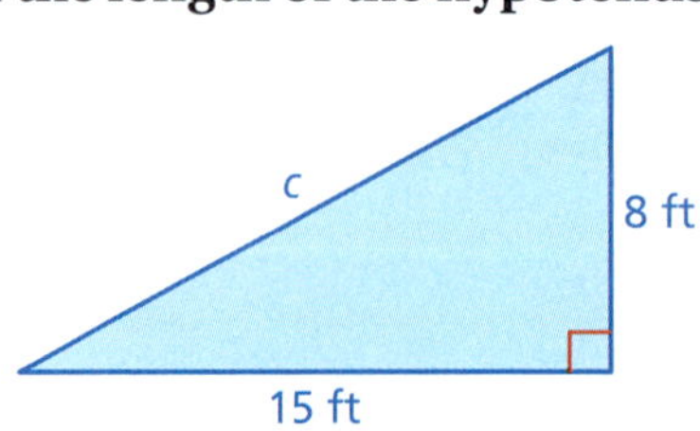

2.

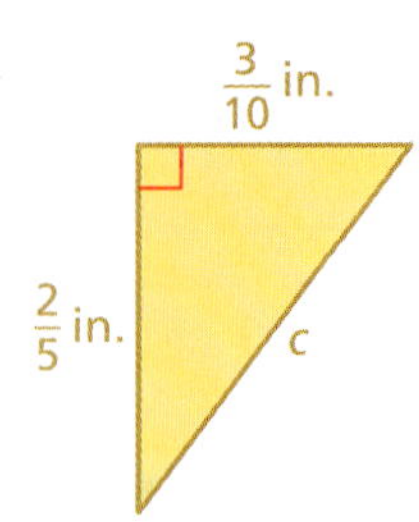

EXAMPLE 2 Finding the Length of a Leg

Find the missing length of the triangle.

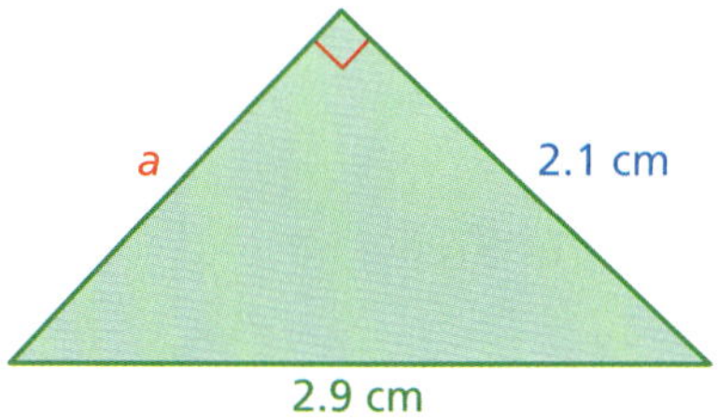

$a^2 + b^2 = c^2$	Write the Pythagorean Theorem.
$a^2 + 2.1^2 = 2.9^2$	Substitute 2.1 for *b* and 2.9 for *c*.
$a^2 + 4.41 = 8.41$	Evaluate powers.
$a^2 = 4$	Subtract 4.41 from each side.
$a = 2$	Take positive square root of each side.

∴ The length of the leg is 2 centimeters.

EXAMPLE 3 Standardized Test Practice

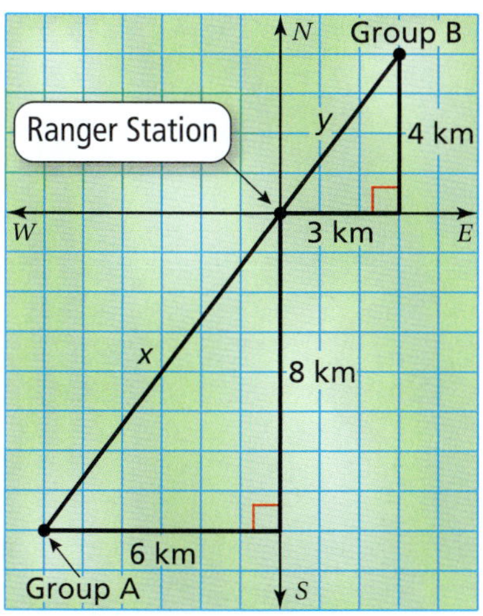

Hiking Group A leaves a ranger station and hikes 8 kilometers south then 6 kilometers west. Group B leaves the station and hikes 3 kilometers east then 4 kilometers north. Using the figure, how far apart are the two groups of hikers?

Ⓐ 5 km Ⓑ 10 km Ⓒ 15 km Ⓓ 21 km

The distance between the groups is the sum of the hypotenuses, *x* and *y*. Use the Pythagorean Theorem to find *x* and *y*.

$a^2 + b^2 = c^2$	Write the Pythagorean Theorem.	$a^2 + b^2 = c^2$
$6^2 + 8^2 = x^2$	Substitute.	$3^2 + 4^2 = y^2$
$36 + 64 = x^2$	Evaluate powers.	$9 + 16 = y^2$
$100 = x^2$	Add.	$25 = y^2$
$10 = x$	Take positive square root of each side.	$5 = y$

∴ The distance between the groups of hikers is $10 + 5 = 15$ kilometers. So, the correct answer is Ⓒ.

On Your Own

Now You're Ready
Exercises 3–8

Find the missing length of the triangle.

3.

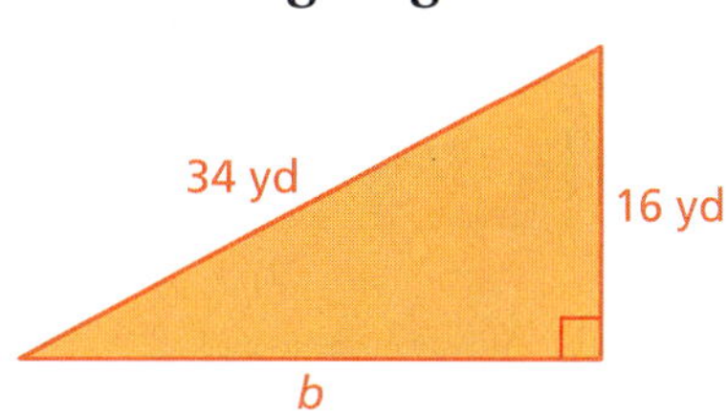

4.

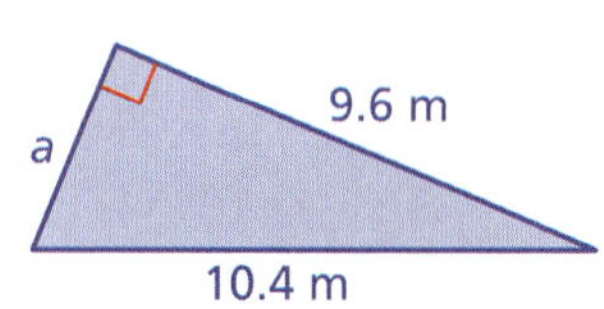

5. **WHAT IF?** In Example 3, Group A hikes 12 kilometers south and 9 kilometers west. How far apart are the hikers?

6.2 Exercises

Check It Out
Help with Homework
BigIdeasMath.com

Vocabulary and Concept Check

1. **VOCABULARY** In a right triangle, how can you tell which sides are the legs and which side is the hypotenuse?

2. **DIFFERENT WORDS, SAME QUESTION** Which is different? Find "both" answers.

   Which side is the hypotenuse?

   Which side is the longest?

   Which side is a leg?

   Which side is opposite the right angle?

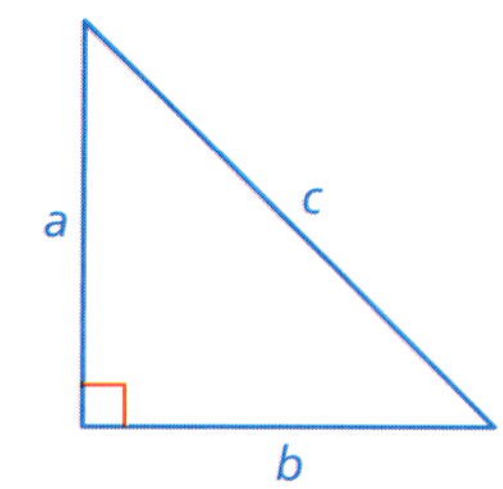

Practice and Problem Solving

Find the missing length of the triangle.

1 2 3.
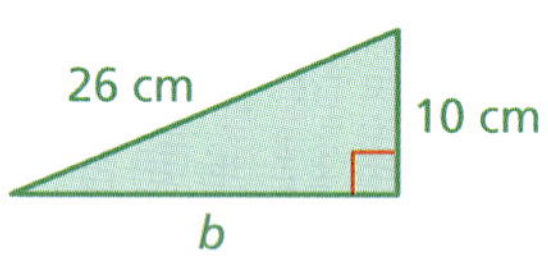

4.
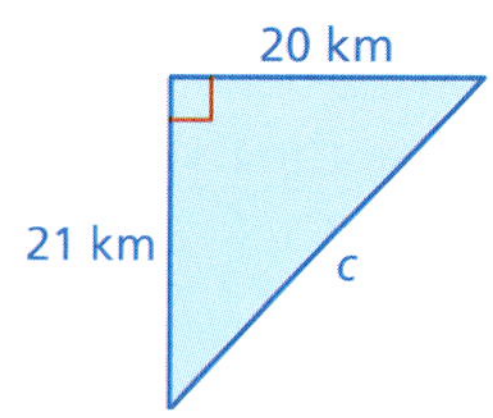

5.
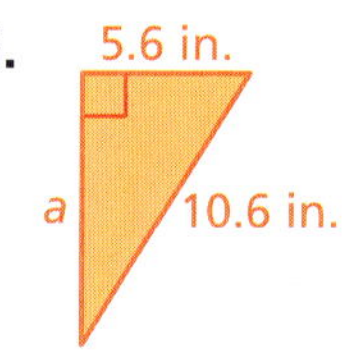

6.
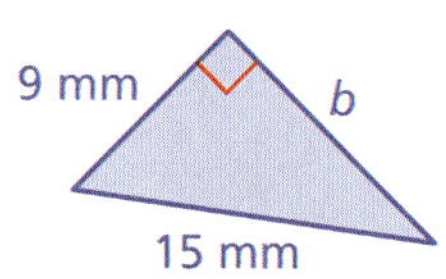

7.
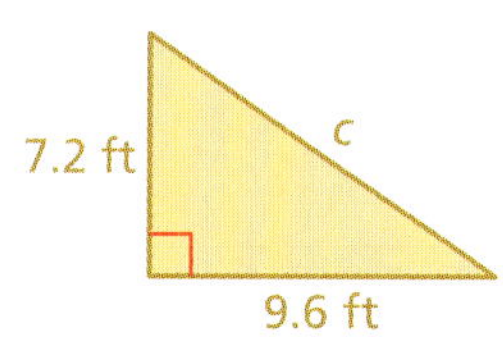

8.
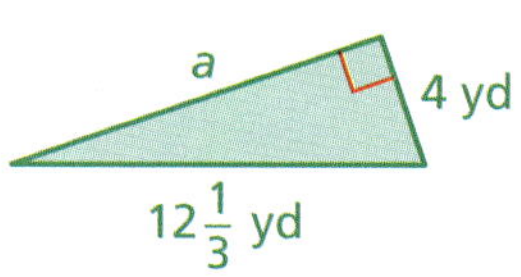

9. **ERROR ANALYSIS** Describe and correct the error in finding the missing length of the triangle.

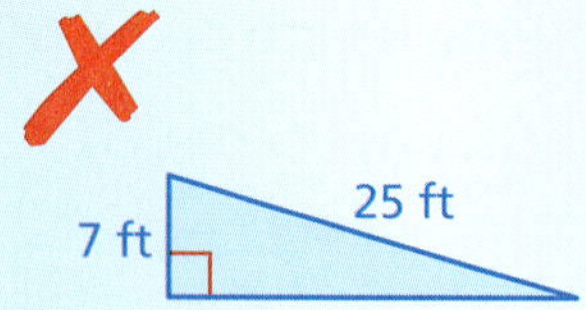

$$a^2 + b^2 = c^2$$
$$7^2 + 25^2 = c^2$$
$$674 = c^2$$
$$\sqrt{674} = c$$

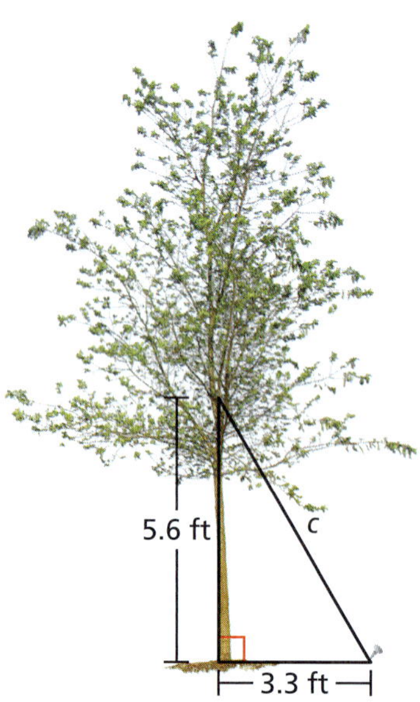

10. **TREE SUPPORT** How long is the wire that supports the tree?

Find the value of x.

11.

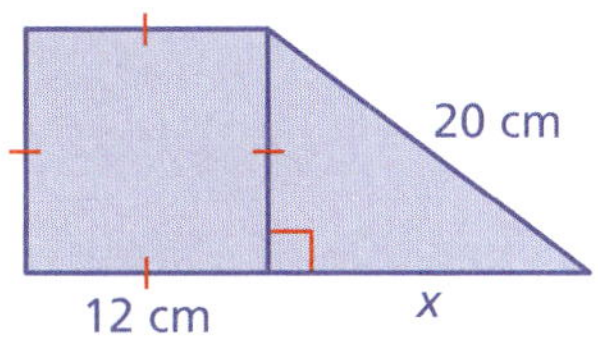

12.

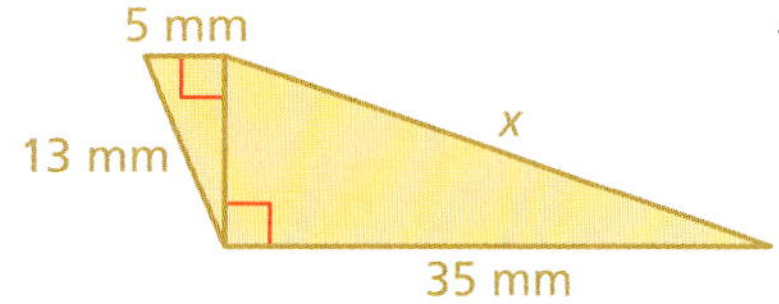

13.

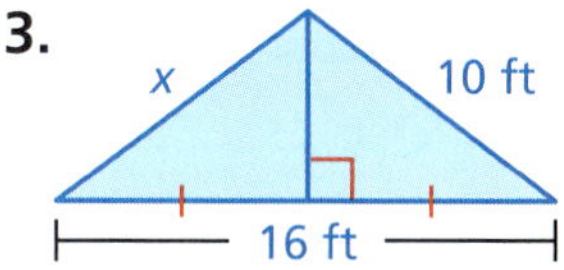

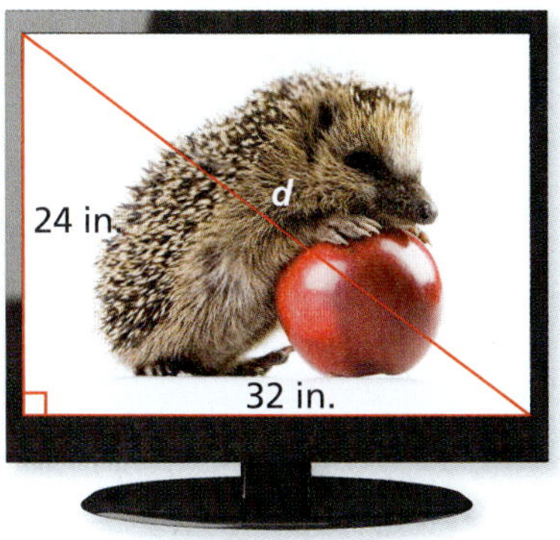

14. FLAT SCREEN Televisions are advertised by the lengths of their diagonals. A store has a sale on televisions 40 inches and larger. Is the television on sale? Explain.

15. BUTTERFLY Approximate the wingspan of the butterfly.

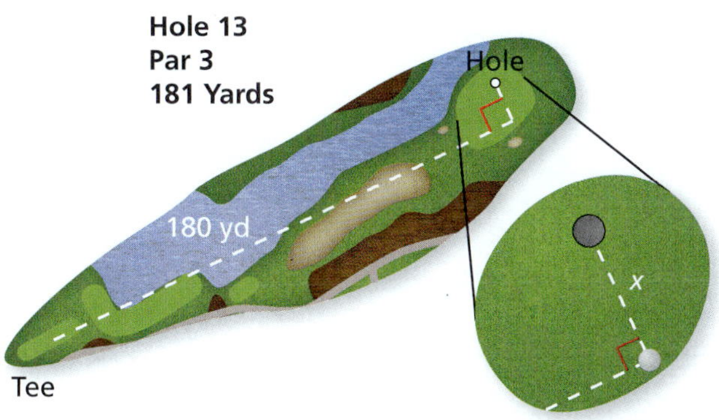

16. PGA The Players Championship is played at TPC Sawgrass golf course in Ponte Vedra Beach. The figure shows the location of the golf ball after a tee shot. How many feet from the hole is the ball?

17. SNOWBALLS You and a friend stand back-to-back. You run 20 feet forward then 15 feet to your right. At the same time, your friend runs 16 feet forward then 12 feet to her right. She stops and hits you with a snowball.

a. Draw the situation in a coordinate plane.

b. How far does your friend throw the snowball?

18. Algebra The legs of a right triangle have lengths of 28 meters and 21 meters. The hypotenuse has a length of $5x$ meters. What is the value of x?

Fair Game Review What you learned in previous grades & lessons

Find the square root(s). *(Section 6.1)*

19. $\pm\sqrt{36}$ **20.** $-\sqrt{121}$ **21.** $\sqrt{169}$ **22.** $-\sqrt{225}$

23. MULTIPLE CHOICE Which type of triangle can have an obtuse angle? *(Section 5.2)*

Ⓐ equiangular Ⓑ right Ⓒ isosceles Ⓓ equilateral

6 Study Help

Check It Out
Graphic Organizer
BigIdeasMath.com

You can use a **summary triangle** to explain a topic. Here is an example of a summary triangle for finding the length of the hypotenuse of a triangle.

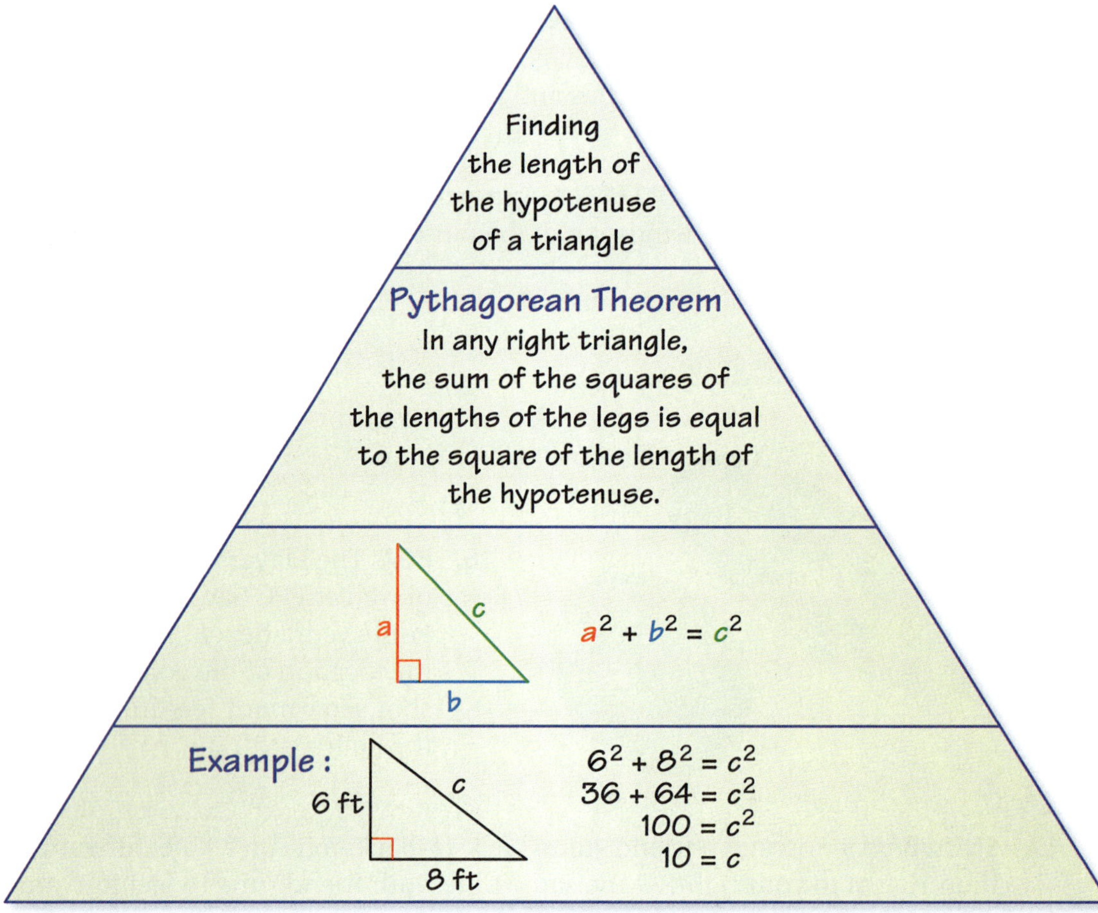

On Your Own

Make a summary triangle to help you study these topics.

1. finding square roots
2. evaluating expressions involving square roots
3. finding the length of a leg of a right triangle

After you complete this chapter, make summary triangles for the following topics.

4. approximating square roots
5. simplifying square roots

"What do you call a cheese summary triangle that isn't yours?"

6.1–6.2 Quiz

Check It Out
Progress Check
BigIdeasMath.com

Find the two square roots of the number. *(Section 6.1)*

1. 196 **2.** 49 **3.** 400

Find the square root(s). *(Section 6.1)*

4. $-\sqrt{4}$ **5.** $\sqrt{\frac{16}{25}}$ **6.** $\pm\sqrt{6.25}$

Evaluate the expression. *(Section 6.1)*

7. $3\sqrt{49} + 5$ **8.** $10 - 4\sqrt{16}$ **9.** $\frac{1}{4} + \sqrt{\frac{100}{4}}$

Find the missing length of the triangle. *(Section 6.2)*

10.

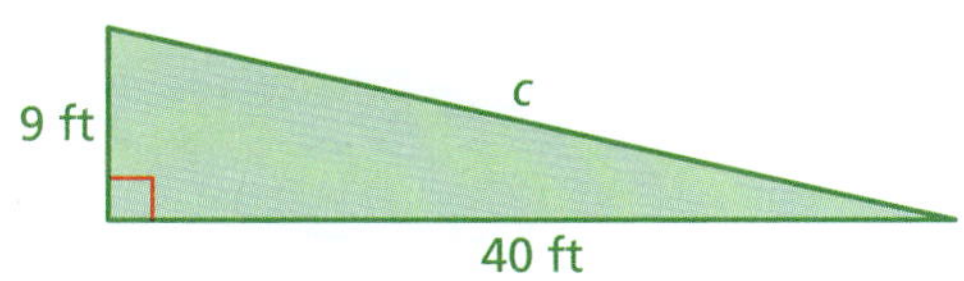

11.

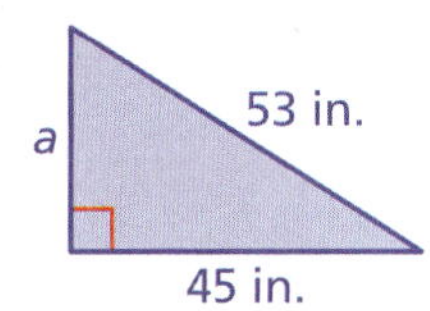

12.

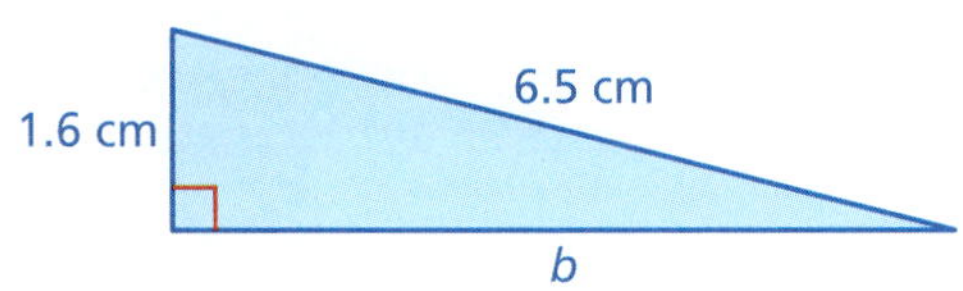

13.

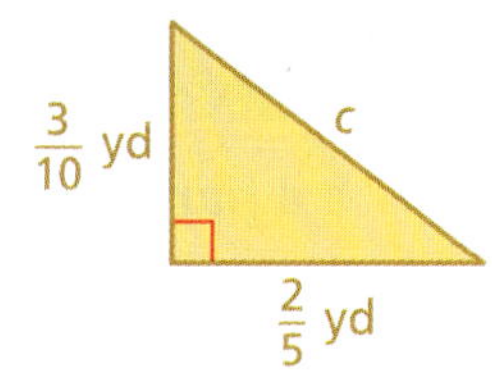

14. POOL The area of a circular pool cover is 314 square feet. Write and solve an equation to find the diameter of the pool cover. Use 3.14 for π. *(Section 6.1)*

15. LAND A square parcel of land has an area of 1 million square feet. What is the length of one side of the parcel? *(Section 6.1)*

16. FABRIC You are cutting a rectangular piece of fabric in half along the diagonal. The fabric measures 28 inches wide and $1\frac{1}{4}$ yards long. What is the length (in inches) of the diagonal? *(Section 6.2)*

6.3 Approximating Square Roots

STATE STANDARDS
MA.8.A.6.2

Essential Question How can you find decimal approximations of square roots that are irrational?

You already know that a rational number is a number that can be written as the ratio of two integers. Numbers that cannot be written as the ratio of two integers are called **irrational**.

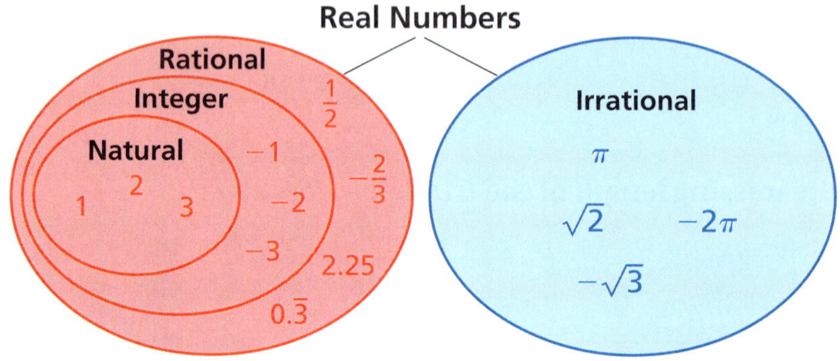

1 ACTIVITY: Approximating Square Roots

Work with a partner.

Archimedes was a Greek mathematician, physicist, engineer, inventor, and astronomer.

Archimedes
(c. 287 B.C.–c. 212 B.C.)

a. Archimedes tried to find a rational number whose square is 3. Here are two that he tried.

$$\frac{265}{153} \text{ and } \frac{1351}{780}$$

Are either of these numbers equal to $\sqrt{3}$? How can you tell?

b. Use a calculator with a square root key to approximate $\sqrt{3}$.

Write the number on a piece of paper. Then enter it into the calculator and square it. Then subtract 3. Do you get 0? Explain.

c. Calculators did not exist in the time of Archimedes. How do you think he might have approximated $\sqrt{3}$?

2 ACTIVITY: Approximating Square Roots Geometrically

Work with a partner.

a. Use grid paper and the given scale to draw a horizontal line segment 1 unit in length. Label this segment *AC*.

b. Draw a vertical line segment 2 units in length. Label this segment *DC*.

c. Set the point of a compass on *A*. Set the compass to 2 units. Swing the compass to intersect segment *DC*. Label this intersection as *B*.

d. Use the Pythagorean Theorem to show that the length of segment *BC* is $\sqrt{3}$ units.

e. Use the grid paper to approximate $\sqrt{3}$.

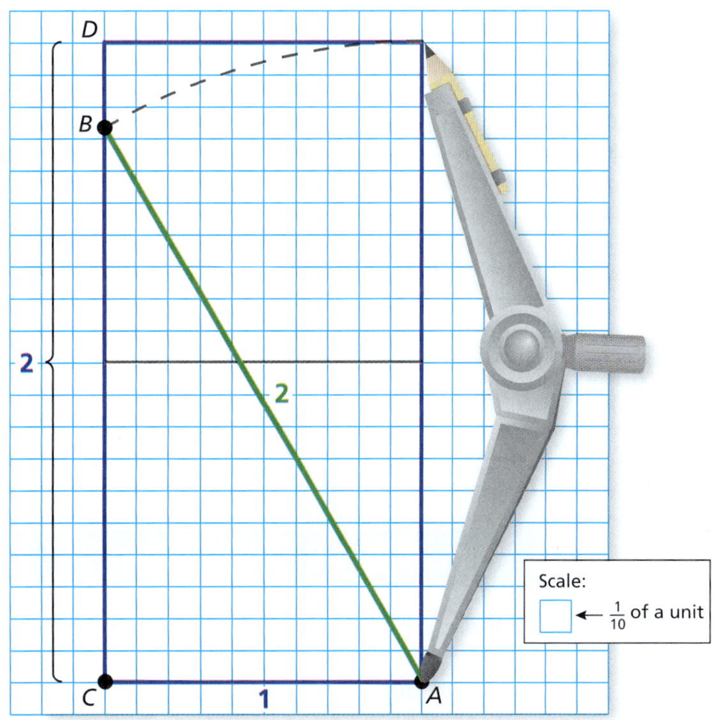

What Is Your Answer?

3. Repeat Activity 2 for a triangle in which segment *CA* is 2 units and segment *BA* is 3 units. Use the Pythagorean Theorem to show that segment *BC* is $\sqrt{5}$ units. Use the grid paper to approximate $\sqrt{5}$.

4. **IN YOUR OWN WORDS** How can you find decimal approximations of square roots that are irrational?

Practice Use what you learned about approximating square roots to complete Exercises 5–8 on page 249.

6.3 Lesson

Check It Out
Lesson Tutorials
BigIdeasMath.com

Key Vocabulary
irrational number, *p. 246*
real numbers, *p. 246*

A rational number is a number that can be written as the ratio of two integers. An **irrational number** cannot be written as the ratio of two integers.

- The square root of any whole number that is not a perfect square is irrational.
- The decimal form of an irrational number neither terminates nor repeats.

Key Idea

Real Numbers

Rational numbers and irrational numbers together form the set of **real numbers**.

Remember
Decimals that *terminate* or *repeat* are rational.

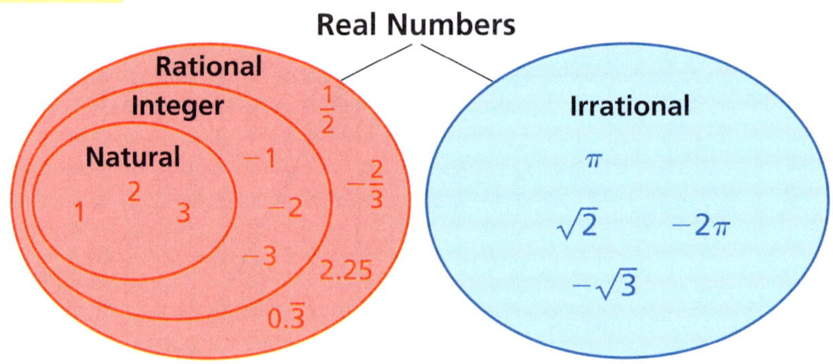

EXAMPLE 1 Classifying Real Numbers

Tell whether the number is *rational* or *irrational*. Explain.

	Number	Rational or Irrational	Reasoning
a.	$\sqrt{12}$	Irrational	12 is not a perfect square.
b.	$-0.36\overline{4}$	Rational	$-0.36\overline{4}$ is a repeating decimal.
c.	$-1\frac{3}{7}$	Rational	$-1\frac{3}{7}$ can be written as $\frac{-10}{7}$.
d.	0.85	Rational	0.85 can be written as $\frac{17}{20}$.

On Your Own

Now You're Ready
Exercises 9–14

Tell whether the number is *rational* or *irrational*. Explain.

1. 0.121221222. . . **2.** $-\sqrt{196}$ **3.** $\sqrt{2}$

EXAMPLE 2 Approximating Square Roots

Estimate $\sqrt{52}$ to the nearest integer.

Use a number line and the square roots of the perfect squares nearest to the radicand. The nearest perfect square less than 52 is 49. The nearest perfect square greater than 52 is 64.

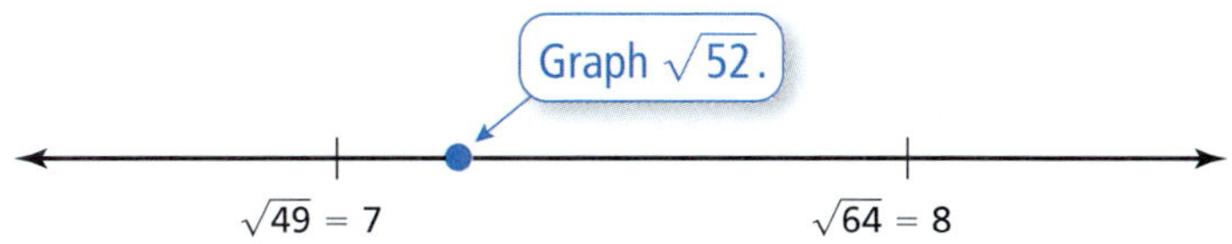

Because 52 is closer to 49 than to 64, $\sqrt{52}$ is closer to 7 than to 8.

So, $\sqrt{52} \approx 7$.

On Your Own

Now You're Ready
Exercises 18–23

Estimate to the nearest integer.

4. $\sqrt{33}$ **5.** $\sqrt{85}$ **6.** $\sqrt{190}$ **7.** $-\sqrt{7}$

EXAMPLE 3 Comparing Real Numbers

a. Which is greater, $\sqrt{5}$ or $2\frac{3}{4}$?

Graph the numbers on a number line.

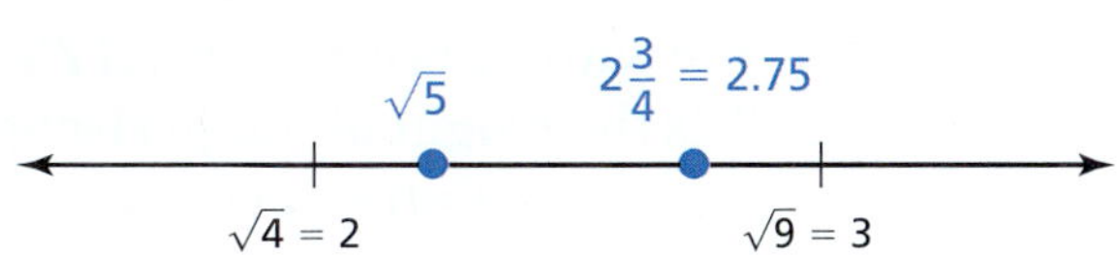

$2\frac{3}{4}$ is to the right of $\sqrt{5}$. So, $2\frac{3}{4}$ is greater.

b. Which is greater, $0.\overline{6}$ or $\sqrt{0.36}$?

Graph the numbers on a number line.

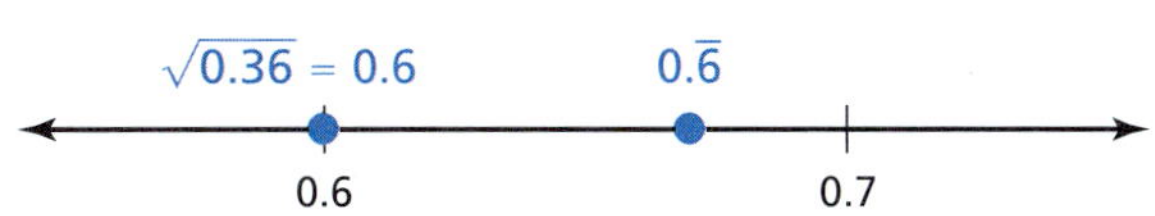

$0.\overline{6}$ is to the right of $\sqrt{0.36}$. So, $0.\overline{6}$ is greater.

On Your Own

Now You're Ready
Exercises 25–30

Which number is greater? Explain.

8. $4\frac{1}{5}, \sqrt{23}$ **9.** $\sqrt{10}, -\sqrt{5}$ **10.** $-\sqrt{2}, -2$

EXAMPLE 4 Approximating an Expression

The radius of a circle with area A is approximately $\sqrt{\frac{A}{3}}$. The area of a circular mouse pad is 51 square inches. Estimate its radius.

$$\sqrt{\frac{A}{3}} = \sqrt{\frac{51}{3}} \qquad \text{Substitute 51 for } A.$$

$$= \sqrt{17} \qquad \text{Divide.}$$

The nearest perfect square less than 17 is 16. The nearest perfect square greater than 17 is 25.

Because 17 is closer to 16 than to 25, $\sqrt{17}$ is closer to 4 than to 5.

∴ The radius is about 4 inches.

On Your Own

11. **WHAT IF?** The area of a circular mouse pad is 64 square inches. Estimate its radius.

EXAMPLE 5 Real-Life Application

The distance (in nautical miles) you can see with a periscope is $1.17\sqrt{h}$, where h is the height of the periscope above the water. Can a periscope that is 6 feet above the water see twice as far as a periscope that is 3 feet above the water? Explain.

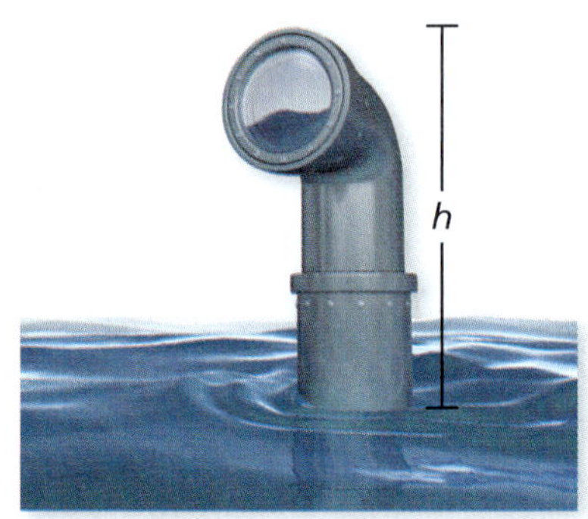

Use a calculator to find the distances.

3 feet above water		*6 feet above water*
$1.17\sqrt{h} = 1.17\sqrt{3}$	Substitute for h.	$1.17\sqrt{h} = 1.17\sqrt{6}$
≈ 2.03	Use a calculator.	≈ 2.87

You can see $\frac{2.87}{2.03} \approx 1.41$ times farther with the periscope that is 6 feet above the water than with the periscope that is 3 feet above the water.

∴ No, the periscope that is 6 feet above the water cannot see twice as far.

On Your Own

12. You use a periscope that is 10 feet above the water. Can you see farther than 4 nautical miles? Explain.

6.3 Exercises

Check It Out
Help with Homework
BigIdeasMath.com

Vocabulary and Concept Check

1. **VOCABULARY** What is the difference between a rational number and an irrational number?
2. **WRITING** Describe a method of approximating $\sqrt{32}$.
3. **VOCABULARY** What are real numbers? Give three examples.
4. **WHICH ONE DOESN'T BELONG?** Which number does *not* belong with the other three? Explain your reasoning.

$-\frac{11}{12}$ 25.075 $\sqrt{8}$ $-3.\overline{3}$

Practice and Problem Solving

Tell whether the rational number is a reasonable approximation of the square root.

5. $\frac{559}{250}, \sqrt{5}$
6. $\frac{3021}{250}, \sqrt{11}$
7. $\frac{678}{250}, \sqrt{28}$
8. $\frac{1677}{250}, \sqrt{45}$

Tell whether the number is *rational* or *irrational*. Explain.

1 9. $3.66666\overline{6}$
10. $\frac{\pi}{6}$
11. $-\sqrt{7}$
12. -1.125
13. $-3\frac{8}{9}$
14. $\sqrt{15}$

15. **ERROR ANALYSIS** Describe and correct the error in classifying the number.

✗ $\sqrt{144}$ is irrational.

16. **SCRAPBOOKING** You cut a picture into a right triangle for your scrapbook. The lengths of the legs of the triangle are 4 inches and 6 inches. Is the length of the hypotenuse a rational number? Explain.

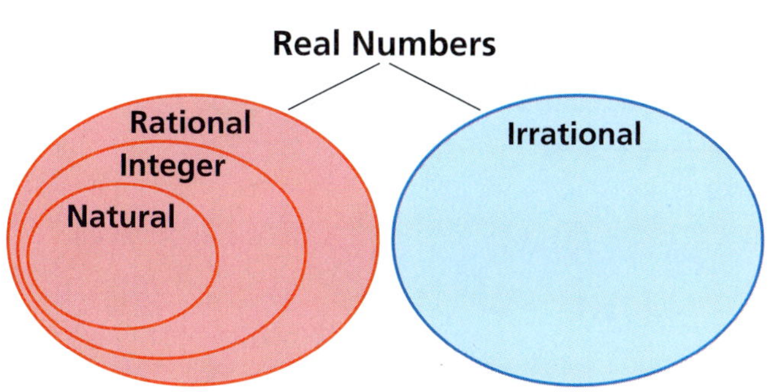

17. **VENN DIAGRAM** Place each number in the correct area of the Venn Diagram.
    a. Your age
    b. The square root of any prime number
    c. The ratio of the circumference of a circle to its diameter

Estimate to the nearest integer.

2 **18.** $\sqrt{24}$ **19.** $\sqrt{685}$ **20.** $-\sqrt{61}$

21. $-\sqrt{105}$ **22.** $\sqrt{\frac{27}{4}}$ **23.** $-\sqrt{\frac{335}{2}}$

24. CHECKERS A checkerboard is 8 squares long and 8 squares wide. The area of each square is 14 square centimeters. Estimate the perimeter of the checkerboard.

Which number is greater? Explain.

3 **25.** $\sqrt{20}$, 10 **26.** $\sqrt{15}$, -3.5 **27.** $\sqrt{133}$, $10\frac{3}{4}$

28. $\frac{2}{3}$, $\sqrt{\frac{16}{81}}$ **29.** $-\sqrt{0.25}$, -0.25 **30.** $-\sqrt{182}$, $-\sqrt{192}$

31. FOUR SQUARE The area of a four square court is 66 square feet. Estimate the length s of one of the sides of the court.

32. RADIO SIGNAL The maximum distance (in nautical miles) that a radio transmitter signal can be sent is represented by the expression $1.23\sqrt{h}$, where h is the height (in feet) above the transmitter.

Estimate the maximum distance x (in nautical miles) between the plane that is receiving the signal and the transmitter. Round your answer to the nearest tenth.

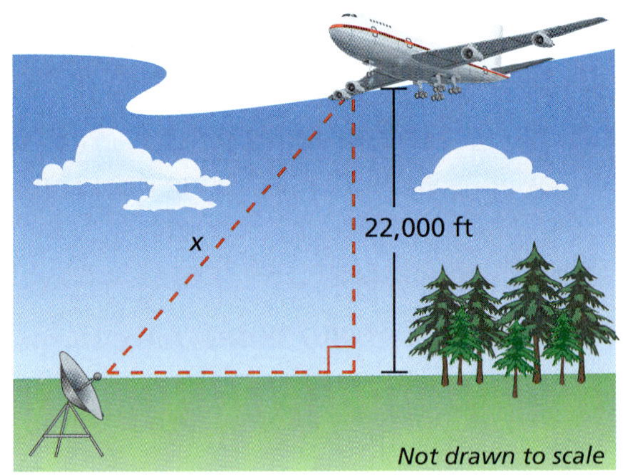

33. OPEN-ENDED Find two numbers a and b that satisfy the diagram.

Estimate to the nearest tenth.

34. $\sqrt{0.39}$ **35.** $\sqrt{1.19}$ **36.** $\sqrt{1.52}$

37. ROLLER COASTER The velocity v (in meters per second) of a roller coaster is represented by the equation $v = 3\sqrt{6r}$, where r is the radius of the loop. Estimate the velocity of a car going around the loop. Round your answer to the nearest tenth.

38. Is $\sqrt{\frac{1}{4}}$ a rational number? Is $\sqrt{\frac{3}{16}}$ a rational number? Explain.

39. WATER BALLOON The time t (in seconds) it takes a water balloon to fall d meters is represented by the equation $t = \sqrt{\frac{d}{4.9}}$. Estimate the time it takes the balloon to fall to the ground from a window that is 14 meters above the ground. Round your answer to the nearest tenth.

40. Number Sense Determine if the statement is *sometimes, always,* or *never* true. Explain your reasoning and give an example of each.

a. A rational number multiplied by a rational number is rational.

b. A rational number multiplied by an irrational number is rational.

c. An irrational number multiplied by an irrational number is rational.

Fair Game Review What you learned in previous grades & lessons

Simplify the expression. *(Skills Review Handbook)*

41. $2x + 3y - 5x$ **42.** $3\pi + 8(t - \pi) - 4t$ **43.** $17k - 9 + 23k$

44. MULTIPLE CHOICE What is the ratio (red to blue) of the corresponding side lengths of the similar triangles? *(Skills Review Handbook)*

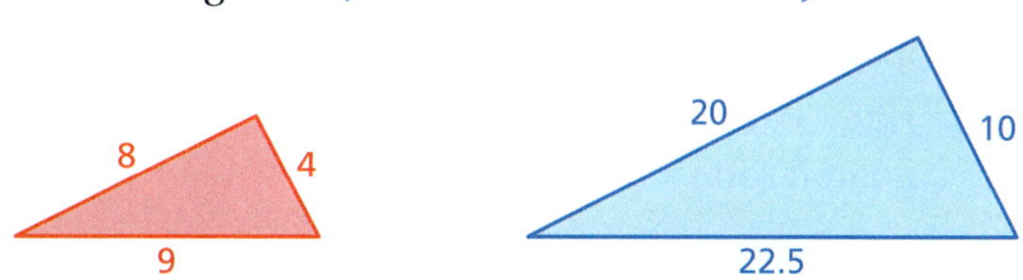

Ⓐ 1 : 3 Ⓑ 5 : 2 Ⓒ 3 : 4 Ⓓ 2 : 5

6.4 Simplifying Square Roots

STATE STANDARDS
MA.8.A.6.2
MA.8.A.6.4

Essential Question How can you use a square root to describe the golden ratio?

Two quantities are in the *golden ratio* if the ratio between the sum of the quantities and the larger quantity is the same as the ratio between the larger quantity and the smaller quantity.

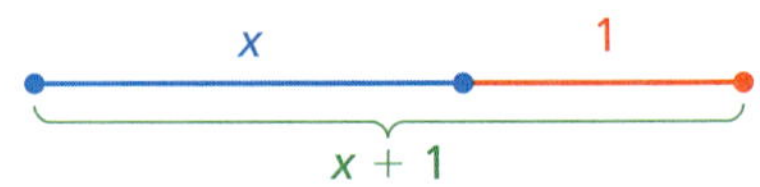

$$\frac{x+1}{x} = \frac{x}{1}$$

In a future algebra course, you will be able to prove that the golden ratio is

$$\frac{1+\sqrt{5}}{2} \qquad \text{Golden ratio.}$$

1 ACTIVITY: Constructing a Golden Ratio

Work with a partner.

a. Use grid paper and the given scale to draw a square that is 1 unit by 1 unit (blue).

b. Draw a line from midpoint *C* of one side of the square to the opposite corner *D*, as shown.

c. Use the Pythagorean Theorem to find the length of segment *CD*.

d. Set the point of a compass on *C*. Set the compass radius to the length of segment *CD*. Swing the compass to intersect line *BC* at point *E*.

e. The rectangle *ABEF* is called a *golden rectangle* because the ratio of its side lengths is the golden ratio.

f. Use a calculator to find a decimal approximation of the golden ratio. Round your answer to two decimal places.

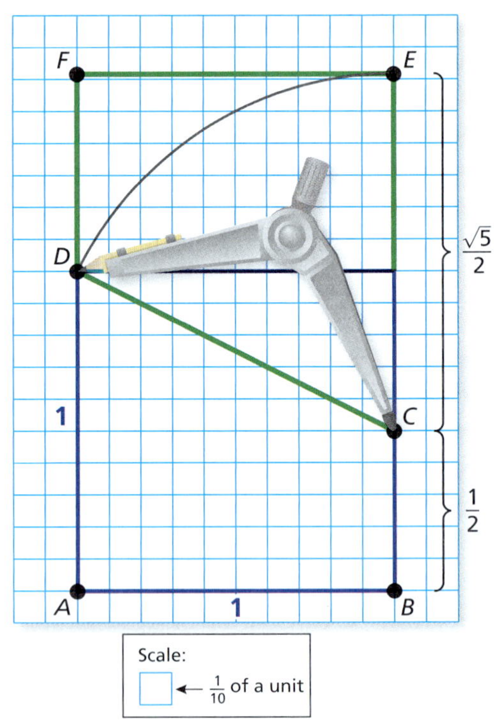

2 ACTIVITY: The Golden Ratio and the Human Body

Work with a partner.

Leonardo da Vinci was one of the first to notice that there are several ratios in the human body that approximate the golden ratio.

a. Use a tape measure or two yardsticks to measure the lengths shown in the diagram for both you and your partner. (Take your shoes off before measuring.)

b. Copy the tables below. Record your results in the first two columns.

c. Calculate the ratios shown in the tables.

d. Leonardo da Vinci stated that for many people, the ratios are close to the golden ratio. How close are your ratios?

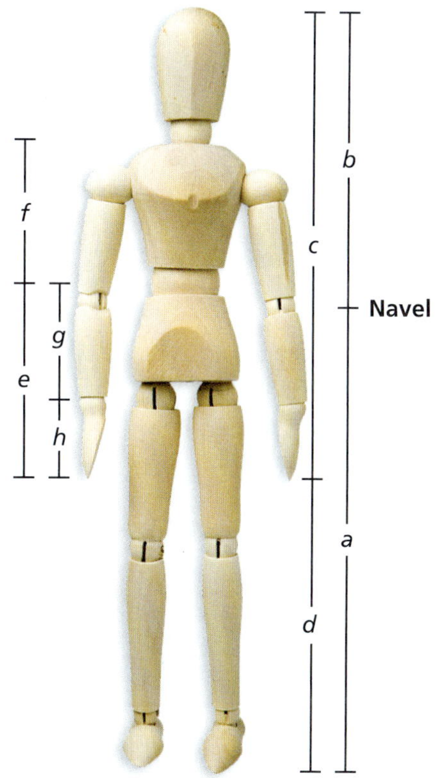

You		
$a =$	$b =$	$\frac{a}{b} =$
$c =$	$d =$	$\frac{c}{d} =$
$e =$	$f =$	$\frac{e}{f} =$
$g =$	$h =$	$\frac{g}{h} =$

Partner		
$a =$	$b =$	$\frac{a}{b} =$
$c =$	$d =$	$\frac{c}{d} =$
$e =$	$f =$	$\frac{e}{f} =$
$g =$	$h =$	$\frac{g}{h} =$

What Is Your Answer?

3. **IN YOUR OWN WORDS** How can you use a square root to describe the golden ratio? Use the Internet or some other reference to find examples of the golden ratio in art and architecture.

Practice Use what you learned about square roots to complete Exercises 3–5 on page 256.

6.4 Lesson

Check It Out
Lesson Tutorials
BigIdeasMath.com

You can add or subtract radical expressions the same way you combine like terms, such as $5x + 4x = 9x$.

EXAMPLE 1 Adding and Subtracting Square Roots

Reading

Do not assume that radicals that have different radicands cannot be simplified.

An expression such as $2\sqrt{4} + \sqrt{1}$ can easily be simplified.

a. Simplify $5\sqrt{2} + 4\sqrt{2}$.

$5\sqrt{2} + 4\sqrt{2} = (5 + 4)\sqrt{2}$ — Use the Distributive Property.

$= 9\sqrt{2}$ — Simplify.

b. Simplify $2\sqrt{3} - 7\sqrt{3}$.

$2\sqrt{3} - 7\sqrt{3} = (2 - 7)\sqrt{3}$ — Use the Distributive Property.

$= -5\sqrt{3}$ — Simplify.

On Your Own

Now You're Ready
Exercises 6–14

Simplify the expression.

1. $\sqrt{5} + \sqrt{5}$ **2.** $6\sqrt{10} + 4\sqrt{10}$ **3.** $2\sqrt{7} - \sqrt{7}$

To simplify square roots that are not perfect squares, use the following property.

Key Idea

Product Property of Square Roots

Algebra $\sqrt{xy} = \sqrt{x} \cdot \sqrt{y}$, where $x, y \geq 0$

Numbers $\sqrt{4 \cdot 3} = \sqrt{4} \cdot \sqrt{3} = 2\sqrt{3}$

EXAMPLE 2 Simplifying Square Roots

Study Tip

A square root is simplified when the radicand has no perfect square factors other than 1.

Simplify $\sqrt{50}$.

$\sqrt{50} = \sqrt{25 \cdot 2}$ — Factor using the greatest perfect square factor.

$= \sqrt{25} \cdot \sqrt{2}$ — Use the Product Property of Square Roots.

$= 5\sqrt{2}$ — Simplify.

On Your Own

Now You're Ready
Exercises 16–20

Simplify the expression.

4. $\sqrt{24}$ **5.** $\sqrt{45}$ **6.** $\sqrt{98}$

Key Idea

Quotient Property of Square Roots

Algebra $\sqrt{\frac{x}{y}} = \frac{\sqrt{x}}{\sqrt{y}}$, where $x \geq 0$ and $y > 0$

Numbers $\sqrt{\frac{7}{9}} = \frac{\sqrt{7}}{\sqrt{9}} = \frac{\sqrt{7}}{3}$

EXAMPLE 3 Simplifying Square Roots

Simplify $\sqrt{\frac{11}{16}}$.

$\sqrt{\frac{11}{16}} = \frac{\sqrt{11}}{\sqrt{16}}$ Use the Quotient Property of Square Roots.

$= \frac{\sqrt{11}}{4}$ Simplify.

EXAMPLE 4 Finding a Volume

Remember

The volume V of a rectangular prism is the product of the area of its base B and its height h.

$V = Bh$

Find the volume of the rectangular prism.

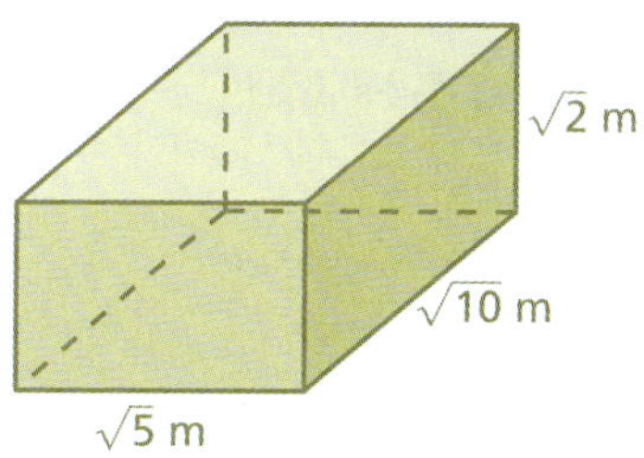

$V = Bh$ Write formula for volume.

$= (\sqrt{5})(\sqrt{10})(\sqrt{2})$ Substitute.

$= \sqrt{5 \cdot 10 \cdot 2}$ Use the Product Property of Square Roots.

$= \sqrt{100}$ Multiply.

$= 10$ Simplify.

The volume is 10 cubic meters.

On Your Own

Now You're Ready
Exercises 21–24

Simplify the expression.

7. $\sqrt{\frac{35}{36}}$ **8.** $\sqrt{\frac{13}{4}}$ **9.** $\sqrt{\frac{5}{b^2}}$

10. WHAT IF? In Example 4, the height of the rectangular prism is $\sqrt{8}$ meters. Find the volume of the prism.

6.4 Exercises

Check It Out
Help with Homework
BigIdeasMath.com

Vocabulary and Concept Check

1. **WRITING** Describe how combining like terms is similar to adding and subtracting square roots.
2. **WRITING** How are the Product Property of Square Roots and the Quotient Property of Square Roots similar?

Practice and Problem Solving

Find the ratio of the side lengths. Is the ratio close to the golden ratio?

3.

4.

5.

Simplify the expression.

1 6. $\frac{\sqrt{2}}{9} + \frac{1}{9}$

7. $\frac{\sqrt{7}}{3} + \frac{1}{3}$

8. $\frac{1}{4} + \frac{\sqrt{13}}{4}$

9. $2\sqrt{3} + 4\sqrt{3}$

10. $6\sqrt{7} - 2\sqrt{7}$

11. $\frac{3}{4}\sqrt{5} + \frac{5}{4}\sqrt{5}$

12. $\sqrt{6} - 4\sqrt{6}$

13. $1.5\sqrt{15} - 9.2\sqrt{15}$

14. $\frac{7}{8}\sqrt{11} + \frac{3}{8}\sqrt{11}$

15. **ERROR ANALYSIS** Describe and correct the error in simplifying the expression.

✗ $4\sqrt{5} + 3\sqrt{5} = 7\sqrt{10}$

Simplify the expression.

2 3 16. $\sqrt{18}$

17. $\sqrt{200}$

18. $\sqrt{12}$

19. $\sqrt{48}$

20. $\sqrt{125}$

21. $\sqrt{\frac{23}{64}}$

22. $\sqrt{\frac{65}{121}}$

23. $\sqrt{\frac{17}{49}}$

24. $\sqrt{\frac{22}{c^2}}$

25. **RAIN GUTTER** A rain gutter is made from a single sheet of metal. What is the length of the red cross-section?

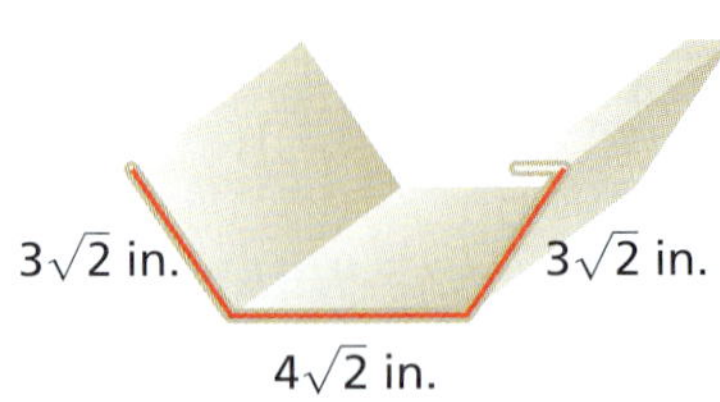

Simplify the expression.

26. $3\sqrt{5} - \sqrt{45}$

27. $\sqrt{24} + 4\sqrt{6}$

28. $\frac{4}{3}\sqrt{7} + \sqrt{28}$

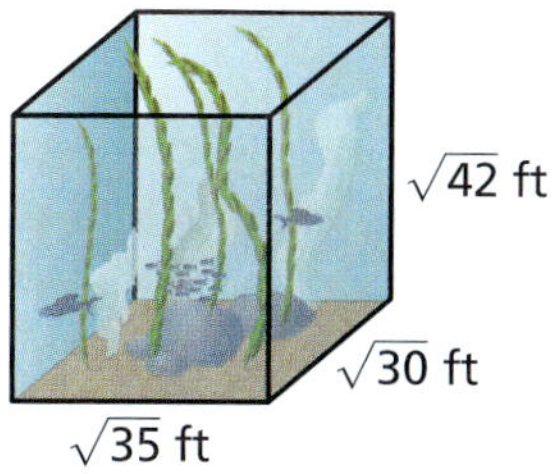

29. **VOLUME** What is the volume of the aquarium (in cubic feet)?

30. **RATIO** The ratio $3:x$ is equivalent to the ratio $x:5$. What are the possible values of x?

31. **GEOGRAPHY** DeSoto County is in the approximate shape of a rectangle.

    a. What is the perimeter of DeSoto County?

    b. What is the area of DeSoto County?

32. **MT. FUJI** Mt. Fuji is in the shape of a cone with a volume of about 475π cubic kilometers. What is the radius of the base of Mt. Fuji?

The height of Mt. Fuji is 3.8 kilometers.

33. **Geometry** A block of ice is in the shape of a square prism. You want to put the block of ice in a cylindrical cooler. The equation $s^2 = 2r^2$ represents the minimum radius r needed for the block of ice with side length s to fit in the cooler.

    a. Solve the equation for r.

    b. Use the equation in part (a) to find the minimum radius needed when the side length of the block of ice is $\sqrt{98}$ inches.

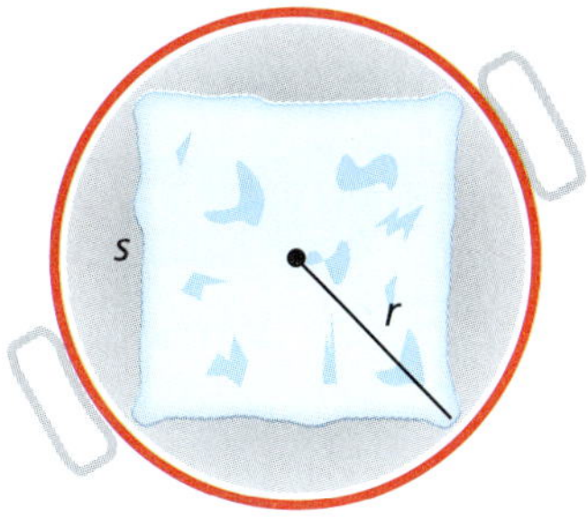

Fair Game Review What you learned in previous grades & lessons

Find the missing length of the triangle. *(Section 6.2)*

34.

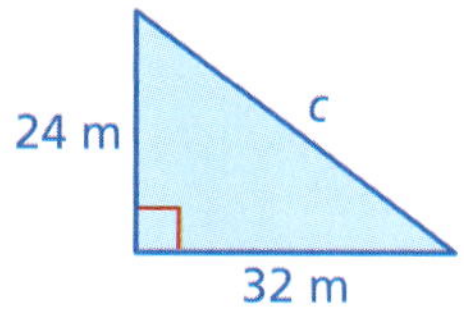

35.

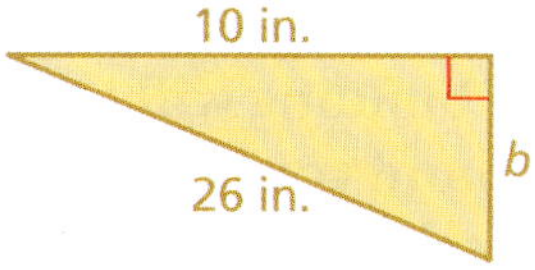

36.

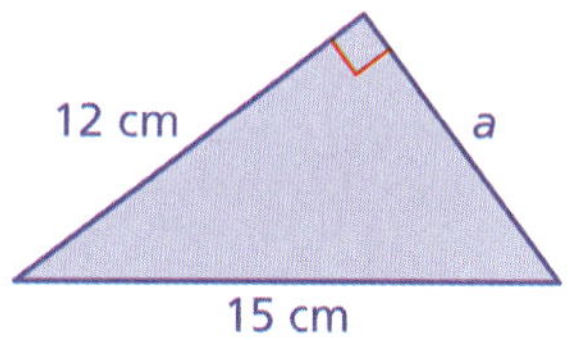

37. **MULTIPLE CHOICE** Where is $-\sqrt{110}$ on a number line? *(Section 6.3)*

    Ⓐ Between -9 and -10 Ⓑ Between 9 and 10

    Ⓒ Between -10 and -11 Ⓓ Between 10 and 11

6.5 Using the Pythagorean Theorem

STATE STANDARDS
MA.8.G.2.4
MA.8.A.6.2
MA.8.A.6.4

Essential Question How can you use the Pythagorean Theorem to solve real-life problems?

Share Your Work at... My.BigIdeasMath.com

1 ACTIVITY: Using the Pythagorean Theorem

Work with a partner.

a. A baseball player throws a ball from second base to home plate. How far does the player throw the ball? Include a diagram showing how you got your answer. Decide how many decimal points of accuracy are reasonable. Explain your reasoning.

b. The distance from the pitcher's mound to home plate is 60.5 feet. Does this form a right triangle with first base? Explain your reasoning.

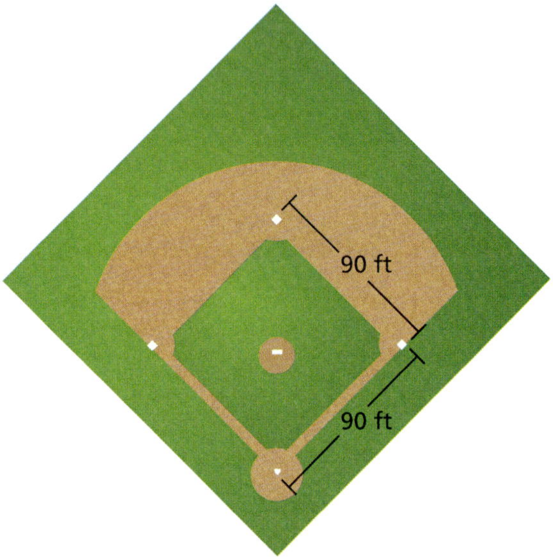

2 ACTIVITY: Firefighting and Ladders

Work with a partner.

The recommended angle for a firefighting ladder is 75°.

When a 110-foot ladder is put up against a building at this angle, the base of the ladder is about 28 feet from the building.

The base of the ladder is 8 feet above the ground.

How high on the building will the ladder reach? Round your answer to the nearest tenth.

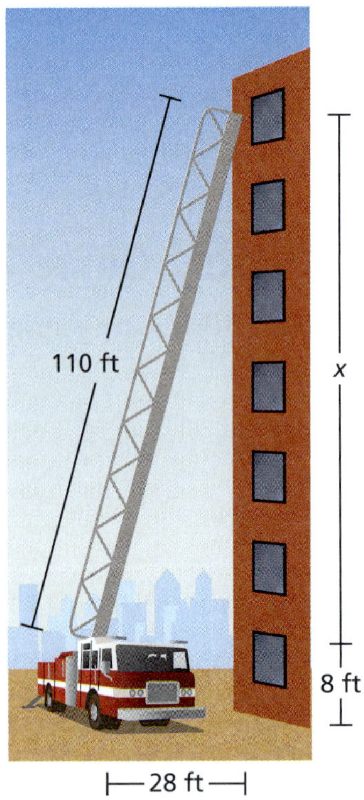

3 ACTIVITY: Finding Perimeters

Work with a partner.

Find the perimeter of each figure. Round your answer to the nearest tenth. Did you use the Pythagorean Theorem? If so, explain.

a. Right triangle

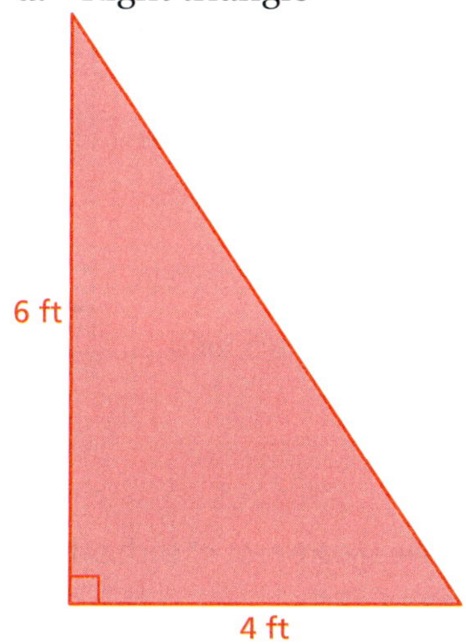

b. Trapezoid

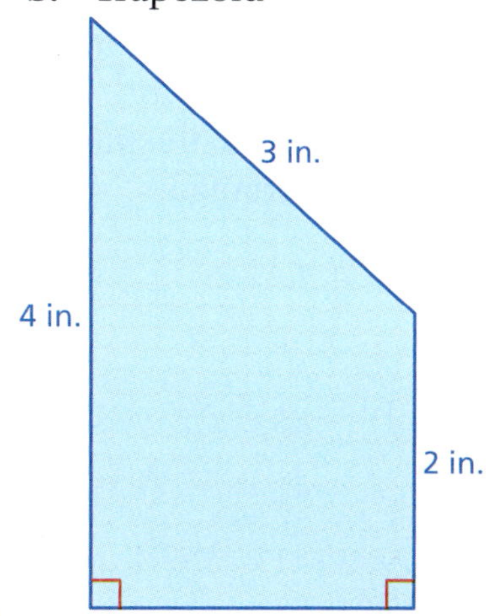

c. Parallelogram

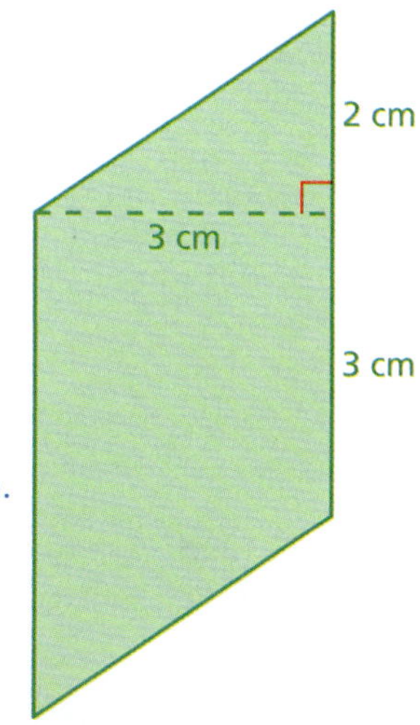

4 ACTIVITY: Writing a Formula

Work with a partner.

a. Write a formula for the area of an equilateral triangle with side length *s*.

b. Use your formula to find the area of an equilateral triangle with a side length of 10 inches.

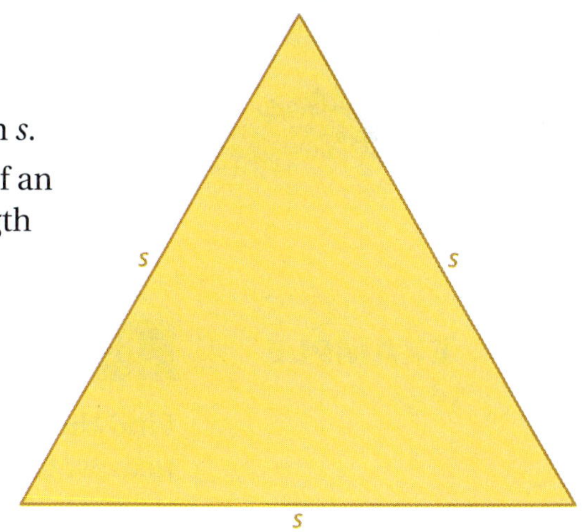

What Is Your Answer?

5. **IN YOUR OWN WORDS** How can you use the Pythagorean Theorem to solve real-life problems?

6. Describe a situation in which you could use the Pythagorean Theorem to help make decisions. Give an example of a real-life problem.

Practice Use what you learned about using the Pythagorean Theorem to complete Exercises 3–5 on page 262.

6.5 Lesson

Check It Out
Lesson Tutorials
BigIdeasMath.com

EXAMPLE 1 Finding a Distance in a Coordinate Plane

Key Vocabulary
Pythagorean triple, *p. 261*

The park is 5 miles east of your home. The library is 4 miles north of the park. How far is your home from the library? Round your answer to the nearest tenth.

Plot a point for your home at the origin in a coordinate plane. Then plot points for the park and the library to form a right triangle.

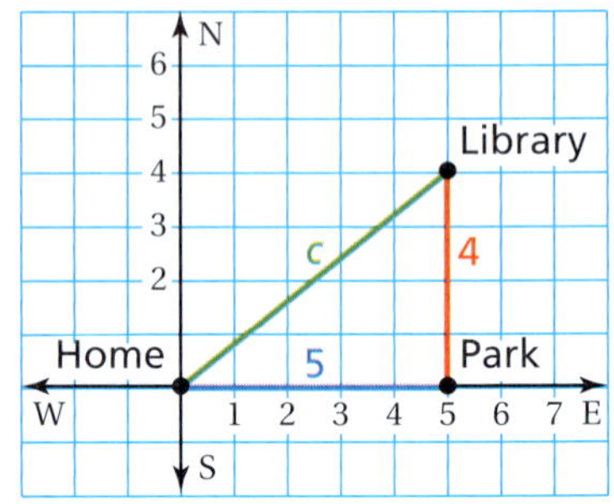

$a^2 + b^2 = c^2$	Write the Pythagorean Theorem.
$4^2 + 5^2 = c^2$	Substitute 4 for *a* and 5 for *b*.
$16 + 25 = c^2$	Evaluate powers.
$41 = c^2$	Add.
$\sqrt{41} = \sqrt{c^2}$	Take positive square root of each side.
$6.4 \approx c$	Use a calculator.

Your home is about 6.4 miles from the library.

On Your Own

Now You're Ready
Exercises 6–8

1. The post office is 3 miles west of your home. Your school is 2 miles north of the post office. How far is your home from your school? Round your answer to the nearest tenth.

EXAMPLE 2 Real-Life Application

Find the height of the firework. Round your answer to the nearest tenth.

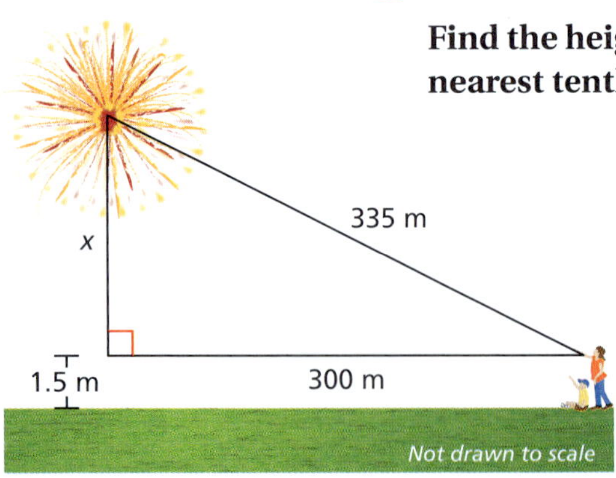

$a^2 + b^2 = c^2$	Write the Pythagorean Theorem.
$x^2 + 300^2 = 335^2$	Substitute.
$x^2 + 90{,}000 = 112{,}225$	Evaluate powers.
$x^2 = 22{,}225$	Subtract 90,000 from each side.
$\sqrt{x^2} = \sqrt{22{,}225}$	Take positive square root of each side.
$x \approx 149.1$	Use a calculator.

The height of the firework is about $149.1 + 1.5 = 150.6$ meters.

On Your Own

Now You're Ready
Exercises 9–11

2. **WHAT IF?** In Example 2, the distance between you and the firework is 350 meters. Find the height of the firework. Round your answer to the nearest tenth.

A **Pythagorean triple** is a set of three positive integers a, b, and c where $a^2 + b^2 = c^2$.

Key Idea

Converse of the Pythagorean Theorem

If the equation $a^2 + b^2 = c^2$ is true for the side lengths of a triangle, then the triangle is a right triangle.

When using the converse of the Pythagorean Theorem, always substitute the length of the longest side for c.

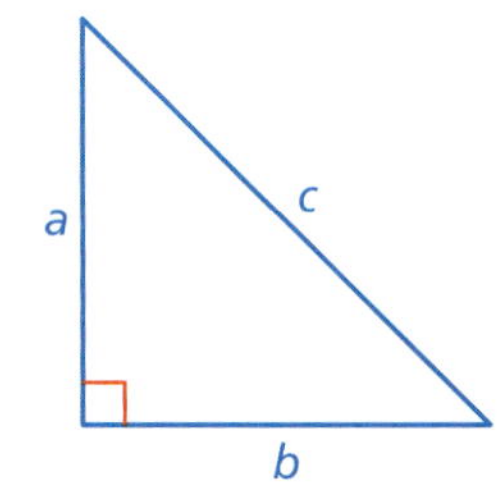

EXAMPLE 3 Identifying a Right Triangle

Tell whether the given triangle is a right triangle.

a.

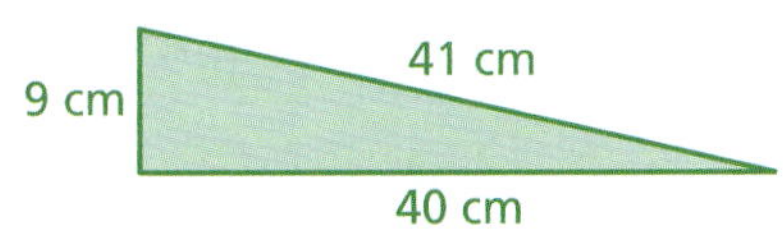

$$a^2 + b^2 = c^2$$

$$9^2 + 40^2 \stackrel{?}{=} 41^2$$

$$81 + 1600 \stackrel{?}{=} 1681$$

$$1681 = 1681 \ ✓$$

∴ It *is* a right triangle.

b.

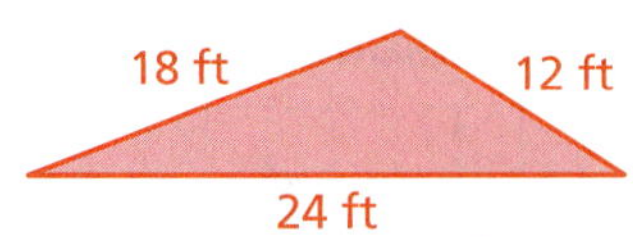

$$a^2 + b^2 = c^2$$

$$12^2 + 18^2 \stackrel{?}{=} 24^2$$

$$144 + 324 \stackrel{?}{=} 576$$

$$468 \neq 576 \ ✗$$

∴ It is *not* a right triangle.

On Your Own

Now You're Ready
Exercises 13–18

Tell whether the triangle with the given side lengths is a right triangle.

3.

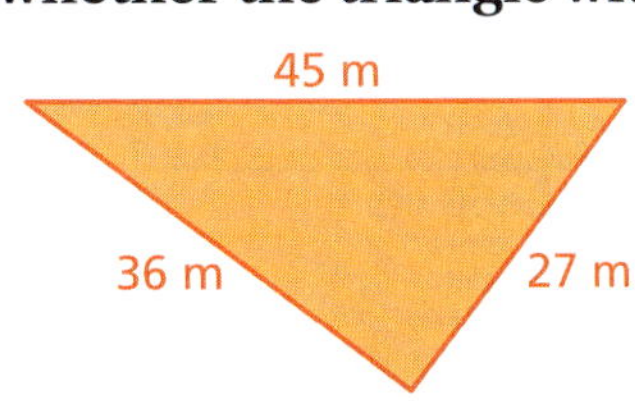

4.

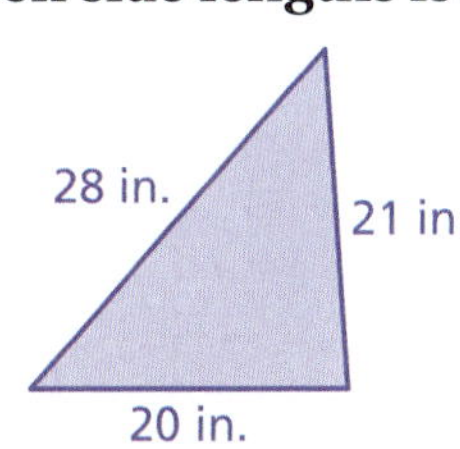

5. $1\frac{1}{2}$ yd, $2\frac{1}{2}$ yd, $3\frac{1}{2}$ yd

6. 1.25 mm, 1 mm, 0.75 mm

6.5 Exercises

Check It Out
Help with Homework
BigIdeasMath.com

Vocabulary and Concept Check

1. **WRITING** How can the Pythagorean Theorem be used to find distances in a coordinate plane?

2. **WHICH ONE DOESN'T BELONG?** Which set of numbers does *not* belong with the other three? Explain your reasoning.

3, 6, 8	6, 8, 10	5, 12, 13	7, 24, 25

Practice and Problem Solving

Find the perimeter of the figure. Round your answer to the nearest tenth.

3. Right triangle

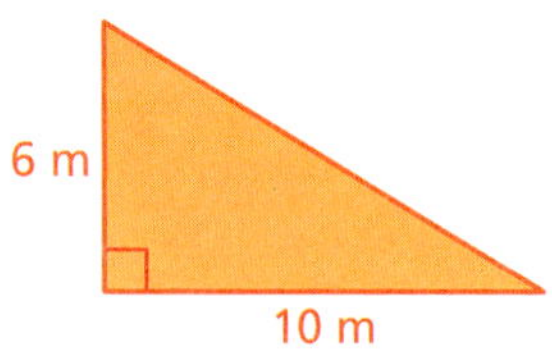

4. Parallelogram

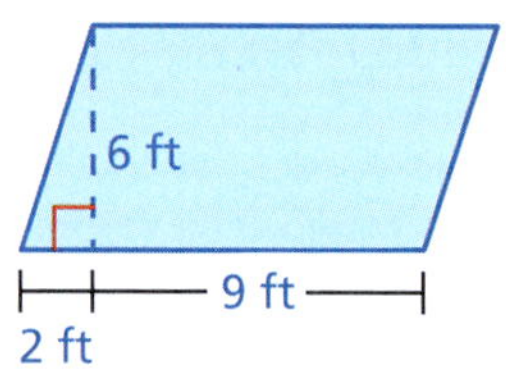

5. Square

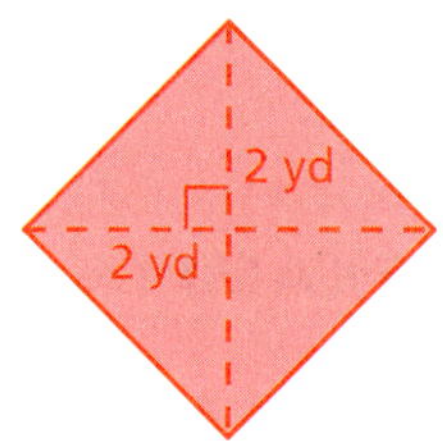

Find the distance d. Round your answer to the nearest tenth.

1 6.

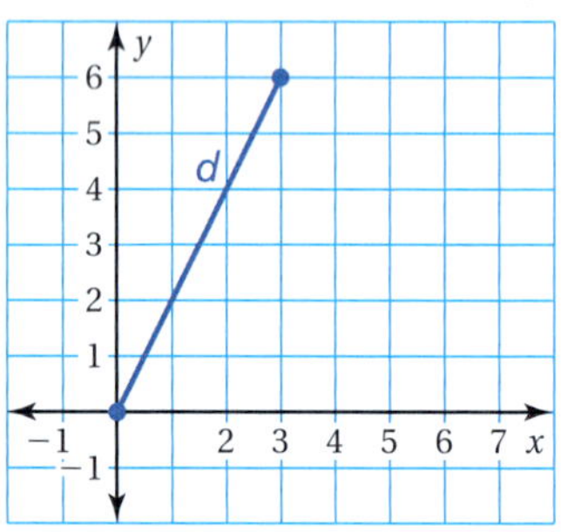

7.

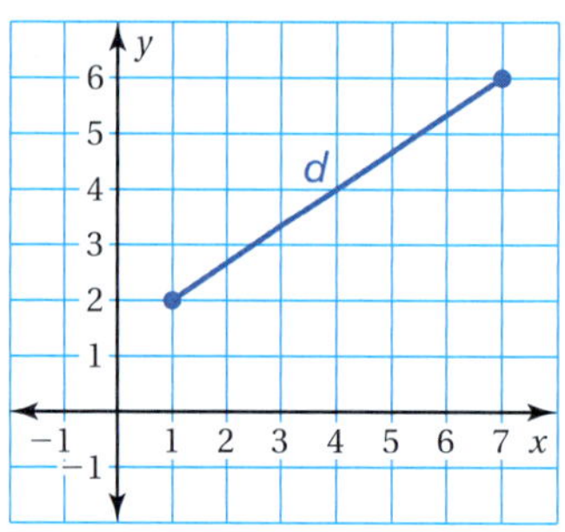

8.

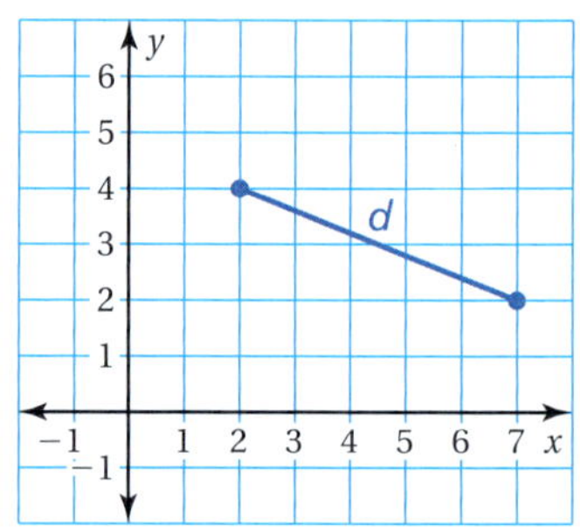

Find the height x. Round your answer to the nearest tenth.

2 9.

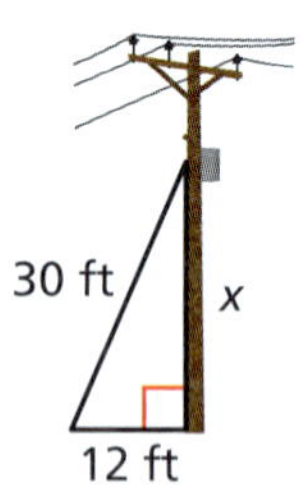

10.

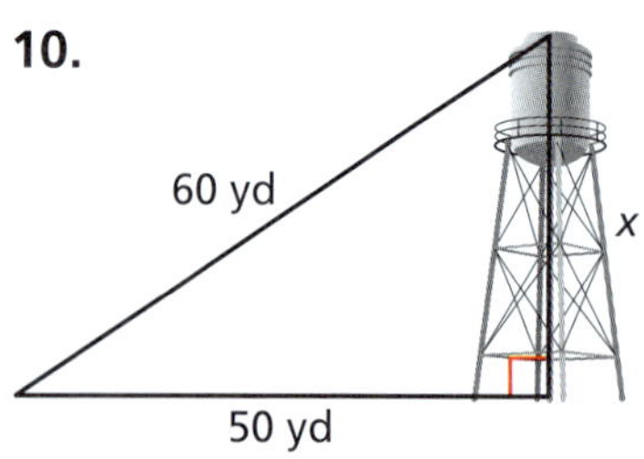

11.

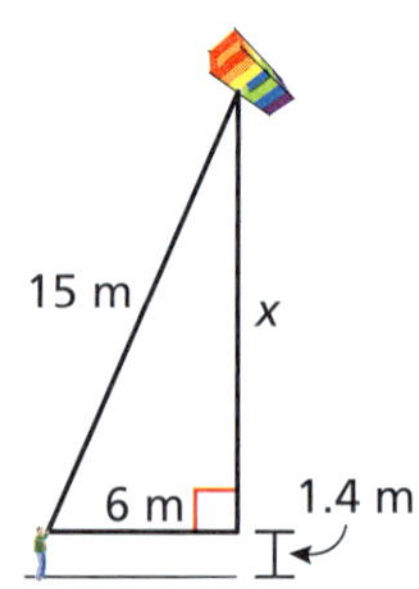

12. **BICYCLE** You ride your bicycle along the outer edge of a park. Then you take a shortcut back to where you started. Find the length of the shortcut. Round your answer to the nearest tenth.

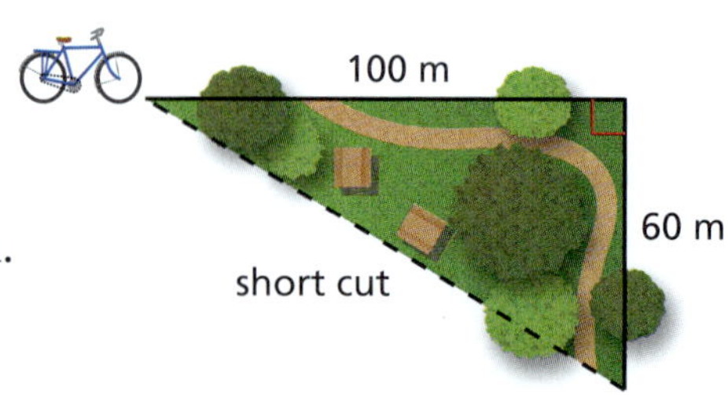

Tell whether the triangle with the given side lengths is a right triangle.

3 **13.**

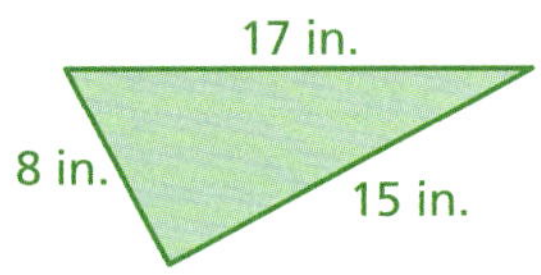

14.

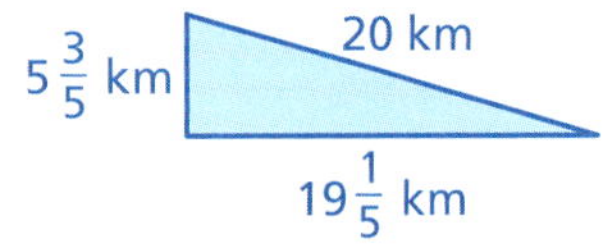

15.

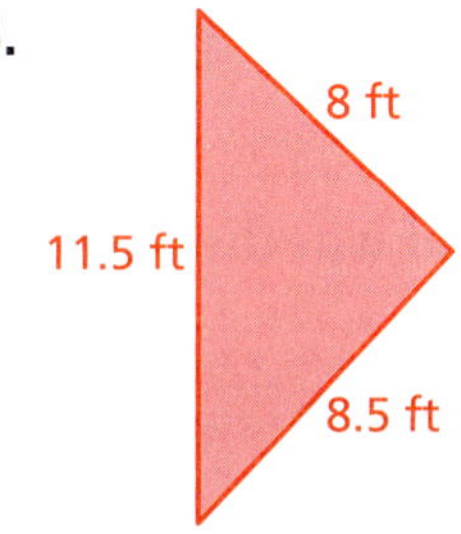

16. 14 mm, 19 mm, 23 mm

17. $\frac{9}{10}$ mi, $1\frac{1}{5}$ mi, $1\frac{1}{2}$ mi

18. 1.4 m, 4.8 m, 5 m

19. **STAIRS** There are 12 steps in the staircase. Find the distance from point A to point B (in feet). Round your answer to the nearest tenth.

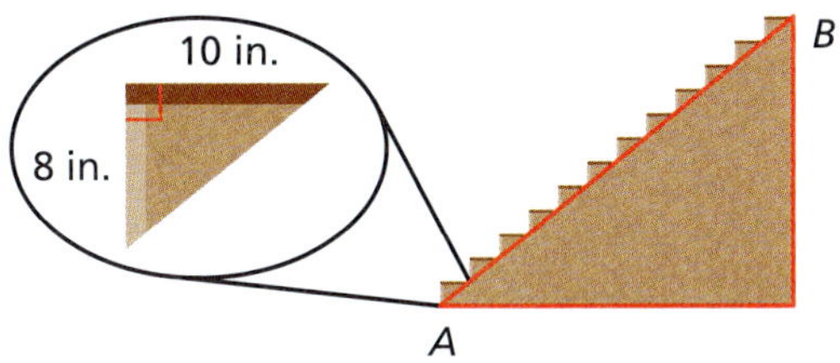

20. **AIRPORT** Which plane is closer to the tower? Explain.

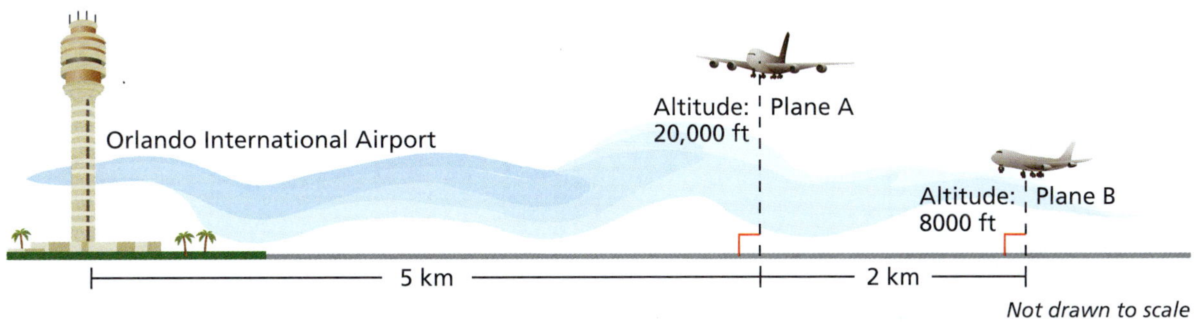

21. **PROJECT** Find a shoebox or some other small box.

a. Measure the dimensions of the box.

b. Without measuring, find length BC and length AB.

c. Use a piece of string and a ruler to check the lengths you found in part (b).

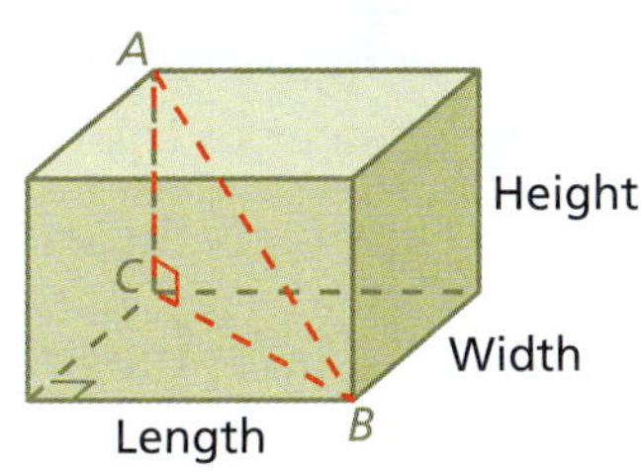

22. Critical Thinking Plot the points $(-1, -2)$, $(2, 1)$, and $(-3, 6)$ in a coordinate plane. Are the points the vertices of a right triangle? Explain.

Fair Game Review What you learned in previous grades & lessons

Find the mean, median, and mode of the data. *(Skills Review Handbook)*

23. 12, 9, 17, 15, 12, 13

24. 21, 32, 16, 27, 22, 19, 10

25. 67, 59, 34, 71, 59

26. **MULTIPLE CHOICE** What is the sum of the angle measures of an octagon? *(Section 5.3)*

Ⓐ 720° Ⓑ 1080° Ⓒ 1440° Ⓓ 1800°

6.3–6.5 Quiz

Check It Out
Progress Check
BigIdeasMath.com

Tell whether the number is *rational* or *irrational*. Explain. *(Section 6.3)*

1. $-\sqrt{225}$

2. $-1\frac{1}{9}$

3. $\sqrt{41}$

Estimate to the nearest integer. *(Section 6.3)*

4. $\sqrt{38}$

5. $-\sqrt{99}$

6. $\sqrt{172}$

Which number is greater? Explain. *(Section 6.3)*

7. $\sqrt{11}, 3\frac{3}{5}$

8. $\sqrt{1.44}, 1.1\overline{8}$

Simplify the expression. *(Section 6.4)*

9. $\sqrt{2} + 2\sqrt{2}$

10. $3\sqrt{15} - 7\sqrt{15}$

11. $\sqrt{\frac{6}{25}}$

Find the volume of the rectangular prism. *(Section 6.4)*

12.

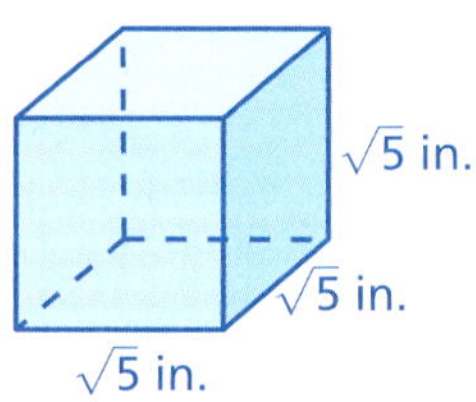

13.

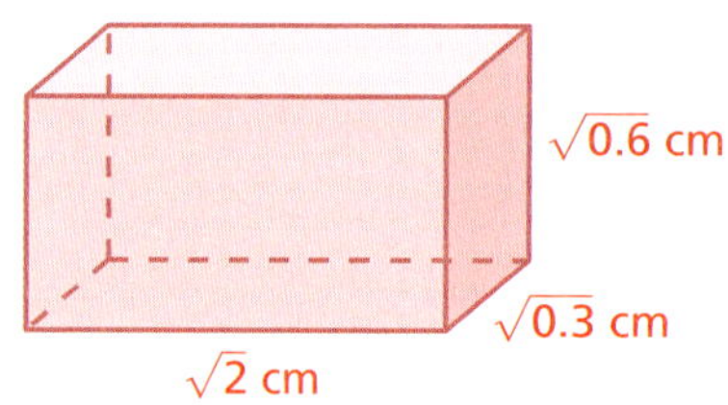

Use the figure to answer Exercises 14–17. Round your answer to the nearest tenth. *(Section 6.5)*

14. How far is the cabin from the peak?

15. How far is the fire tower from the lake?

16. How far is the lake from the peak?

17. You are standing at $(-5, -6)$. How far are you from the lake?

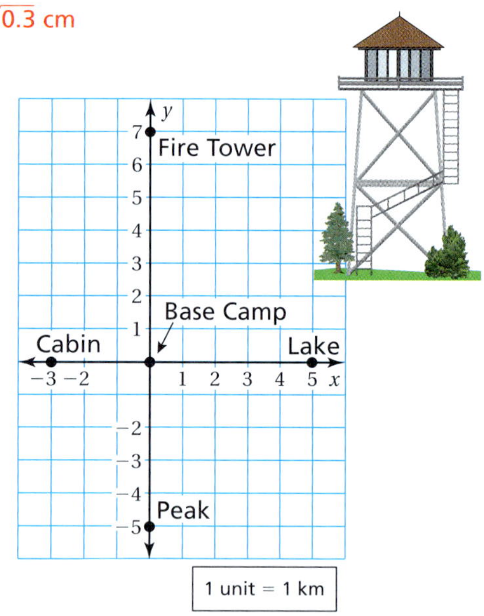

Tell whether the triangle with the given side lengths is a right triangle. *(Section 6.5)*

18.

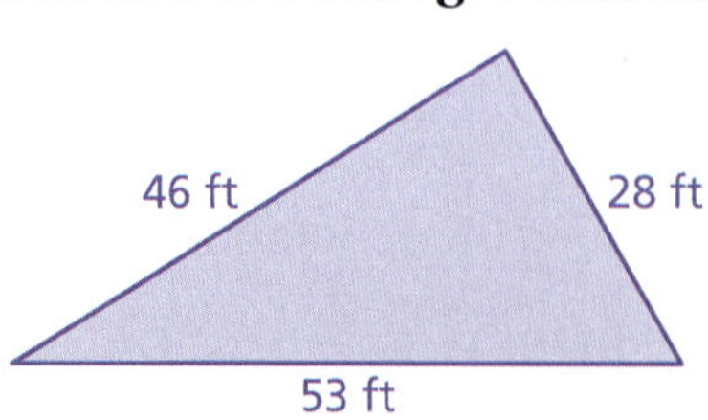

19.

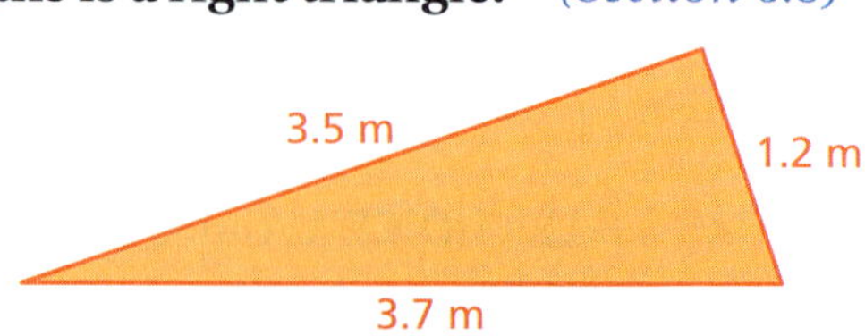

6 Chapter Review

Check It Out
Vocabulary Help
BigIdeasMath.com

Review Key Vocabulary

square root, *p. 232*
perfect square, *p. 232*
radical sign, *p. 232*
radicand, *p. 232*
theorem, *p. 236*
legs, *p. 238*
hypotenuse, *p. 238*
Pythagorean Theorem, *p. 238*
irrational number, *p. 246*
real numbers, *p. 246*
Pythagorean triple, *p. 261*

Review Examples and Exercises

6.1 Finding Square Roots *(pp. 230–235)*

Find the square root(s).

a. $-\sqrt{36}$

$-\sqrt{36}$ represents the *negative* square root.

Because $6^2 = 36$, $-\sqrt{36} = -\sqrt{6^2} = -6$.

b. $\sqrt{1.96}$

$\sqrt{1.96}$ represents the *positive* square root.

Because $1.4^2 = 1.96$, $\sqrt{1.96} = \sqrt{1.4^2} = 1.4$.

c. $\pm\sqrt{\frac{16}{81}}$

$\pm\sqrt{\frac{16}{81}}$ represents both the *positive and negative* square roots.

Because $\left(\frac{4}{9}\right)^2 = \frac{16}{81}$, $\pm\sqrt{\frac{16}{81}} = \pm\sqrt{\left(\frac{4}{9}\right)^2} = \frac{4}{9}$ and $-\frac{4}{9}$.

Exercises

Find the two square roots of the number.

1. 16 **2.** 900 **3.** 2500

Find the square root(s).

4. $\sqrt{1}$ **5.** $-\sqrt{\frac{9}{25}}$ **6.** $\pm\sqrt{1.96}$

Evaluate the expression.

7. $15 - 4\sqrt{16}$ **8.** $\sqrt{\frac{54}{6}} + \frac{2}{3}$ **9.** $10(\sqrt{81} - 12)$

6.2 The Pythagorean Theorem *(pp. 236–241)*

Find the length of the hypotenuse of the triangle.

$a^2 + b^2 = c^2$	Write the Pythagorean Theorem.
$7^2 + 24^2 = c^2$	Substitute.
$49 + 576 = c^2$	Evaluate powers.
$625 = c^2$	Add.
$\sqrt{625} = \sqrt{c^2}$	Take positive square root of each side.
$25 = c$	Simplify.

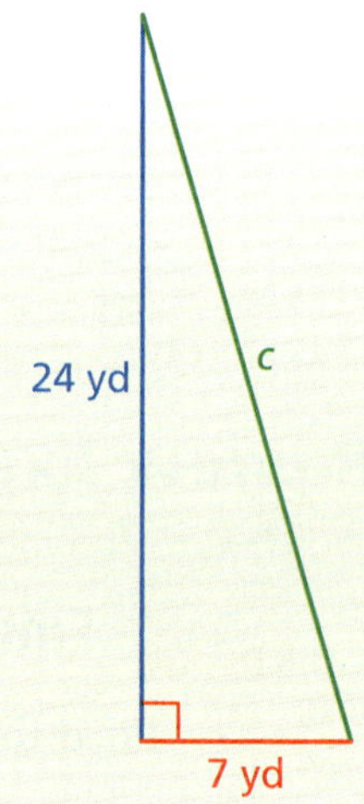

The length of the hypotenuse is 25 yards.

Exercises

Find the missing length of the triangle.

10.

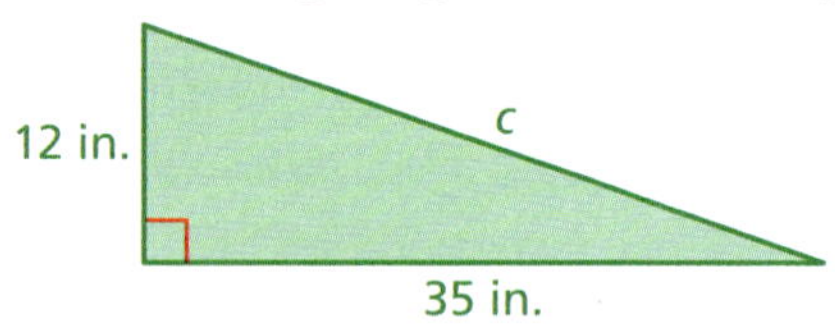

11.

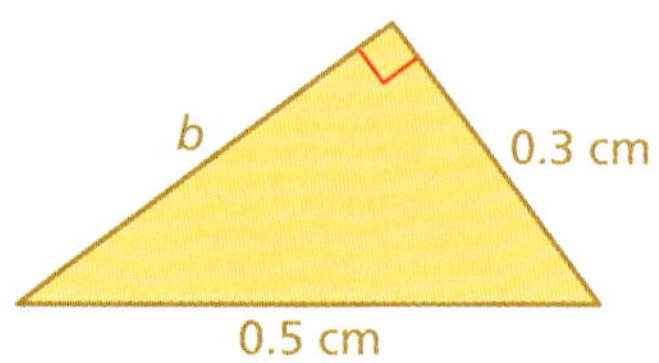

6.3 Approximating Square Roots *(pp. 244–251)*

Estimate $\sqrt{34}$ to the nearest integer.

Use a number line and the square roots of the perfect squares nearest to the radicand. The nearest perfect square less than 34 is 25. The nearest perfect square greater than 34 is 36.

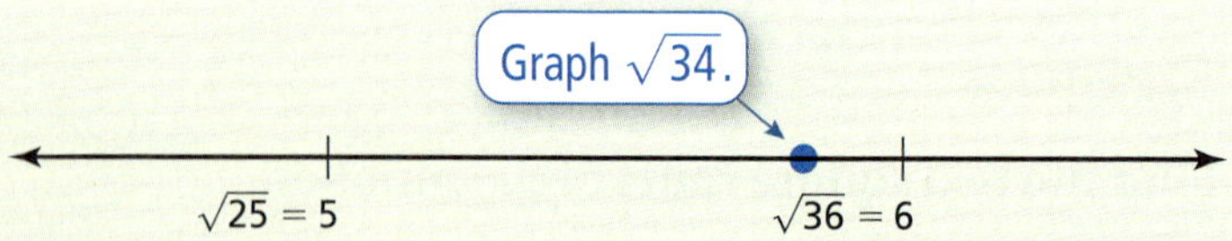

Because 34 is closer to 36 than to 25, $\sqrt{34}$ is closer to 6 than to 5.

So, $\sqrt{34} \approx 6$.

Exercises

Estimate to the nearest integer.

12. $\sqrt{14}$ **13.** $\sqrt{90}$ **14.** $\sqrt{175}$

6.4 Simplifying Square Roots (pp. 252–257)

Simplify $\sqrt{28}$.

$$
\begin{aligned}
\sqrt{28} &= \sqrt{4 \cdot 7} && \text{Factor using the greatest perfect square factor.}\\
&= \sqrt{4} \cdot \sqrt{7} && \text{Use the Product Property of Square Roots.}\\
&= 2\sqrt{7} && \text{Simplify.}
\end{aligned}
$$

Simplify $\sqrt{\dfrac{13}{64}}$.

$$
\begin{aligned}
\sqrt{\frac{13}{64}} &= \frac{\sqrt{13}}{\sqrt{64}} && \text{Use the Quotient Property of Square Roots.}\\
&= \frac{\sqrt{13}}{8} && \text{Simplify.}
\end{aligned}
$$

Exercises

Simplify the expression.

15. $\sqrt{\dfrac{99}{100}}$ **16.** $\sqrt{96}$ **17.** $\sqrt{75}$

6.5 Using the Pythagorean Theorem (pp. 258–263)

Find the height of the stilt walker. Round your answer to the nearest tenth.

$$
\begin{aligned}
a^2 + b^2 &= c^2 && \text{Write the Pythagorean Theorem.}\\
6^2 + x^2 &= 13^2 && \text{Substitute.}\\
36 + x^2 &= 169 && \text{Evaluate powers.}\\
x^2 &= 133 && \text{Subtract 36 from each side.}\\
\sqrt{x^2} &= \sqrt{133} && \text{Take positive square root of each side.}\\
x &\approx 11.5 && \text{Use a calculator.}
\end{aligned}
$$

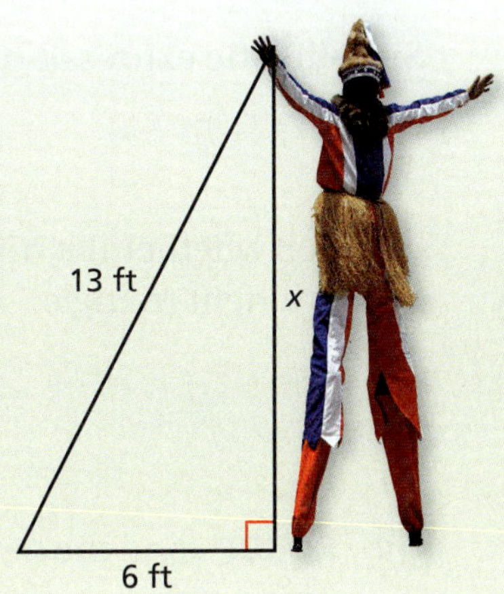

The height of the stilt walker is about 11.5 feet.

Exercises

Find the height x. Round your answer to the nearest tenth, if necessary.

18.

19.

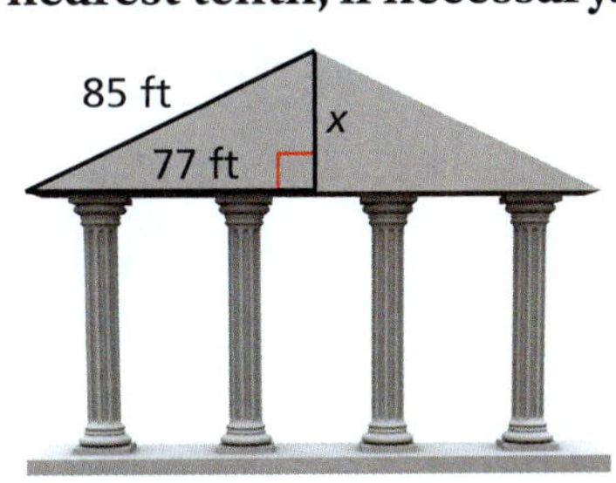

Check It Out
Test Practice
BigIdeasMath.com

Find the square root(s).

1. $-\sqrt{1600}$
2. $\sqrt{\frac{25}{49}}$
3. $\pm\sqrt{\frac{100}{9}}$

Evaluate the expression.

4. $12 + 8\sqrt{16}$
5. $\frac{1}{2} + \sqrt{\frac{72}{2}}$
6. Find the missing length of the triangle.

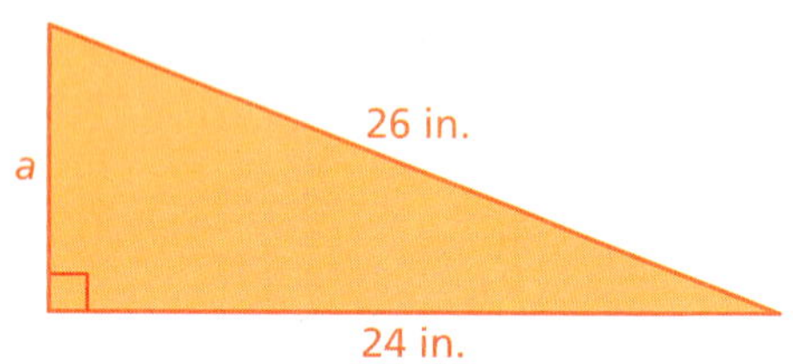

Tell whether the number is *rational* or *irrational*. Explain.

7. 16π
8. $-\sqrt{49}$

Which number is greater? Explain.

9. $\sqrt{0.16}, \frac{1}{2}$
10. $\sqrt{45}, 6.\overline{3}$

Simplify the expression.

11. $6\sqrt{5} + 5\sqrt{5}$
12. $\sqrt{250}$
13. Tell whether the triangle is a right triangle.

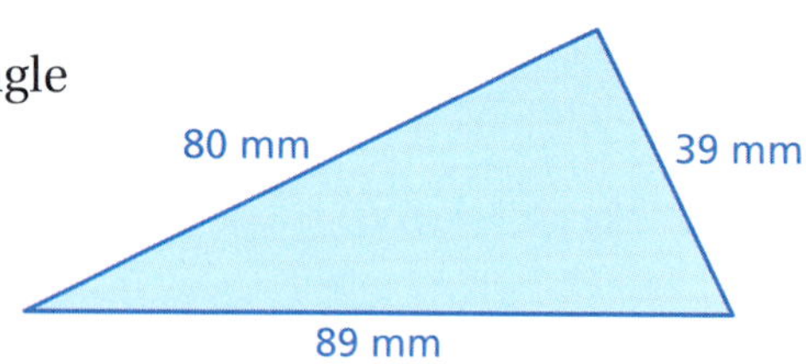

14. **ROBOT** Find the height of the dinosaur robot.

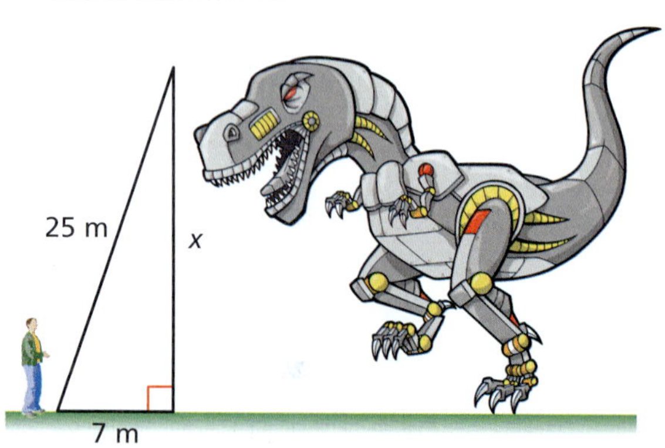

15. **SUPERHERO** Find the altitude of the superhero balloon.

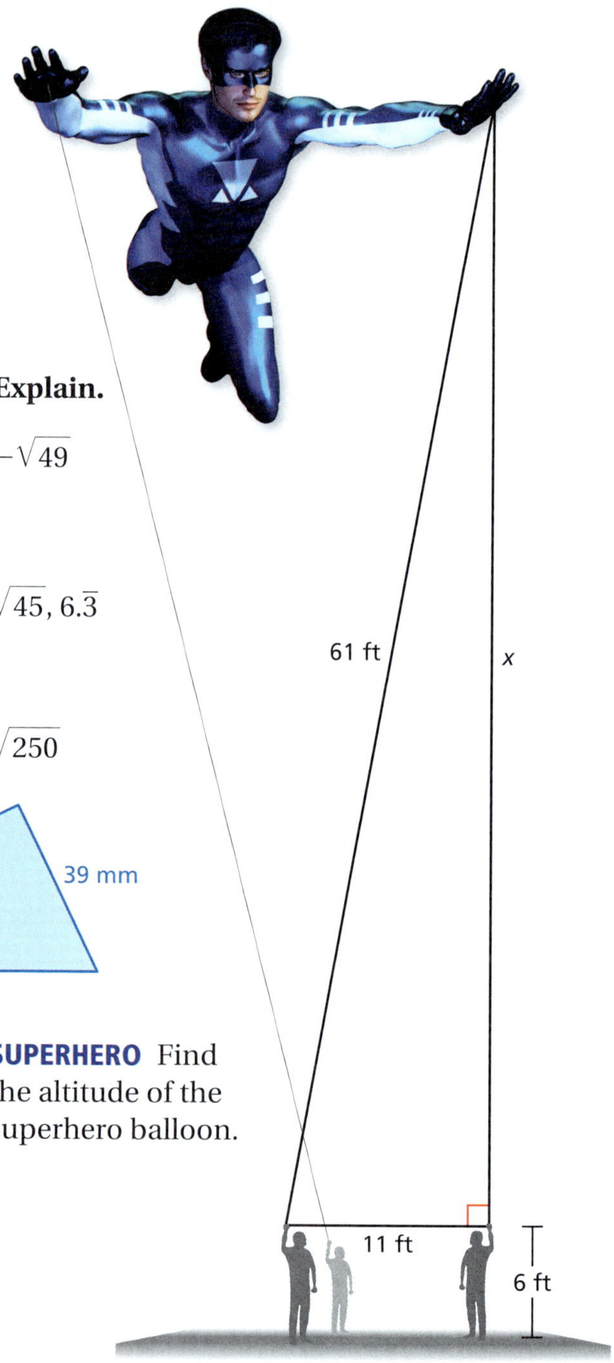

6 Standardized Test Practice

1. The period T of a pendulum is the time, in seconds, it takes the pendulum to swing back and forth. The period can be found using the formula $T = 1.1\sqrt{L}$, where L is the length, in feet, of the pendulum. A pendulum has a length of 4 feet. Find its period.

   A. 5.1 sec **C.** 3.1 sec

   B. 4.4 sec **D.** 2.2 sec

Test-Taking Strategy

Answer Easy Questions First

"Scan the test and answer the easy questions first. You know the square root of 4 is 2."

2. The steps Pat took to write the equation in slope-intercept form are shown below. What should Pat change in order to correctly rewrite the equation in slope-intercept form?

$$3x - 6y = 1$$
$$3x = 6y + 1$$
$$x = 2y + \frac{1}{3}$$

   F. Use the formula $m = \frac{\text{rise}}{\text{run}}$.

   G. Use the formula $m = \frac{\text{run}}{\text{rise}}$.

   H. Subtract $3x$ from both sides of the equation and divide every term by -6.

   I. Subtract 1 from both sides of the equation and divide every term by 3.

3. A housing community started with 60 homes. In each of the following years, 8 more homes were built. Let y represent the number of years that have passed since the first year and let n represent the number of homes. Which equation describes the relationship between n and y?

   A. $n = 8y + 60$ **C.** $n = 60y + 8$

   B. $n = 68y$ **D.** $n = 60 + 8 + y$

4. The domain of a function is 0, 1, 2, 3, 4, 5. What can you conclude?

   F. The domain is continuous. **H.** The function is linear.

   G. The domain is discrete. **I.** The range is 0, 1, 2, 3, 4, 5.

5. A football field is 40 yards wide and 120 yards long. Find the distance between opposite corners of the football field. Show your work and explain your reasoning.

Think Solve Explain

6. A computer consultant charges \$50 plus \$40 for each hour she works. The consultant charged \$650 for one job. This can be represented by the equation below, where h represents the number of hours worked.

$$40h + 50 = 650$$

How many hours did the consultant work?

7. The formula below can be used to find the number S of degrees in a polygon with n sides. Solve the formula for n.

$$S = 180(n - 2)$$

A. $n = 180(S - 2)$

B. $n = \frac{S}{180} + 2$

C. $n = \frac{S}{180} - 2$

D. $n = \frac{S}{180} + \frac{1}{90}$

8. The table below shows a linear pattern. Which linear function relates y to x?

x	1	2	3	4	5
y	4	2	0	−2	−4

F. $y = 2x + 2$

G. $y = 4x$

H. $y = -2x + 2$

I. $y = -2x + 6$

9. What is the value of x in the right triangle shown?

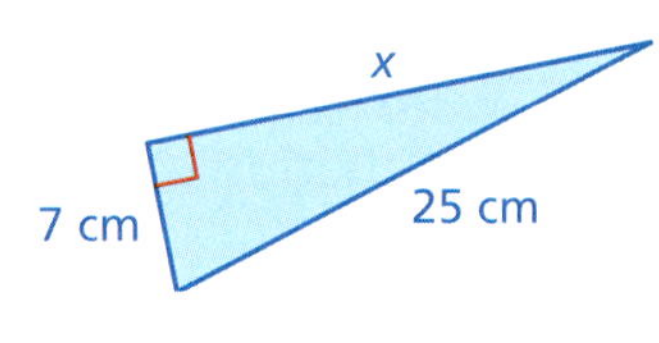

A. 16 cm

B. 18 cm

C. 24 cm

D. $\sqrt{674}$ cm

10. Find the height of the tree in the diagram.

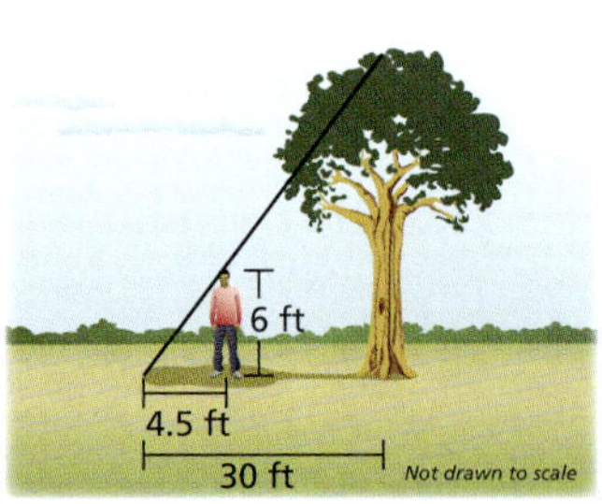

F. 22.5 ft

G. 31.5 ft

H. 35 ft

I. 40 ft

11. Which expression is equivalent to $12\sqrt{24}$?

A. $48\sqrt{6}$

B. $24\sqrt{12}$

C. $24\sqrt{6}$

D. 6

12. The measure of an angle is x degrees. What is the measure of its complement?

F. $(90 - x)°$

G. $(180 - x)°$

H. $(x - 90)°$

I. $(x - 180)°$

13. You fill up the gas tank of your car and begin driving on the interstate. You drive at an average speed of 60 miles per hour. The amount g, in gallons, of gas left in your car can be estimated. Use the formula shown below, where h is the number of hours you have been driving.

$$g = 18 - 2.5h$$

You will fill up when you have 3 gallons of gas left in the gas tank. How long after you start driving will you fill up again?

A. about 36 min

B. about 6.0 h

C. about 7.2 h

D. about 8.4 h

14. An airplane flies 56 miles due north and then 33 miles due east. How many miles is the plane from its starting point?

15. Which graph represents the linear equation $y = -2x - 2$?

F.

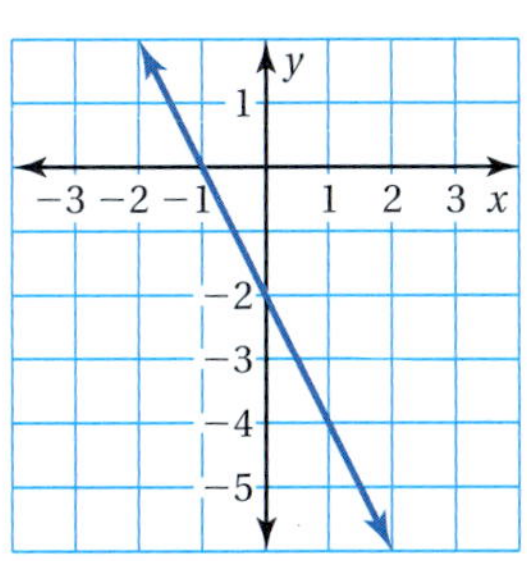

H.

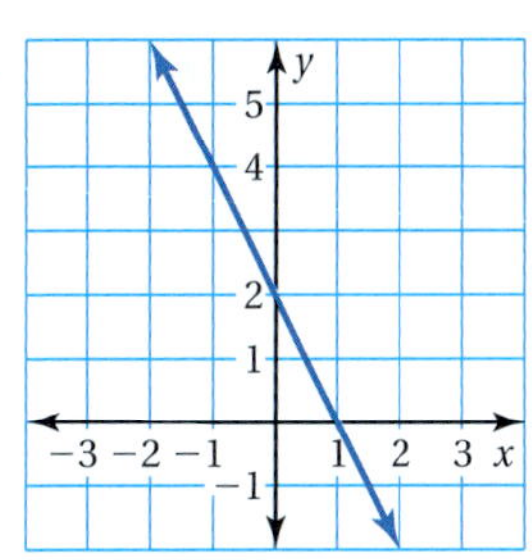

G.

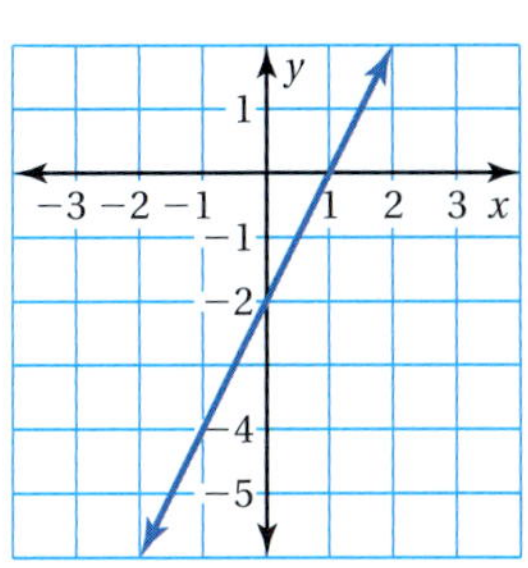

I.

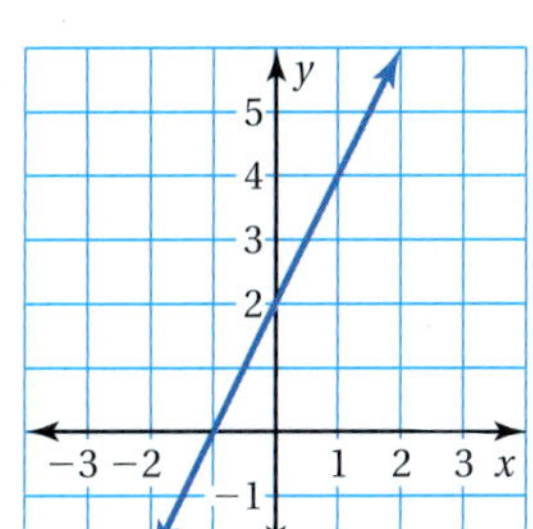

7 Data Analysis and Displays

7.1 Measures of Central Tendency

7.2 Box-and-Whisker Plots

7.3 Scatter Plots and Lines of Best Fit

7.4 Choosing a Data Display

"Wow. The number of minutes I can dog paddle is growing like crazy!"

"Please hold still. I am trying to find the mean of 6, 8, and 10 by dividing their sum into three equal piles."

What You Learned Before

Displaying Data (Grade 7)

Example 1 The table shows the results of a survey. Display the data in a circle graph.

Class Trip Location	Water park	Museum	Zoo	Other
Students	25	11	5	4

A total of 45 students took the survey.

Water park:

$\frac{25}{45} \cdot 360^\circ = 200^\circ$

Museum:

$\frac{11}{45} \cdot 360^\circ = 88^\circ$

Zoo:

$\frac{5}{45} \cdot 360^\circ = 40^\circ$

Other:

$\frac{4}{45} \cdot 360^\circ = 32^\circ$

Class Trip Locations

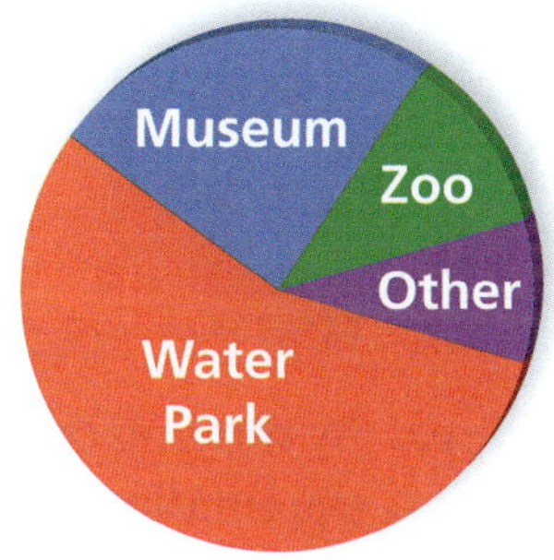

Example 2 The frequency table shows the number of books that 12 people read last month. Display the data in a histogram.

Books Read Last Month	Frequency
0–1	6
2–3	4
4–5	0
6–7	2

Try It Yourself

1. Conduct a survey to determine the after-school activities of students in your class. Display the results in a circle graph.
2. Conduct a survey to determine the number of pets owned by students in your class. Display the results in a histogram.

7.1 Measures of Central Tendency

STATE STANDARDS
MA.8.S.3.2

Essential Question How can you use measures of central tendency to distribute an amount evenly among a group of people?

1 ACTIVITY: Exploring Mean, Median, and Mode

Work with a partner. Forty-five coins are arranged in nine stacks.

a. Record the number of coins in each stack in a table.

Stack	1	2	3	4	5	6	7	8	9
Coins									

b. Find the mean, median, and mode of the number of coins in each stack.

c. By moving coins from one stack to another, can you change the mean? the median? the mode? Explain.

d. Is it possible to arrange the coins in stacks so that the median is 6? 8? Explain.

2 EXAMPLE: Drawing a Line Plot

Work with a partner.

a. Draw a number line. Label the tick marks from 1 to 10.

b. Place each stack of coins in Activity 1 above the number of coins in the stack.

c. Draw an × to represent each stack. This graph is called a *line plot*.

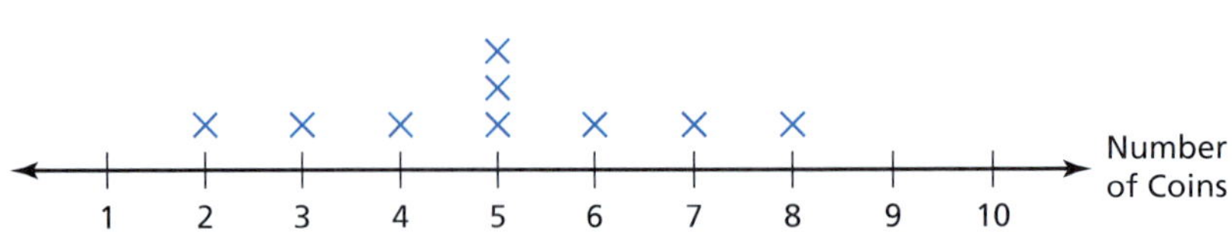

3 ACTIVITY: Fair and Unfair Distributions

Work with a partner.

A distribution of coins to nine people is considered *fair* if each person has the same number of coins.

- Distribute the 45 coins into 9 stacks using a fair distribution. How is this distribution related to the mean?
- Draw a line plot for each distribution. Which distributions seem most fair? Which distributions seem least fair? Explain your reasoning.

a.

b.

c.

d.

e.

f.

What Is Your Answer?

4. **IN YOUR OWN WORDS** How can you use measures of central tendency to distribute an amount evenly among a group of people?
5. Use the Internet or some other reference to find examples of mean or median incomes of groups of people. Describe possible distributions that could produce the given means or medians.

Practice Use what you learned about measures of central tendency to complete Exercise 4 on page 278.

7.1 Lesson

Check It Out
Lesson Tutorials
BigIdeasMath.com

Key Vocabulary
measure of central tendency, *p. 276*

A **measure of central tendency** is a measure that represents the center of a data set. The *mean, median,* and *mode* are measures of central tendency.

Key Ideas

Mean

The *mean* of a data set is the sum of the data divided by the number of data values.

Median

Order the data. For a set with an odd number of values, the *median* is the middle value. For a set with an even number of values, the *median* is the mean of the two middle values.

Mode

The *mode* of a data set is the value or values that occur most often.

Remember

Data can have one mode, more than one mode, or no mode. When each value occurs only once, there is no mode.

EXAMPLE 1 Finding the Mean, Median, and Mode

Students' Hourly Wages	
\$3.87	\$7.25
\$8.75	\$8.45
\$8.25	\$7.25
\$6.99	\$7.99

An amusement park hires students for the summer. The students' hourly wages are given in the table. Find the mean, median, and mode of the hourly wages.

Mean: $\frac{58.8}{8} = 7.35$ (sum of the data / number of values)

Median: 3.87, 6.99, 7.25, 7.25, 7.99, 8.25, 8.45, 8.75 Order the data.

$\frac{15.24}{2} = 7.62$ Mean of two middle values

Mode: 3.87, 6.99, 7.25, 7.25, 7.99, 8.25, 8.45, 8.75

The value 7.25 occurs most often.

∴ The mean is \$7.35, the median is \$7.62, and the mode is \$7.25.

On Your Own

Now You're Ready
Exercises 5–8

1. **WHAT IF?** In Example 1, the park hires another student at an hourly wage of \$6.99. How does this additional value affect the mean, median, and mode? Explain.

Multi-Language Glossary at BigIdeasMath.com.

EXAMPLE 2 Removing an Outlier

Remember

An *outlier* is a data value that is much greater or much less than the other values.

Identify the outlier in Example 1. How does the outlier affect the mean, median, and mode?

The value \$3.87 is low compared to the other wages. It is the outlier.

Find the mean, median, and mode without the outlier.

Mean: $\frac{54.93}{7} \approx 7.85$

Median: 6.99, 7.25, 7.25, 7.99, 8.25, 8.45, 8.75 — The middle value, 7.99, is the median.

Mode: 6.99, 7.25, 7.25, 7.99, 8.25, 8.45, 8.75 — The mode is 7.25.

By removing the outlier, the mean increases \$7.85 − \$7.35 = \$0.50, the median increases \$7.99 − \$7.62 = \$0.37, and the mode is the same.

EXAMPLE 3 Changing the Values of a Data Set

In Example 1, each hourly wage increases \$0.40. How does this increase affect the mean, median, and mode?

Make a new table by adding \$0.40 to each hourly wage.

Students' Hourly Wages	
\$4.27	\$7.65
\$9.15	\$8.85
\$8.65	\$7.65
\$7.39	\$8.39

Mean: $\frac{62}{8} = 7.75$

Median: 4.27, 7.39, 7.65, 7.65, 8.39, 8.65, 8.85, 9.15 — Order the data.

$\frac{16.04}{2} = 8.02$ — Mean of two middle values

Mode: 4.27, 7.39, 7.65, 7.65, 8.39, 8.65, 8.85, 9.15 — The mode is 7.65.

By increasing each hourly wage \$0.40, the mean, median, and mode all increase \$0.40.

On Your Own

Now You're Ready
Exercises 16 and 17

The figure shows the altitudes of several airplanes.

2. Identify the outlier. How does the outlier affect the mean, median, and mode? Explain.

3. Each airplane increases its altitude $1\frac{1}{2}$ miles. How does this affect the mean, median and mode? Explain.

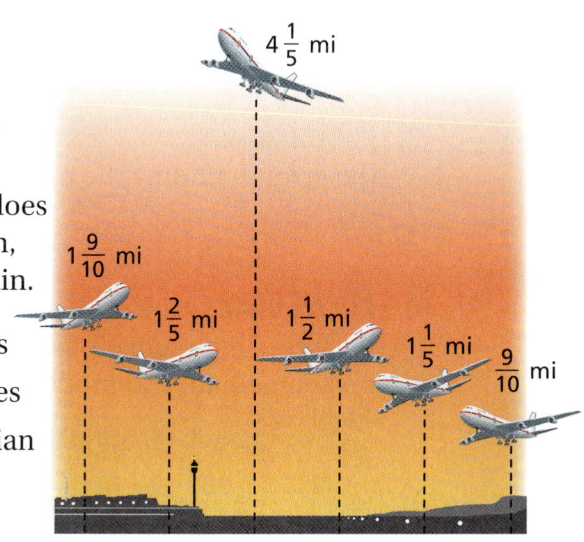

7.1 Exercises

Check It Out
Help with Homework
BigIdeasMath.com

Vocabulary and Concept Check

1. **VOCABULARY** Can a data value be an outlier *and* a measure of central tendency of the same data set? Explain.
2. **OPEN-ENDED** Create a data set that has more than one mode.
3. **WRITING** Describe how removing an outlier from a data set affects the mean of the data set.

Practice and Problem Solving

4. Draw a line plot of the data. Then find the mean, median, and mode of the data.

Bag	1	2	3	4	5	6	7	8	9
Strawberries	10	13	11	15	8	14	7	11	12

Find the mean, median, and mode of the data.

1 5.

Golf Scores		
3	−2	1
6	4	−1
−3	−1	2

6.

Changes in Stock Value (dollars)			
1.05	2.03	−1.78	−2.41
−2.64	0.67	4.02	1.39
0.66	−0.38	−3.01	2.20

7.

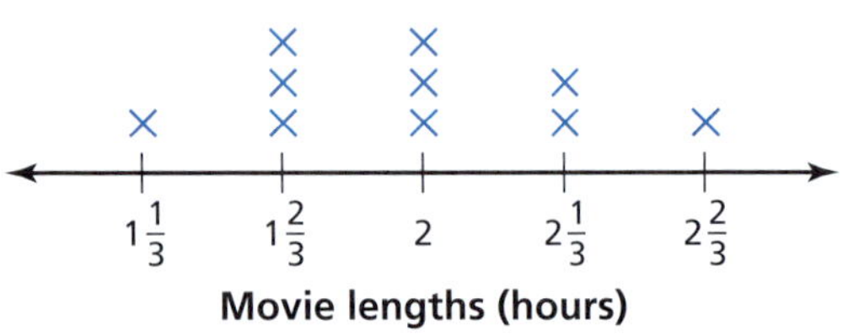

8. **Available Memory**

Stem	Leaf
6	5
7	0 5 5
8	0 4 5
9	4

Key: 7 | 5 = 75 megabytes

9. **ERROR ANALYSIS** Describe and correct the error in finding the median.

✗ Test scores: 98, 90, 80, 80, 90, 90

The median is $\frac{528}{6} = 88$.

10. **POLAR BEARS** The table shows the masses of eight polar bears. Find the mean, median, and mode of the masses.

Masses (kilograms)			
455	262	471	358
364	553	352	467

Find the value of x.

11. Mean is 6; 2, 8, 9, 7, 6, x

12. Mean is 0; 11.5, 12.5, -10, -7.5, x

13. Median is 14; 9, 10, 12, x, 20, 25

14. Median is 51; 30, 45, x, 100

(2) **15. TEMPERATURES** An environmentalist records the average temperatures of five regions.

a. Identify the outlier.

b. Which measure of central tendency will be most affected by removing the outlier?

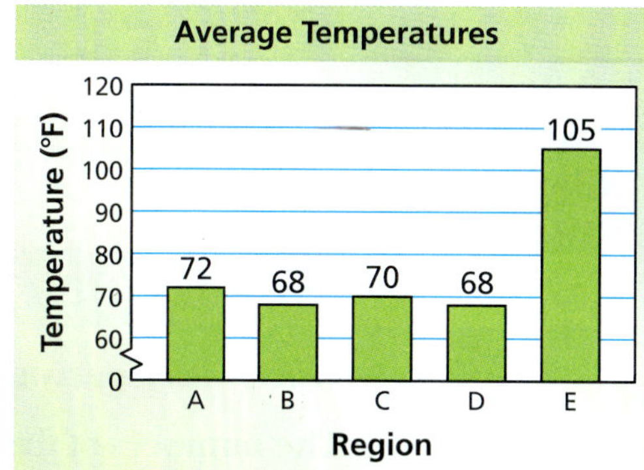

16. RIVER The Suwannee River Trail has 11 segments in Florida. The lengths (in miles) of the segments are 8.9, 9.7, 6, 2.2, 19.6, 15, 7.6, 0.5, 14.2, 14.6, and 22.4.

a. Find the mean, median, and mode of the lengths.

b. A canoeist skips the first segment of the trail. How does this affect the mean, median, and mode of the remaining lengths? Explain.

(3) **17. REASONING** The value of each stock in Exercise 6 decreases \$0.05. How does this affect the mean, median, and mode? Explain.

18. Critical Thinking The circle graph shows the ages of 200 students in a college psychology class.

a. Find the mean, median, and mode of the students' ages.

b. Identify the outliers. How do the outliers affect the mean, median, and mode?

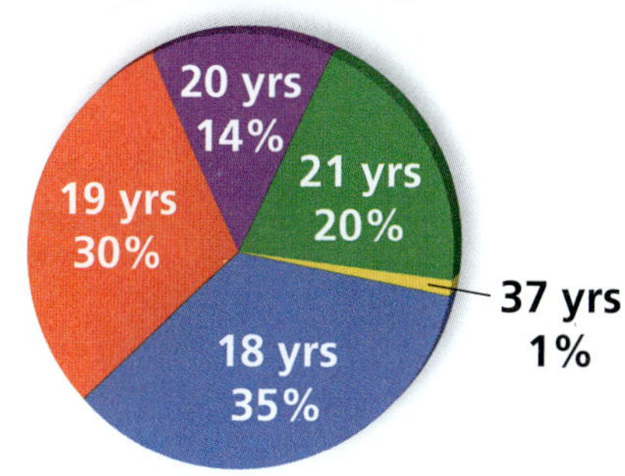

Fair Game Review What you learned in previous grades & lessons

Order the values from least to greatest. *(Skills Review Handbook)*

19. 1, -3, -8, 4, 7, -5

20. 1.2, -2.8, $\frac{3}{2}$, 5.4, -4.7, $-\frac{2}{3}$

21. MULTIPLE CHOICE Which equation represents a linear function? *(Section 4.4)*

Ⓐ $y = x^2$ Ⓑ $y = 2x$ Ⓒ $y = \frac{2}{x}$ Ⓓ $xy = 2$

7.2 Box-and-Whisker Plots

STATE STANDARDS
MA.8.S.3.1

Essential Question How can you use a box-and-whisker plot to describe a population?

1 ACTIVITY: Drawing a Box-and-Whisker Plot

Work with a partner.

The numbers of first cousins of each student in an eighth-grade class are shown.

A box-and-whisker plot uses a number line to represent the data visually.

Numbers of First Cousins

3	10	18	8
9	3	0	32
23	19	13	8
6	3	3	10
12	45	1	5
13	24	16	14

a. Order the data set and write it on a strip of grid paper with 24 equally spaced boxes.

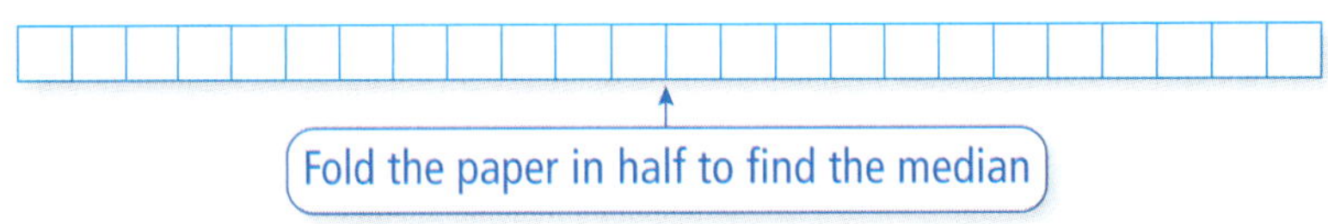

b. Fold the paper in half again to divide the data into four groups. Because there are 24 numbers in the data set, each group should have six numbers.

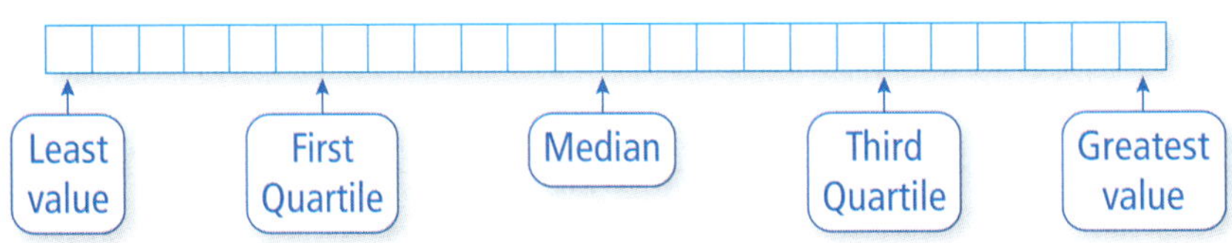

c. Draw a number line that includes the least value and the greatest value in the data set. Plot the five numbers that you found in part (b).

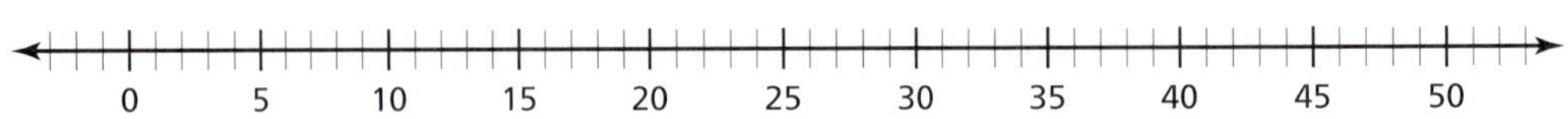

d. Explain how the box-and-whisker plot shown below represents the data set.

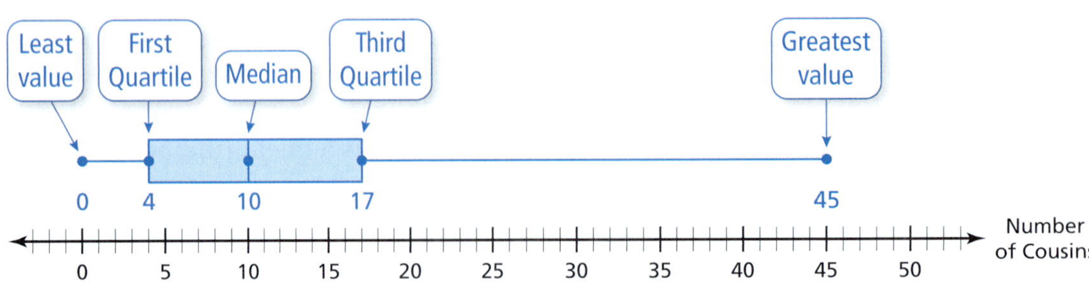

2 ACTIVITY: Conducting a Survey

Conduct a survey in your class. Ask each student to write the number of his or her first cousins on a piece of paper. Collect the pieces of paper and write the data on the chalkboard.

Now, work with a partner to draw a box-and-whisker plot of the data.

Two people are first cousins if they share at least one grandparent, but do not share a parent.

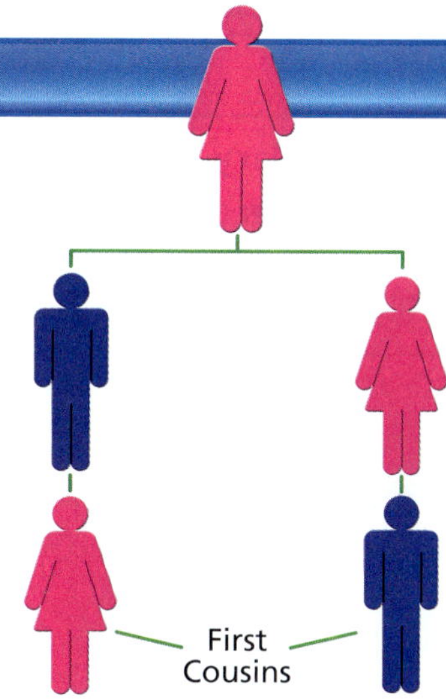

3 ACTIVITY: Reading a Box-and-Whisker Plot

Work with a partner. Collier and Hendry are neighboring counties in Florida. However, their age distributions are quite different.

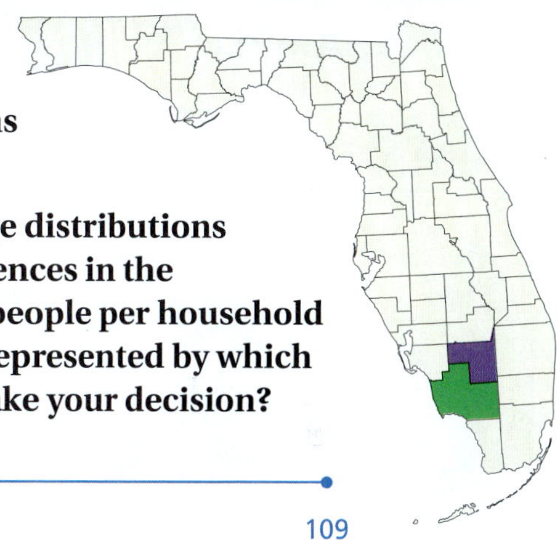

The box-and-whisker plots show the age distributions of the two counties. Describe the differences in the populations. Collier County has fewer people per household than Hendry County. Which county is represented by which box-and-whisker plot? How did you make your decision?

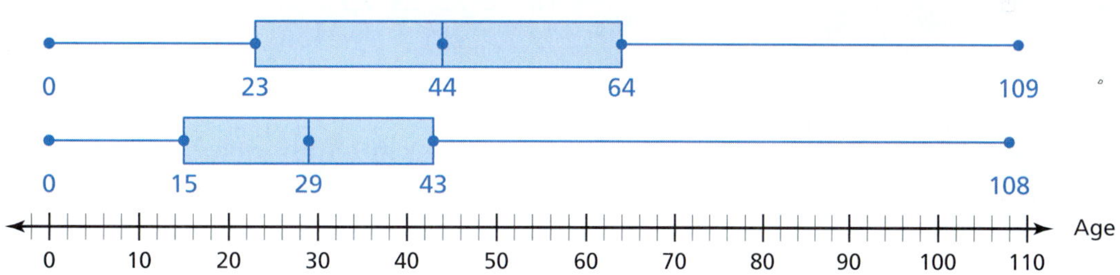

What Is Your Answer?

4. **IN YOUR OWN WORDS** How can you use a box-and-whisker plot to describe a population?

5. Describe some careers in which age distributions (like those shown in Activity 3) are important. Explain why it is important to know the age distribution in a county.

Practice Use what you learned about box-and-whisker plots to complete Exercise 4 on page 284.

7.2 Lesson

Check It Out
Lesson Tutorials
BigIdeasMath.com

Key Vocabulary
box-and-whisker plot, *p. 282*
quartiles, *p. 282*

Key Idea

Box-and-Whisker Plot

A **box-and-whisker plot** displays a data set along a number line using medians. **Quartiles** divide the data set into four equal parts. The median (second quartile) divides the data set into two halves. The median of the lower half is the first quartile. The median of the upper half is the third quartile.

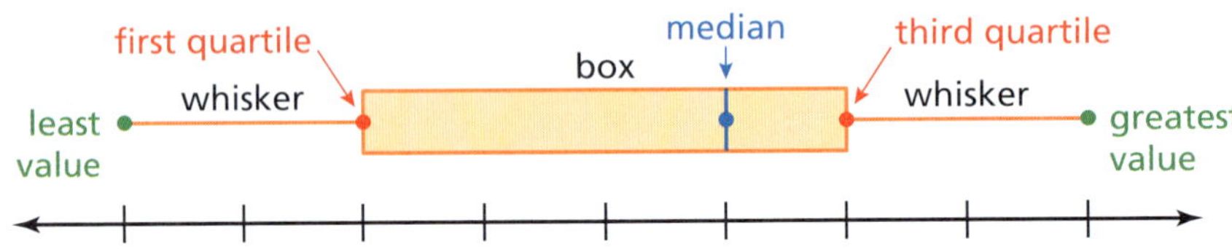

Study Tip

A box-and-whisker plot shows the *variability* of a data set.

EXAMPLE 1 Making a Box-and-Whisker Plot

Make a box-and-whisker plot for the ages of the members of the 2008 U.S. women's wheelchair basketball team.

24, 30, 30, 22, 25, 22, 18, 25, 28, 30, 25, 27

Step 1: Order the data. Find the median and the quartiles.

lower half: 18 22 22 24 25 25 | upper half: 25 27 28 30 30 30

least value → 18; greatest value → 30

first quartile, 23; median, 25; third quartile, 29

Step 2: Draw a number line that includes the least and greatest values. Graph points above the number line for the least value, greatest value, median, first quartile, and third quartile.

Step 3: Draw a box using the quartiles. Draw a line through the median. Draw whiskers from the box to the least and greatest values.

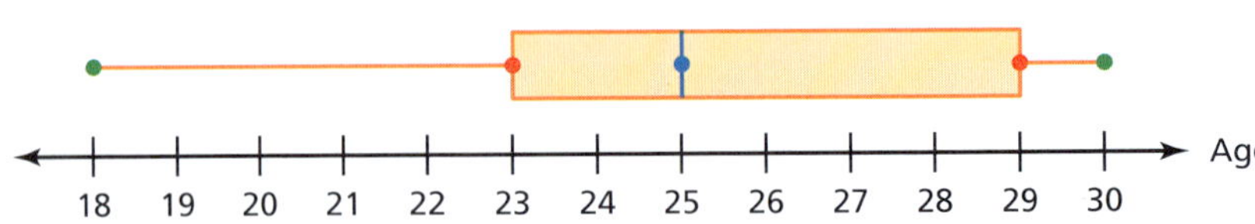

On Your Own

Now You're Ready
Exercises 5–7

1. A basketball player scores 14, 16, 20, 5, 22, 30, 16, and 28 points during a tournament. Make a box-and-whisker plot for the points scored by the player.

Multi-Language Glossary at BigIdeasMath.com.

EXAMPLE 2 Interpreting a Box-and-Whisker Plot

Study Tip

A long whisker or box indicates data is more spread out.

What does the box-and-whisker plot tell you about the data?

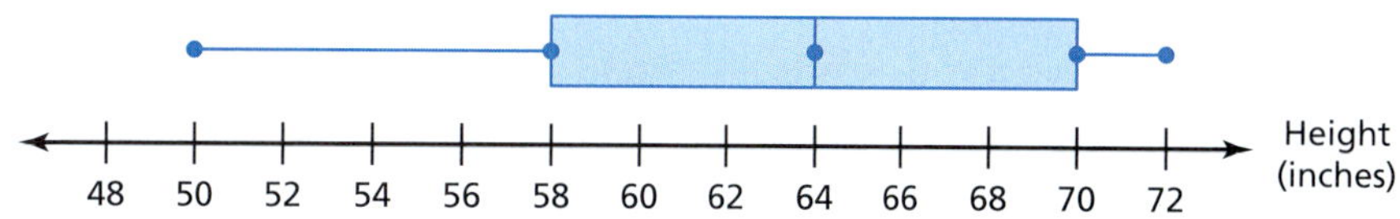

- The left whisker is longer than the right whisker. So, the data are more spread out below the first quartile than above the third quartile.
- The range of the data is $72 - 50 = 22$ inches.

EXAMPLE 3 Standardized Test Practice

Which statement is true about the double box-and-whisker plot?

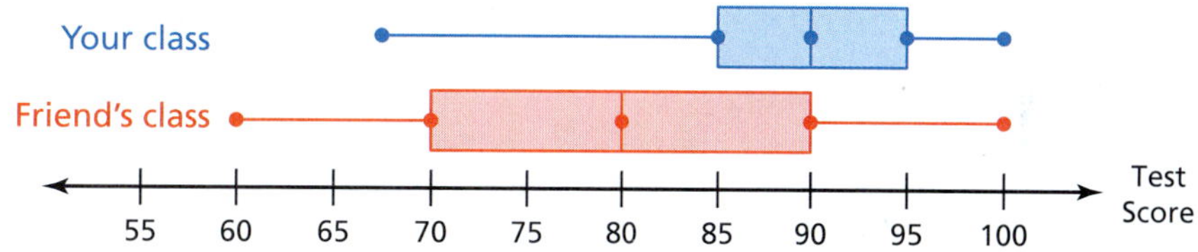

(A) Half of the test scores in your class are between 85 and 100.

(B) 25% of the test scores in your friend's class are 80 or above.

(C) The medians are the same for both classes.

(D) The test scores in your friend's class are more spread out than the test scores in your class.

The range of the test scores in your class is less than the range in your friend's class. Also, the box for your friend's class is longer than the box for your class. So, the test scores in your friend's class are more spread out than the test scores in your class.

∴ The correct answer is (D).

On Your Own

Now You're Ready
Exercise 10

2. Compare the surfboard prices of Shop A and Shop B. What are three conclusions you can make from the double box-and-whisker plot?

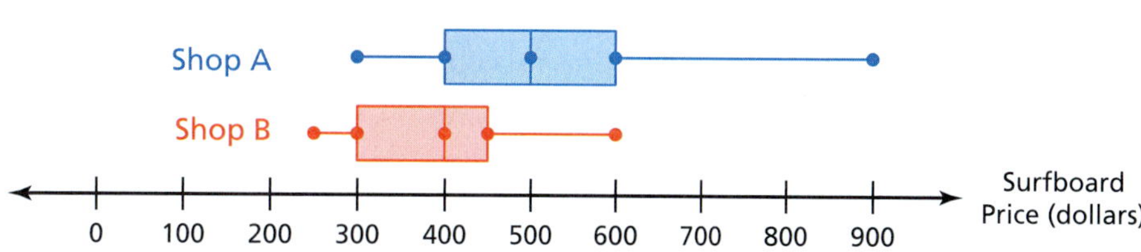

7.2 Exercises

Check It Out
Help with Homework
BigIdeasMath.com

Vocabulary and Concept Check

1. **VOCABULARY** In a box-and-whisker plot, what percent of the data is represented by each whisker? the box?

2. **WRITING** Describe how to find the first quartile of a data set.

3. **NUMBER SENSE** What does the length of the box-and-whisker plot tell you about the data?

Practice and Problem Solving

4. The box-and-whisker plots show the monthly car sales for a year for two sales representatives. Compare and contrast the sales of the two representatives.

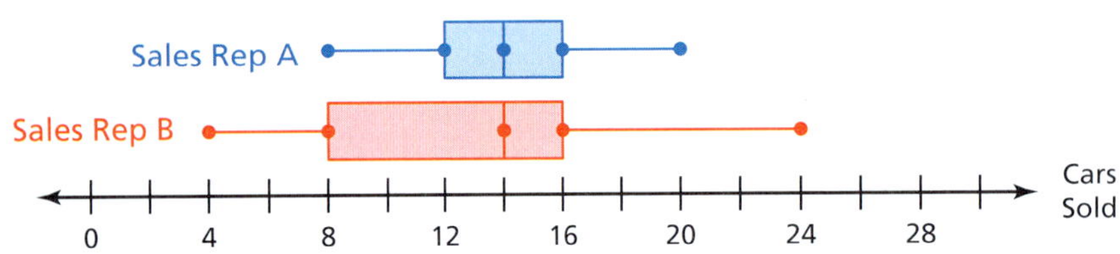

Make a box-and-whisker plot for the data.

1 5. Hours of television watched: 0, 3, 4, 5, 3, 4, 6, 5

6. Lengths (in inches) of cats: 16, 18, 20, 25, 17, 22, 23, 21

7. Elevations (in feet): −2, 0, 5, −4, 1, −3, 2, 0, 2, −3, 6, −1

8. **ERROR ANALYSIS** Describe and correct the error in making a box-and-whisker plot for the data.

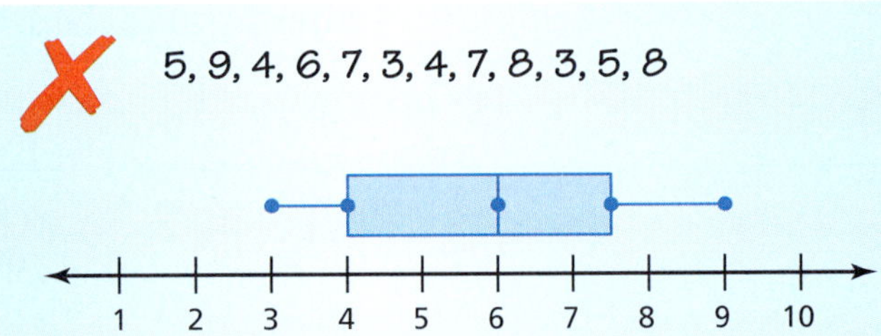

9. **FISH** The lengths (in inches) of the fish caught on a fishing trip are 9, 10, 12, 8, 13, 10, 12, 14, 7, 14, 8, and 14. Make a box-and-whisker plot for the data. What is the range of the data?

2 10. **INCHWORM** The table shows the lengths of 12 inchworms. Make a box-and-whisker plot for the data. What does the box-and-whisker plot tell you about the data?

Length (cm)	2.5	2.4	2.3	2.5	2.7	2.1	2.8	2.6	2.1	2.6	2.9	2.0

11. **CALORIES** The table shows the number of calories burned per hour for nine activities.

a. Make a box-and-whisker plot for the data.

b. Identify the outlier.

c. Make another box-and-whisker plot without the outlier.

d. **WRITING** Describe how the outlier affects the whiskers, the box, and the quartiles of the box-and-whisker plot.

Calories Burned per Hour	
Fishing	207
Mowing the lawn	325
Canoeing	236
Bowling	177
Hunting	295
Fencing	354
Bike racing	944
Horseback riding	236
Dancing	266

12. **CELL PHONES** The double box-and-whisker plot compares the battery life (in hours) of two brands of cell phones.

a. What is the range of the upper 75% of each brand?

b. Which battery has a longer battery life? Explain.

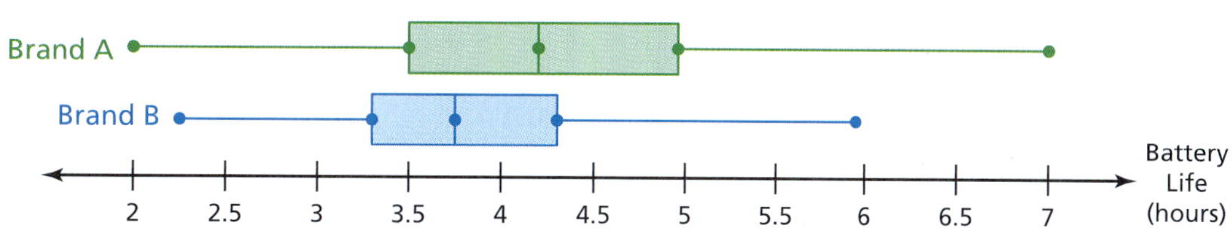

Critical Thinking **Create a set of data values whose box-and-whisker plot has the given characteristic(s).**

13. The least value, greatest value, quartiles, and median are all equally spaced.

14. Both whiskers are the same length as the box.

15. The box between the median and the first quartile is three times as long as the box between the median and the third quartile.

16. There is no right whisker.

Fair Game Review What you learned in previous grades & lessons

Write an equation of the line that passes through the points. *(Section 3.3)*

17. $(-4, -10), (2, 8)$

18. $(-3, 3), (0, -1)$

19. $(-4, 1), (4, -1)$

20. $(6, 7), (8, 8)$

21. **MULTIPLE CHOICE** You run 10 feet per second. What is this rate in miles per hour? *(Section 1.5)*

Ⓐ 0.11 mi/h Ⓑ 6.82 mi/h Ⓒ 10.23 mi/h Ⓓ 14.67 mi/h

7 Study Help

Check It Out
Graphic Organizer
BigIdeasMath.com

You can use a **word magnet** to organize information associated with a vocabulary word. Here is an example of a word magnet for measures of central tendency.

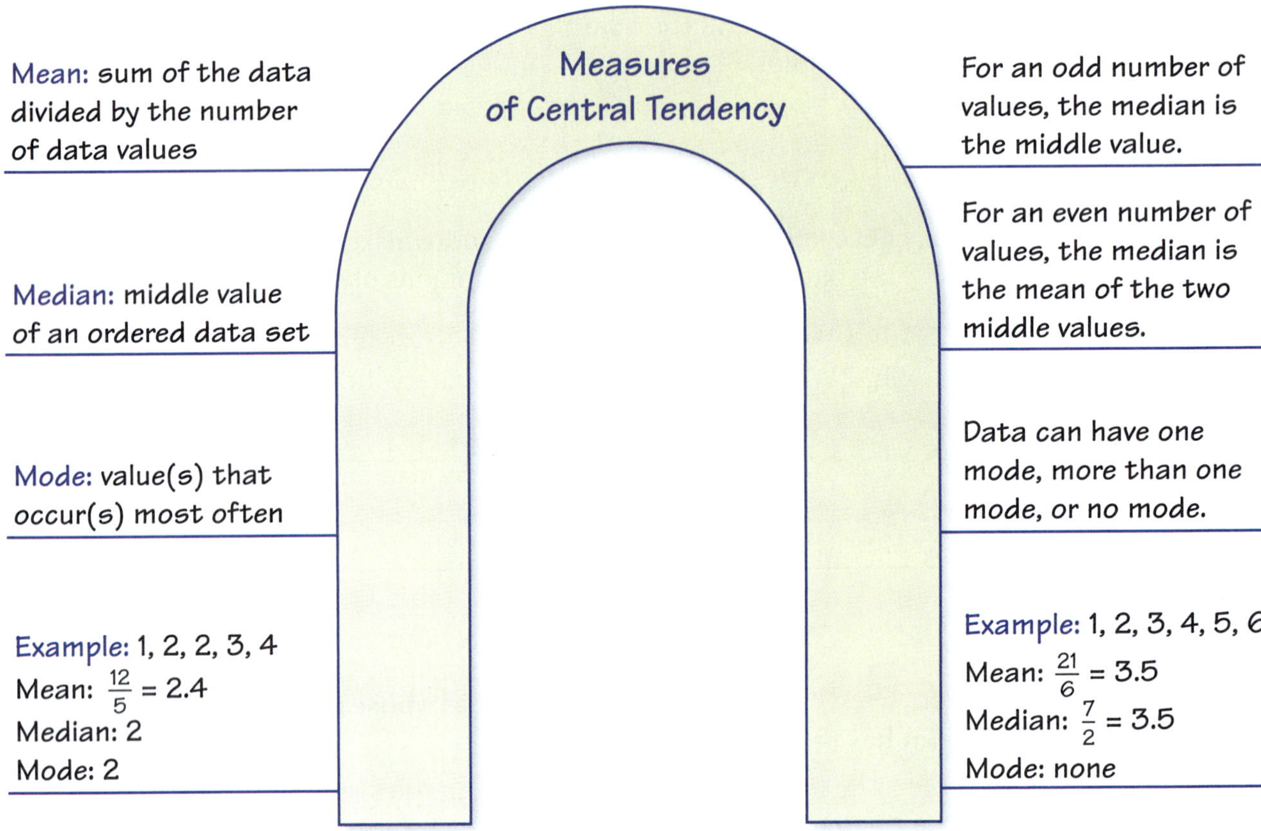

On Your Own

Make a word magnet to help you study these topics.

1. outliers
2. box-and-whisker plots

After you complete this chapter, make word magnets for the following topics.

3. scatter plots
4. lines of best fit
5. data displays

"How do you like the word magnet I made for 'Beagle'?"

7.1–7.2 Quiz

Check It Out
Progress Check
BigIdeasMath.com

Find the mean, median, and mode of the data. *(Section 7.1)*

1.

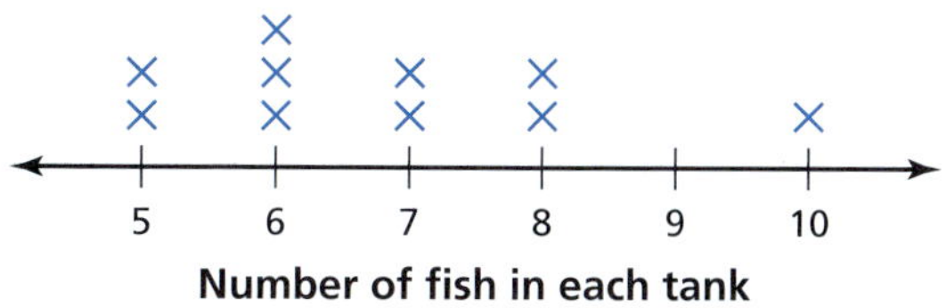

2.

Checkbook Balances (dollars)		
40	10	−20
0	−10	40
30	40	50

3.

Hours Spent on Project		
$3\frac{1}{2}$	5	$2\frac{1}{2}$
3	$3\frac{1}{2}$	$\frac{1}{2}$

4. Students in a Grade

Stem	Leaf
9	4 9
10	1 2 6
11	3 3
12	0

Key: 10 | 6 = 106 students

Make a box-and-whisker plot for the data. *(Section 7.2)*

5. Hours spent on each babysitting job: 2, 4, 7, 5, 4, 1, 7, 4

6. Minutes of violin practice: 20, 50, 60, 40, 40, 30, 60, 40, 50, 20, 20, 35

7. Players' scores at end of first round: 200, −100, 100, 350, −50, 0, −50, 300

8. The table shows the prices of eight acoustic guitars at a music store. *(Section 7.1)*

Prices of Acoustic Guitars (dollars)			
650	225	320	615
595	495	200	110

a. Find the mean, median, and mode of the prices.

b. The store gets a ninth guitar in stock. The guitar costs \$615. How does this additional value affect the mean, median, and mode of the data?

9. ANOLES The table shows the lengths of 12 green anoles. Make a box-and-whisker plot for the data. What does the box-and-whisker plot tell you about the data? *(Section 7.2)*

Length (cm)	17.5	17.3	16.5	16.8	17.0	16.5	17.0	16.7	16.5	17.0	17.4	17.1

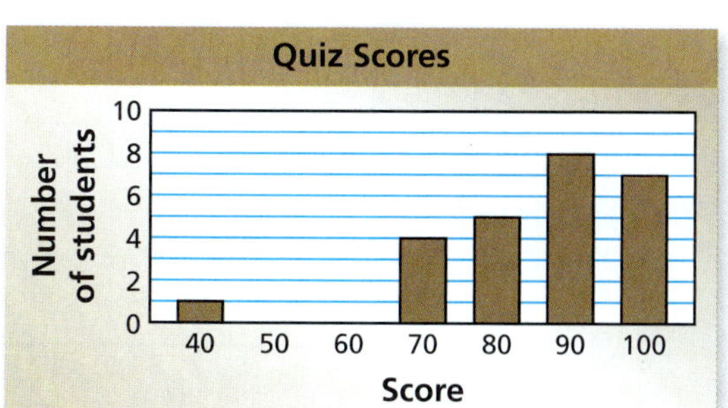

10. QUIZ SCORES The graph shows the quiz scores of students in a class. *(Section 7.1)*

a. Identify the outlier.

b. Which measure of central tendency will be most affected by removing the outlier? Explain.

7.3 Scatter Plots and Lines of Best Fit

STATE STANDARDS
MA.8.S.3.1

Essential Question How can you use data to predict an event?

Share Your Work at... My.BigIdeasMath.com

1 ACTIVITY: Representing Data by a Linear Equation

Work with a partner. You have been working on a science project for 8 months. Each month, you have measured the length of a baby alligator.

The table shows your measurements.

September (Month 0) — April (Month 7)

Month, x	0	1	2	3	4	5	6	7
Length (in.), y	22.0	22.5	23.5	25.0	26.0	27.5	28.5	29.5

Use the following steps to predict the baby alligator's length next September.

a. Graph the data in the table.

b. Draw the straight line that you think best approximates the points.

c. Write an equation of the line you drew.

d. Use the equation to predict the baby alligator's length next September.

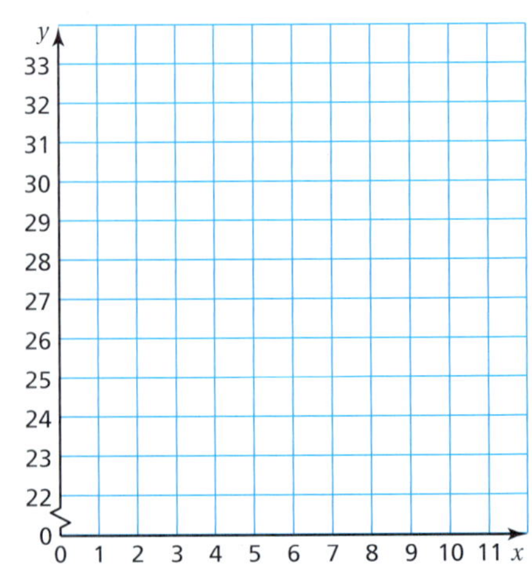

2 ACTIVITY: Representing Data by a Linear Equation

Work with a partner. You work for the Bureau of Tourism in Miami.

You are asked to predict the number of tourists from overseas who will visit Miami during the next 2 years.

To start, you find the number of overseas tourists who visited Miami during the past 8 years.

The table shows the results of your research.

	7 years ago							this year
Year, x	0	1	2	3	4	5	6	7
Tourists (thousands), y	3270	3060	2990	2700	2540	2320	2150	1970

Use the following steps to predict the number of overseas tourists who will visit Miami during the next 2 years.

a. Graph the data in the table.

b. Draw the straight line that you think best approximates the points.

c. Write an equation of the line you drew.

d. Use the equation to predict the number of tourists during the next 2 years.

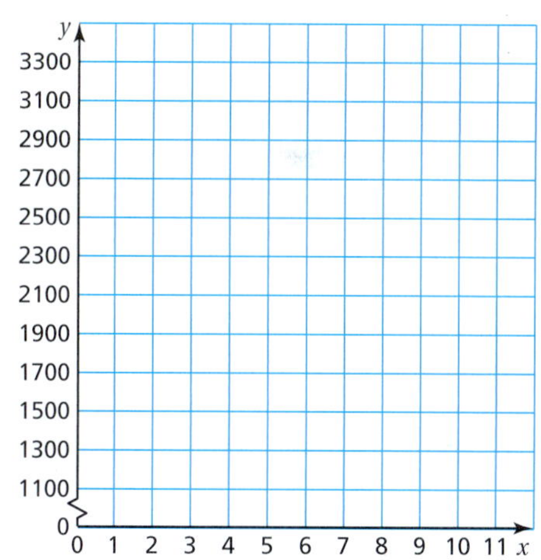

What Is Your Answer?

3. **IN YOUR OWN WORDS** How can you use data to predict an event?

4. Use the Internet or some other reference to find data that appear to have a linear pattern. List the data in a table and graph the data. Use an equation that is based on the data to predict a future event.

Practice Use what you learned about scatter plots and lines of best fit to complete Exercise 3 on page 293.

7.3 Lesson

Check It Out
Lesson Tutorials
BigIdeasMath.com

Key Vocabulary
scatter plot, *p. 290*
line of best fit, *p. 292*

Key Idea

Scatter Plot

A **scatter plot** is a graph that shows the relationship between two data sets. The two sets of data are graphed as ordered pairs in a coordinate plane.

EXAMPLE 1 Interpreting a Scatter Plot

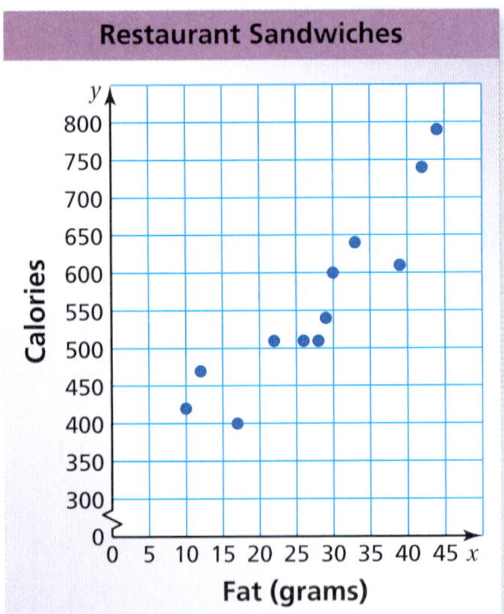

The scatter plot at the left shows the total fat (in grams) and the total calories in 12 restaurant sandwiches.

a. How many calories are in the sandwich that contains 17 grams of fat?

Draw a horizontal line from the point that has an x-value of 17. It crosses the y-axis at 400.

So, the sandwich has 400 calories.

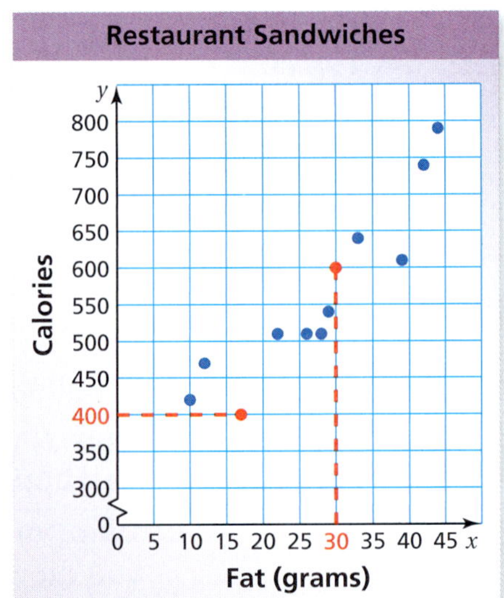

b. How many grams of fat are in the sandwich that contains 600 calories?

Draw a vertical line from the point that has a y-value of 600. It crosses the x-axis at 30.

So, the sandwich has 30 grams of fat.

c. What tends to happen to the number of calories as the number of grams of fat increases?

Looking at the graph, the plotted points go up from left to right.

So, as the number of grams of fat increases, the number of calories increases.

On Your Own

Now You're Ready
Exercises 4 and 5

1. **WHAT IF?** A sandwich has 650 calories. Based on the scatter plot in Example 1, how many grams of fat would you expect the sandwich to have? Explain your reasoning.

Multi-Language Glossary at BigIdeasMath.com.

A scatter plot can show that a relationship exists between two data sets.

Positive Relationship	***Negative Relationship***	***No Relationship***
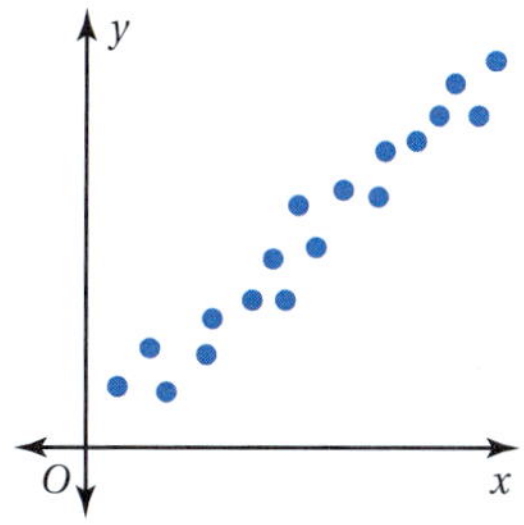	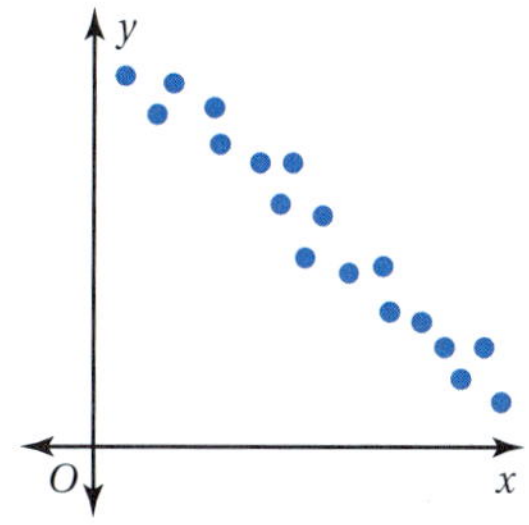	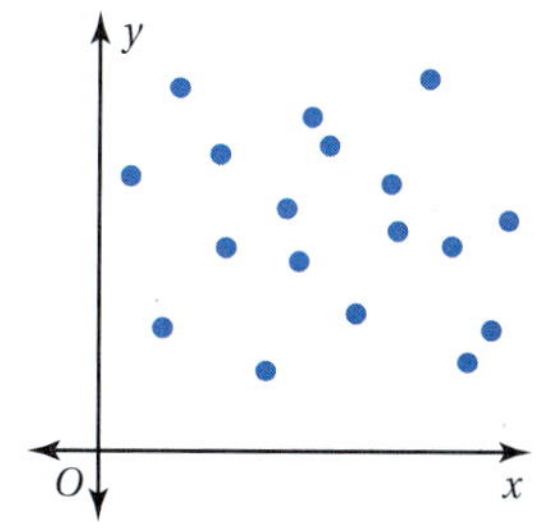
As x increases, y increases.	As x increases, y decreases.	The points show no pattern.

EXAMPLE 2 Identifying a Relationship

Tell whether the data show a *positive*, a *negative*, or *no* relationship.

a. Television size and price

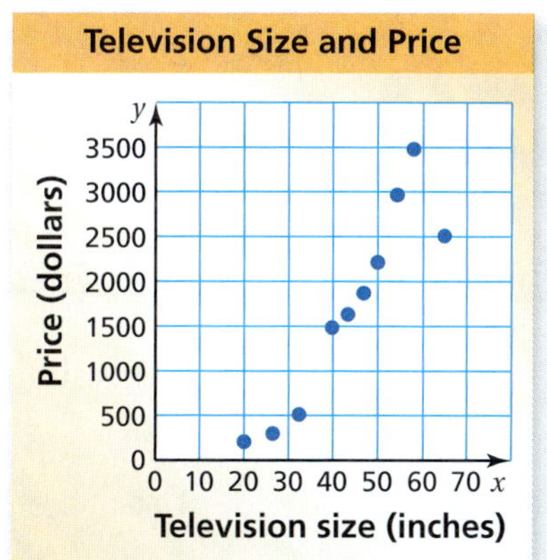

As the size of the television increases, the price increases.

So, the scatter plot shows a positive relationship.

b. Age and number of pets owned

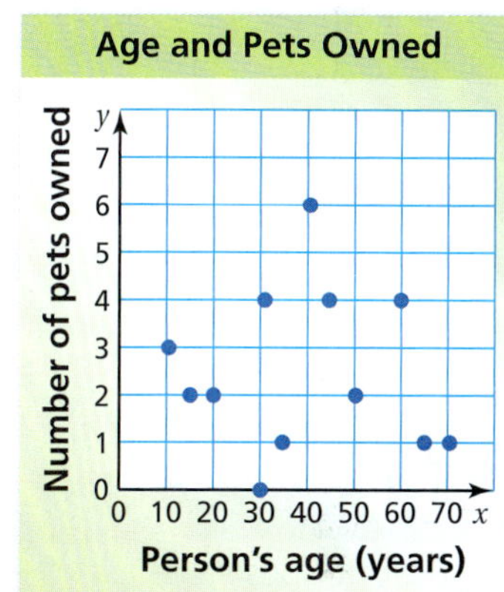

The number of pets owned does not depend on a person's age.

So, the scatter plot shows no relationship.

On Your Own

Now You're Ready
Exercises 6–8

Make a scatter plot of the data. Tell whether the data show a *positive*, a *negative*, or *no* relationship.

2.

Study Time (min), x	30	20	60	90	45	10	30	75	120	80
Test Score, y	87	74	92	97	85	62	83	90	95	91

3.

Age of a Car (years), x	1	2	3	4	5	6	7	8
Value (thousands), y	\$24	\$21	\$19	\$18	\$15	\$12	\$8	\$7

A **line of best fit** is a line drawn on a scatter plot that is close to most of the data points. It can be used to estimate data on a graph.

EXAMPLE 3 Finding a Line of Best Fit

Week, x	Sales (millions), y
1	\$19
2	\$15
3	\$13
4	\$11
5	\$10
6	\$8
7	\$7
8	\$5

The table shows the weekly sales of a DVD and the number of weeks since its release. (a) Make a scatter plot of the data. (b) Draw a line of best fit. (c) Write an equation of the line of best fit. (d) Predict the sales in week 9.

a. Plot the points in a coordinate plane. The scatter plot shows a negative relationship.

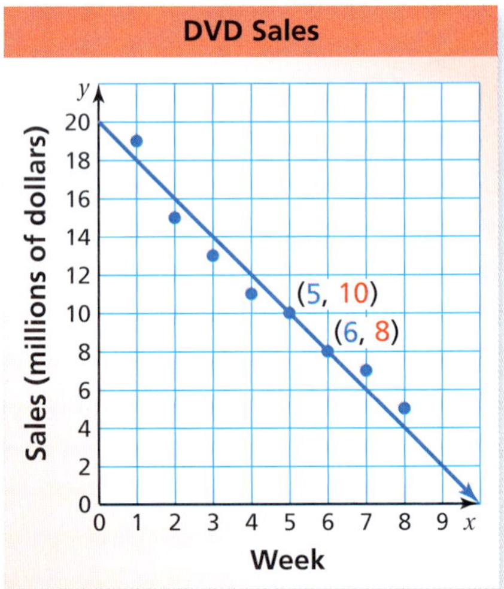

b. Draw a line that is close to the data points. Try to have as many points above the line as below it.

c. The line passes through (5, 10) and (6, 8).

$$\text{slope} = \frac{\text{rise}}{\text{run}} = \frac{-2}{1} = -2$$

Because the line crosses the y-axis at (0, 20), the y-intercept is 20.

∴ So, the equation of the line of best fit is $y = -2x + 20$.

d. To predict the sales for week 9, substitute 9 for x in the equation of the line of best fit.

$y = -2x + 20$ — Line of best fit

$= -2(9) + 20$ — Substitute 9 for x.

$= 2$ — Evaluate.

∴ The sales in week 9 should be about \$2 million.

Study Tip

A line of best fit does not need to pass through any of the data points.

On Your Own

Now You're Ready
Exercise 11

4. The table shows the number of people who have attended a neighborhood festival over an 8-year period.

Year, x	1	2	3	4	5	6	7	8
Attendance, y	420	500	650	900	1100	1500	1750	2400

a. Make a scatter plot of the data.

b. Draw a line of best fit.

c. Write an equation of the line of best fit.

d. Predict the number of people who will attend the festival in year 10.

7.3 Exercises

Check It Out
Help with Homework
BigIdeasMath.com

Vocabulary and Concept Check

1. **VOCABULARY** What type of data are needed to make a scatter plot? Explain.
2. **WRITING** Explain why a line of best fit is helpful when analyzing data.

Practice and Problem Solving

3. **BLUEBERRIES** The table shows the weights y of x pints of blueberries.

Number of Pints, x	0	1	2	3	4	5
Weight (pounds), y	0	0.8	1.50	2.20	3.0	3.75

   a. Graph the data in the table.
   b. Draw the straight line that you think best approximates the points.
   c. Write an equation of the line you drew.
   d. Use the equation to predict the weight of 10 pints of blueberries.
   e. Blueberries cost $2.25 per pound. How much do 10 pints of blueberries cost?

1 4. **SUVS** The scatter plot shows the number of sport utility vehicles sold in a city from 2005 to 2010.

   a. In what year were 1000 SUVs sold?
   b. About how many SUVs were sold in 2009?
   c. Describe the relationship shown by the data.

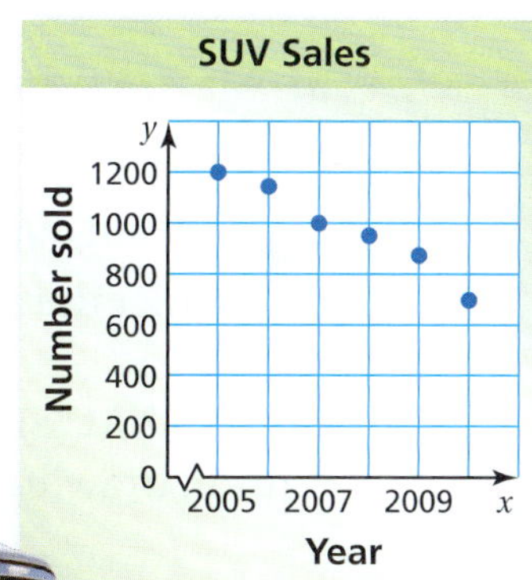

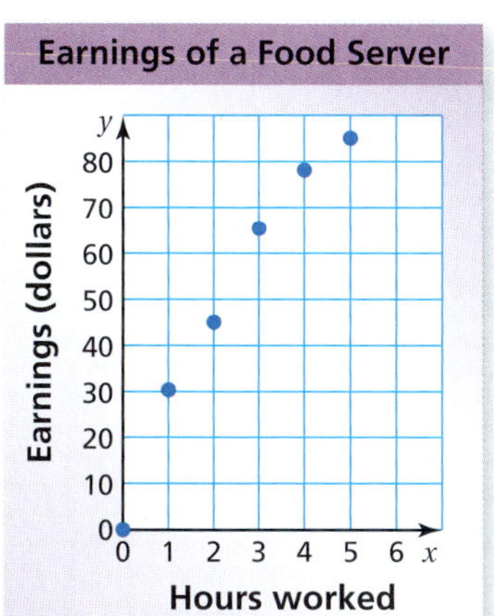

5. **EARNINGS** The scatter plot shows the total earnings (wages and tips) of a food server during 1 day.

   a. About how many hours must the server work to earn $70?
   b. About how much did the server earn for 5 hours of work?
   c. Describe the relationship shown by the data.

Tell whether the data show a *positive*, a *negative*, or *no* relationship.

② **6.**

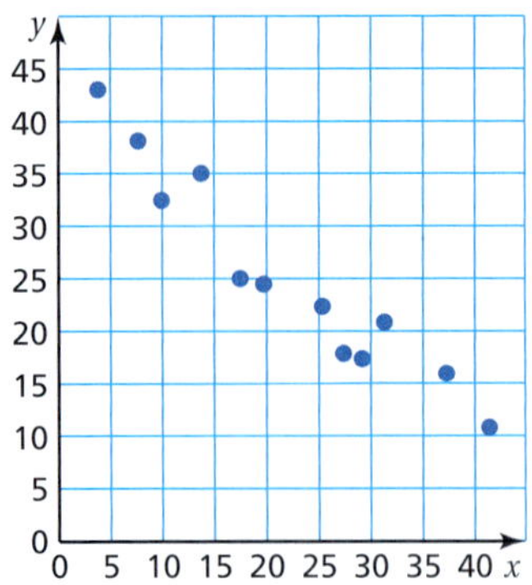

7.

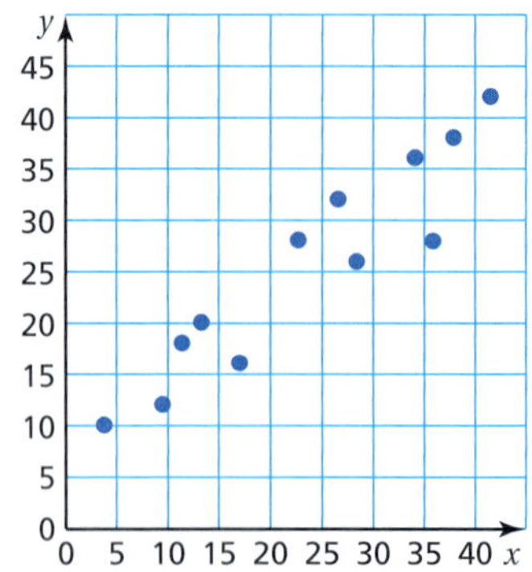

8.

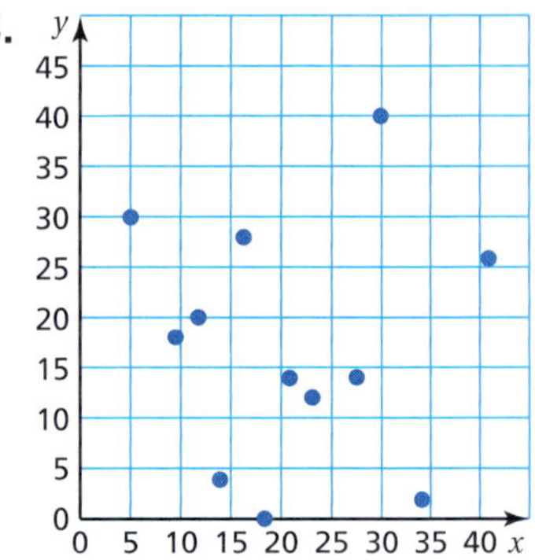

9. HONEYBEES The table shows the number of honeybee colonies in the United States from 2003 to 2006. What type of relationship do the data show?

Year, x	2003	2004	2005	2006
Honeybee Colonies (millions), y	2.599	2.556	2.413	2.392

10. OPEN-ENDED Describe a set of real-life data that has a positive relationship.

③ **11. VACATION** The table shows the distance you travel over a 6-hour period.

a. Make a scatter plot of the data.

b. Draw a line of best fit.

c. Write an equation of the line of best fit.

d. Predict the distance you will travel in 7 hours.

Hours, x	Distance (miles), y
1	62
2	123
3	188
4	228
5	280
6	344

12. ERROR ANALYSIS Describe and correct the error in drawing the line of best fit.

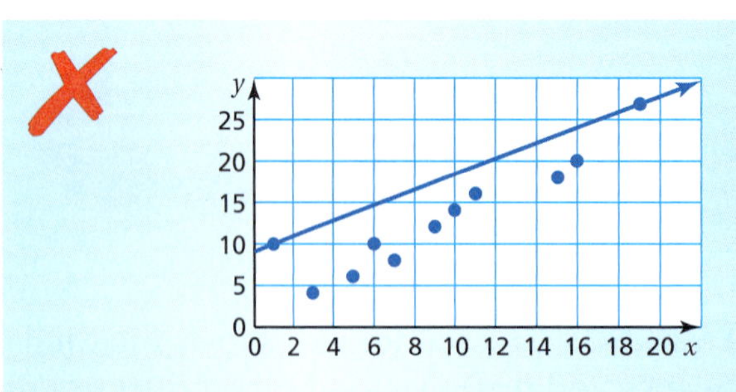

13. TEST SCORES The scatter plot shows the relationship between the number of minutes spent studying and the test scores for a science class.

a. What type of relationship does the data show?

b. Interpret the relationship.

Study Time and Test Scores

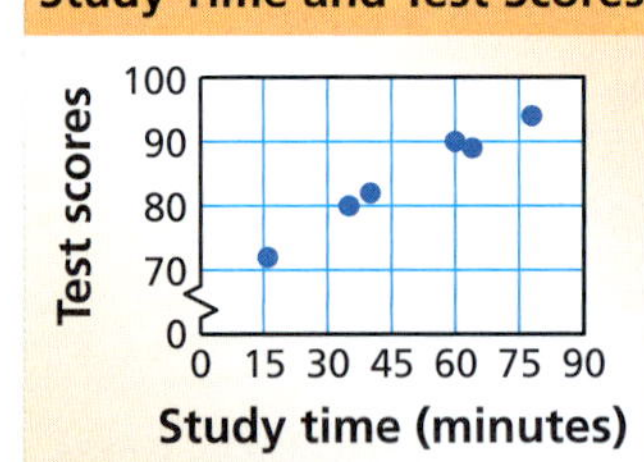

14. REASONING A data set has no relationship. Is it possible to find the line of best fit for the data? Explain.

15. PROJECT Use a ruler or a yardstick to find the height and arm span of three people.

a. Make a scatter plot using the data you collected. Then draw the line of best fit for the data.

b. Use your height and the line of best fit to predict your arm span.

c. Measure your arm span. Compare the result with your prediction in part (b).

d. Is there a relationship between a person's height x and arm span y? Explain.

16. Critical Thinking The table shows the price of admission to a local theater and the yearly attendance for several years.

Price of Admission (dollars), x	Yearly Attendance, y
19.50	50,000
21.95	48,000
23.95	47,500
24.00	40,000
24.50	45,000
25.00	43,500

a. Identify the outlier.

b. How does the outlier affect the line of best fit? Explain.

c. Make a scatter plot of the data and draw the line of best fit.

d. Use the line of best fit to predict the attendance when the admission cost is \$27.

Fair Game Review What you learned in previous grades & lessons

Use a graph to solve the equation. Check your solution. *(Section 2.7)*

17. $5x = 2x + 6$

18. $7x + 3 = 9x - 13$

19. $\frac{2}{3}x = -\frac{1}{3}x - 4$

20. MULTIPLE CHOICE The circle graph shows the super powers chosen by a class. What percent of the students want strength as their super power? *(Skills Review Handbook)*

Ⓐ 10.5% Ⓑ 12.5%

Ⓒ 15% Ⓓ 25%

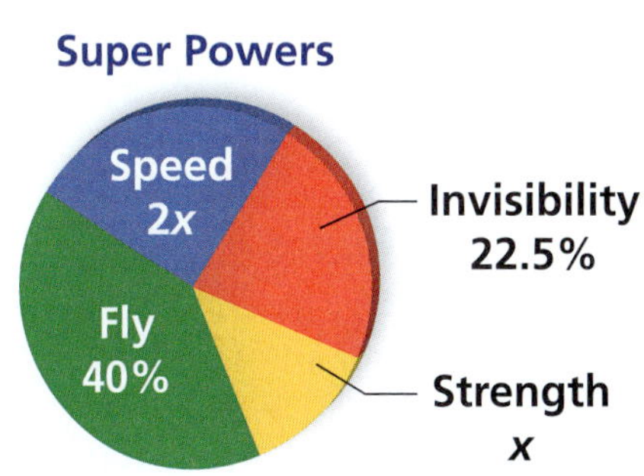

7.4 Choosing a Data Display

STATE STANDARDS
MA.8.S.3.1

Essential Question How can you display data in a way that helps you make decisions?

Share Your Work at... My.BigIdeasMath.com

1 ACTIVITY: Displaying Data

Work with a partner. Analyze and display each data set in a way that best describes the data. Explain your choice of display.

a. **ROAD KILL** A 1-year study of a 2-mile length of U.S. Highway 441 (through Payne's Prairie State Park) in Florida found a total of 3365 dead animals.

1291 snakes	1333 frogs
374 turtles	265 birds
72 mammals	29 alligators
1 lizard	

b. **BLACK BEAR ROAD KILL** The data below show the number of black bears killed on Florida roads from 1987 to 2006.

1987	30	1994	47	2001	99
1988	37	1995	49	2002	129
1989	46	1996	61	2003	111
1990	33	1997	74	2004	127
1991	43	1998	88	2005	141
1992	35	1999	82	2006	135
1993	43	2000	109		

c. **RACCOON ROAD KILL** A 1-week study of a 4-mile road in Santa Rosa county found the following weights (in pounds) of raccoons that had been killed by vehicles.

13.4	14.8	17.0	12.9
21.3	21.5	16.8	14.8
15.2	18.7	18.6	17.2
18.5	9.4	19.4	15.7
14.5	9.5	25.4	21.5
17.3	19.1	11.0	12.4
20.4	13.6	17.5	18.5
21.5	14.0	13.9	19.0

d. What do you think can be done to minimize the number of animals killed by vehicles in Florida?

2 ACTIVITY: Statistics Project

ENDANGERED SPECIES PROJECT Use the Internet or some other reference to write a report about an animal species in Florida that is (or has been) endangered. Include graphical displays of the data you have gathered.

Sample: Florida Key Deer

In 1939, Florida banned the hunting of Key deer. The numbers of Key deer fell to about 100 in the 1940s.

In 1947, public sentiment was stirred by 11-year-old Glenn Allen from Miami. Allen organized Boy Scouts and others in a letter-writing campaign that led to the establishment of the National Key Deer Refuge in 1957. The approximately 8600-acre refuge includes 2280 acres of designated wilderness.

Key Deer Refuge has increased the population of Key deer. A recent study estimated the total Key deer population to be between 700 and 800.

About half of Key deer deaths are due to vehicles.

One of two Key deer wildlife underpasses on Big Pine Key

What Is Your Answer?

3. **IN YOUR OWN WORDS** How can you display data in a way that helps you make decisions? Use the Internet or some other reference to find examples of the following types of data displays.
   - Bar graph
   - Circle graph
   - Scatter plot
   - Stem-and-leaf plot
   - Box-and-whisker plot

Practice Use what you learned about choosing data displays to complete Exercise 3 on page 300.

Check It Out
Lesson Tutorials
BigIdeasMath.com

Key Idea

Data Display	What does it do?	
Pictograph	shows data using pictures	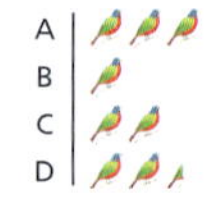
Bar Graph	shows data in specific categories	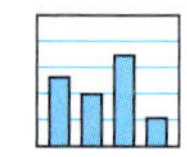
Circle Graph	shows data as parts of a whole	
Line Graph	shows how data change over time	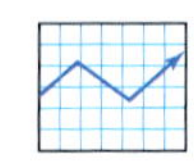
Histogram	shows frequencies of data values in intervals of the same size	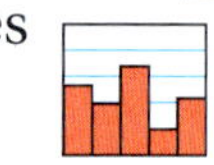
Stem-and-Leaf Plot	orders numerical data and shows how they are distributed	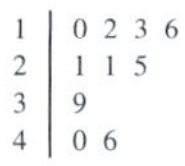
Box-and-Whisker Plot	shows the variability of a data set using quartiles	
Line Plot	shows the number of times each value occurs in a data set	
Scatter Plot	shows the relationship between two data sets using ordered pairs in a coordinate plane	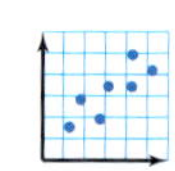

EXAMPLE 1 Choosing an Appropriate Data Display

Choose an appropriate data display for the situation. Explain your reasoning.

a. the number of students in a marching band each year

A line graph shows change over time. So, a line graph is an appropriate data display.

b. comparison of people's shoe sizes and their heights

You want to compare two different data sets. So, a scatter plot is an appropriate data display.

On Your Own

Now You're Ready
Exercises 4–7

Choose an appropriate data display for the situation. Explain your reasoning.

1. the population of Florida divided into age groups
2. the percents of students in your school who speak Spanish, French, or Haitian Creole

EXAMPLE 2 Identifying a Misleading Data Display

Which line graph is misleading? Explain.

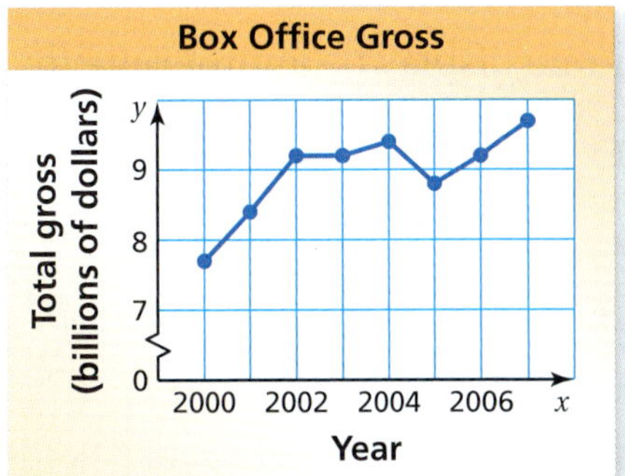

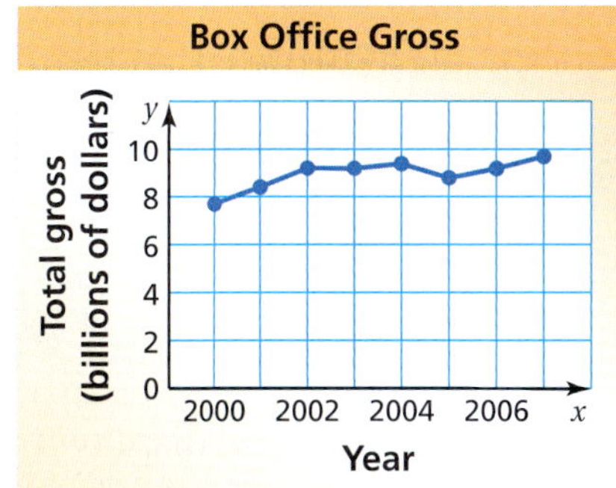

The vertical axis of the line graph on the left has a break (⦚) and begins at 7. This graph makes it appear that the total gross increased rapidly from 2000 to 2004. The graph on the right has an unbroken axis. It is more honest and shows that the total gross increased slowly.

∴ So, the graph on the left is misleading.

EXAMPLE 3 Analyzing a Misleading Data Display

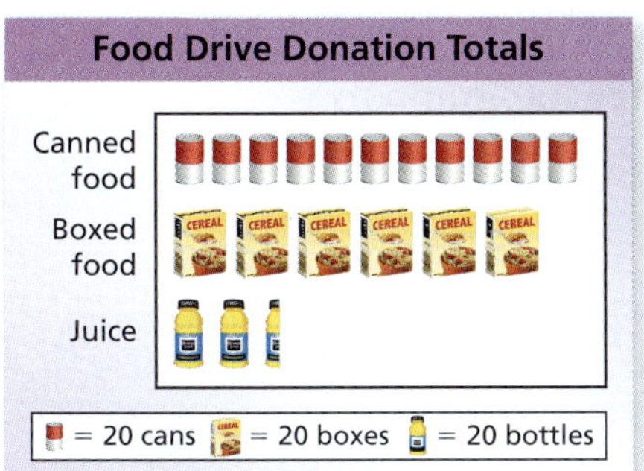

A volunteer concludes that the number of cans of food and boxes of food donated were about the same. Is this conclusion accurate? Explain.

Each icon represents the same number of items. Because the box icon is larger than the can icon, it looks like the number of boxes is about the same as the number of cans, but the number of boxes is actually about half of the number of cans.

∴ So, the conclusion is not accurate.

On Your Own

Now You're Ready
Exercises 9–12

Explain why the data display is misleading.

3.

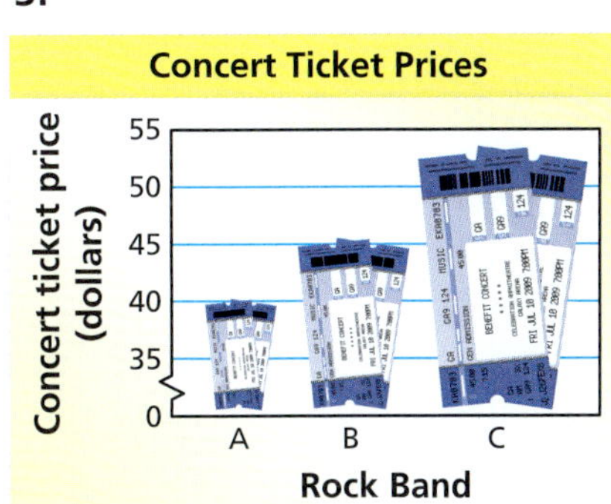

4.

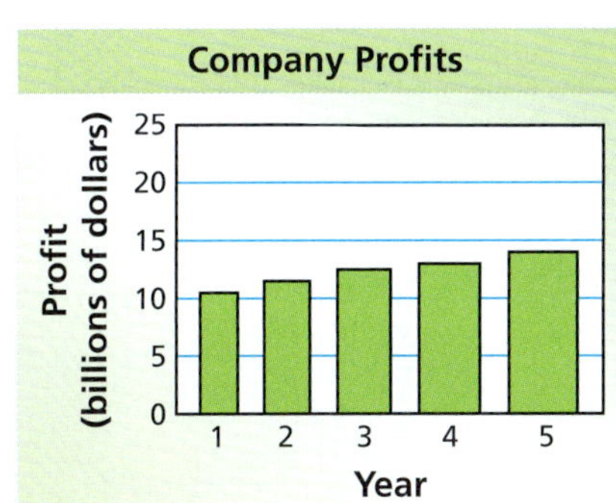

7.4 Exercises

Check It Out
Help with Homework
BigIdeasMath.com

Vocabulary and Concept Check

1. **REASONING** Can more than one display be appropriate for a data set? Explain.
2. **OPEN-ENDED** Describe how a histogram can be misleading.

Practice and Problem Solving

3. Analyze and display the data in a way that best describes the data. Explain your choice of display.

Notebooks Sold in One Week				
192 red	170 green	203 black	183 pink	230 blue
165 yellow	210 purple	250 orange	179 white	218 other

Choose an appropriate data display for the situation. Explain your reasoning.

1 4. a student's test scores and how the scores are spread out

5. the distance a person drives each month
6. the outcome of rolling a number cube
7. homework problems assigned each day
8. **WRITING** When would you choose a histogram instead of a bar graph to display data?

Explain why the data display is misleading.

2 3 9.

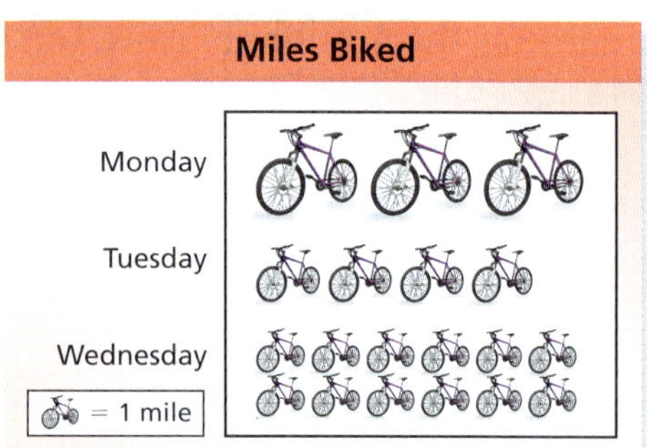

10.

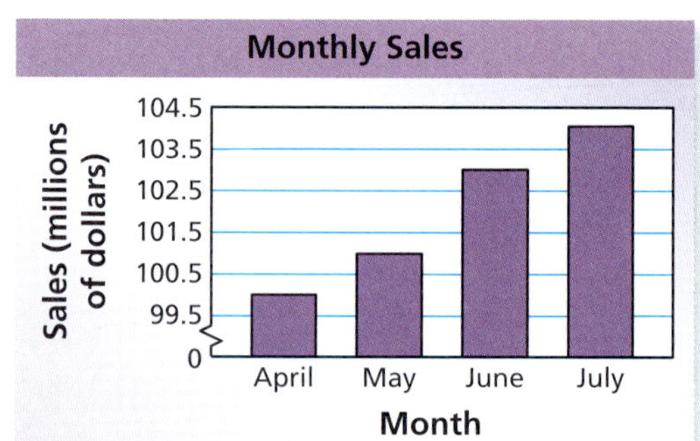

11.

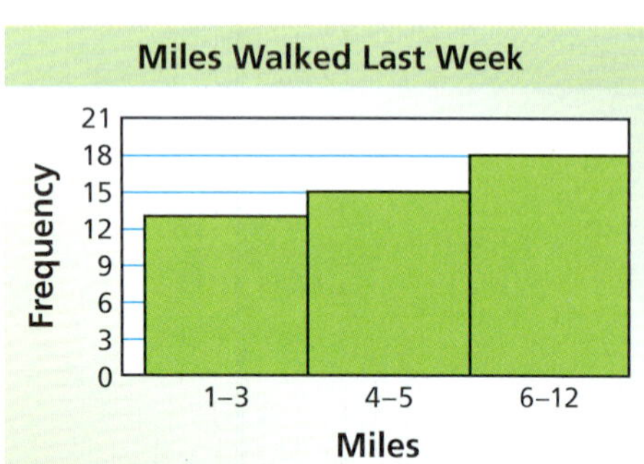

12.

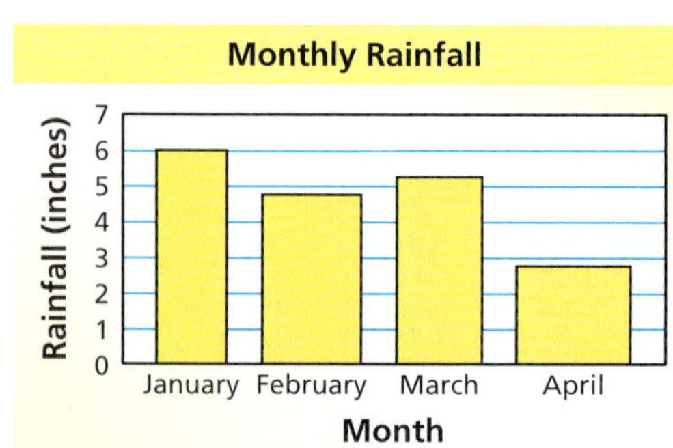

13. **VEGETABLES** A nutritionist wants to use a data display to show the favorite vegetables of the students at a school. Choose an appropriate data display for the situation. Explain your reasoning.

14. **CHEMICALS** A scientist gathers data about a decaying chemical compound. The results are shown in the scatter plot. Is the data display misleading? Explain.

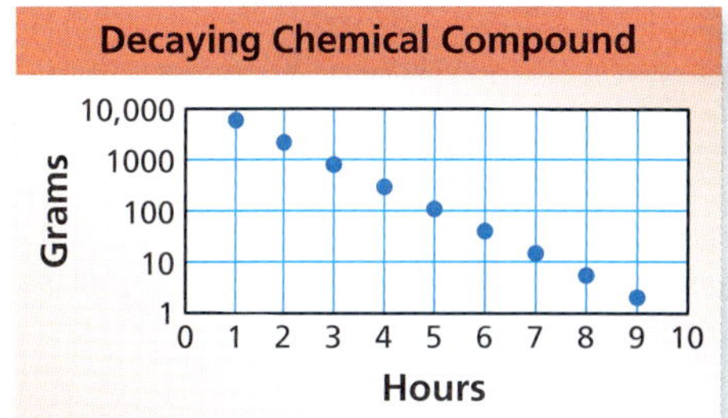

15. **REASONING** What type of data display is appropriate for showing the mode of a data set?

16. **SPORTS** A survey asked 100 students to choose their favorite sports. The results are shown in the circle graph.

a. Explain why the graph is misleading.

b. What type of data display would be more appropriate for the data? Explain.

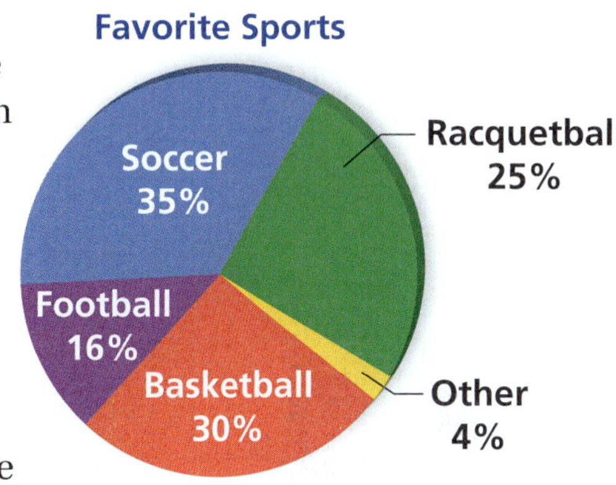

17. **Critical Thinking** With the help of computers, mathematicians have computed and analyzed billions of digits of the irrational number π. One of the things they analyze is the frequency of each of the numbers 0 through 9. The table shows the frequency of each number in the first 100,000 digits of π.

a. Display the data in a bar graph.

b. Display the data in a circle graph.

c. Which data display is more appropriate? Explain.

d. Describe the distribution.

Number	0	1	2	3	4	5	6	7	8	9
Frequency	9999	10,137	9908	10,025	9971	10,026	10,029	10,025	9978	9902

Fair Game Review What you learned in previous grades & lessons

Write the verbal statement as an equation. *(Skills Review Handbook)*

18. A number plus 3 is 5.

19. 8 times a number is 24.

20. **MULTIPLE CHOICE** What is 20% of 25% of 400? *(Skills Review Handbook)*

Ⓐ 20　Ⓑ 200　Ⓒ 240　Ⓓ 380

7.3–7.4 Quiz

Check It Out
Progress Check
BigIdeasMath.com

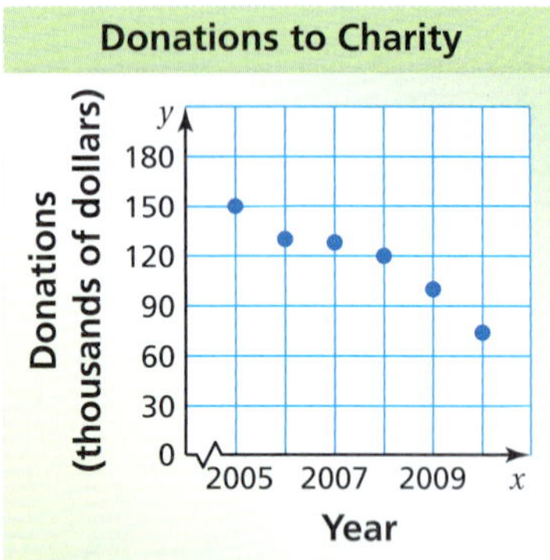

1. The scatter plot shows the number of donations a charity received from 2005 to 2010. *(Section 7.3)*

a. In what year did the charity receive $150,000?

b. How much did the charity receive in 2008?

c. Describe the relationship shown by the data.

Tell whether the data show a *positive*, a *negative*, or *no* relationship. *(Section 7.3)*

2.

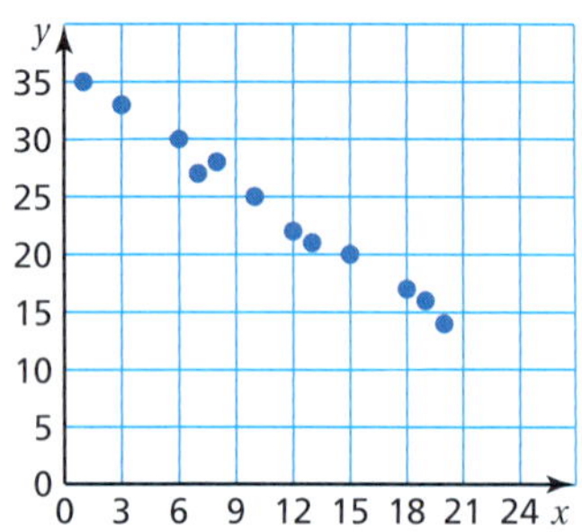

3.

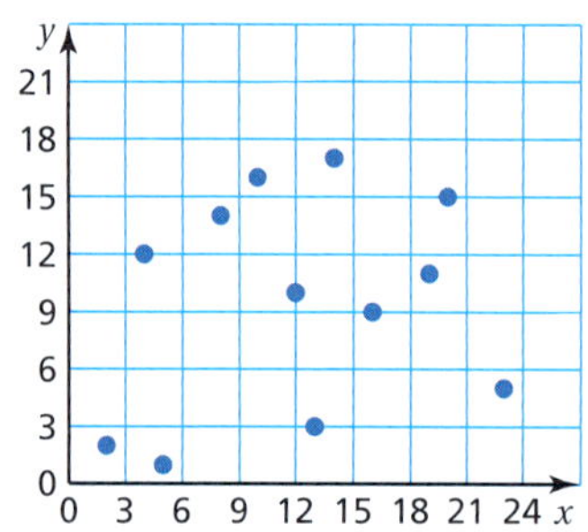

4.

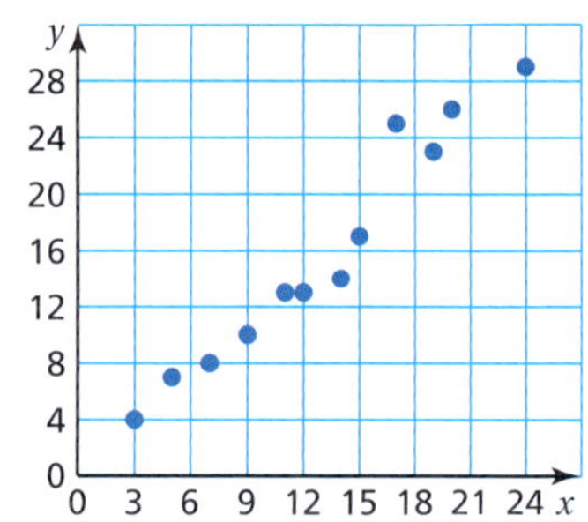

Choose an appropriate data display for the situation. Explain your reasoning. *(Section 7.4)*

5. percent of band students in each section

6. company's profit for each week

7. **FUNDRAISER** The graph shows the amount of money that the eighth-grade students at a school raised each month to pay for the class trip. Is the data display misleading? Explain. *(Section 7.4)*

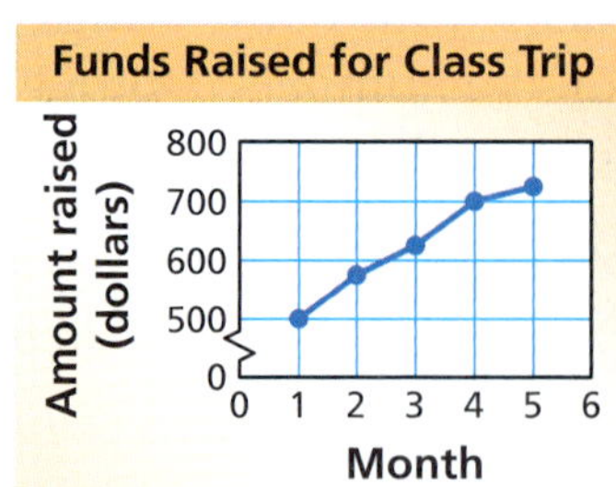

8. **CATS** The table shows the number of cats adopted from an animal shelter each month. *(Section 7.3)*

a. Make a scatter plot of the data.

b. Draw a line of best fit.

c. Write an equation of the line of best fit.

d. Predict how many cats will be adopted in month 10.

Month	1	2	3	4	5	6	7	8	9
Cats	3	6	7	11	13	14	15	18	19

7 Chapter Review

Check It Out
Vocabulary Help
BigIdeasMath.com

Review Key Vocabulary

measure of central tendency, *p. 276*
box-and-whisker plot, *p. 282*
quartiles, *p. 282*
scatter plot, *p. 290*
line of best fit, *p. 292*

Review Examples and Exercises

7.1 Measures of Central Tendency *(pp. 274–279)*

The table shows the number of kilometers you ran each day for the past 10 days. Find the mean, median, and mode of the distances.

Kilometers Run	
3.5	4.1
4.0	4.3
4.4	4.5
3.9	2.0
4.3	5.0

Mean: $\frac{40}{10} = 4$ (sum of the data / number of values)

Median: 2.0, 3.5, 3.9, 4.0, 4.1, 4.3, 4.3, 4.4, 4.5, 5.0 — Order the data.

$\frac{8.4}{2} = 4.2$ — Mean of two middle values

Mode: 2.0, 3.5, 3.9, 4.0, 4.1, 4.3, 4.3, 4.4, 4.5, 5.0

The value 4.3 occurs most often.

∴ The mean is 4 kilometers, the median is 4.2 kilometers, and the mode is 4.3 kilometers.

Exercises

1. Use the data in the example above. You run 4.0 miles on day 11. How does this additional value affect the mean, median, and mode? Explain.

Find the mean, median, and mode of the data.

2.

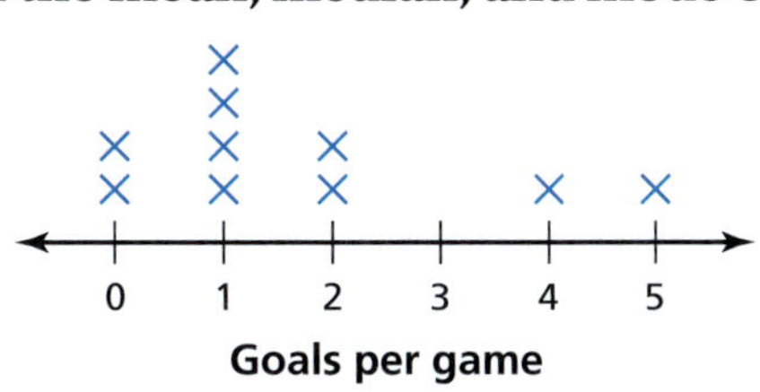

3.

Ski Resort Temperatures (°F)		
11	3	3
0	−9	−2
10	10	10

7.2 Box-and-Whisker Plots (pp. 280–285)

Make a box-and-whisker plot for the weights (in pounds) of pumpkins sold at a market.

16, 20, 14, 15, 12, 8, 8, 19, 14, 10, 8, 16

Step 1: Order the data. Find the median and the quartiles.

lower half | upper half

least value → 8 8 8 10 12 14 14 15 16 16 19 20 ← greatest value

first quartile, 9 | median, 14 | third quartile, 16

Step 2: Draw a number line that includes the least and greatest values. Graph points above the number line for the least value, greatest value, median, first quartile, and third quartile.

Step 3: Draw a box using the quartiles. Draw a line through the median. Draw whiskers from the box to the least and greatest values.

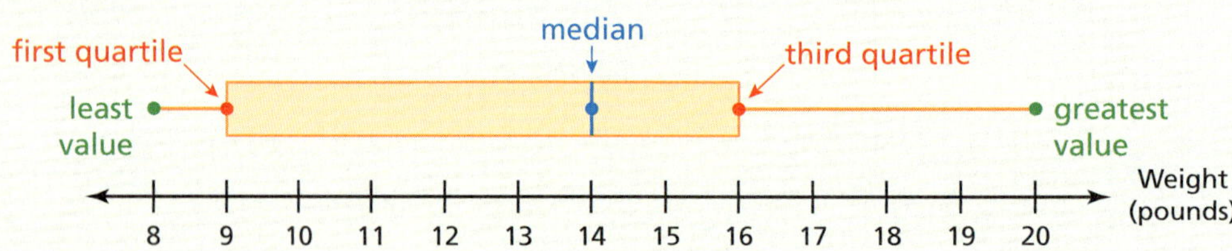

Exercises

Make a box-and-whisker plot for the data.

4. Ages of volunteers at a hospital: 14, 17, 20, 16, 17, 14, 21, 18

5. Masses (in kilograms) of lions: 120, 200, 180, 150, 200, 200, 230, 160

7.3 Scatter Plots and Lines of Best Fit (pp. 288–295)

Your school is ordering custom T-shirts. The scatter plot shows the number of T-shirts ordered and the cost per shirt. What tends to happen to the cost per shirt as the number of T-shirts ordered increases?

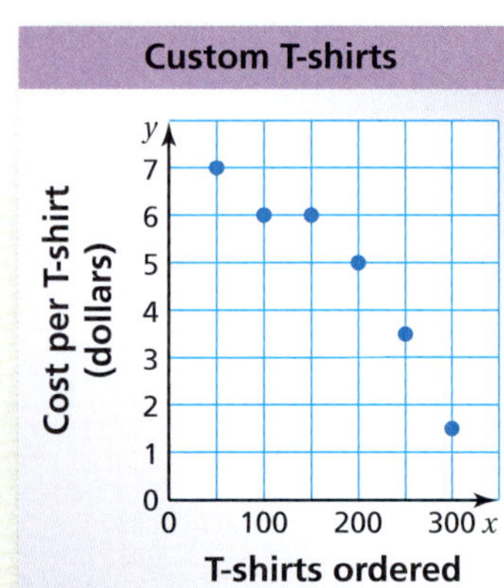

Looking at the graph, the plotted points go down from left to right.

∴ So, as the number of T-shirts ordered increases, the cost per shirt decreases.

Exercises

6. The scatter plot shows the number of geese that migrated to a park each season.

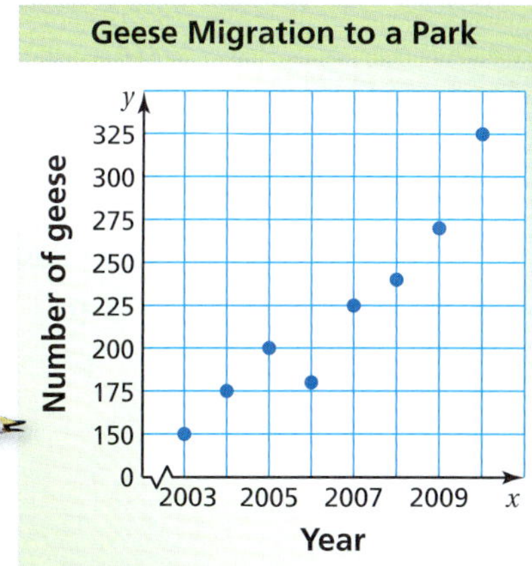

   a. In what year did 270 geese migrate?

   b. How many geese migrated in 2007?

   c. Describe the relationship shown by the data.

Tell whether the data show a *positive*, a *negative*, or *no* relationship.

7. 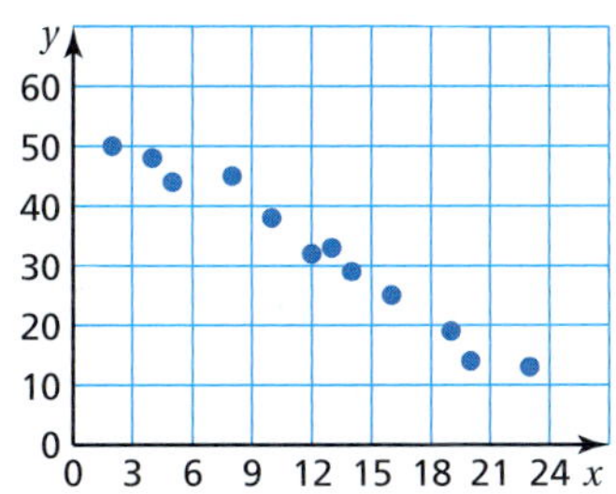

8. 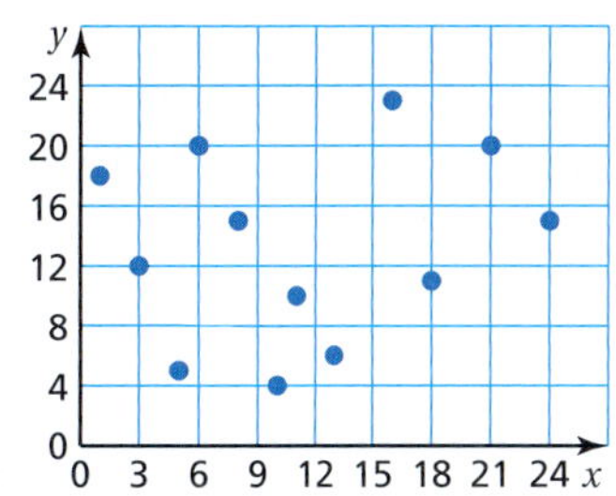

7.4 Choosing a Data Display (pp. 296–301)

Choose an appropriate data display for the situation. Explain your reasoning.

a. the percent of votes that each candidate received in an election

A circle graph shows data as parts of a whole. So, a circle graph is an appropriate data display.

b. the distribution of the ages of U.S. presidents

A stem-and-leaf plot orders numerical data and shows how they are distributed. So, a stem-and-leaf plot is an appropriate data display.

Exercises

Choose an appropriate data display for the situation. Explain your reasoning.

9. the number of pairs of shoes sold by a store each week
10. the outcomes of spinning a spinner with 3 equal sections numbered 1, 2, and 3
11. comparison of the number of cans of food donated by each eighth-grade class
12. comparison of the heights of brothers and sisters

7 Chapter Test

Check It Out
Test Practice
BigIdeasMath.com

Find the mean, median, and mode of the data.

1.

Distances (feet) Above or Below Water Level in Pool		
−3	0	−3
3	10	0
11	−6	−3

2. Cooking Time (minutes)

Stem	Leaf
3	5 8
4	0 1 8
5	0 4 4 4 5 9
6	0

Key: 4 | 1 = 41 minutes

Make a box-and-whisker plot for the data.

3. Ages (in years) of dogs at a vet's office: 1, 3, 5, 11, 5, 7, 5, 9

4. Lengths (in inches) of fish in a pond: 12, 13, 7, 8, 14, 6, 13, 10

5. Hours practiced each week: 7, 6, 5, 4.5, 3.5, 7, 7.5, 2, 8, 7, 7.5, 6.5

6. **POPULATION** The graph shows the population (in millions) of Florida from 1960 to 2000.

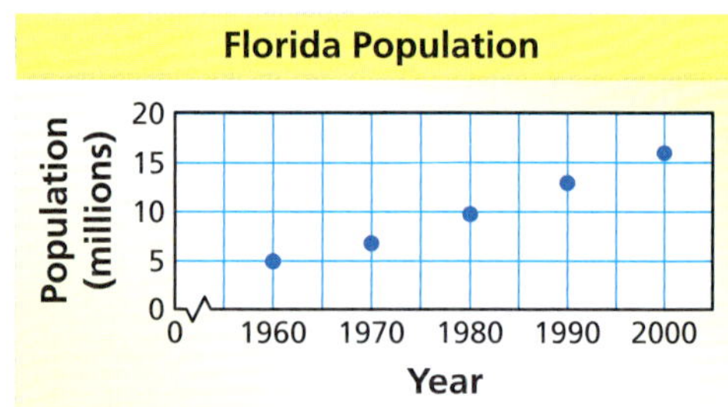

a. In what year was the population of Florida about 5 million?

b. What was the approximate population of Florida in 1990?

c. Describe the relationship shown by the data.

Choose an appropriate data display for the situation. Explain your reasoning.

7. magazine sales grouped by price

8. distance a person hikes each week

9. **ALLIGATORS** The table shows the lengths of 12 alligators. Make a box-and-whisker plot of the data. What does the box-and-whisker plot tell you about the data?

Length (meters)	2.0	1.9	2.2	2.8	3.0	2.0	2.2	3.0	2.5	1.8	2.1	3.0

10. **REASONING** Name two types of data displays that are appropriate for showing the median of a data set.

11. **NEWBORNS** The table shows the lengths and weights of several newborn babies.

Length (inches)	Weight (pounds)
19	6
19.5	7
20	7.75
20.25	8.5
20.5	8.5
22.5	11

a. Make a scatter plot of the data.

b. Draw the line of best fit.

c. Write an equation of the line of best fit.

d. Use the equation to predict the weight of a newborn that is 19.75 inches long.

7 Standardized Test Practice

Test-Taking Strategy
Read Question Before Answering

"Be sure to read the question before choosing your answer. You may find a word that changes the meaning."

1. Research scientists are measuring the number of days lettuce seeds take to germinate. In a study, 500 seeds were planted. Of these, 473 seeds germinated. The box-and-whisker plot summarizes the number of days it took the seeds to germinate. What can you conclude from the box-and-whisker plot?

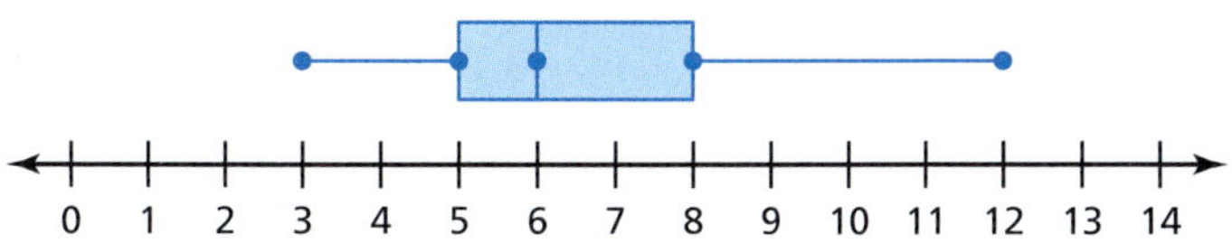

A. The median number of days for the seeds to germinate is 12.

B. 50% of the seeds took more than 8 days to germinate.

C. 50% of the seeds took less than 5 days to germinate.

D. The median number of days for the seeds to germinate was 6.

2. An object dropped from a height will fall under the force of gravity. The time t, in seconds, it takes to fall a distance d, in feet, can be found using the formula below.

$$t = \frac{\sqrt{d}}{4}$$

A ball is dropped from the top of a building that is 40 feet tall. Approximately how many seconds will it take for the ball to reach the ground?

F. 1.6 sec

G. 3.2 sec

H. 5 sec

I. 6.3 sec

3. A plumber charges a fixed amount for a house call plus an amount based on the number of hours worked. A job lasting 1 hour costs \$95 and a job lasting 2 hours costs \$140. What is the fixed amount charged by the plumber?

A. \$45

B. \$50

C. \$70

D. \$95

4. The diagram below shows parallel lines cut by a transversal. Which angle is the corresponding angle for $\angle 6$?

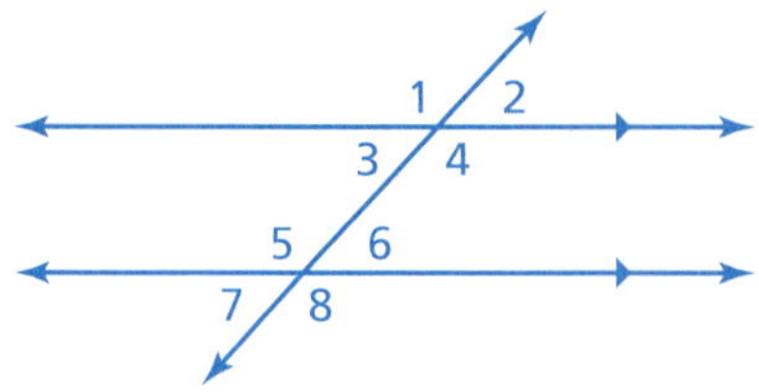

F. $\angle 2$

G. $\angle 3$

H. $\angle 4$

I. $\angle 8$

5. Which value of x makes the equation below true?

$$3x - 9 = 2(x + 4)$$

6. As part of a probability experiment, students were asked to roll two number cubes and find the sum of the numbers obtained. One group of students did this 600 times and obtained the results shown in the table. Which sum was the mode for this group's 600 rolls?

Sum	2	3	4	5	6	7	8	9	10	11	12
Number of Rolls	16	30	51	66	83	98	93	64	47	35	17

A. 4

B. 7

C. 12

D. 98

7. Which expression is equivalent to $\frac{\sqrt{32}}{\sqrt{18}}$?

F. $\sqrt{14}$

G. $\frac{4}{3}$

H. $\frac{8}{3}$

I. $\frac{16}{9}$

8. Which point lies on the graph of the line given by $y = -\frac{1}{2}x + 7$?

A. $(5, 4)$

B. $(-4, 5)$

C. $(20, 3)$

D. $(40, -13)$

9. Which scatter plot shows a negative relationship between x and y?

F.

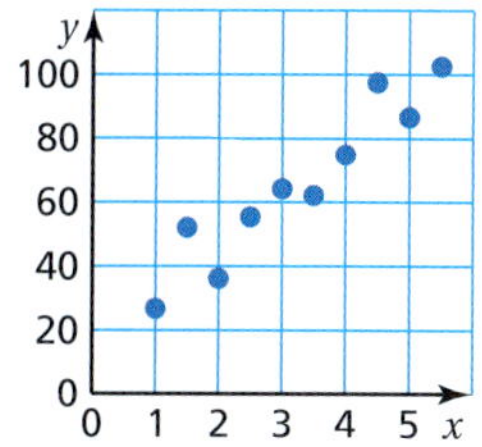

H.

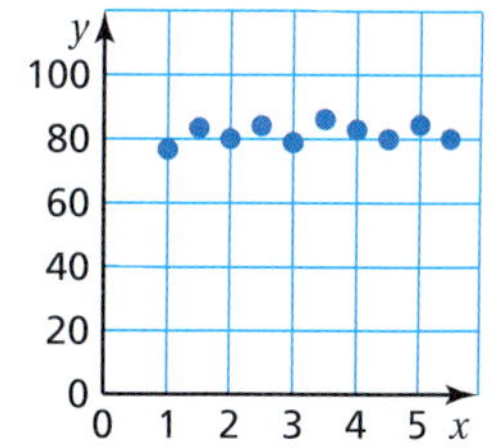

G.

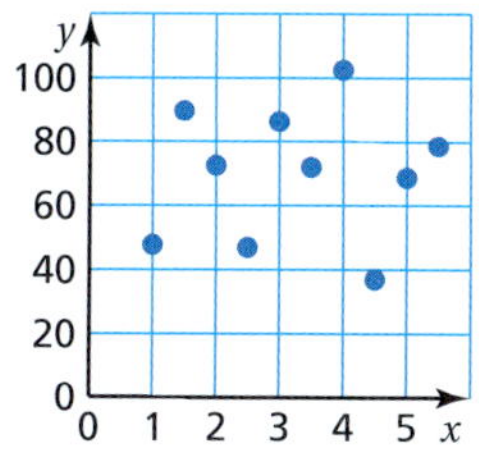

I.

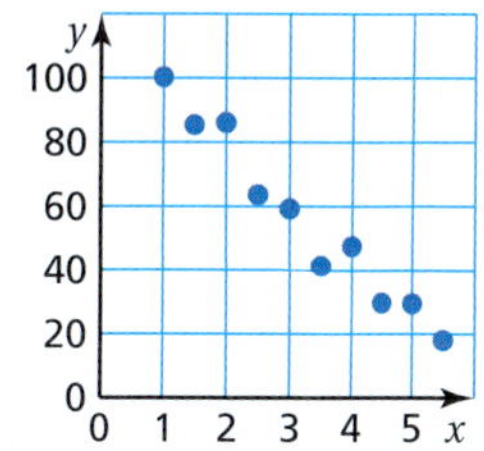

10. The legs of a right triangle have the lengths of 8 centimeters and 15 centimeters. What is the length of the hypotenuse, in centimeters?

11. The 16 members of a camera club have the ages listed below.

40, 22, 24, 58, 30, 31, 37, 25, 62, 40, 39, 37, 28, 28, 51, 44

Part A Order the ages from least to greatest.

Part B Find the median of the ages.

Part C Make a box-and-whisker plot for the ages of the camera club members.

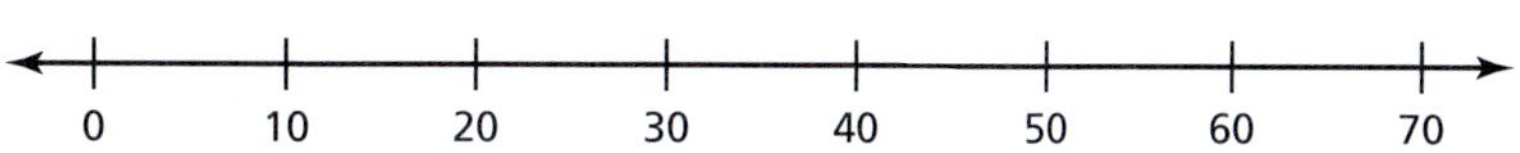

8 Linear Inequalities

"Here is a math quiz, Descartes. Tell me about these symbols."

"Just think of the Addition Property of Inequality in this way. If Fluffy has more cat treats than you have ..."

"... and you each get 2 more cat treats, then Fluffy will STILL have more cat treats than you have!"

What You Learned Before

"Some people remember which is bigger by thinking that < is the mouth of a hungry alligator who is trying to eat the LARGER number."

Comparing Real Numbers (Grade 8, Chapter 6)

Complete the number sentence with <, >, or =.

Example 1 $\frac{1}{3} \;\square\; 0.3$

$$\frac{1}{3} = \frac{10}{30},\quad 0.3 = \frac{3}{10} = \frac{9}{30}$$

Because $\frac{10}{30}$ is greater than $\frac{9}{30}$, $\frac{1}{3}$ is greater than 0.3.

∴ So, $\frac{1}{3} > 0.3$.

Example 2 $\sqrt{6} \;\square\; 6$

Use a calculator to estimate $\sqrt{6}$.

$$\sqrt{6} \approx 2.45$$

Because 2.45 is less than 6, $\sqrt{6}$ is less than 6.

∴ So, $\sqrt{6} < 6$.

Try It Yourself

Complete the number sentence with <, >, or =.

1. $\frac{1}{4} \;\square\; 0.25$ **2.** $0.1 \;\square\; \frac{1}{9}$ **3.** $\pi \;\square\; \sqrt{10}$

Graphing Inequalities (Grade 6)

Example 3 **Graph $x \geq 3$.**

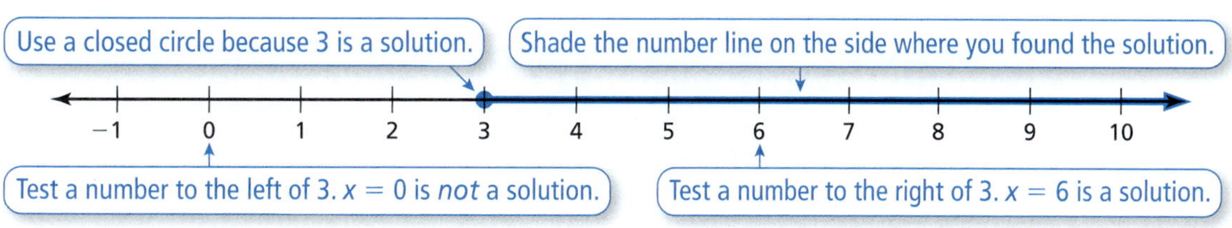

Example 4 **Graph $x < -2$.**

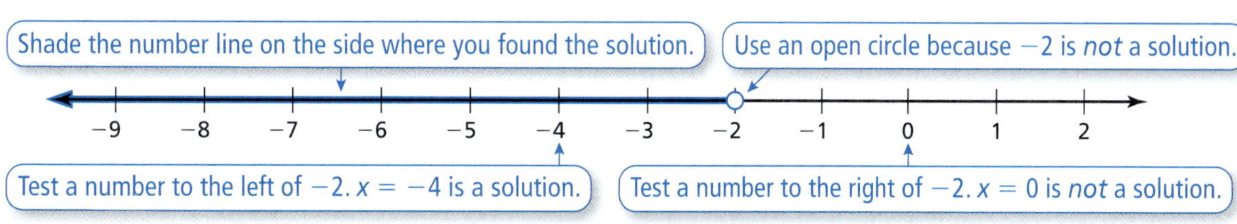

Try It Yourself

Graph the inequality.

4. $x \geq 0$ **5.** $x < 6$ **6.** $x \leq -4$ **7.** $x > -10$

8.1 Writing and Graphing Inequalities

STATE STANDARDS
MA.8.A.4.2

Essential Question How can you use an inequality to describe a real-life statement?

1 ACTIVITY: Writing and Graphing Inequalities

Work with a partner. Write an inequality for the statement. Then sketch the graph of all the numbers that make the inequality true.

a. **Statement:** The temperature t in Minot, North Dakota has never been below -36°F.

Inequality:

Graph:

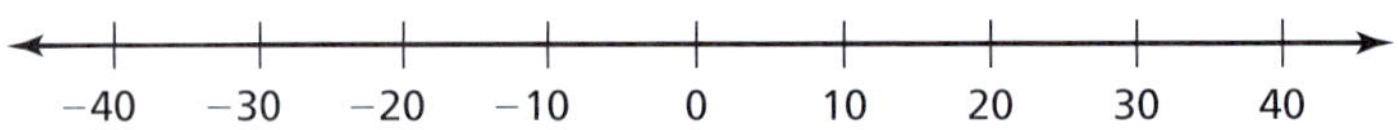

b. **Statement:** The elevation e in Florida is at most 345 feet above sea level.

Inequality:

Graph:

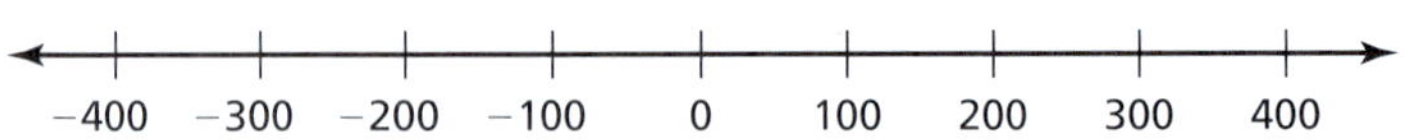

2 ACTIVITY: Writing and Graphing Inequalities

Work with a partner. Write an inequality for the graph. Then, in words, describe all the values of *x* that make the inequality true.

a.

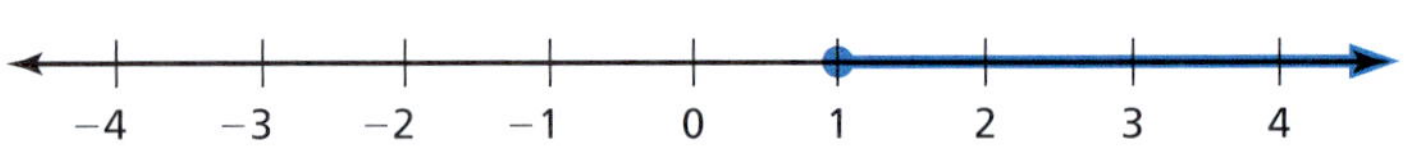

b.

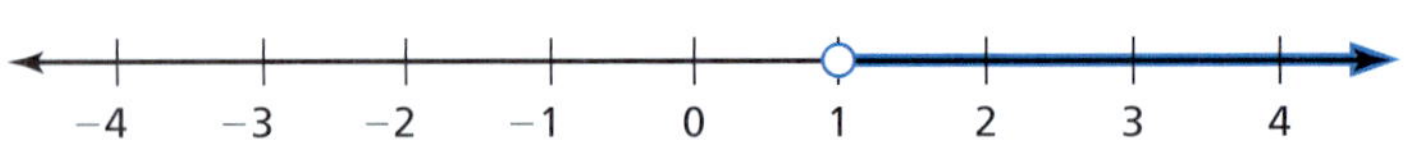

c.

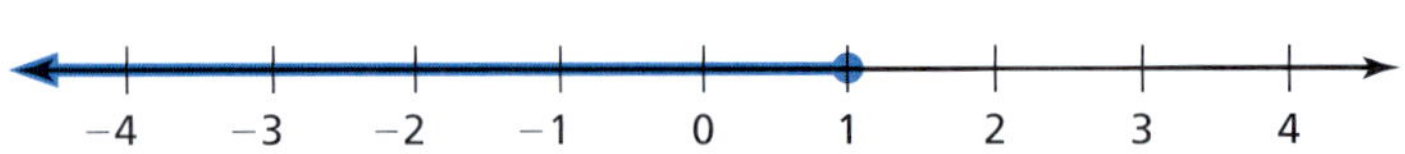

d.

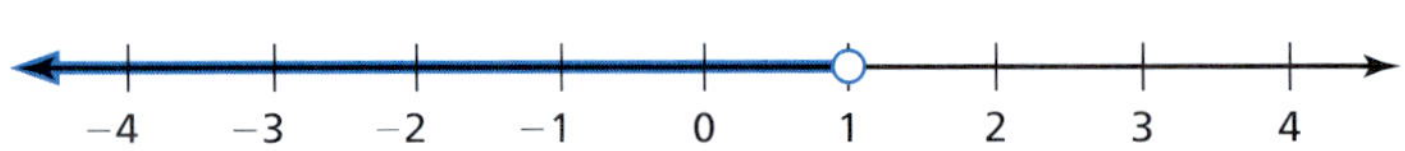

3 ACTIVITY: Triangle Inequality

Work with a partner. Use 8 to 10 pieces of spaghetti.

- Break one piece of spaghetti into three parts that can be used to form a triangle.
- Form a triangle and use a centimeter ruler to measure each side. Round the side lengths to the nearest tenth.
- Record the side lengths in a table.

Side Lengths That Form a Triangle			
Small	Medium	Large	$S + M$

- Repeat the process with two other pieces of spaghetti.
- Repeat the experiment by breaking pieces of spaghetti into three pieces that *do not* form a triangle. Record the lengths in a table.

Side Lengths That Do Not Form a Triangle			
Small	Medium	Large	$S + M$

- **INDUCTIVE REASONING** Write a rule that uses an inequality to compare the lengths of three sides of a triangle.
- Use your rule to decide whether the following triangles are possible. Explain.

a.

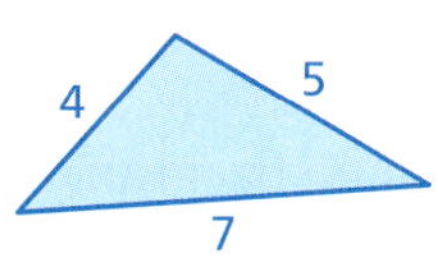

b.

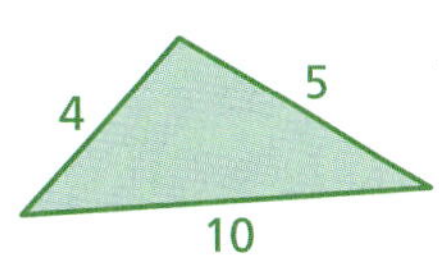

c.

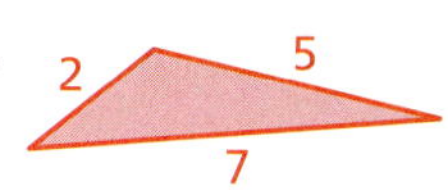

What Is Your Answer?

4. **IN YOUR OWN WORDS** How can you use an inequality to describe a real-life statement? Give two examples of real-life statements that can be represented by inequalities.

Practice Use what you learned about writing and graphing inequalities to complete Exercises 4 and 5 on page 316.

8.1 Lesson

Check It Out
Lesson Tutorials
BigIdeasMath.com

Key Vocabulary
inequality, *p. 314*
solution of an inequality, *p. 314*
solution set, *p. 314*
graph of an inequality, *p. 315*

An **inequality** is a mathematical sentence that compares expressions. It contains the symbols $<$, $>$, $\leq$, or $\geq$. To write an inequality, look for the following phrases to determine where to place the inequality symbol.

Inequality Symbols				
Symbol	$<$	$>$	$\leq$	$\geq$
Key Phrases	• is less than • is fewer than	• is greater than • is more than	• is less than or equal to • is at most • is no more than	• is greater than or equal to • is at least • is no less than

EXAMPLE 1 Writing an Inequality

A number w minus 3.5 is less than or equal to -2. Write this sentence as an inequality.

A number w minus 3.5 is less than or equal to -2.

$$w - 3.5 \quad \leq \quad -2$$

∴ An inequality is $w - 3.5 \leq -2$.

On Your Own

Now You're Ready
Exercises 6–9

Write the word sentence as an inequality.

1. A number b is fewer than 30.4.
2. Twice a number k is at least $-\frac{7}{10}$.

A **solution of an inequality** is a value that makes the inequality true. An inequality can have more than one solution. The set of all solutions of an inequality is called the **solution set**.

Value of x	$x + 5 \geq -2$	Is the inequality true?
-6	$-6 + 5 \overset{?}{\geq} -2$ $-1 \geq -2$ ✓	yes
-7	$-7 + 5 \overset{?}{\geq} -2$ $-2 \geq -2$ ✓	yes
-8	$-8 + 5 \overset{?}{\geq} -2$ $-3 \ngeq -2$ ✗	no

Reading

The symbol $\ngeq$ means "is not greater than or equal to."

Multi-Language Glossary at BigIdeasMath.com.

EXAMPLE 2 Checking Solutions

Tell whether -4 is a solution of the inequality.

a. $x + 8 < -3$

$x + 8 < -3$	Write the inequality.	$-4.5x > -21$
$-4 + 8 \overset{?}{<} -3$	Substitute -4 for x.	$-4.5(-4) \overset{?}{>} -21$
$4 \not< -3$ ✗	Simplify.	$18 > -21$ ✓

4 is *not* less than -3.

So, -4 is *not* a solution of the inequality.

b. $-4.5x > -21$

18 is greater than -21.

So, -4 is a solution of the inequality.

On Your Own

Now You're Ready
Exercises 11–16

Tell whether -6 is a solution of the inequality.

3. $c + 4 < -1$ **4.** $5 - m \le 10$ **5.** $21 \div x \ge -3.5$

The **graph of an inequality** shows all of the solutions of the inequality on a number line. An open circle ○ is used when a number is *not* a solution. A closed circle ● is used when a number is a solution. An arrow to the left or right shows that the graph continues in that direction.

EXAMPLE 3 Graphing an Inequality

Graph $y \le -3$.

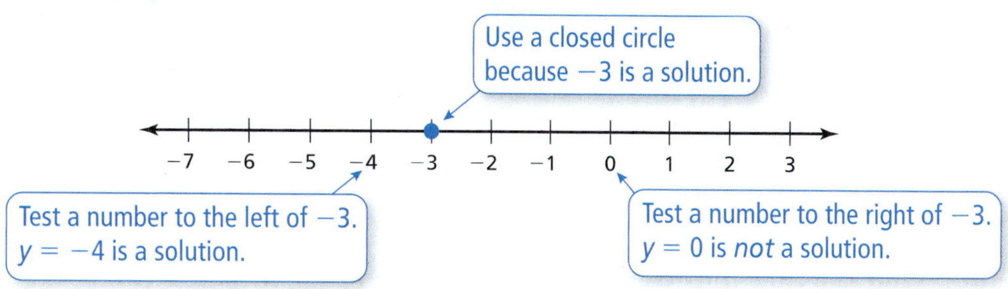

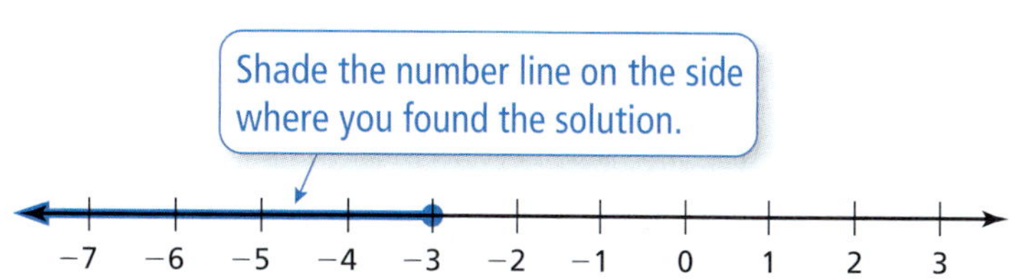

On Your Own

Now You're Ready
Exercises 17–20

Graph the inequality on a number line.

6. $b > -8$ **7.** $g \le 1.4$ **8.** $r < -\frac{1}{2}$ **9.** $v \ge \sqrt{0.09}$

8.1 Exercises

Check It Out
Help with Homework
BigIdeasMath.com

Vocabulary and Concept Check

1. **VOCABULARY** Would an open circle or a closed circle be used in the graph of the inequality $k < 250$? Explain.

2. **DIFFERENT WORDS, SAME QUESTION** Which is different? Write "both" inequalities.

w is greater than or equal to -7.	w is no less than -7.
w is no more than -7.	w is at least -7.

3. **REASONING** Do $x \geq -9$ and $-9 \geq x$ represent the same inequality? Explain.

Practice and Problem Solving

Write an inequality for the graph. Then, in words, describe all the values of x that make the inequality true.

4. 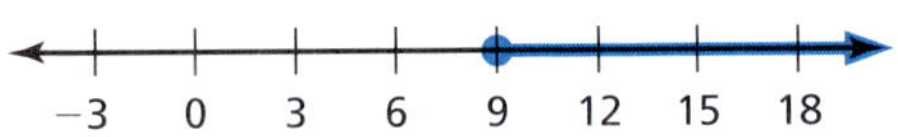

5. 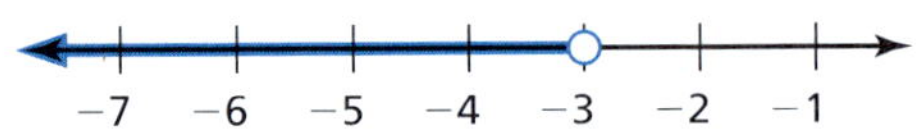

Write the word sentence as an inequality.

1 6. A number x is no less than -4.

7. A number y added to 5.2 is less than 23.

8. A number b multiplied by -5 is at most $-\frac{3}{4}$.

9. A number k minus 8.3 is greater than 48.

10. **ERROR ANALYSIS** Describe and correct the error in writing the word sentence as an inequality.

> ✗ Twice a number c is at least $-\frac{4}{9}$.
>
> $2c \leq -\frac{4}{9}$

Tell whether the given value is a solution of the inequality.

2 11. $s + 6 \leq 12$; $s = 4$

12. $15n > -3$; $n = -2$

13. $a - 2.5 \leq 1.6$; $a = 4.1$

14. $-3.3q > -13$; $q = 4.6$

15. $\frac{4}{5}h \geq -4$; $h = -15$

16. $\frac{1}{12} - p < \frac{1}{3}$; $p = \frac{1}{6}$

Graph the inequality on a number line.

3 17. $g \geq -6$

18. $q > 1.25$

19. $z < 11\frac{1}{4}$

20. $w \leq -\sqrt{289}$

21. **DRIVING** When you are driving with a learner's license, a licensed driver who is 21 years of age or older must be with you. Write an inequality that represents this situation.

Tell whether the given value is a solution of the inequality.

22. $3p > 5 + p;\ p = 4$

23. $\frac{y}{2} \geq y - 11;\ y = 18$

24. VIDEO GAME RATINGS Each rating is matched with the inequality that represents the recommended ages of players. Your friend is old enough to play "E 10+" games. Is your friend old enough to play "T" games? Explain.

$x \geq 3$ | $x \geq 6$ | $x \geq 10$ | $x \geq 13$ | $x \geq 17$

The ESRB rating icons are registered trademarks of the Entertainment Software Association.

25. SCUBA DIVING Three requirements for a scuba diving training course are shown.

a. Write and graph three inequalities that represent the requirements.

b. You can swim 10 lengths of a 25-yard pool. Do you satisfy the swimming requirement of the course? Explain.

26. LUGGAGE On an airplane, the maximum sum of the length, width, and height of a carry-on bag is 45 inches. Find three different sets of dimensions that are reasonable for a carry-on bag.

27. Critical Thinking A number m is less than another number n. The number n is less than or equal to a third number p.

a. Write two inequalities representing these relationships.

b. Describe the relationship between m and p.

c. Can m be equal to p? Explain.

Fair Game Review What you learned in previous grades & lessons

Solve the equation. Check your solution. *(Section 1.1)*

28. $r - 12 = 3$

29. $4.2 + p = 2.5$

30. $n - 3\pi = 7\pi$

31. MULTIPLE CHOICE Which linear function relates y to x? *(Section 4.3)*

Ⓐ $y = -0.5x - 3$ Ⓑ $y = 2x + 3$

Ⓒ $y = 0.5x - 3$ Ⓓ $y = 2x - 3$

x	-1	0	1	2
y	-5	-3	-1	1

8.2 Solving Inequalities Using Addition or Subtraction

STATE STANDARDS
MA.8.A.4.2

Essential Question How can you use addition or subtraction to solve an inequality?

1 ACTIVITY: Quarterback Passing Efficiency

Work with a partner. The National Collegiate Athletic Association (NCAA) uses the following formula to rank the passing efficiency P of quarterbacks.

$$P = \frac{8.4Y + 100C + 330T - 200N}{A}$$

Y = total length of all completed passes (in Yards)

C = Completed passes

T = passes resulting in a Touchdown

N = iNtercepted passes

A = Attempted passes

M = incoMplete passes

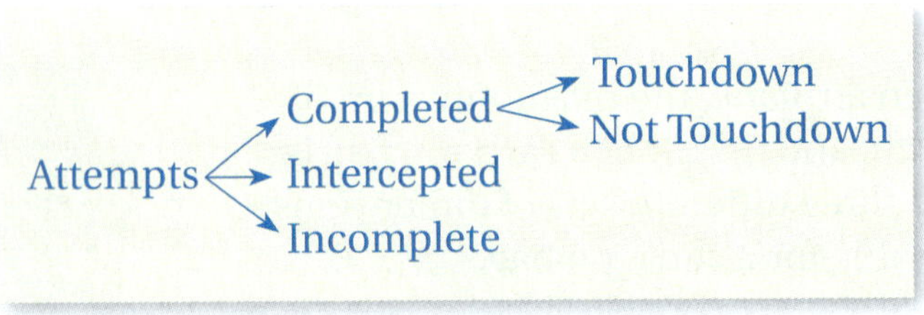

Which of the following equations or inequalities are true relationships among the variables? Explain your reasoning.

a. $C + N < A$ **b.** $C + N \leq A$ **c.** $T < C$ **d.** $T \leq C$

e. $N < A$ **f.** $A > T$ **g.** $A - C \geq M$ **h.** $A = C + N + M$

2 ACTIVITY: Quarterback Passing Efficiency

Work with a partner. Which of the following quarterbacks has a passing efficiency rating that satisfies the inequality $P > 100$? Show your work.

Player	Attempts	Completions	Yards	Touchdowns	Interceptions
A	149	88	1065	7	9
B	400	205	2000	10	3
C	426	244	3105	30	9
D	188	89	1167	6	15

3 ACTIVITY: Finding Solutions of Inequalities

Work with a partner. Use the passing efficiency formula to create a passing record that makes the inequality true. Then describe the values of *P* that make the inequality true.

a. $P < 0$

Attempts	Completions	Yards	Touchdowns	Interceptions

b. $P + 100 \geq 250$

Attempts	Completions	Yards	Touchdowns	Interceptions

c. $180 < P - 50$

Attempts	Completions	Yards	Touchdowns	Interceptions

d. $P + 30 \geq 120$

Attempts	Completions	Yards	Touchdowns	Interceptions

e. $P - 250 > -80$

Attempts	Completions	Yards	Touchdowns	Interceptions

What Is Your Answer?

4. Write a rule that describes how to solve inequalities like those in Activity 3. Then use your rule to solve each of the inequalities in Activity 3.

5. **IN YOUR OWN WORDS** How can you use addition or subtraction to solve an inequality?

6. How is solving the inequality $x + 3 < 4$ similar to solving the equation $x + 3 = 4$? How is it different?

Practice Use what you learned about solving inequalities using addition or subtraction to complete Exercises 3–5 on page 322.

8.2 Lesson

Check It Out
Lesson Tutorials
BigIdeasMath.com

Key Ideas

Study Tip

You can solve inequalities the same way you solve equations. Use inverse operations to get the variable by itself.

Addition Property of Inequality

Words If you add the same number to each side of an inequality, the inequality remains true.

Numbers
$$\begin{array}{rcr} -3 & < & 2 \\ +4 & & +4 \\ \hline 1 & < & 6 \end{array}$$

Algebra
$$\begin{array}{rcr} x - 3 & > & -10 \\ +3 & & +3 \\ \hline x & > & -7 \end{array}$$

Subtraction Property of Inequality

Words If you subtract the same number from each side of an inequality, the inequality remains true.

Numbers
$$\begin{array}{rcr} -3 & < & 1 \\ -5 & & -5 \\ \hline -8 & < & -4 \end{array}$$

Algebra
$$\begin{array}{rcr} x + 7 & > & -20 \\ -7 & & -7 \\ \hline x & > & -27 \end{array}$$

These properties are also true for $\leq$ and $\geq$.

EXAMPLE 1 Solving an Inequality Using Addition

Solve $x - 6 \geq -10$. Graph the solution.

$$x - 6 \geq -10$$ Write the inequality.

Undo the subtraction. → $$+6 \quad +6$$ Add 6 to each side.

$$x \geq -4$$ Simplify.

∴ The solution is $x \geq -4$.

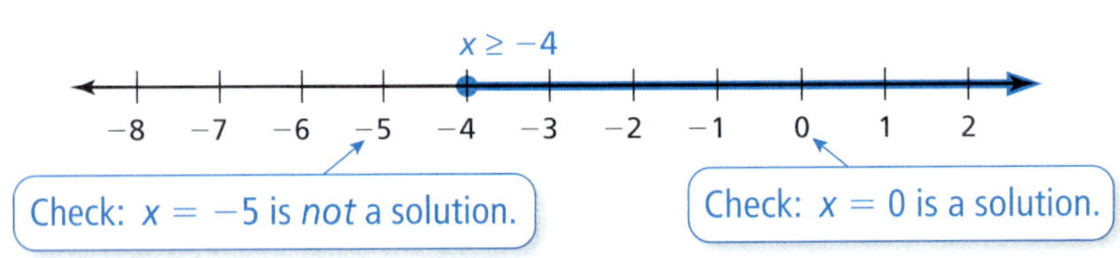

Study Tip

To check a solution, you check some numbers that are solutions and some that are not.

On Your Own

Solve the inequality. Graph the solution.

1. $b - 2 > -9$
2. $m - 3.8 \leq 5$
3. $\frac{1}{4} > y - \frac{1}{4}$

EXAMPLE 2 Solving an Inequality Using Subtraction

Solve $-8 > 1.4 + x$. Graph the solution.

$$
\begin{aligned}
-8 &> 1.4 + x && \text{Write the inequality.}\\
\underline{-\,1.4} &\quad \underline{-\,1.4} && \text{Subtract 1.4 from each side.}\\
-9.4 &> x && \text{Simplify.}
\end{aligned}
$$

Undo the addition.

∴ The solution is $x < -9.4$.

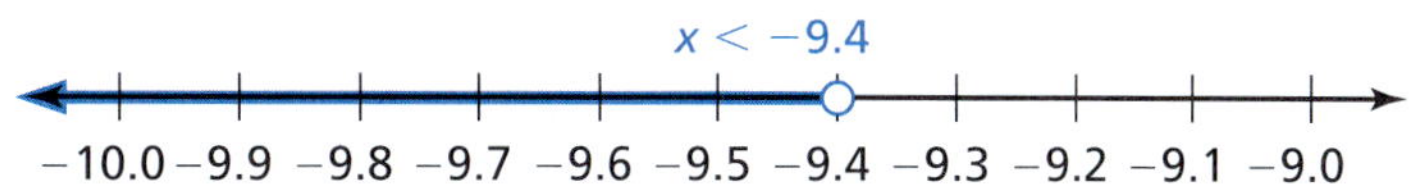

Reading

The inequality $-9.4 > x$ is the same as $x < -9.4$.

On Your Own

Now You're Ready
Exercises 6–17

Solve the inequality. Graph the solution.

4. $k + 5 \le -3$

5. $\frac{5}{6} \le z + \frac{2}{3}$

6. $p + 0.7 > -2.3$

EXAMPLE 3 Real-Life Application

On a train, carry-on bags can weigh no more than 50 pounds. Your bag weighs 24.8 pounds. Write and solve an inequality that represents the amount of weight you can add to your bag.

Words	Weight of your bag	plus	amount of weight you can add	is no more than	the weight limit.
Variable	Let w be the possible weight you can add.				
Inequality	24.8	$+$	w	$\le$	50

$$
\begin{aligned}
24.8 + w &\le 50 && \text{Write the inequality.}\\
\underline{-\,24.8} &\quad \underline{-\,24.8} && \text{Subtract 24.8 from each side.}\\
w &\le 25.2 && \text{Simplify.}
\end{aligned}
$$

∴ You can add no more than 25.2 pounds to your bag.

On Your Own

7. **WHAT IF?** Your carry-on bag weighs 32.5 pounds. Write and solve an inequality that represents the possible weight you can add to your bag.

8.2 Exercises

Check It Out
Help with Homework
BigIdeasMath.com

Vocabulary and Concept Check

1. **REASONING** Is the inequality $r - 5 \le 8$ the same as $8 \le r - 5$? Explain.

2. **WHICH ONE DOESN'T BELONG?** Which inequality does *not* belong with the other three? Explain your reasoning.

$c + \frac{7}{2} \le \frac{3}{2}$ $\quad$ $c + \frac{7}{2} \ge \frac{3}{2}$ $\quad$ $\frac{3}{2} \ge c + \frac{7}{2}$ $\quad$ $c - \frac{3}{2} \le -\frac{7}{2}$

Practice and Problem Solving

Use the formula in Activity 1 to create a passing record that makes the inequality true.

3. $P \ge 180$

4. $P + 40 < 110$

5. $280 \le P - 20$

Solve the inequality. Graph the solution.

1 2

6. $y - 3 \ge 7$

7. $t - 8 > -4$

8. $n + 11 \le 20$

9. $a + 7 > -1$

10. $5 < v - \frac{1}{2}$

11. $\frac{1}{5} > d + \frac{4}{5}$

12. $-\frac{2}{3} \le g - \frac{1}{3}$

13. $m + \frac{7}{4} \le \frac{11}{4}$

14. $11.2 \le k + 9.8$

15. $h - 1.7 < -3.2$

16. $0 > s + \pi$

17. $5 \ge u - 4.5$

18. **ERROR ANALYSIS** Describe and correct the error in graphing the solution of the inequality.

✗ $5 \ge x - 5$
$10 \ge x$

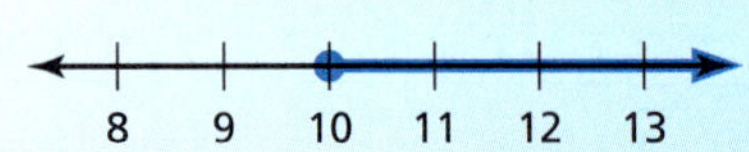

19. **PELICAN** The maximum volume of a great white pelican's bill is about 700 cubic inches.

 a. A pelican scoops up 100 cubic inches of water. Write and solve an inequality that represents the additional volume the bill can contain.

 b. A pelican's stomach can contain about one-third the maximum amount that its bill can contain. Write an inequality that represents the volume of the pelican's stomach.

Write and solve an inequality that represents the value of x.

20. The perimeter is less than 16 feet.

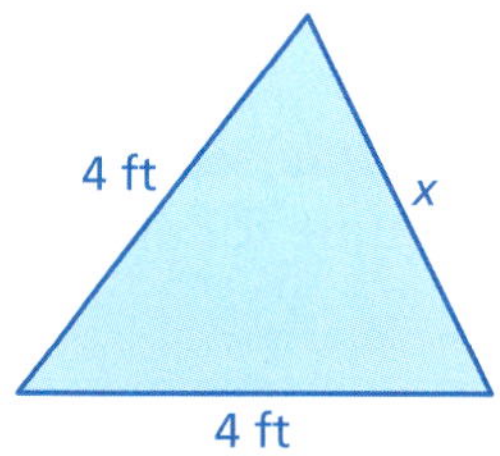

21. The base is greater than the height.

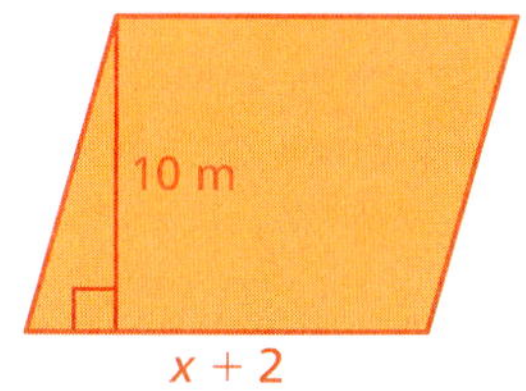

22. The perimeter is less than or equal to 5 feet.

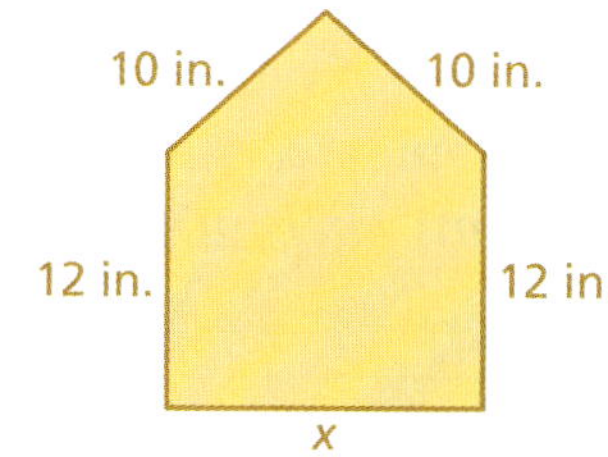

23. REASONING The solution of $w + c \leq 8$ is $w \leq 3$. What is the value of c?

24. FENCE The hole for a fence post is 2 feet deep. The top of the fence post needs to be at least 4 feet above the ground. Write and solve an inequality that represents the required length of the fence post.

25. VIDEO GAME You need at least 12,000 points to advance to the next level of a video game.

a. Write and solve an inequality that represents the number of points you need to advance.

b. You find a treasure chest that increases your score by 60%. How does this change the inequality?

26. POWER A circuit overloads at 1800 watts of electricity. A microwave that uses 1100 watts of electricity is plugged into the circuit.

a. Write and solve an inequality that represents the additional number of watts you can plug in without overloading the circuit.

b. In addition to the microwave, what two appliances in the table can you plug in without overloading the circuit?

Appliance	Watts
Clock radio	50
Blender	300
Hot plate	1200
Toaster	800

27. Critical Thinking The maximum surface area of the solid is 15π square millimeters. Write and solve an inequality that represents the height of the cylinder.

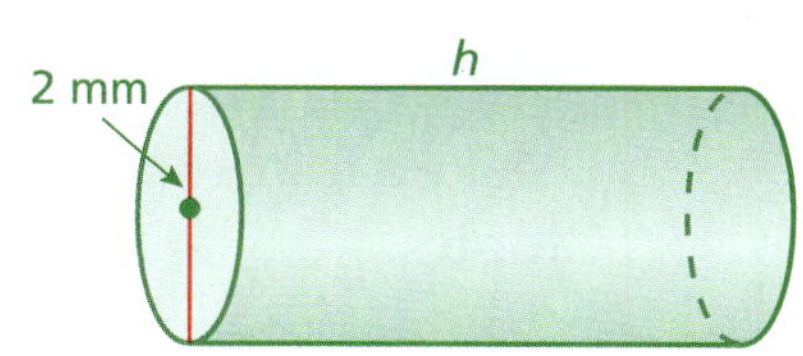

Fair Game Review What you learned in previous grades & lessons

Solve the equation. *(Section 1.1)*

28. $6 = 3x$

29. $\frac{r}{5} = 2$

30. $4c = 15$

31. $8 = \frac{2}{3}b$

32. MULTIPLE CHOICE Which fraction is equivalent to 3.8? *(Skills Review Handbook)*

Ⓐ $\frac{5}{19}$ Ⓑ $\frac{19}{5}$ Ⓒ $\frac{12}{15}$ Ⓓ $\frac{12}{5}$

8 Study Help

Check It Out
Graphic Organizer
BigIdeasMath.com

You can use a **four square** to organize information about a topic. Each of the four squares can be a category, such as *definition, vocabulary, example, non-example, words, algebra, table, numbers, visual, graph,* or *equation*. Here is an example of a four square for an inequality.

Definition:

A mathematical sentence that uses an inequality symbol (<).

Vocabulary:

Key phrases that can represent an inequality:

- is less than
- is fewer than

INEQUALITY (<)

Example:

Words: A number x is less than 2.
Symbols: $x < 2$

Graph:

To graph $x < 2$, draw an open circle at $x = 2$. Then draw an arrow pointing to the left.

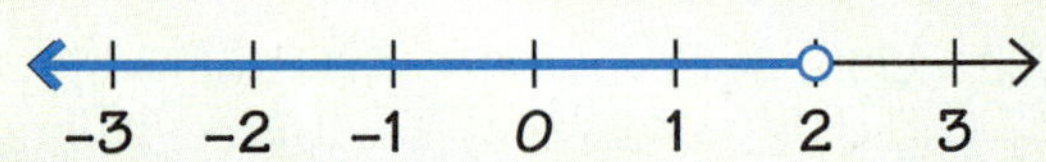

On Your Own

Make a four square to help you study these topics.

1. inequality (>)
2. inequality (≤)
3. inequality (≥)
4. solving an inequality using addition
5. solving an inequality using subtraction

After you complete this chapter, make four squares for the following topics.

6. solving an inequality using multiplication
7. solving an inequality using division

"Sorry, but I have limited space in my four square. I needed pet names with only three letters."

8.1–8.2 Quiz

Check It Out
Progress Check
BigIdeasMath.com

Write the word sentence as an inequality. *(Section 8.1)*

1. A number x plus 1 is less than -13.
2. A number t minus 1.6 is at most 9.

Tell whether the given value is a solution of the inequality. *(Section 8.1)*

3. $12n < -2;\ n = -1$
4. $y + 4 < -3;\ y = -7$

Graph the inequality on a number line. *(Section 8.1)*

5. $x > -10$
6. $y \leq \frac{3}{5}$
7. $w < 6.8$

Solve the inequality. *(Section 8.2)*

8. $x - 2 < 4$
9. $g + 14 \geq 30$
10. $h - 1 \leq -9$
11. $s + 3 > -7$
12. $v - \frac{3}{4} < 0$
13. $\frac{3}{2} < p + \frac{1}{2}$

14. **WATERCRAFT** You must be at least 14 years old to operate a personal watercraft in Florida. Write an inequality that represents this situation. *(Section 8.1)*

15. **REASONING** The solution of $x - a > 4$ is $x > 11$. What is the value of a? *(Section 8.2)*

16. **MP3 PLAYER** Your MP3 player can store up to 8 gigabytes of media. You transfer 3.5 gigabytes of media to the MP3 player. Write and solve an inequality that represents the amount of memory available on the MP3 player. *(Section 8.2)*

17. **LIFEGUARD** Three requirements for a lifeguard training course are shown. *(Section 8.1)*

    a. Write and graph three inequalities that represent the requirements.

    b. You can swim 350 feet. Do you satisfy the swimming requirement of the course? Explain.

8.3 Solving Inequalities Using Multiplication or Division

STATE STANDARDS
MA.8.A.4.2

Essential Question How can you use multiplication or division to solve an inequality?

1 ACTIVITY: Using a Table to Solve an Inequality

Work with a partner.

- **Copy and complete the table.**
- **Decide which graph represents the solution of the inequality.**
- **Write the solution of the inequality.**

a. $3x \le 6$

x	-1	0	1	2	3	4	5
$3x$							
$3x \overset{?}{\le} 6$							

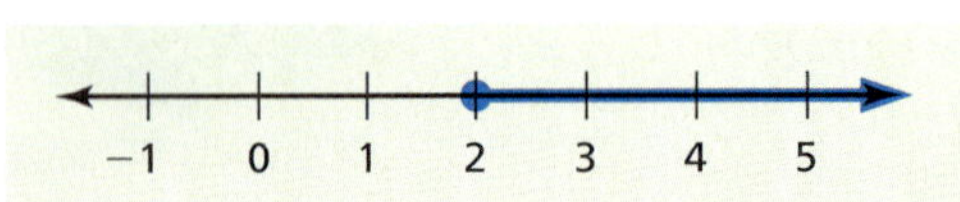

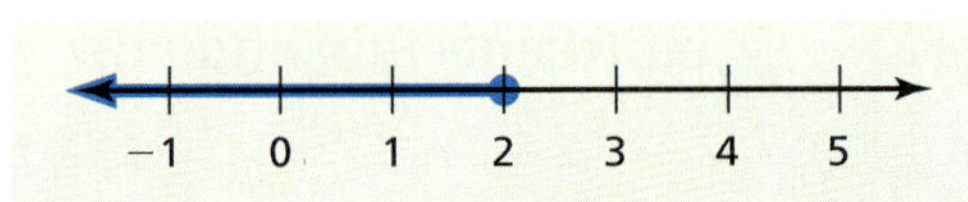

b. $-2x > 4$

x	-5	-4	-3	-2	-1	0	1
$-2x$							
$-2x \overset{?}{>} 4$							

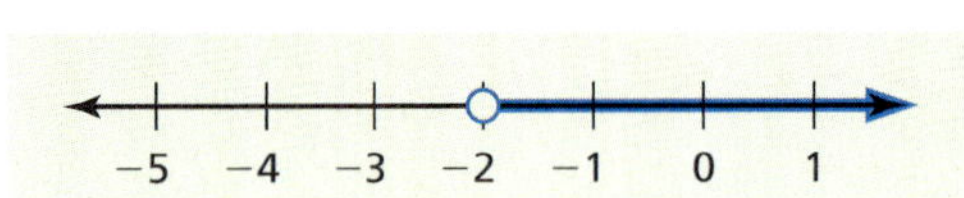

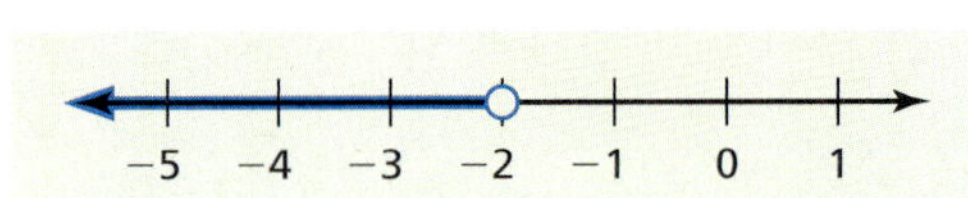

2 ACTIVITY: Writing a Rule

Work with a partner. Use a table to solve each inequality.

a. $3x > 3$ **b.** $4x \le 4$ **c.** $-2x \ge 6$ **d.** $-5x < 10$

Write a rule that describes how to solve inequalities like those in Activity 1. Then use your rule to solve each of the four inequalities above.

3 ACTIVITY: Using a Table to Solve an Inequality

Work with a partner.

- **Copy and complete the table.**
- **Decide which graph represents the solution of the inequality.**
- **Write the solution of the inequality.**

a. $\frac{x}{2} \geq 1$

x	-1	0	1	2	3	4	5
$\frac{x}{2}$							
$\frac{x}{2} \overset{?}{\geq} 1$							

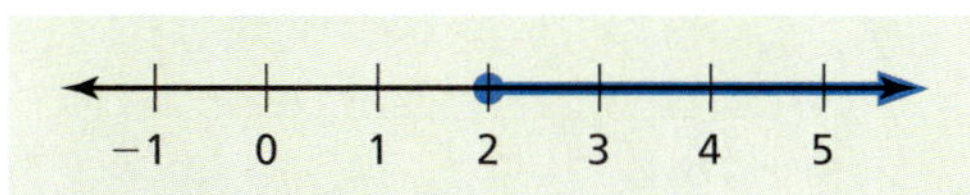

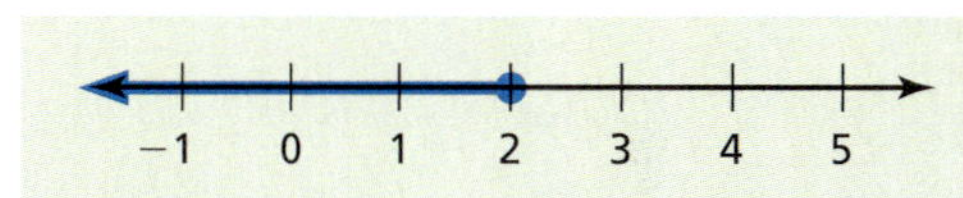

b. $\frac{x}{-3} < \frac{2}{3}$

x	-5	-4	-3	-2	-1	0	1
$\frac{x}{-3}$							
$\frac{x}{-3} \overset{?}{<} \frac{2}{3}$							

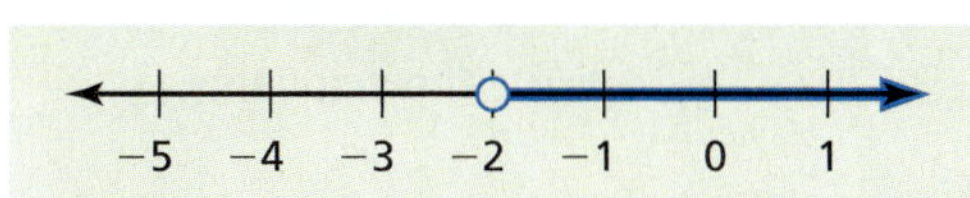

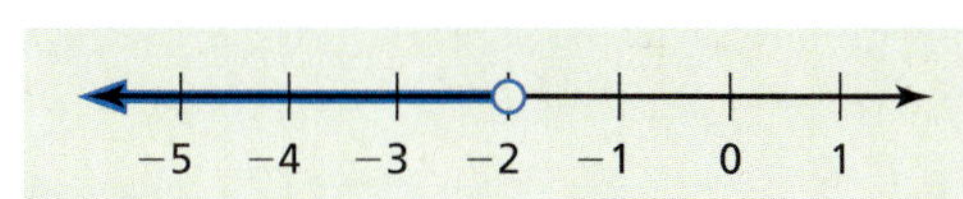

4 ACTIVITY: Writing a Rule

Work with a partner. Use a table to solve each inequality.

a. $\frac{x}{4} \geq 1$ **b.** $\frac{x}{2} < \frac{3}{2}$ **c.** $\frac{x}{-2} > 2$ **d.** $\frac{x}{-5} \leq \frac{1}{5}$

Write a rule that describes how to solve inequalities like those in Activity 3. Then use your rule to solve each of the four inequalities above.

What Is Your Answer?

5. **IN YOUR OWN WORDS** How can you use multiplication or division to solve an inequality?

Practice Use what you learned about solving inequalities using multiplication or division to complete Exercises 4–9 on page 331.

8.3 Lesson

Check It Out
Lesson Tutorials
BigIdeasMath.com

Remember

Multiplication and division are inverse operations.

Key Idea

Multiplication and Division Properties of Inequality (Case 1)

Words If you multiply or divide each side of an inequality by the same *positive* number, the inequality remains true.

Numbers

$-6 < 8$	$6 > -8$
$2 \cdot (-6) < 2 \cdot 8$	$\frac{6}{2} > \frac{-8}{2}$
$-12 < 16$	$3 > -4$

Algebra

$\frac{x}{2} < -9$	$4x > -12$
$2 \cdot \frac{x}{2} < 2 \cdot (-9)$	$\frac{4x}{4} > \frac{-12}{4}$
$x < -18$	$x > -3$

These properties are also true for $\leq$ and $\geq$.

EXAMPLE 1 Solving an Inequality Using Multiplication

Solve $\frac{x}{8} > -5$. Graph the solution.

$\frac{x}{8} > -5$ Write the inequality.

Undo the division. → $8 \cdot \frac{x}{8} > 8 \cdot (-5)$ Multiply each side by 8.

$x > -40$ Simplify.

∴ The solution is $x > -40$.

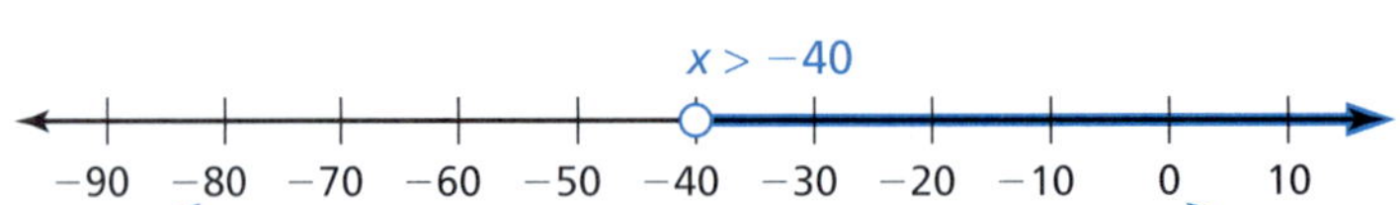

Check: $x = -80$ is *not* a solution.

Check: $x = 0$ is a solution.

On Your Own

Solve the inequality. Graph the solution.

1. $a \div 2 < 4$
2. $\frac{n}{7} \geq -1$
3. $-6.4 \geq \frac{w}{5}$

EXAMPLE 2 Solving an Inequality Using Division

Solve $3x \le -24$. Graph the solution.

	$3x \le -24$	Write the inequality.
Undo the multiplication.	$\dfrac{3x}{3} \le \dfrac{-24}{3}$	Divide each side by 3.
	$x \le -8$	Simplify.

The solution is $x \le -8$.

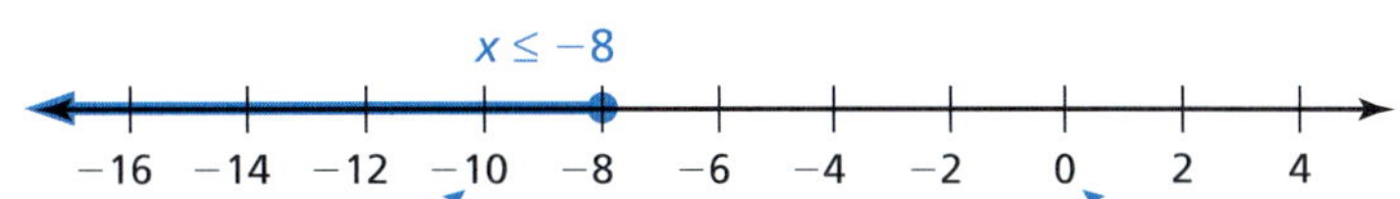

Check: $x = -10$ is a solution.

Check: $x = 0$ is *not* a solution.

On Your Own

Now You're Ready
Exercises 10–18

Solve the inequality. Graph the solution.

4. $4b \ge 36$ **5.** $2k > -10$ **6.** $-18 > 1.5q$

Key Idea

Multiplication and Division Properties of Inequality (Case 2)

Words If you multiply or divide each side of an inequality by the same *negative* number, the direction of the inequality symbol must be reversed for the inequality to remain true.

Numbers

$$-6 < 8 \qquad\qquad 6 > -8$$
$$(-2) \cdot (-6) > (-2) \cdot 8 \qquad\qquad \frac{6}{-2} < \frac{-8}{-2}$$
$$12 > -16 \qquad\qquad -3 < 4$$

Algebra

$$\frac{x}{-6} < 3 \qquad\qquad -5x > 30$$
$$-6 \cdot \frac{x}{-6} > -6 \cdot 3 \qquad\qquad \frac{-5x}{-5} < \frac{30}{-5}$$
$$x > -18 \qquad\qquad x < -6$$

These properties are also true for $\le$ and $\ge$.

Common Error

A negative sign in an inequality does not necessarily mean you must reverse the inequality symbol.

Only reverse the inequality symbol when you multiply or divide both sides by a negative number.

EXAMPLE 3 Solving an Inequality Using Multiplication

Solve $\frac{y}{-3} > 2$. Graph the solution.

	$\frac{y}{-3} > 2$	Write the inequality.
Undo the division. →	$-3 \cdot \frac{y}{-3} < -3 \cdot 2$	Multiply each side by −3. Reverse the inequality symbol.
	$y < -6$	Simplify.

The solution is $y < -6$.

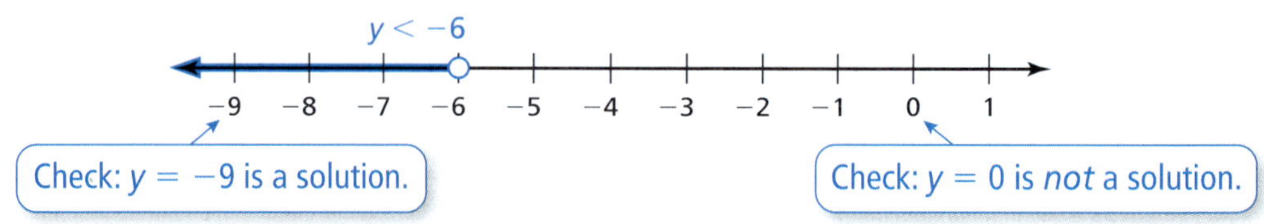

EXAMPLE 4 Solving an Inequality Using Division

Solve $-7y \le -35$. Graph the solution.

	$-7y \le -35$	Write the inequality.
Undo the multiplication. →	$\frac{-7y}{-7} \ge \frac{-35}{-7}$	Divide each side by −7. Reverse the inequality symbol.
	$y \ge 5$	Simplify.

The solution is $y \ge 5$.

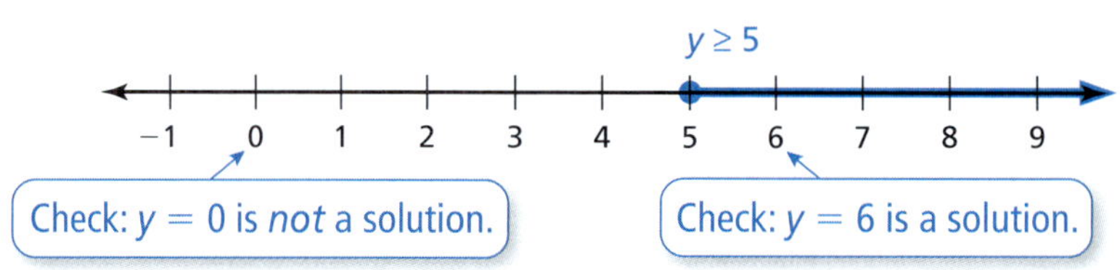

On Your Own

Now You're Ready
Exercises 27–35

Solve the inequality. Graph the solution.

7. $\frac{p}{-4} < 7$

8. $\frac{x}{-5} \le -5$

9. $1 \ge -\frac{1}{10}z$

10. $-9m > 63$

11. $-2r \ge -22$

12. $-0.4y \ge -12$

8.3 Exercises

Check It Out
Help with Homework
BigIdeasMath.com

Vocabulary and Concept Check

1. **VOCABULARY** Explain how to solve $\frac{x}{6} < -5$.
2. **WRITING** Explain how solving $2x < -8$ is different from solving $-2x < 8$.
3. **OPEN-ENDED** Write an inequality that is solved using the Division Property of Inequality where the inequality symbol needs to be reversed.

Practice and Problem Solving

Use a table to solve the inequality.

4. $4x < 4$
5. $-2x \leq 2$
6. $-5x > 15$
7. $\frac{x}{-3} \geq 1$
8. $\frac{x}{-2} > \frac{5}{2}$
9. $\frac{x}{4} \leq \frac{3}{8}$

Solve the inequality. Graph the solution.

1 2

10. $3n > 18$
11. $\frac{c}{4} \leq -9$
12. $1.2m < 12$
13. $-14 > x \div 2$
14. $\frac{w}{5} \geq -2.6$
15. $5 < 2.5k$
16. $4x \leq -\frac{3}{2}$
17. $2.6y \leq -10.4$
18. $10.2 > \frac{b}{3.4}$
19. **ERROR ANALYSIS** Describe and correct the error in solving the inequality.

✗

$$\frac{x}{2} < -5$$
$$2 \cdot \frac{x}{2} > 2 \cdot (-5)$$
$$x > -10$$

Write the word sentence as an inequality. Then solve the inequality.

20. The quotient of a number and 3 is at most 4.
21. A number divided by 8 is less than -2.
22. Four times a number is at least -12.
23. The product of 5 and a number is greater than 20.

24. **CAMERA** You earn \$9.50 per hour at your summer job. Write and solve an inequality that represents the number of hours you need to work in order to buy a digital camera that costs \$247.

25. **COPIES** You have \$3.65 to make copies. Write and solve an inequality that represents the number of copies you can make.

26. **SPEED LIMIT** The maximum speed limit for a school bus is 55 miles per hour. Write and solve an inequality that represents the number of hours it takes to travel 165 miles in a school bus.

Solve the inequality. Graph the solution.

3 4 **27.** $-2n \le 10$ **28.** $-5w > 30$ **29.** $\frac{h}{-6} \ge 7$

30. $-8 < -\frac{1}{3}x$ **31.** $-2y < -11$ **32.** $-7d \ge 56$

33. $2.4 > -\frac{m}{5}$ **34.** $\frac{k}{-0.5} \le 18$ **35.** $-2.5 > \frac{b}{-1.6}$

36. **ERROR ANALYSIS** Describe and correct the error in solving the inequality.

$$-4m \ge 16$$
$$\frac{-4m}{-4} \ge \frac{16}{-4}$$
$$m \ge -4$$

37. **CRITICAL THINKING** Are all numbers greater than zero solutions of $-x > 0$? Explain.

38. **TRUCKING** By law, the maximum height (including freight) of a vehicle in Florida is 13.5 feet.

a. Write and solve an inequality that represents the number of crates that can be stacked vertically on the bed of the truck.

b. Five crates are stacked vertically on the bed of the truck. Is this legal? Explain.

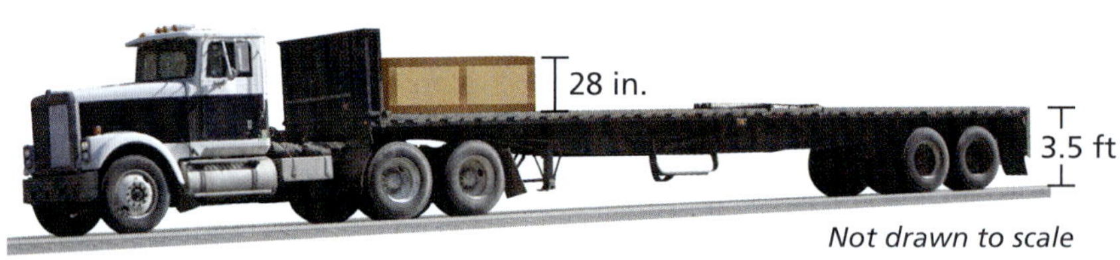

Not drawn to scale

Write and solve an inequality that represents the value of x.

39. Area $\ge$ 102 cm^2

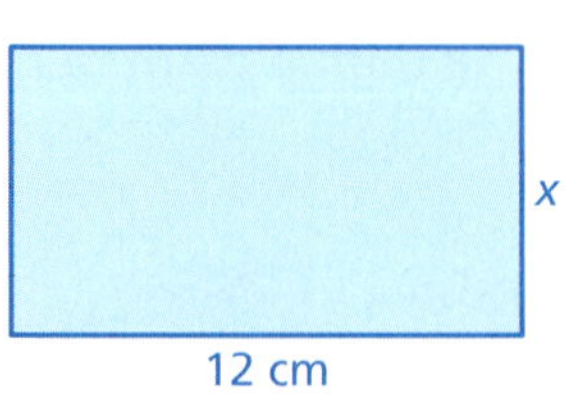

40. Area $<$ 30 ft^2

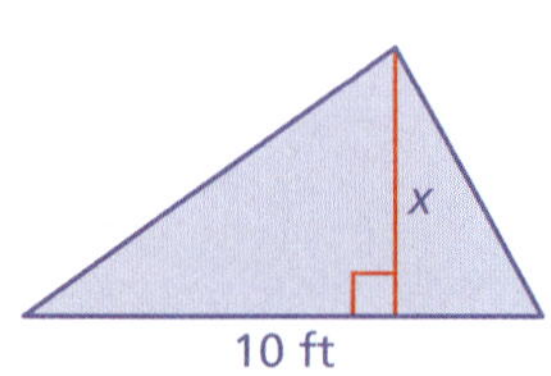

41. **TRIP** You and three friends are planning a trip. You want to keep the cost below \$80 per person. Write and solve an inequality that represents the total cost of the trip.

42. **REASONING** Explain why the direction of the inequality symbol must be reversed when multiplying or dividing by the same negative number.

43. **PROJECT** Choose two musical artists to research.

 a. Use the Internet or a magazine to complete the table.

 b. Find the average number of copies sold per month for each CD.

 c. Use the release date to write and solve an inequality that represents the minimum average number of copies sold per month for each CD.

 d. In how many months do you expect the number of copies of the second top selling CD to surpass the current number of copies of the top selling CD?

Artist	Name of CD	Release Date	Current Number of Copies Sold
1.			
2.			

Number Sense **Describe all numbers that satisfy *both* inequalities. Include a graph with your description.**

44. $3m > -12$ and $2m < 12$

45. $\frac{n}{2} \geq -3$ and $\frac{n}{-4} \geq 1$

46. $2x \geq -4$ and $2x \geq 4$

47. $\frac{m}{-4} > -5$ and $\frac{m}{4} < 10$

Fair Game Review What you learned in previous grades & lessons

Solve the equation. *(Section 1.2)*

48. $-4w + 5 = -11$

49. $4(x - 3) = 21$

50. $\frac{v}{6} - 7 = 4$

51. $\frac{m + 300}{4} = 96$

52. **MULTIPLE CHOICE** Which measure can have more than one value for a given data set? *(Section 7.1)*

Ⓐ mean Ⓑ median Ⓒ mode Ⓓ range

8.4 Solving Multi-Step Inequalities

STATE STANDARDS
MA.8.A.4.2

Essential Question How can you use an inequality to describe the area and perimeter of a composite figure?

1 ACTIVITY: Areas and Perimeters of Composite Figures

Work with a partner.

a. For what values of x will the area of the blue region be greater than 12 square units?

b. For what values of x will the sum of the inner and outer perimeters of the blue region be greater than 20 units?

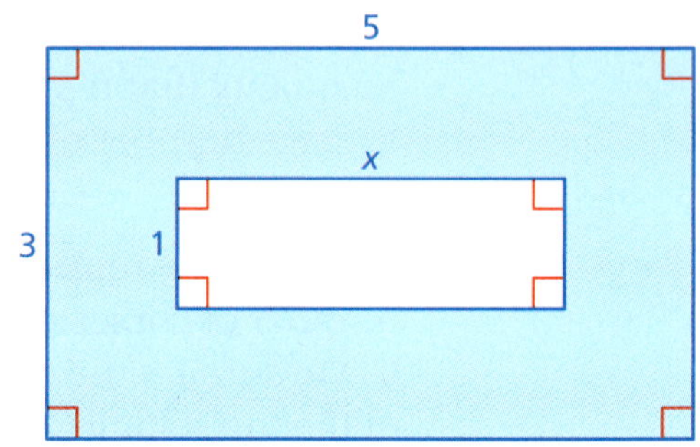

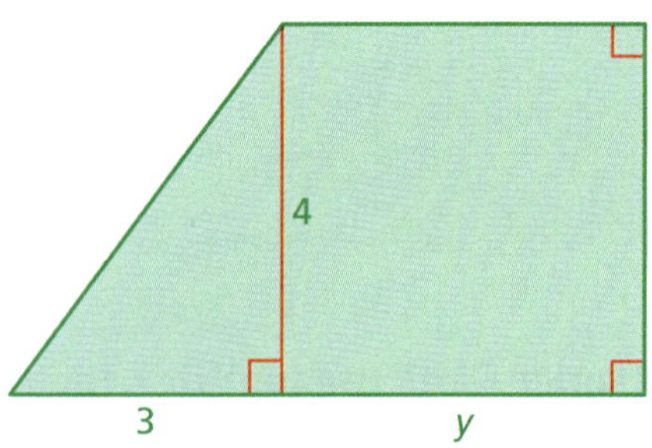

c. For what values of y will the area of the trapezoid be less than or equal to 10 square units?

d. For what values of y will the perimeter of the trapezoid be less than or equal to 16 units?

e. For what values of w will the area of the red region be greater than or equal to 36 square units?

f. For what values of w will the sum of the inner and outer perimeters of the red region be greater than 47 units?

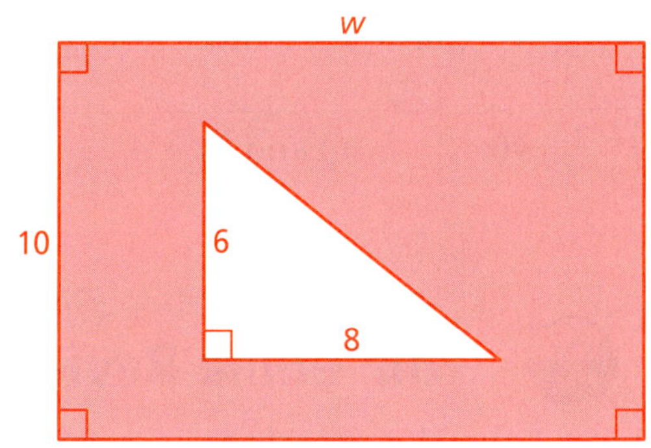

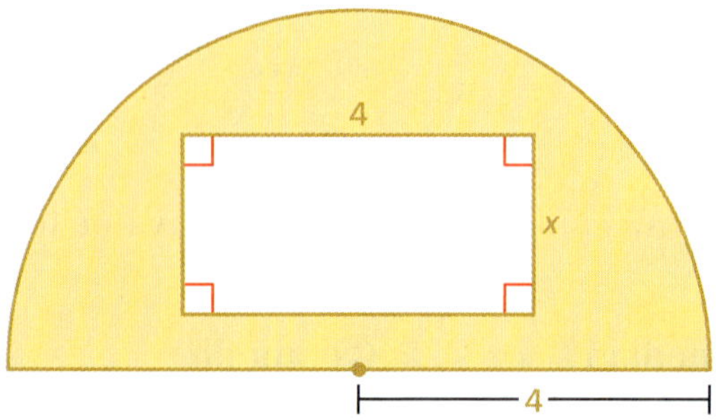

g. For what values of x will the area of the yellow region be less than 4π square units?

h. For what values of x will the sum of the inner and outer perimeters of the yellow region be less than $4\pi + 20$ units?

2 ACTIVITY: Volume and Surface Area of a Composite Solid

Work with a partner.

a. For what values of x will the volume of the solid be greater than or equal to 42 cubic units?

b. For what values of x will the surface area of the solid be greater than 72 square units?

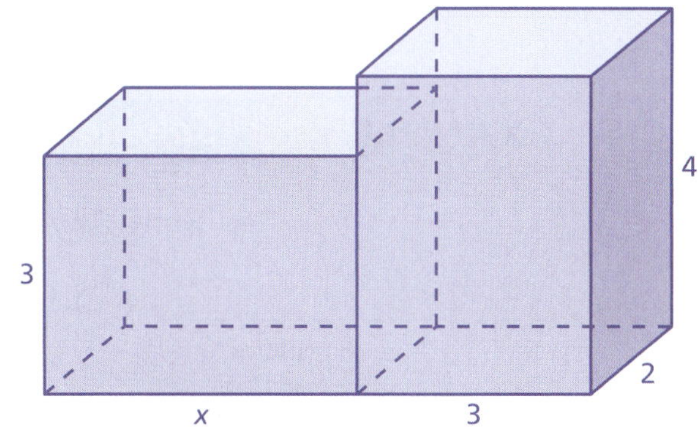

3 ACTIVITY: Planning a Budget

Work with a partner.

You are building a patio. You want to cover the patio with Spanish tile that costs \$5 per square foot. Your budget for the tile is \$1700. How wide can you make the patio without going over your budget?

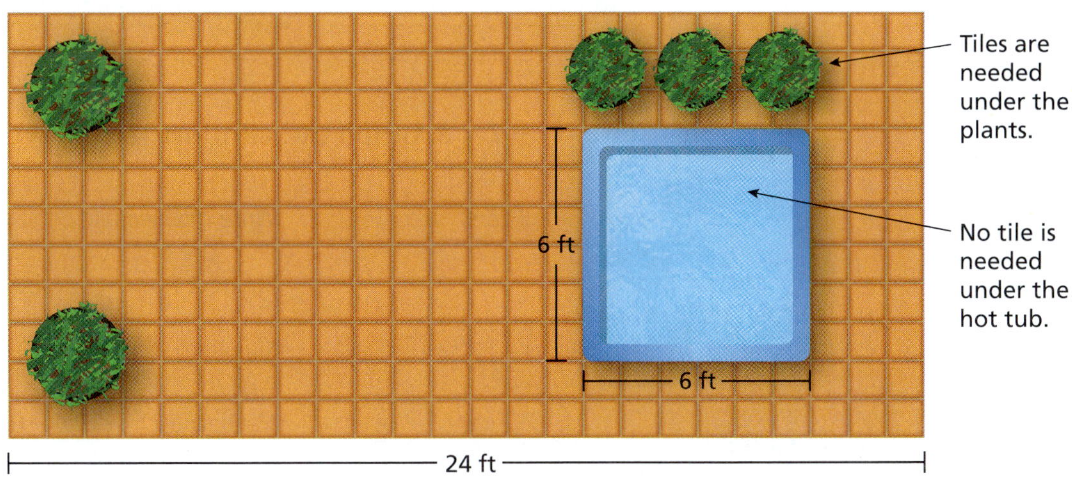

What Is Your Answer?

4. **IN YOUR OWN WORDS** How can you use an inequality to describe the area and perimeter of a composite figure? Give an example. Include a diagram with your example.

Practice Use what you learned about solving multi-step inequalities to complete Exercises 3 and 4 on page 338.

8.4 Lesson

Check It Out
Lesson Tutorials
BigIdeasMath.com

You can solve multi-step inequalities the same way you solve multi-step equations.

EXAMPLE 1 Solving Two-Step Inequalities

a. Solve $5x - 4 \geq 11$. Graph the solution.

$5x - 4 \geq 11$ Write the inequality.

Step 1: Undo the subtraction. → $+4 \quad +4$ Add 4 to each side.

$5x \geq 15$ Simplify.

Step 2: Undo the multiplication. → $\frac{5x}{5} \geq \frac{15}{5}$ Divide each side by 5.

$x \geq 3$ Simplify.

The solution is $x \geq 3$.

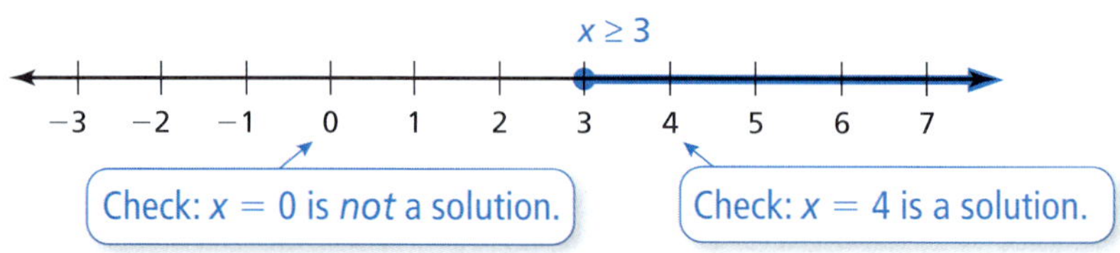

b. Solve $\frac{y}{-6} + 7 < 9$. Graph the solution.

$\frac{y}{-6} + 7 < 9$ Write the inequality.

$-7 \quad -7$ Subtract 7 from each side.

$\frac{y}{-6} < 2$ Simplify.

$-6 \cdot \frac{y}{-6} > -6 \cdot 2$ Multiply each side by -6. Reverse the inequality symbol.

$y > -12$ Simplify.

The solution is $y > -12$.

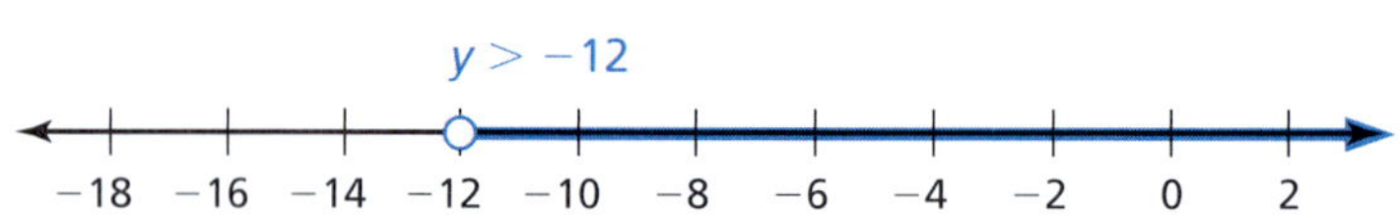

On Your Own

Now You're Ready
Exercises 5–10

Solve the inequality. Graph the solution.

1. $4b - 1 < 7$
2. $8 + 9c \geq -28$
3. $\frac{n}{-2} + 11 > 12$

EXAMPLE 2 Standardized Test Practice

Which graph represents the solution of $-7(x + 3) \le 28$?

Ⓐ 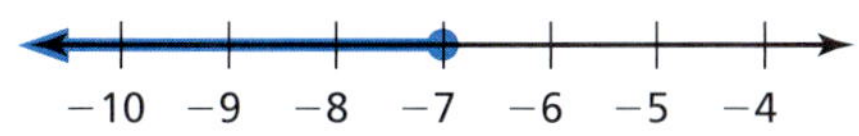

Ⓑ 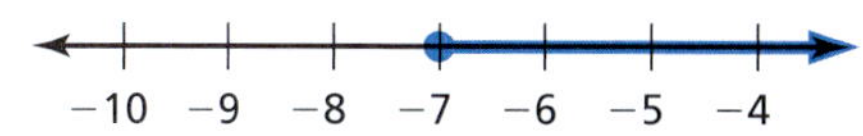

Ⓒ 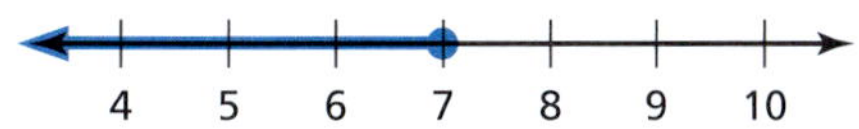

Ⓓ 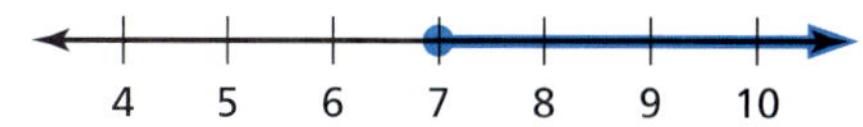

$-7(x + 3) \le 28$	Write the inequality.
$-7x - 21 \le 28$	Use Distributive Property.
$+21 \quad\quad +21$	Add 21 to each side.
$-7x \le 49$	Simplify.
$\dfrac{-7x}{-7} \ge \dfrac{49}{-7}$	Divide each side by -7. Reverse the inequality symbol.
$x \ge -7$	Simplify.

∴ The correct answer is Ⓑ.

EXAMPLE 3 Real-Life Application

You need a mean score of at least 90 to advance to the next round of the trivia game. What score do you need on the fifth game to advance?

Use the definition of mean to write and solve an inequality. Let x be the score on the fifth game.

The phrase "at least" means greater than or equal to.

$\dfrac{95 + 91 + 77 + 89 + x}{5} \ge 90$	
$\dfrac{352 + x}{5} \ge 90$	Simplify.
$5 \cdot \dfrac{352 + x}{5} \ge 5 \cdot 90$	Multiply each side by 5.
$352 + x \ge 450$	Simplify.
$-352 \quad\quad -352$	Subtract 352 from each side.
$x \ge 98$	Simplify.

∴ You need at least 98 points to advance to the next round.

Remember

The mean in Example 3 is equal to the sum of the game scores divided by the number of games.

On Your Own

Now You're Ready
Exercises 12–17

Solve the inequality. Graph the solution.

4. $2(k - 5) < 6$ **5.** $-4(n - 10) < 32$ **6.** $-3 \le 0.5(8 + y)$

7. WHAT IF? In Example 3, you need a mean score of at least 88 to advance to the next round of the trivia game. What score do you need on the fifth game to advance?

8.4 Exercises

Check It Out
Help with Homework
BigIdeasMath.com

Vocabulary and Concept Check

1. **WRITING** Compare and contrast solving multi-step inequalities and solving multi-step equations.

2. **OPEN-ENDED** Describe how to solve the inequality $3(a + 5) < 9$.

Practice and Problem Solving

3. For what values of k will the perimeter of the octagon be less than or equal to 64 units?

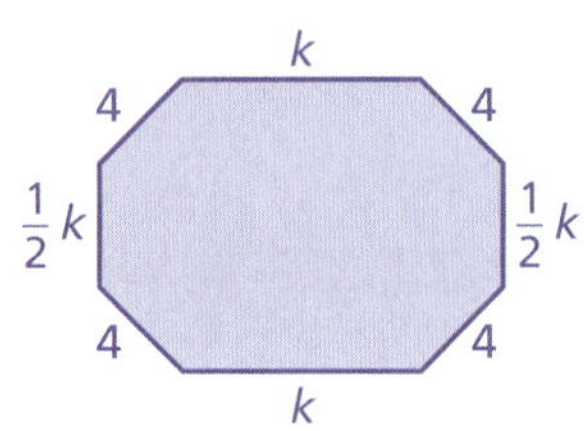

4. For what values of h will the surface area of the solid be greater than 46 square units?

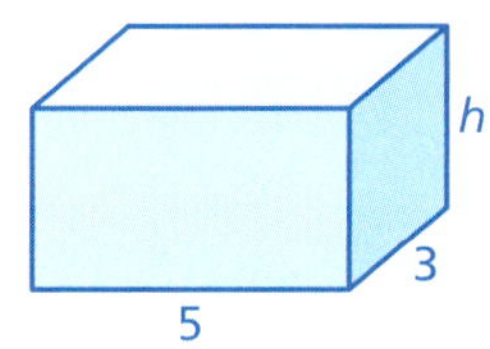

Solve the inequality. Graph the solution.

(1) 5. $7b + 4 \geq 11$

6. $2v - 4 < 8$

7. $1 - \frac{m}{3} \leq 6$

8. $\frac{4}{5} < 3w - \frac{11}{5}$

9. $1.8 < 0.5 - 1.3p$

10. $-2.4r + 9.6 \geq 4.8$

11. **ERROR ANALYSIS** Describe and correct the error in solving the inequality.

✗

$$\frac{x}{4} + 6 \geq 3$$
$$x + 6 \geq 12$$
$$x \geq 6$$

Solve the inequality. Graph the solution.

(2) 12. $6(g + 2) \leq 18$

13. $2(y - 5) \leq 16$

14. $-10 \geq \frac{5}{3}(h - 3)$

15. $-\frac{1}{3}(u + 2) > 5$

16. $2.7 > 0.9(n - 1.7)$

17. $10 > -2.5(z - 3.1)$

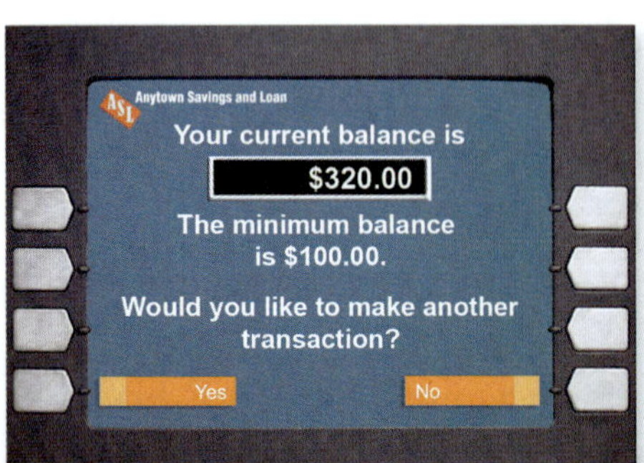

18. **ATM** Write and solve an inequality that represents the number of \$20 bills you can withdraw from the account without going below the minimum balance.

Solve the inequality. Graph the solution.

19. $5x - 2x + 7 \leq 15 + 10$

20. $7b - 12b + 1.4 > 8.4 - 22$

21. TYPING One line of text on a page uses about $\frac{3}{16}$ of an inch. There are 1-inch margins at the top and bottom of a page. Write and solve an inequality to find the number of lines that can be typed on a page that is 11 inches long.

22. WOODWORKING A woodworker builds a cabinet in 20 hours. The cabinet is sold at a store for \$500. Write and solve an inequality that represents the hourly wage the store can pay the woodworker and still make a profit of at least \$100.

23. FIRE TRUCK The height of one story of a building is about 10 feet. The bottom of the ladder on the fire truck must be at least 24 feet away from the building. Write and solve an inequality to find the number of stories the ladder can reach.

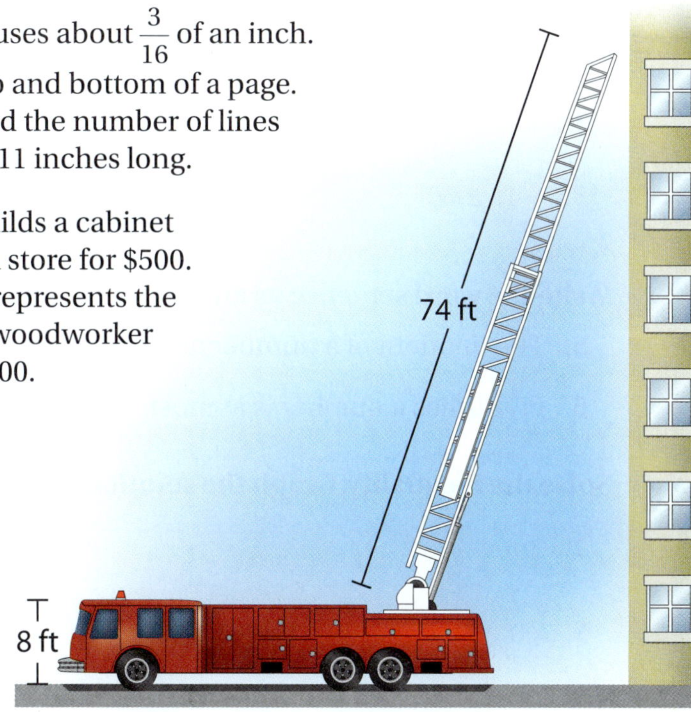

24. DRIVE-IN A drive-in movie theater charges \$3.50 per car. The drive-in has already admitted 100 cars. Write and solve an inequality to find the number of cars the drive-in needs to admit to make at least \$500.

25. Challenge For what values of r will the area of the shaded region be greater than or equal to $9(\pi - 2)$?

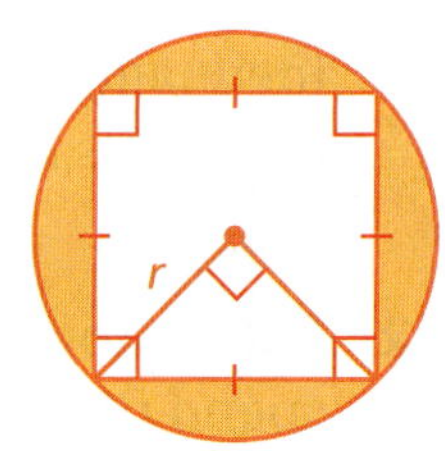

Fair Game Review What you learned in previous grades & lessons

Find the area of the circle. *(Skills Review Handbook)*

26.

27.

28.

29. MULTIPLE CHOICE What is the volume of the cube? *(Skills Review Handbook)*

(A) 8 ft^3 (B) 16 ft^3

(C) 24 ft^3 (D) 32 ft^3

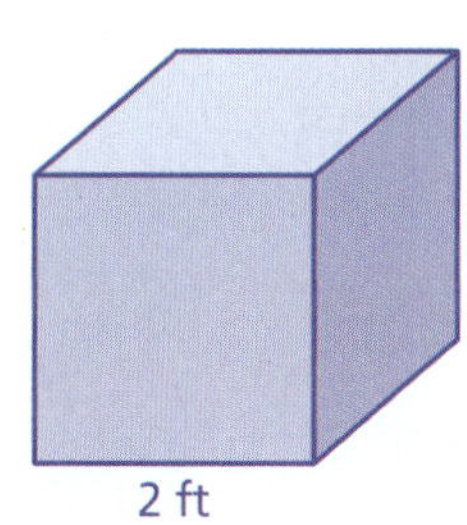

8.3–8.4 Quiz

Check It Out
Progress Check
BigIdeasMath.com

Solve the inequality. Graph the solution. *(Section 8.3)*

1. $x \div 4 > 12$

2. $\frac{n}{-6} \geq -2$

3. $-4y \geq 60$

4. $-2.3 \geq \frac{p}{5}$

Write the word sentence as an inequality. Then solve the inequality. *(Section 8.3)*

5. The quotient of a number and 6 is more than 9.

6. Five times a number is at most -10.

Solve the inequality. Graph the solution. *(Section 8.4)*

7. $2m + 1 \geq 7$

8. $\frac{n}{6} - 8 \leq 2$

9. $2 - \frac{j}{5} > 7$

10. $\frac{5}{4} > -3w - \frac{7}{4}$

11. FLOWERS A soccer team needs to raise \$200 for new uniforms. The team earns \$0.50 for each flower sold. Write and solve an inequality to find the number of flowers it must sell to meet or exceed its fundraising goal. *(Section 8.3)*

12. PARTY You buy lunch for guests at a party. You can spend no more than \$100. You will spend \$20 on beverages and \$10 per guest on sandwiches. Write and solve an inequality to find the number of guests you can invite to the party. *(Section 8.4)*

13. BOOKS You have a gift card worth \$50. You want to buy several paperback books that cost \$6 each. Write and solve an inequality to find the number of books you can buy and still have at least \$20 on the gift card. *(Section 8.4)*

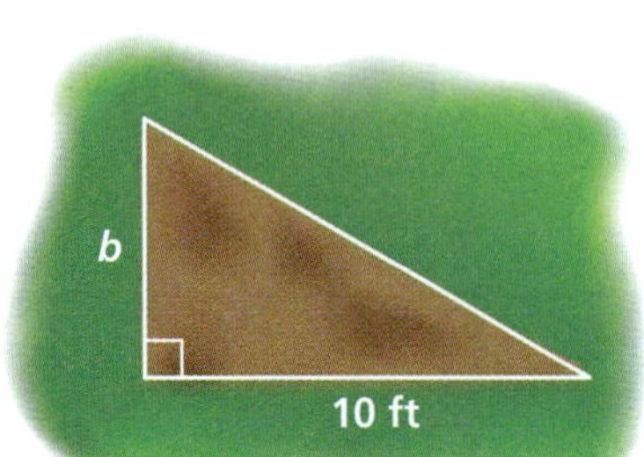

14. GARDEN The area of the triangular garden must be less than 35 square feet. Write and solve an inequality that represents the value of b. *(Section 8.3)*

8 Chapter Review

Check It Out
Vocabulary Help
BigIdeasMath.com

Review Key Vocabulary

inequality, *p. 314*
solution of an inequality, *p. 314*
solution set, *p. 314*
graph of an inequality, *p. 315*

Review Examples and Exercises

8.1 Writing and Graphing Inequalities *(pp. 312–317)*

a. Four plus a number w is at least $-\frac{1}{2}$. Write this sentence as an inequality.

Four plus a number w | is at least | $-\frac{1}{2}$.

$4 + w$ | $\geq$ | $-\frac{1}{2}$

An inequality is $4 + w \geq -\frac{1}{2}$.

b. Graph $m > 4$.

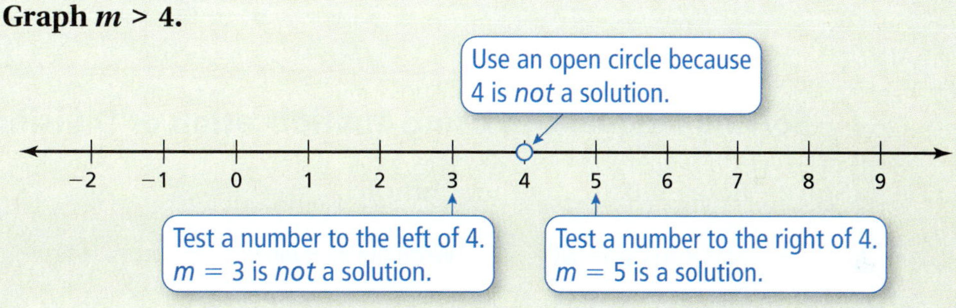

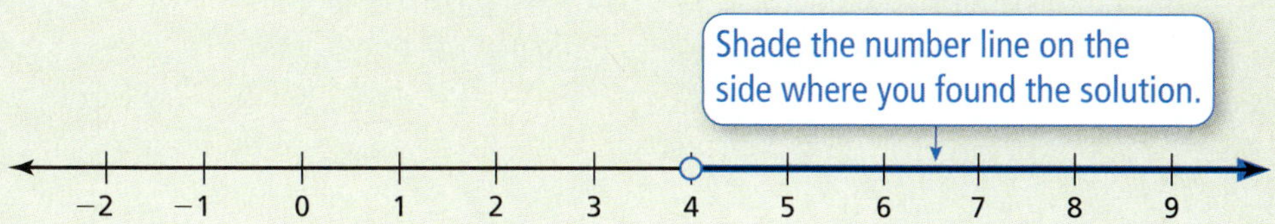

Exercises

Write the word sentence as an inequality.

1. A number v is less than -2.
2. A number x minus $\frac{1}{4}$ is no more than $-\frac{3}{4}$.

Tell whether the given value is a solution of the inequality.

3. $10 - q < 3$; $q = 6$
4. $12 \div m \geq -4$; $m = -3$

Graph the inequality on a number line.

5. $p < 1.2$
6. $n > 10\frac{1}{4}$

8.2 Solving Inequalities Using Addition or Subtraction (pp. 318–323)

Solve $-4 < n - 3$. Graph the solution.

Undo the subtraction.

$-4 < n - 3$	Write the inequality.
$\underline{+3} \quad \underline{+3}$	Add 3 to each side.
$-1 < n$	Simplify.

The solution is $n > -1$.

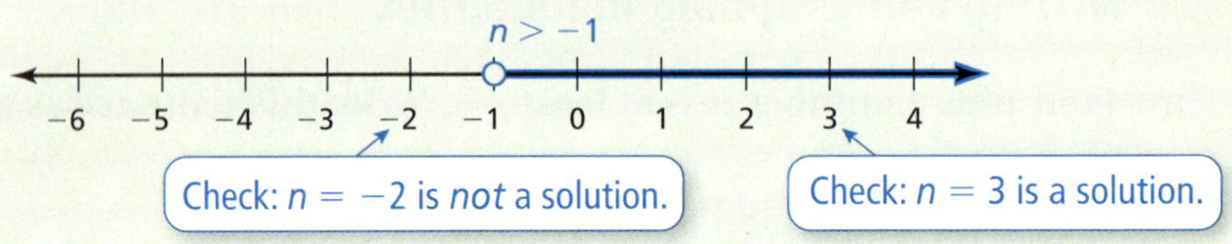

Exercises

Solve the inequality. Graph the solution.

7. $b + 13 < 18$ **8.** $x - 3 \le 10$ **9.** $y + 1 \ge -2$

8.3 Solving Inequalities Using Multiplication or Division (pp. 326–333)

Solve $-8a \ge -48$. Graph the solution.

Undo the multiplication.

$-8a \ge -48$	Write the inequality.
$\dfrac{-8a}{-8} \le \dfrac{-48}{-8}$	Divide each side by -8. Reverse the inequality symbol.
$a \le 6$	Simplify.

The solution is $a \le 6$.

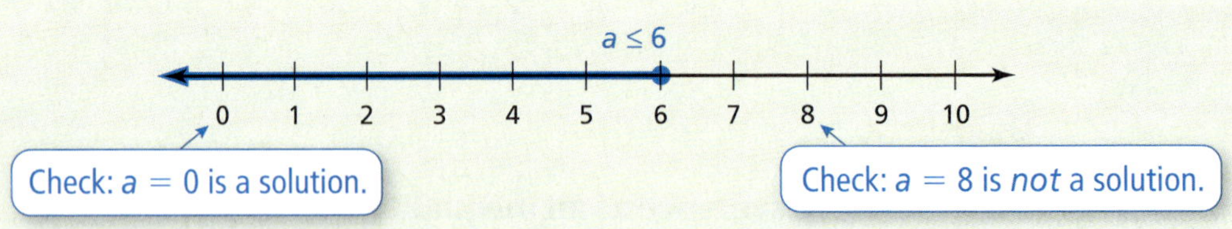

Exercises

Solve the inequality. Graph the solution.

10. $\dfrac{x}{2} \ge 4$ **11.** $4z < -44$ **12.** $-2q \ge -18$

8.4 Solving Multi-Step Inequalities (pp. 334–339)

a. Solve $2x - 3 \le -9$. Graph the solution.

	$2x - 3 \le -9$	Write the inequality.
Step 1: Undo the subtraction.	$+3 \quad +3$	Add 3 to each side.
	$2x \le -6$	Simplify.
Step 2: Undo the multiplication.	$\frac{2x}{2} \le \frac{-6}{2}$	Divide each side by 2.
	$x \le -3$	Simplify.

The solution is $x \le -3$.

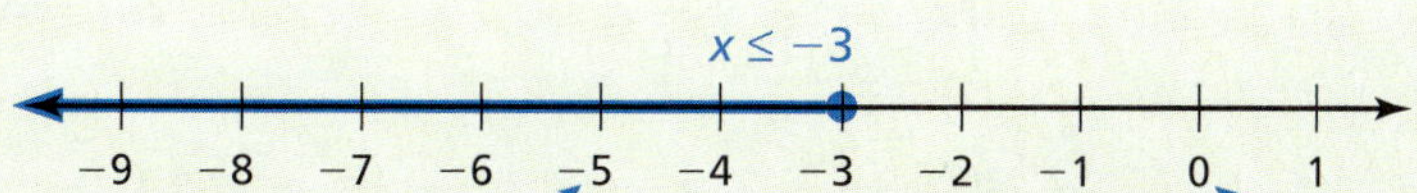

Check: $x = -5$ is a solution.

Check: $x = 0$ is *not* a solution.

b. Solve $\frac{t}{-3} + 4 > 7$. Graph the solution.

	$\frac{t}{-3} + 4 > 7$	Write the inequality.
Step 1: Undo the addition.	$-4 \quad -4$	Subtract 4 from each side.
	$\frac{t}{-3} > 3$	Simplify.
Step 2: Undo the division.	$-3 \cdot \frac{t}{-3} < -3 \cdot 3$	Multiply each side by −3. Reverse the inequality symbol.
	$t < -9$	Simplify.

The solution is $t < -9$.

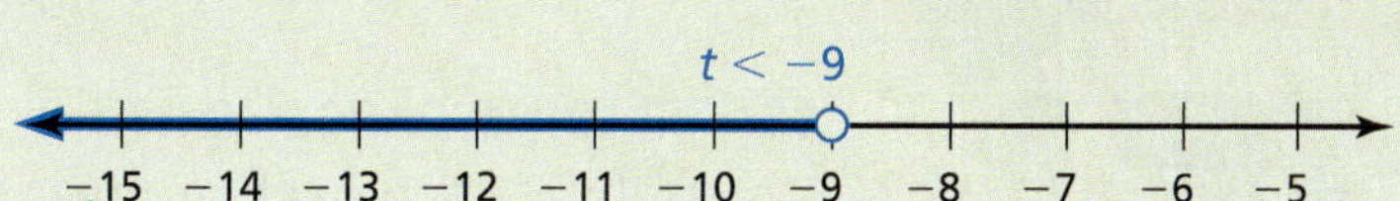

Check: $t = -15$ is a solution.

Check: $t = -6$ is *not* a solution.

Exercises

Solve the inequality. Graph the solution.

13. $4x + 3 < 11$

14. $\frac{z}{-4} - 3 \le 1$

15. $-3w - 4 > 8$

16. $8(q + 2) < 40$

17. $-\frac{1}{2}(p + 4) \le 18$

18. $1.5(k + 3.2) \ge 6.9$

8 Chapter Test

Check It Out
Test Practice
BigIdeasMath.com

Write the word sentence as an inequality.

1. A number j plus 20.5 is greater than or equal to 50.
2. A number r multiplied by $\frac{1}{7}$ is less than -14.

Tell whether the given value is a solution of the inequality.

3. $v - 2 \le 7;\ v = 9$
4. $\frac{3}{10}p < 0;\ p = 10$
5. $-3n \ge 6;\ n = -3$

Solve the inequality. Graph the solution.

6. $n - 3 > -3$
7. $x - \frac{7}{8} \le \frac{9}{8}$
8. $-6b \ge -30$
9. $\frac{y}{-4} \ge 13$
10. $3v - 7 \ge -13.3$
11. $-5(t + 11) < -60$

12. **VOTING** U.S. citizens must be at least 18 years of age on Election Day to vote. Write an inequality that represents this situation.

13. **GARAGE** The vertical clearance for a hotel parking garage is 10 feet. Write and solve an inequality that represents the height (in feet) of the vehicle.

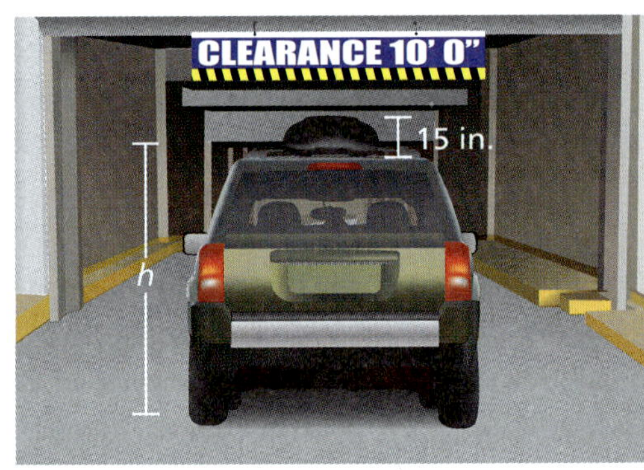

14. **LUNCH BILL** A lunch bill, including tax, is divided equally among you and five friends. Everyone pays less than $8.75. Write and solve an inequality that describes the total amount of the bill.

15. **TRADING CARDS** You have $25 to buy trading cards online. Each pack of cards costs $4.50. Shipping costs $2.95. Write and solve an inequality to find the number of packs of trading cards you can buy.

16. **SCIENCE QUIZZES** The table shows your scores on four science quizzes. What score do you need on the fifth quiz to have a mean score of at least 80?

Test	1	2	3	4	5
Score (%)	76	87	73	72	?

8 Standardized Test Practice

Test-Taking Strategy
After Answering Easy Questions, Relax

"After answering the easy questions, relax and try the harder ones. For this, $x = \$946$. So, it's D."

1. The perimeter of the triangle shown below is greater than 50 centimeters. Which inequality represents this algebraically?

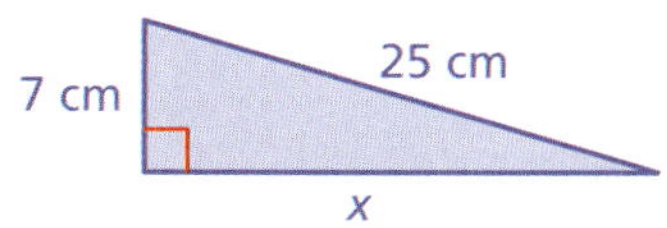

A. $\frac{1}{2}(7x) < 50$ **C.** $\frac{1}{2}(7x) > 50$

B. $x + 32 < 50$ **D.** $x + 32 > 50$

2. A store has recorded total dollar sales each month for the past three years. Which type of graph would best show how sales have increased over this time period?

F. circle graph **H.** box-and-whisker plot

G. line graph **I.** stem-and-leaf plot

3. What is the length d in the coordinate plane?

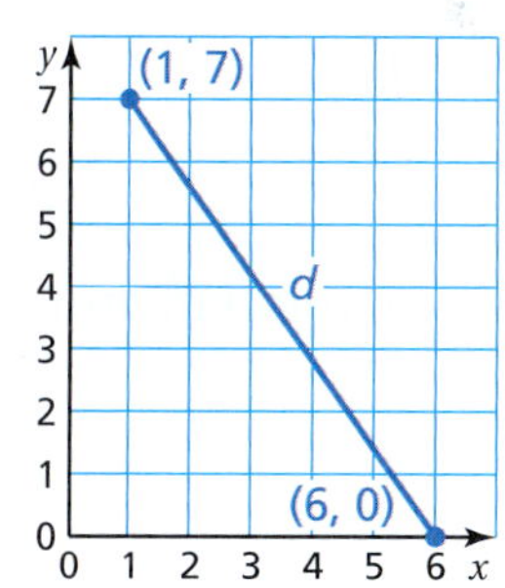

A. 74 **C.** 12

B. $\sqrt{74}$ **D.** $\sqrt{12}$

4. Which system of equations has infinitely many solutions?

F. $x + y = 1$
$x + y = 2$

H. $4x - 2y = 9$
$y = 2x - 4.5$

G. $y = x$
$y = -x$

I. $3x + 4y = 12$
$y = \frac{3}{4}x + 3$

5. Which is equivalent to $4\sqrt{25}$?

A. 20

B. $\sqrt{29}$

C. 100

D. $\sqrt{100}$

6. The triangles shown below are similar. What is the value of x?

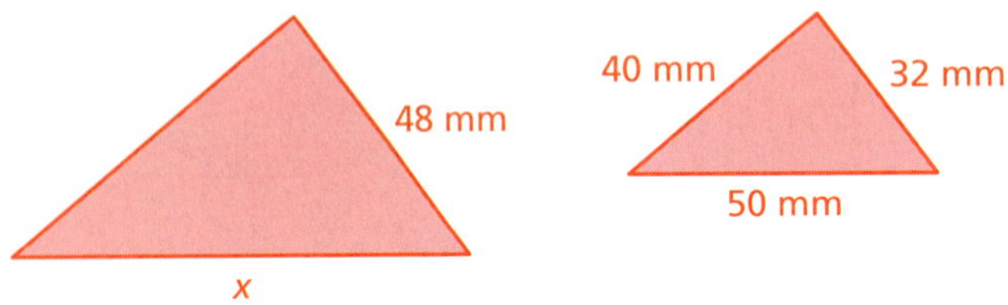

7. The table lists the mean, median, and mode salaries at a company. Suppose a new worker is hired at a salary of $70,000. Which statement is true?

Mean	Median	Mode
$62,000	$58,000	$54,000

F. The mean annual salary must increase.

G. The median annual salary must increase.

H. The median annual salary must remain the same.

I. The mode annual salary must increase.

8. Does squaring a number always make it bigger? Is the inequality shown below true for all numbers?

$$x^2 > x$$

Show your work and explain your reasoning.

9. Which graph represents the inequality below?

$$-2x + 3 < 1$$

A. 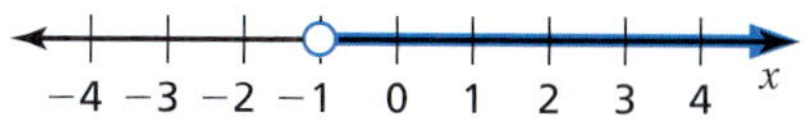

C. 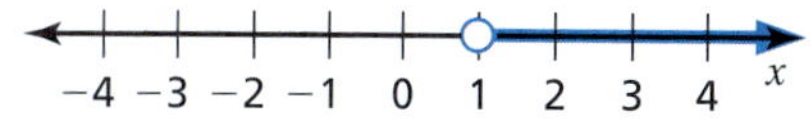

B. 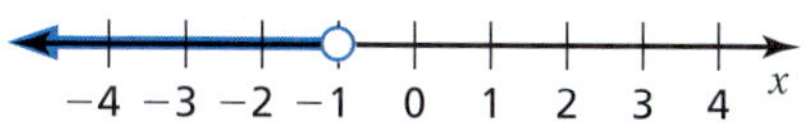

D. 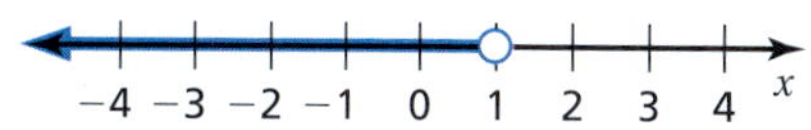

10. The function $y = 29.95x$ represents the total cost y of purchasing x day passes to a water park. Which statement is true?

F. The domain represents day passes and it is continuous.

G. The domain represents day passes and it is discrete.

H. The domain represents total cost and it is continuous.

I. The domain represents total cost and it is discrete.

11. A television screen measures 36 inches across and 27 inches high. What is the length, in inches, of the television screen's diagonal?

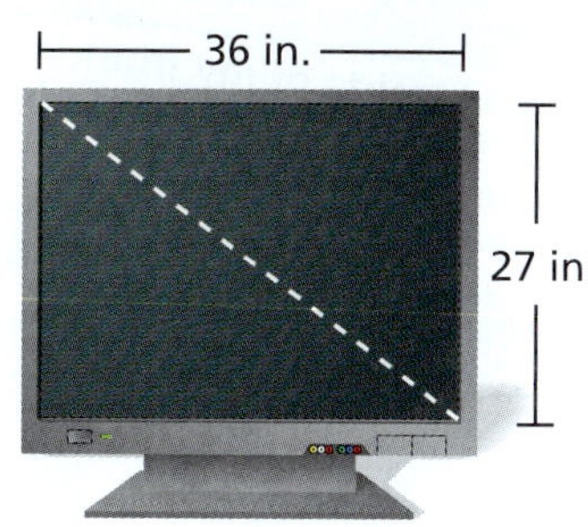

9 Exponents and Scientific Notation

What You Learned Before

"It's called the Power of Negative One, Descartes!"

Adding and Subtracting Decimals (Grade 5)

Example 1 Find 2.65 + 5.012.

$$\begin{array}{r} 2.650 \\ +\ 5.012 \\ \hline 7.662 \end{array}$$

Example 2 Find 3.7 − 0.48.

$$\begin{array}{r} \overset{6}{\cancel{7}}\ \overset{10}{\cancel{0}} \\ 3.\cancel{7}\,\cancel{0} \\ -\ 0.4\,8 \\ \hline 3.2\,2 \end{array}$$

Try It Yourself

Find the sum or difference.

1. 2.73 + 1.007 **2.** 3.4 − 1.27 **3.** 0.35 + 0.749 **4.** 1.019 + 0.09

5. 6.03 − 1.008 **6.** 4.21 − 0.007 **7.** 0.228 + 1.205 **8.** 3.003 − 1.9

Multiplying and Dividing Decimals (Grade 6)

Example 3 Find 2.1 • 0.35.

$$\begin{array}{rl} 2.1 & \leftarrow \text{1 decimal place} \\ \times\ 0.3\,5 & \leftarrow \underline{+\ \text{2 decimal places}} \\ \hline 1\,0\,5 & \\ 6\,3\ \ & \\ \hline 0.7\,3\,5 & \leftarrow \text{3 decimal places} \end{array}$$

Example 4 Find 1.08 ÷ 0.9.

$0.9\overline{)1.08}$ Multiply each number by 10.

$$\begin{array}{r} 1.2 \\ 9\overline{)10.8} \\ -\ 9\ \ \\ \hline 1\,8 \\ -\ 1\,8 \\ \hline 0 \end{array}$$

Place the decimal point above the decimal point in the dividend 10.8.

Try It Yourself

Find the product or quotient.

9. 1.75 • 0.2 **10.** 1.4 • 0.6 **11.** $\begin{array}{r} 7.03 \\ \times\ 4.3 \\ \hline \end{array}$ **12.** $\begin{array}{r} 0.894 \\ \times\ \ 0.2 \\ \hline \end{array}$

13. 5.40 ÷ 0.09 **14.** 4.17 ÷ 0.3 **15.** $0.15\overline{)3.6}$ **16.** $0.004\overline{)7.2}$

9.1 Exponents

STATE STANDARDS
MA.8.A.6.1
MA.8.A.6.4

Essential Question How can you use exponents to write numbers?

The expression 3^5 is called a **power**. The **base** is 3. The **exponent** is 5.

Base → 3^5 ← Exponent

1 ACTIVITY: Using Exponent Notation

Work with a partner.

a. Copy and complete the table.

Power	Repeated Multiplication Form	Value
$(-3)^1$	-3	-3
$(-3)^2$	$(-3) \cdot (-3)$	9
$(-3)^3$		
$(-3)^4$		
$(-3)^5$		
$(-3)^6$		
$(-3)^7$		

b. Describe what is meant by the expression $(-3)^n$. How can you find the value of $(-3)^n$?

2 ACTIVITY: Using Exponent Notation

Work with a partner.

a. The cube at the right has \$3 in each of its small cubes. Write a single power that represents the total amount of money in the large cube.

b. Evaluate the power to find the total amount of money in the large cube.

3 ACTIVITY: Writing Powers as Whole Numbers

Work with a partner. Write each distance as a whole number. Which numbers do you know how to write in words? For instance, in words, 10^3 is equal to *one thousand*.

a. 10^{26} meters:

Diameter of observable universe

b. 10^{21} meters:

Diameter of Milky Way Galaxy

c. 10^{16} meters:

Diameter of Solar System

d. 10^{7} meters:

Diameter of Earth

e. 10^{6} meters:

Length of Florida

f. 10^{5} meters:

Length of Miami-Dade County

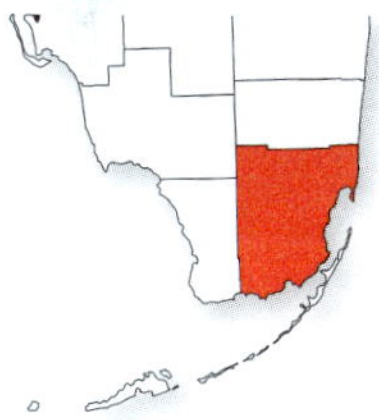

4 ACTIVITY: Writing a Power

Work with a partner. Write the number of kits, cats, sacks, and wives as a power.

As I was going to St. Ives
I met a man with seven wives
And every wife had seven sacks
And every sack had seven cats
And every cat had seven kits
Kits, cats, sacks, wives
How many were going to St. Ives?

Nursery Rhyme, 1730

What Is Your Answer?

5. **IN YOUR OWN WORDS** How can you use exponents to write numbers? Give some examples of how exponents are used in real life.

Practice Use what you learned about exponents to complete Exercises 3–5 on page 354.

9.1 Lesson

Check It Out
Lesson Tutorials
BigIdeasMath.com

Key Vocabulary
power, *p. 352*
base, *p. 352*
exponent, *p. 352*

A **power** is a product of repeated factors. The **base** of a power is the common factor. The **exponent** of a power indicates the number of times the base is used as a factor.

base → $\left(\frac{1}{2}\right)^5$ ← exponent

$$\left(\frac{1}{2}\right)^5 = \frac{1}{2} \cdot \frac{1}{2} \cdot \frac{1}{2} \cdot \frac{1}{2} \cdot \frac{1}{2}$$

power — $\frac{1}{2}$ is used as a factor 5 times.

EXAMPLE 1 Writing Expressions Using Exponents

Study Tip
Use parentheses to write powers with negative bases.

Write each product using exponents.

a. $(-7) \cdot (-7) \cdot (-7)$

Because -7 is used as a factor 3 times, its exponent is 3.

So, $(-7) \cdot (-7) \cdot (-7) = (-7)^3$.

b. $\pi \cdot \pi \cdot r \cdot r \cdot r$

Because π is used as a factor 2 times, its exponent is 2. Because r is used as a factor 3 times, its exponent is 3.

So, $\pi \cdot \pi \cdot r \cdot r \cdot r = \pi^2 r^3$.

On Your Own

Now You're Ready
Exercises 3–10

Write the product using exponents.

1. $\frac{1}{4} \cdot \frac{1}{4} \cdot \frac{1}{4} \cdot \frac{1}{4} \cdot \frac{1}{4}$

2. $0.3 \cdot 0.3 \cdot 0.3 \cdot 0.3 \cdot x \cdot x$

EXAMPLE 2 Evaluating Expressions

Evaluate the expression.

a. $(-2)^4$

$(-2)^4 = (-2) \cdot (-2) \cdot (-2) \cdot (-2)$ — Write as repeated multiplication. (The factor is -2.)

$= 16$ — Simplify.

b. -2^4

$-2^4 = -(2 \cdot 2 \cdot 2 \cdot 2)$ — Write as repeated multiplication. (The factor is 2.)

$= -16$ — Simplify.

Multi-Language Glossary at BigIdeasMath.com.

EXAMPLE 3 Using Order of Operations

Evaluate the expression.

a. $3 + 2 \cdot 3^4$

$$3 + 2 \cdot 3^4 = 3 + 2 \cdot 81 \quad \text{Evaluate the power.}$$
$$= 3 + 162 \quad \text{Multiply.}$$
$$= 165 \quad \text{Add.}$$

b. $3^3 - 8^2 \div 2$

$$3^3 - 8^2 \div 2 = 27 - 64 \div 2 \quad \text{Evaluate the powers.}$$
$$= 27 - 32 \quad \text{Divide.}$$
$$= -5 \quad \text{Subtract.}$$

On Your Own

Now You're Ready
Exercises 11–16 and 21–26

Evaluate the expression.

3. -5^4 **4.** $\left(-\frac{1}{6}\right)^3$ **5.** $\left|-3^3 \div 27\right|$ **6.** $9 - 2^5 \cdot 0.5$

EXAMPLE 4 Real-Life Application

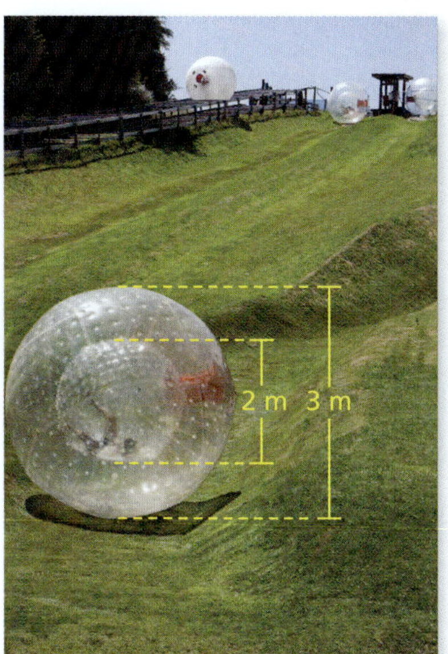

In sphering, a person is secured inside a small, hollow sphere that is surrounded by a larger sphere. The space between the spheres is inflated with air. What is the volume of the inflated space? (The volume V of a sphere is $V = \frac{4}{3}\pi r^3$. Use 3.14 for π.)

Outer sphere		*Inner sphere*
$V = \frac{4}{3}\pi r^3$	Write formula.	$V = \frac{4}{3}\pi r^3$
$= \frac{4}{3}\pi(1.5)^3$	Substitute.	$= \frac{4}{3}\pi(1)^3$
$= \frac{4}{3}\pi(3.375)$	Evaluate the power.	$= \frac{4}{3}\pi(1)$
≈ 14.13	Multiply.	≈ 4.19

So, the volume of the inflated space is about $14.13 - 4.19$, or 9.94 cubic meters.

On Your Own

7. **WHAT IF?** In Example 4, the diameter of the inner sphere is 1.8 meters. What is the volume of the inflated space?

9.1 Exercises

Check It Out
Help with Homework
BigIdeasMath.com

Vocabulary and Concept Check

1. **VOCABULARY** Describe the difference between an exponent and a power. Can the two words be used interchangeably?

2. **WHICH ONE DOESN'T BELONG?** Which one does *not* belong with the other three? Explain your reasoning.

5^3 The exponent is 3.	5^3 The power is 5.	5^3 The base is 5.	5^3 Five is used as a factor 3 times.

Practice and Problem Solving

Write the product using exponents.

1 3. $3 \cdot 3 \cdot 3 \cdot 3$

4. $(-6) \cdot (-6)$

5. $\left(-\frac{1}{2}\right) \cdot \left(-\frac{1}{2}\right) \cdot \left(-\frac{1}{2}\right)$

6. $\frac{1}{3} \cdot \frac{1}{3} \cdot \frac{1}{3}$

7. $\pi \cdot \pi \cdot \pi \cdot x \cdot x \cdot x \cdot x$

8. $(-4) \cdot (-4) \cdot (-4) \cdot y \cdot y$

9. $8 \cdot 8 \cdot 8 \cdot 8 \cdot b \cdot b \cdot b$

10. $(-t) \cdot (-t) \cdot (-t) \cdot (-t) \cdot (-t)$

Evaluate the expression.

2 11. 5^2

12. -11^3

13. $(-1)^6$

14. $\left(\frac{1}{2}\right)^6$

15. $\left(-\frac{1}{12}\right)^2$

16. $-\left(\frac{1}{9}\right)^3$

17. **ERROR ANALYSIS** Describe and correct the error in evaluating the expression.

✗ $6^3 = 6 \cdot 3 = 18$

18. **PRIME FACTORIZATION** Write the prime factorization of 675 using exponents.

19. **NUMBER SENSE** Write $-\left(\frac{1}{4} \cdot \frac{1}{4} \cdot \frac{1}{4} \cdot \frac{1}{4}\right)$ using exponents.

20. **RUSSIAN DOLLS** The largest doll is 12 inches tall. The height of each of the other dolls is $\frac{7}{10}$ the height of the next larger doll. Write an expression for the height of the smallest doll. What is the height of the smallest doll?

Evaluate the expression.

3 **21.** $5 + 3 \cdot 2^3$

22. $2 + 7 \cdot (-3)^2$

23. $(13^2 - 12^2) \div 5$

24. $\frac{1}{2}(4^3 - 6 \cdot 3^2)$

25. $\left| \frac{1}{2}(7 + 5^3) \right|$

26. $\left| \left(-\frac{1}{2}\right)^3 \div \left(\frac{1}{4}\right)^2 \right|$

27. **MONEY** You have a part-time job. One day your boss offers to pay you either $2^h - 1$ or 2^{h-1} dollars for each hour h you work that day. Copy and complete the table. Which option should you choose? Explain.

h	1	2	3	4	5
$2^h - 1$					
2^{h-1}					

28. **CARBON-14 DATING** Carbon-14 dating is used by scientists to determine the age of a sample.

a. The amount C (in grams) of a 100-gram sample of carbon-14 remaining after t years is represented by the equation $C = 100(0.99988)^t$. Use a calculator to find the amount of carbon-14 remaining after 4 years.

b. What percent of the carbon-14 remains after 4 years?

29. Critical Thinking The frequency (in vibrations per second) of a note on a piano is represented by the equation $F = 440(1.0595)^n$, where n is the number of notes above A-440. Each black or white key represents one note.

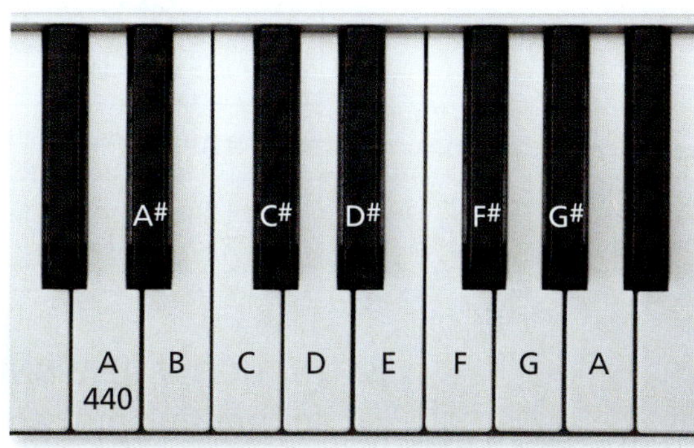

a. How many notes do you take to travel from A-440 to A?

b. What is the frequency of A?

c. Describe the relationship between the number of notes between A-440 and A and the frequency of the notes.

Fair Game Review What you learned in previous grades & lessons

Tell which property is illustrated by the statement. *(Skills Review Handbook)*

30. $8 \cdot x = x \cdot 8$

31. $(2 \cdot 10)x = 2(10 \cdot x)$

32. $3(x \cdot 1) = 3x$

33. **MULTIPLE CHOICE** A cone of yarn has a surface area of 16π square inches. What is the slant height of the cone of yarn? *(Skills Review Handbook)*

Ⓐ 4 in. Ⓑ 6 in.

Ⓒ 8 in. Ⓓ 10 in.

9.2 Product of Powers Property

STATE STANDARDS
MA.8.A.6.1
MA.8.A.6.3
MA.8.A.6.4

Essential Question How can you multiply two powers that have the same base?

1 ACTIVITY: Finding Products of Powers

Work with a partner.

a. Copy and complete the table.

Product	Repeated Multiplication Form	Power
$2^2 \cdot 2^4$	$2 \cdot 2 \cdot 2 \cdot 2 \cdot 2 \cdot 2$	2^6
$(-3)^2 \cdot (-3)^4$	$(-3) \cdot (-3) \cdot (-3) \cdot (-3) \cdot (-3) \cdot (-3)$	$(-3)^6$
$7^3 \cdot 7^2$		
$5.1^1 \cdot 5.1^6$		
$(-4)^2 \cdot (-4)^2$		
$10^3 \cdot 10^5$		
$\left(\frac{1}{2}\right)^5 \cdot \left(\frac{1}{2}\right)^5$		

b. **INDUCTIVE REASONING** Describe the pattern in the table. Then write a rule for multiplying two powers that have the same base.

$$a^m \cdot a^n = a^{\square}$$

c. Use your rule to simplify the products in the first column of the table above. Does your rule give the results in the third column?

2 ACTIVITY: Using a Calculator

Work with a partner.

Some calculators have *exponent keys* that are used to evaluate powers.

Use a calculator with an exponent key to evaluate the products in Activity 1.

3 ACTIVITY: The Penny Puzzle

Work with a partner.

- The rows y and columns x of a chess board are numbered as shown.
- Each position on the chess board has a stack of pennies. (Only the first row is shown.)
- The number of pennies in each stack is

  $2^x \cdot 2^y$.

a. How many pennies are in the stack in location (3, 5)?

b. Which locations have 32 pennies in their stacks?

c. How much money (in dollars) is in the location with the tallest stack?

d. A penny is about 0.06 inch thick. About how tall (in inches) is the tallest stack?

What Is Your Answer?

4. **IN YOUR OWN WORDS** How can you multiply two powers that have the same base? Give two examples of your rule.

Practice Use what you learned about the Product of Powers Property to complete Exercises 3–5 on page 360.

9.2 Lesson

Check It Out
Lesson Tutorials
BigIdeasMath.com

Key Idea

Product of Powers Property

Words To multiply powers with the same base, add their exponents.

Numbers $4^2 \cdot 4^3 = 4^{2+3} = 4^5$ **Algebra** $a^m \cdot a^n = a^{m+n}$

EXAMPLE 1 Multiplying Powers with the Same Base

a. $2^4 \cdot 2^5 = 2^{4+5}$ — The base is 2. Add the exponents.

$= 2^9$ — Simplify.

b. $-5 \cdot (-5)^6 = (-5)^1 \cdot (-5)^6$ — Rewrite -5 as $(-5)^1$.

$= (-5)^{1+6}$ — The base is -5. Add the exponents.

$= (-5)^7$ — Simplify.

c. $x^3 \cdot x^7 = x^{3+7}$ — The base is x. Add the exponents.

$= x^{10}$ — Simplify.

Study Tip

When a number is written without an exponent, its exponent is 1.

On Your Own

Simplify the expression. Write your answer as a power.

1. $6^2 \cdot 6^4$
2. $\left(-\frac{1}{2}\right)^3 \cdot \left(-\frac{1}{2}\right)^6$
3. $z \cdot z^{12}$

EXAMPLE 2 Raising a Power to a Power

a. $(3^4)^3 = 3^4 \cdot 3^4 \cdot 3^4$ — Write as repeated multiplication.

$= 3^{4+4+4}$ — The base is 3. Add the exponents.

$= 3^{12}$ — Simplify.

b. $(w^5)^4 = w^5 \cdot w^5 \cdot w^5 \cdot w^5$ — Write as repeated multiplication.

$= w^{5+5+5+5}$ — The base is w. Add the exponents.

$= w^{20}$ — Simplify.

On Your Own

Now You're Ready
Exercises 3–14

Simplify the expression. Write your answer as a power.

4. $(4^4)^3$
5. $(y^2)^4$
6. $(\pi^3)^3$
7. $((-4)^3)^2$

EXAMPLE 3 Raising a Product to a Power

a. $(2x)^3 = 2x \cdot 2x \cdot 2x$ — Write as repeated multiplication.

$= (2 \cdot 2 \cdot 2) \cdot (x \cdot x \cdot x)$ — Group like bases using properties of multiplication.

$= 2^{1+1+1} \cdot x^{1+1+1}$ — The bases are 2 and x. Add the exponents.

$= 2^3 \cdot x^3 = 8x^3$ — Simplify.

b. $(xy)^2 = xy \cdot xy$ — Write as repeated multiplication.

$= (x \cdot x) \cdot (y \cdot y)$ — Group like bases using properties of multiplication.

$= x^{1+1} \cdot y^{1+1}$ — The bases are x and y. Add the exponents.

$= x^2y^2$ — Simplify.

On Your Own

Now You're Ready
Exercises 17–22

Simplify the expression.

8. $(5y)^4$ **9.** $(0.5n)^2$ **10.** $(ab)^5$

EXAMPLE 4 Standardized Test Practice

Details

Local Disk (C:)
Local Disk

Free Space: 16GB

Total Space: 64GB

A gigabyte (GB) of computer storage space is 2^{30} bytes. The details of a computer are shown. How many bytes of total storage space does the computer have?

(A) 2^{34} (B) 2^{36} (C) 2^{180} (D) 128^{30}

The computer has 64 gigabytes of total storage space. Notice that 64 can be written as a power, 2^6. Use a model to solve the problem.

Total number of bytes = Number of bytes in a gigabyte • Number of gigabytes

$= 2^{30} \cdot 2^6$ — Substitute.

$= 2^{30+6}$ — Add exponents.

$= 2^{36}$ — Simplify.

The computer has 2^{36} bytes of total storage space. The correct answer is (B).

On Your Own

11. How many bytes of free storage space does the computer have?

9.2 Exercises

Vocabulary and Concept Check

1. **REASONING** When should you use the Product of Powers Property?

2. **CRITICAL THINKING** Can you use the Product of Powers Property to multiply powers with different bases? Explain.

Practice and Problem Solving

Simplify the expression. Write your answer as a power.

1 2 **3.** $3^2 \cdot 3^2$ **4.** $8^{10} \cdot 8^4$ **5.** $(-4)^5 \cdot (-4)^7$

6. $a^3 \cdot a^3$ **7.** $h^6 \cdot h$ **8.** $\left(\frac{2}{3}\right)^2 \cdot \left(\frac{2}{3}\right)^6$

9. $\left(-\frac{5}{7}\right)^8 \cdot \left(-\frac{5}{7}\right)^9$ **10.** $(-2.9) \cdot (-2.9)^7$ **11.** $(5^4)^3$

12. $(b^{12})^3$ **13.** $(3.8^3)^4$ **14.** $\left(\left(-\frac{3}{4}\right)^5\right)^2$

ERROR ANALYSIS **Describe and correct the error in simplifying the expression.**

15. ✗ $5^2 \cdot 5^9 = (5 \cdot 5)^{2+9}$
$= 25^{11}$

16. ✗ $(r^6)^4 = r^{6+4}$
$= r^{10}$

Simplify the expression.

3 **17.** $(6g)^3$ **18.** $(-3v)^5$ **19.** $\left(\frac{1}{5}k\right)^2$

20. $(1.2m)^4$ **21.** $(rt)^{12}$ **22.** $\left(-\frac{3}{4}p\right)^3$

23. **CRITICAL THINKING** Is $3^2 + 3^3$ equal to 3^5? Explain.

24. **ARTIFACT** A display case for the artifact is in the shape of a cube. Each side of the display case is three times longer than the width of the artifact.

a. Write an expression for the volume of the case. Write your answer as a power.

b. Simplify the expression.

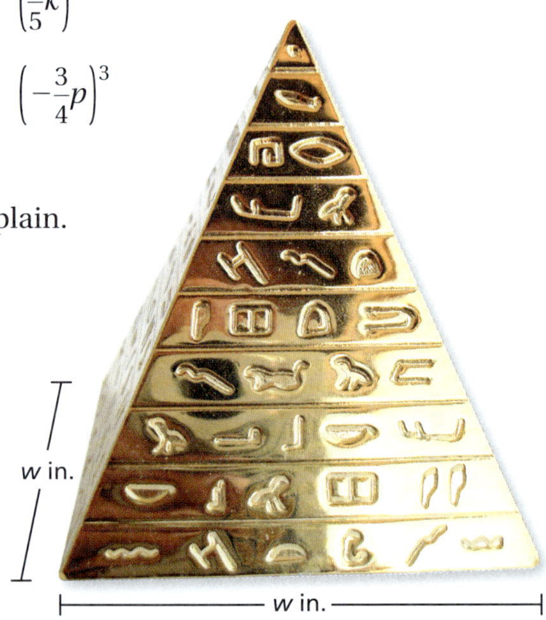

Simplify the expression.

25. $2^4 \cdot 2^5 - (2^2)^2$

26. $16\left(\frac{1}{2}x\right)^4$

27. $5^2(5^3 \cdot 5^2)$

28. **WATER** Florida has about 10 times more miles of rivers, streams, and waterways than miles of coastline. There are approximately 10^3 miles of coastline in Florida. About how many miles of rivers, streams, and waterways are in Florida? Write your answer as a power.

29. **PYTHON EGG** The volume V of a python egg is given by the formula $V = \frac{4}{3}\pi abc$. For the python egg shown, $a = 2$ inches, $b = 2$ inches, and $c = 3$ inches.

a. Find the volume of the python egg.

b. Square the dimensions of the python egg. Then evaluate the formula. How does this volume compare to your answer in part (a)?

30. **PYRAMID** The volume of a square pyramid is $V = \frac{1}{3}b^2h$, where b is the length of one side of the base and h is the height of the pyramid. The length of each side of the base increases by 50%. Write a formula for the volume of the new pyramid.

31. **MAIL** The United States Postal Service delivers about $2^6 \cdot 5^3$ pieces of mail each second. There are $2^8 \cdot 3^4 \cdot 5^2$ seconds in 6 days. How many pieces of mail does the United States Postal Service deliver in 6 days? Write your answer as a power.

32. **Critical Thinking** Find the value of x in the equation without evaluating the power.

a. $2^5 \cdot 2^x = 256$

b. $\left(\frac{1}{3}\right)^2 \cdot \left(\frac{1}{3}\right)^x = \frac{1}{729}$

Fair Game Review What you learned in previous grades & lessons

Simplify. *(Skills Review Handbook)*

33. $\frac{4 \cdot 4}{4}$

34. $\frac{5 \cdot 5 \cdot 5}{5}$

35. $\frac{2 \cdot 3}{2}$

36. $\frac{8 \cdot 6 \cdot 6}{6 \cdot 8}$

37. **MULTIPLE CHOICE** What is the measure of each angle of the regular polygon? *(Section 5.3)*

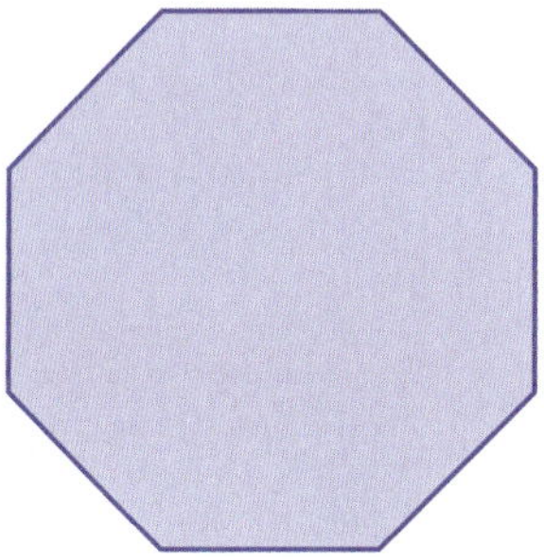

Ⓐ 45° Ⓑ 135°

Ⓒ 1080° Ⓓ 1440°

9.3 Quotient of Powers Property

STATE STANDARDS
MA.8.A.6.1
MA.8.A.6.3
MA.8.A.6.4

Essential Question How can you divide two powers that have the same base?

1 ACTIVITY: Finding Quotients of Powers

Work with a partner.

a. Copy and complete the table.

Quotient	Repeated Multiplication Form	Power
$\frac{2^4}{2^2}$	$\frac{\overset{1}{\cancel{2}} \cdot \overset{1}{\cancel{2}} \cdot 2 \cdot 2}{\underset{1}{\cancel{2}} \cdot \underset{1}{\cancel{2}}}$	2^2
$\frac{(-4)^5}{(-4)^2}$	$\frac{\overset{1}{\cancel{(-4)}} \cdot \overset{1}{\cancel{(-4)}} \cdot (-4) \cdot (-4) \cdot (-4)}{\underset{1}{\cancel{(-4)}} \cdot \underset{1}{\cancel{(-4)}}}$	$(-4)^3$
$\frac{7^7}{7^3}$		
$\frac{8.5^9}{8.5^6}$		
$\frac{10^8}{10^5}$		
$\frac{3^{12}}{3^4}$		
$\frac{(-5)^7}{(-5)^5}$		
$\frac{11^4}{11^1}$		

b. **INDUCTIVE REASONING** Describe the pattern in the table. Then write a rule for dividing two powers that have the same base.

$$\frac{a^m}{a^n} = a^{\square}$$

c. Use your rule to simplify the quotients in the first column of the table above. Does your rule give the results in the third column?

2 ACTIVITY: Comparing Volumes

Work with a partner.

How many of the smaller cubes will fit inside the larger cube? Record your results in the table. Describe the pattern in the table.

a. Sample:

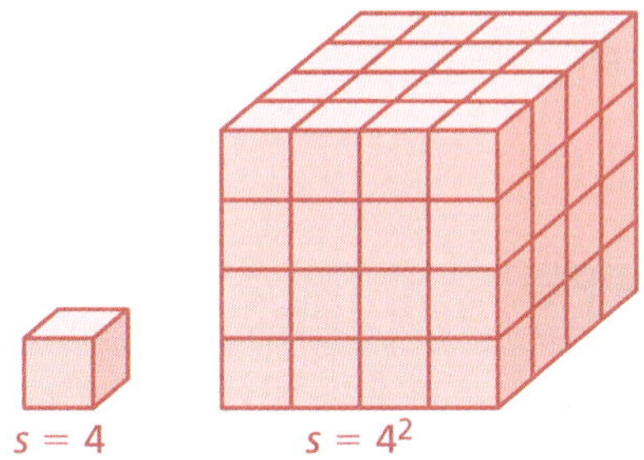

b.

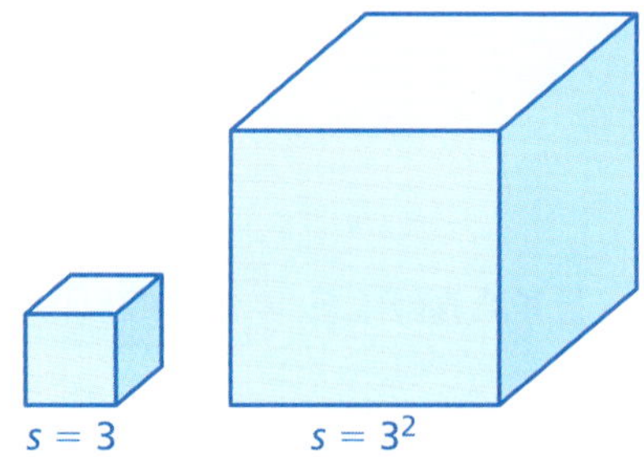

c.

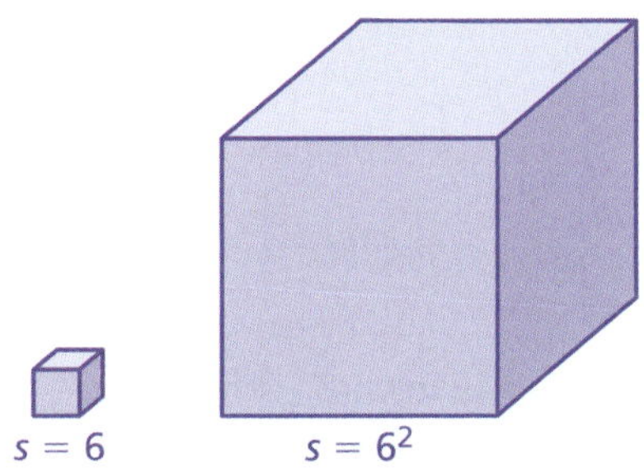

d.

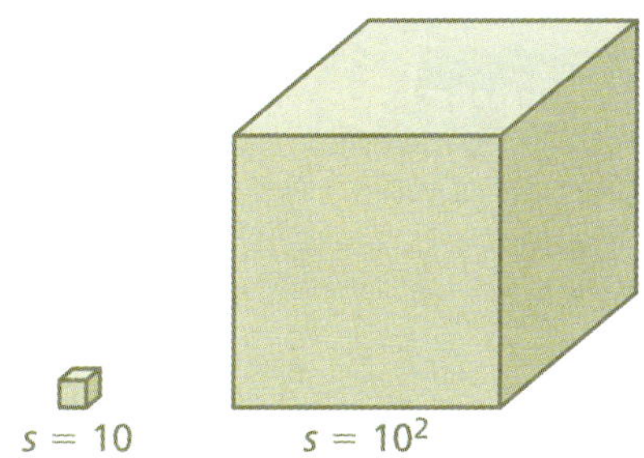

	Volume of Smaller Cube	Volume of Larger Cube	$\frac{\text{Larger Volume}}{\text{Smaller Volume}}$	Answer
a.	4^3	$(4^2)^3 = 4^6$	$\frac{4^6}{4^3}$	4^3
b.				
c.				
d.				

What Is Your Answer?

3. **IN YOUR OWN WORDS** How can you divide two powers that have the same base? Give two examples of your rule.

Practice Use what you learned about the Quotient of Powers Property to complete Exercises 3–6 on page 366.

9.3 Lesson

Check It Out
Lesson Tutorials
BigIdeasMath.com

Key Idea

Quotient of Powers Property

Words To divide powers with the same base, subtract their exponents.

Numbers $\dfrac{4^5}{4^2} = 4^{5-2} = 4^3$ **Algebra** $\dfrac{a^m}{a^n} = a^{m-n}$, where $a \neq 0$

EXAMPLE 1 Dividing Powers with the Same Base

a. $\dfrac{2^6}{2^4} = 2^{6-4}$ The base is 2. Subtract the exponents.

$= 2^2$ Simplify.

b. $\dfrac{(-7)^9}{(-7)^3} = (-7)^{9-3}$ The base is −7. Subtract the exponents.

$= (-7)^6$ Simplify.

c. $\dfrac{h^7}{h^6} = h^{7-6}$ The base is h. Subtract the exponents.

$= h^1 = h$ Simplify.

Common Error

When dividing powers, do not divide the bases.

$\dfrac{2^6}{2^4} = 2^2$, not 1^2.

On Your Own

Now You're Ready
Exercises 7–14

Simplify the expression. Write your answer as a power.

1. $\dfrac{9^7}{9^4}$ 2. $\dfrac{4.2^6}{4.2^5}$ 3. $\dfrac{(-8)^8}{(-8)^4}$ 4. $\dfrac{x^8}{x^3}$

EXAMPLE 2 Simplifying an Expression

Simplify $\dfrac{3^4 \cdot 3^2}{3^3}$. Write your answer as a power.

The numerator is a product of powers.

$\dfrac{3^4 \cdot 3^2}{3^3} = \dfrac{3^{4+2}}{3^3}$ Add the exponents in the numerator.

$= \dfrac{3^6}{3^3}$ Simplify.

$= 3^{6-3}$ The base is 3. Subtract the exponents.

$= 3^3$ Simplify.

EXAMPLE 3 Simplifying an Expression

Study Tip

You can also simplify the expression in Example 3 as follows.

$$\frac{a^{10}}{a^6} \cdot \frac{a^7}{a^4} = \frac{a^{10} \cdot a^7}{a^6 \cdot a^4}$$
$$= \frac{a^{17}}{a^{10}}$$
$$= a^{17-10}$$
$$= a^7$$

Simplify $\frac{a^{10}}{a^6} \cdot \frac{a^7}{a^4}$. Write your answer as a power.

$$\frac{a^{10}}{a^6} \cdot \frac{a^7}{a^4} = a^{10-6} \cdot a^{7-4}$$ Subtract the exponents.

$$= a^4 \cdot a^3$$ Simplify.

$$= a^{4+3}$$ Add the exponents.

$$= a^7$$ Simplify.

On Your Own

Now You're Ready Exercises 16–21

Simplify the expression. Write your answer as a power.

5. $\frac{2^{15}}{2^3 \cdot 2^5}$

6. $\frac{d^5}{d} \cdot \frac{d^9}{d^8}$

EXAMPLE 4 Real-Life Application

The projected population of Florida in 2030 is about $4 \cdot 7.2^8$. Predict the average number of people per square kilometer in 2030.

Use a model to solve the problem.

$$\text{People per square kilometer} = \frac{\text{Population in 2030}}{\text{Land area}}$$

$$= \frac{4 \cdot 7.2^8}{7.2^6}$$ Substitute.

$$= 4 \cdot \frac{7.2^8}{7.2^6}$$ Rewrite.

$$= 4 \cdot 7.2^2$$ Subtract the exponents.

$$= 207.36$$ Evaluate.

There will be about 207 people per square kilometer in Florida in 2030.

On Your Own

Now You're Ready Exercises 23–28

7. The projected population of Alabama in 2020 is about $2.25 \cdot 2^{21}$. The land area of Alabama is about 2^{17} square kilometers. Predict the average number of people per square kilometer in 2020.

9.3 Exercises

Check It Out
Help with Homework
BigIdeasMath.com

Vocabulary and Concept Check

1. **WRITING** Explain in your own words what it means to divide powers.

2. **WHICH ONE DOESN'T BELONG?** Which quotient does *not* belong with the other three? Explain your reasoning.

$\frac{(-10)^7}{(-10)^2}$ $\quad \frac{6^3}{6^2}$ $\quad \frac{(-4)^8}{(-3)^4}$ $\quad \frac{5^6}{5^3}$

Practice and Problem Solving

Simplify the expression. Write your answer as a power.

3. $\frac{6^{10}}{6^4}$
4. $\frac{8^9}{8^7}$
5. $\frac{(-3)^4}{(-3)^1}$
6. $\frac{4.5^5}{4.5^3}$

1 7. $\frac{5^9}{5^3}$
8. $\frac{64^4}{64^3}$
9. $\frac{(-17)^5}{(-17)^2}$
10. $\frac{(-7.9)^{10}}{(-7.9)^4}$

11. $\frac{(-6.4)^8}{(-6.4)^6}$
12. $\frac{\pi^{11}}{\pi^7}$
13. $\frac{b^{24}}{b^{11}}$
14. $\frac{n^{18}}{n^7}$

15. **ERROR ANALYSIS** Describe and correct the error in simplifying the quotient.

✗ $\frac{6^{15}}{6^5} = 6^{\frac{15}{5}}$
$= 6^3$

Simplify the expression. Write your answer as a power.

2 3 16. $\frac{7^5 \cdot 7^3}{7^2}$
17. $\frac{2^{19} \cdot 2^5}{2^{12} \cdot 2^3}$
18. $\frac{(-8.3)^8}{(-8.3)^7} \cdot \frac{(-8.3)^4}{(-8.3)^3}$

19. $\frac{\pi^{30}}{\pi^{18} \cdot \pi^4}$
20. $\frac{c^{22}}{c^8 \cdot c^9}$
21. $\frac{k^{13}}{k^5} \cdot \frac{k^{17}}{k^{11}}$

22. **SOUND INTENSITY** The sound intensity of a normal conversation is 10^6 times greater than the quietest noise a person can hear. The sound intensity of a jet at takeoff is 10^{14} times greater than the quietest noise a person can hear. How many times more intense is the sound of a jet at takeoff than the sound of a normal conversation?

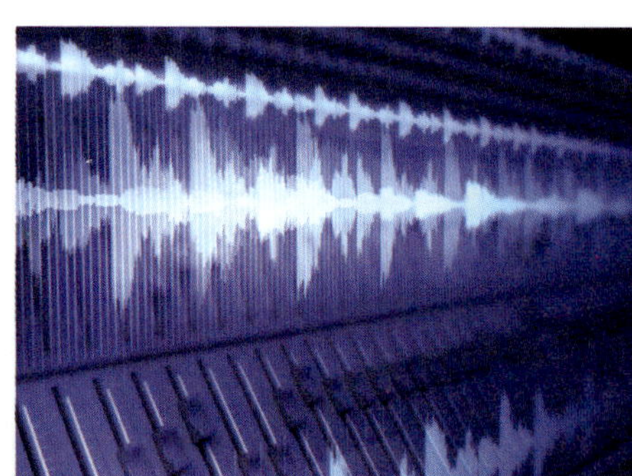

Simplify the expression.

4 23. $\dfrac{x \cdot 4^8}{4^5}$

24. $\dfrac{6^3 \cdot w}{6^2}$

25. $\dfrac{a^3 \cdot b^4 \cdot 5^4}{b^2 \cdot 5}$

26. $\dfrac{5^{12} \cdot c^{10} \cdot d^2}{5^9 \cdot c^9}$

27. $\dfrac{x^{15}y^9}{x^8y^3}$

28. $\dfrac{m^{10}n^7}{m^1n^6}$

MP3 Player	Memory (GB)	Price
A	2^1	\$70
B	2^2	\$120
C	2^3	\$170
D	2^4	\$220
E	2^5	\$270

29. **MEMORY** The memory capacities and prices of five MP3 players are shown in the table.

   a. How many times more memory does MP3 Player D have than MP3 Player B?

   b. Do the differences in price between consecutive sizes reflect a constant rate of change?

30. **CRITICAL THINKING** Consider the equation $\dfrac{9^m}{9^n} = 9^2$.

   a. Find two numbers m and n that satisfy the equation.

   b. Are there any other pairs of numbers that satisfy the equation? Explain.

Milky Way Galaxy
$10 \cdot 10^{10}$ stars

31. **STARS** There are about 10^{24} stars in the Universe. Each galaxy has approximately the same number of stars as the Milky Way Galaxy. About how many galaxies are in the Universe?

32. Number Sense Find the value of x that makes $\dfrac{8^{3x}}{8^{2x+1}} = 8^9$ true. Explain how you found your answer.

Fair Game Review What you learned in previous grades & lessons

Subtract. *(Skills Review Handbook)*

33. $-4 - 5$

34. $-23 - (-15)$

35. $33 - (-28)$

36. $18 - 22$

37. **MULTIPLE CHOICE** What is the value of x? *(Section 5.1)*

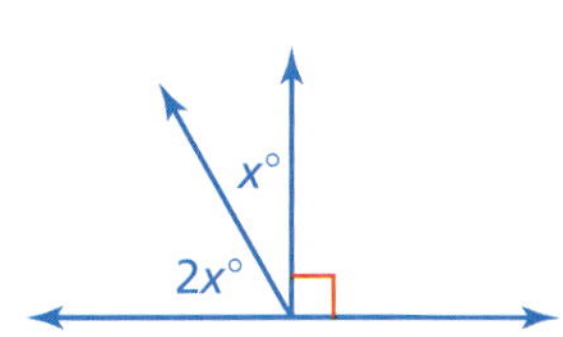

Ⓐ 20 Ⓑ 30

Ⓒ 45 Ⓓ 60

9 Study Help

Check It Out
Graphic Organizer
BigIdeasMath.com

You can use an **information wheel** to organize information about a topic. Here is an example of an information wheel for exponents.

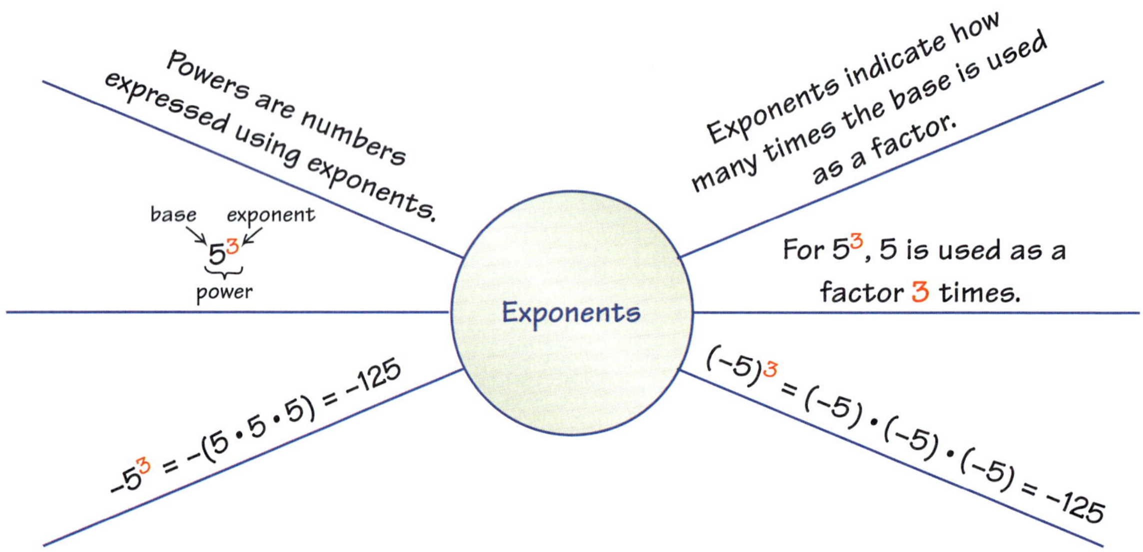

On Your Own

Make an information wheel to help you study these topics.

1. order of operations
2. Product of Powers Property
3. Quotient of Powers Property

After you complete this chapter, make information wheels for the following topics.

4. zero exponents
5. negative exponents
6. writing numbers in scientific notation
7. writing numbers in standard form
8. Choose three other topics you studied earlier in this course. Make an information wheel for each topic to summarize what you know about them.

"My information wheel for Fluffy has matching adjectives and nouns."

9.1–9.3 Quiz

Check It Out
Progress Check
BigIdeasMath.com

Write the product using exponents. *(Section 9.1)*

1. $(-5) \cdot (-5) \cdot (-5) \cdot (-5)$

2. $\frac{1}{6} \cdot \frac{1}{6} \cdot \frac{1}{6} \cdot \frac{1}{6} \cdot \frac{1}{6}$

3. $(-x) \cdot (-x) \cdot (-x) \cdot (-x) \cdot (-x) \cdot (-x)$

4. $7 \cdot 7 \cdot m \cdot m \cdot m$

Evaluate the expression. *(Section 9.1)*

5. 5^4

6. $(-2)^6$

Simplify the expression. Write your answer as a power. *(Section 9.2)*

7. $3^8 \cdot 3$

8. $(a^5)^3$

Simplify the expression. *(Section 9.2)*

9. $(3c)^4$

10. $\left(-\frac{2}{7}p\right)^2$

Simplify the expression. Write your answer as a power. *(Section 9.3)*

11. $\frac{8^7}{8^4}$

12. $\frac{6^3 \cdot 6^7}{6^2}$

13. $\frac{\pi^{15}}{\pi^3 \cdot \pi^9}$

14. $\frac{t^{13}}{t^5} \cdot \frac{t^8}{t^6}$

15. SEQUENCE The nth term of a sequence can be found by evaluating $10^n - 1$. Copy and complete the table to find the first four terms of the sequence. *(Section 9.1)*

n	$10^n - 1$
1	
2	
3	
4	

16. CRITICAL THINKING Is $(ab)^2$ equivalent to ab^2? Explain. *(Section 9.2)*

17. EARTHQUAKES An earthquake of magnitude 3.0 is 10^2 times stronger than an earthquake of magnitude 1.0. An earthquake of magnitude 8.0 is 10^7 times stronger than an earthquake of magnitude 1.0. How many times stronger is an earthquake of magnitude 8.0 than an earthquake of magnitude 3.0? *(Section 9.3)*

9.4 Zero and Negative Exponents

STATE STANDARDS
MA.8.A.6.1
MA.8.A.6.3
MA.8.A.6.4

Essential Question How can you define zero and negative exponents?

1 ACTIVITY: Finding Patterns and Writing Definitions

Work with a partner.

a. Talk about the following notation.

Thousands, Hundreds, Tens, Ones

$$4327 = 4 \cdot 10^3 + 3 \cdot 10^2 + 2 \cdot 10^1 + 7 \cdot 10^{\square}$$

What patterns do you see in the first three exponents?
Continue the pattern to find the fourth exponent.
How would you define 10^0? Explain.

b. Copy and complete the table.

n	5	4	3	2	1	0
2^n						

What patterns do you see in the first six values of 2^n?
How would you define 2^0? Explain.

c. Use the Quotient of Powers Property to complete the table.

$\frac{3^5}{3^2} = 3^{5-2} = 3^3 = 27$
$\frac{3^4}{3^2} = 3^{4-2} = \square = \square$
$\frac{3^3}{3^2} = 3^{3-2} = \square = \square$
$\frac{3^2}{3^2} = 3^{2-2} = \square = \square$

What patterns do you see in the first four rows of the table?
How would you define 3^0? Explain.

2 ACTIVITY: Comparing Volumes

Work with a partner.

The quotients show three ratios of the volumes of the solids. Identify each ratio, find its value, and describe what it means.

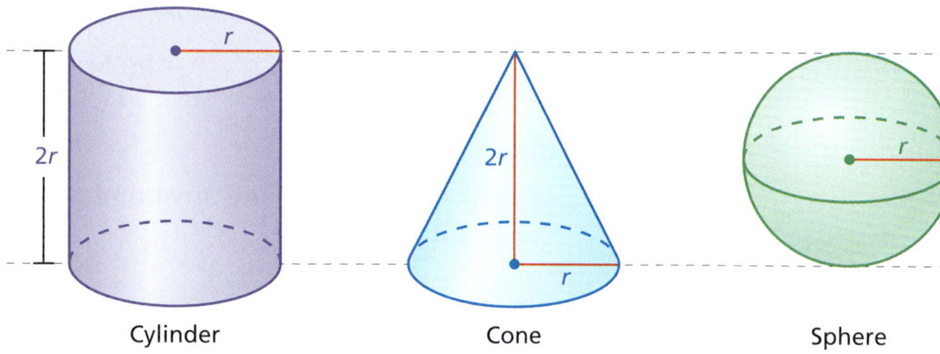

a. $2\pi r^3 \div \frac{2}{3}\pi r^3 = $ ▢

b. $\frac{4}{3}\pi r^3 \div \frac{2}{3}\pi r^3 = $ ▢

c. $2\pi r^3 \div \frac{4}{3}\pi r^3 = $ ▢

3 ACTIVITY: Writing a Definition

Work with a partner.

Compare the two methods used to simplify $\frac{3^2}{3^5}$. Then describe how you can rewrite a power with a negative exponent as a fraction.

Method 1

$$\frac{3^2}{3^5} = \frac{\overset{1}{\cancel{3}} \cdot \overset{1}{\cancel{3}}}{\underset{1}{\cancel{3}} \cdot \underset{1}{\cancel{3}} \cdot 3 \cdot 3 \cdot 3} = \frac{1}{3^3}$$

Method 2

$$\frac{3^2}{3^5} = 3^{2-5} = 3^{-3}$$

What Is Your Answer?

4. **IN YOUR OWN WORDS** How can you define zero and negative exponents? Give two examples of each.

Practice Use what you learned about zero and negative exponents to complete Exercises 5–8 on page 374.

9.4 Lesson

Check It Out
Lesson Tutorials
BigIdeasMath.com

Key Ideas

Zero Exponents

Words Any nonzero number to the zero power is equal to 1. Zero to the zero power, 0^0, is *undefined*.

Numbers $4^0 = 1$

Algebra $a^0 = 1$, where $a \neq 0$

Negative Exponents

Words For any integer n and any number a not equal to 0, a^{-n} is equal to 1 divided by a^n.

Numbers $4^{-2} = \dfrac{1}{4^2}$

Algebra $a^{-n} = \dfrac{1}{a^n}$, where $a \neq 0$

EXAMPLE 1 Evaluating Expressions

a. $3^{-4} = \dfrac{1}{3^4}$ Definition of negative exponent

$= \dfrac{1}{81}$ Evaluate power.

b. $(-8.5)^{-4} \cdot (-8.5)^4 = (-8.5)^{-4+4}$ Add the exponents.

$= (-8.5)^0$ Simplify.

$= 1$ Definition of zero exponent

c. $\dfrac{2^6}{2^8} = 2^{6-8}$ Subtract the exponents.

$= 2^{-2}$ Simplify.

$= \dfrac{1}{2^2}$ Definition of negative exponent

$= \dfrac{1}{4}$ Evaluate power.

On Your Own

Now You're Ready
Exercises 9–16

Evaluate the expression.

1. 4^{-2}
2. $(-2)^{-5}$
3. $6^{-8} \cdot 6^8$
4. $\dfrac{(-3)^5}{(-3)^6}$
5. $\dfrac{1}{5^7} \cdot \dfrac{1}{5^{-4}}$
6. $\dfrac{4^5 \cdot 4^{-3}}{4^2}$

EXAMPLE 2 Simplifying Expressions

a. $-5x^0 = -5(1)$ Definition of zero exponent

$= -5$ Multiply.

b. $\dfrac{9y^{-3}}{y^5} = 9y^{-3-5}$ Subtract the exponents.

$= 9y^{-8}$ Simplify.

$= \dfrac{9}{y^8}$ Definition of negative exponent

On Your Own

Now You're Ready
Exercises 20–27

Simplify. Write the expression using only positive exponents.

7. $8x^{-2}$ **8.** $b^0 \cdot b^{-10}$ **9.** $\dfrac{z^6}{15z^9}$

EXAMPLE 3 Real-Life Application

A drop of water leaks from a faucet every second. How many liters of water leak from the faucet in 1 hour?

Drop of water: 50^{-2} L

Convert 1 hour to seconds.

$$1\cancel{h} \times \frac{60\ \cancel{min}}{1\ \cancel{h}} \times \frac{60\text{ sec}}{1\ \cancel{min}} = 3600\text{ sec}$$

Water leaks from the faucet at a rate of 50^{-2} liter per second. Multiply the time by the rate.

$3600 \cdot 50^{-2} = 3600 \cdot \dfrac{1}{50^2}$ Definition of negative exponent

$= 3600 \cdot \dfrac{1}{2500}$ Evaluate power.

$= \dfrac{3600}{2500}$ Multiply.

$= 1\dfrac{11}{25} = 1.44$ Simplify.

So, 1.44 liters of water leak from the faucet in 1 hour.

On Your Own

10. WHAT IF? In Example 4, the faucet leaks water at a rate of 5^{-5} liter per second. How many liters of water leak from the faucet in 1 hour?

9.4 Exercises

Check It Out
Help with Homework
BigIdeasMath.com

Vocabulary and Concept Check

1. **VOCABULARY** If a is a nonzero number, does the value of a^0 depend on the value of a? Explain.
2. **WRITING** Explain how to evaluate 10^{-3}.
3. **NUMBER SENSE** Without evaluating, order 5^0, 5^4, and 5^{-5} from least to greatest.
4. **DIFFERENT WORDS, SAME QUESTION** Which is different? Find "both" answers.

Rewrite $\frac{1}{3 \cdot 3 \cdot 3}$ using a negative exponent.	Write 3 to the negative third power.
Write $\frac{1}{3}$ cubed as a power.	Write $(-3) \cdot (-3) \cdot (-3)$ as a power.

Practice and Problem Solving

5. Use the Quotient of Powers Property to copy and complete the table.

n	4	3	2	1
$\frac{5^n}{5^2}$				

6. What patterns do you see?
7. How would you define 5^0? Why?
8. How can you rewrite 5^{-1} as a fraction?

Evaluate the expression.

1 9. 6^{-2}
10. 158^0
11. $\frac{4^3}{4^5}$
12. $\frac{-3}{(-3)^2}$
13. $(-2)^{-8} \cdot (-2)^8$
14. $3^{-3} \cdot 3^{-2}$
15. $\frac{1}{5^{-3}} \cdot \frac{1}{5^6}$
16. $\frac{(1.5)^2}{(1.5)^{-2} \cdot (1.5)^4}$

17. **ERROR ANALYSIS** Describe and correct the error in evaluating the expression.

$$(4)^{-3} = (-4)(-4)(-4) = -64$$

18. **SAND** The mass of a grain of sand is about 10^{-3} gram. About how many grains of sand are in the bag of sand?

19. **CRITICAL THINKING** How can you write the number 1 as 2 to a power? 10 to a power?

Simplify. Write the expression using only positive exponents.

20. $6y^{-4}$

21. $8^{-2} \cdot a^7$

22. $\dfrac{9c^3}{c^{-4}}$

23. $\dfrac{5b^{-2}}{b^{-3}}$

24. $\dfrac{8x^3}{2x^9}$

25. $3d^{-4} \cdot 4d^4$

26. $m^{-2} \cdot n^3$

27. $\dfrac{3^{-2} \cdot k^0 \cdot w^0}{w^{-6}}$

METRIC UNITS **In Exercises 28–31, use the table.**

Unit of Length	Length
decimeter	10^{-1} m
centimeter	10^{-2} m
millimeter	10^{-3} m
micrometer	10^{-6} m
nanometer	10^{-9} m

28. How many millimeters are in a decimeter?

29. How many micrometers are in a centimeter?

30. How many nanometers are in a millimeter?

31. How many micrometers are in a meter?

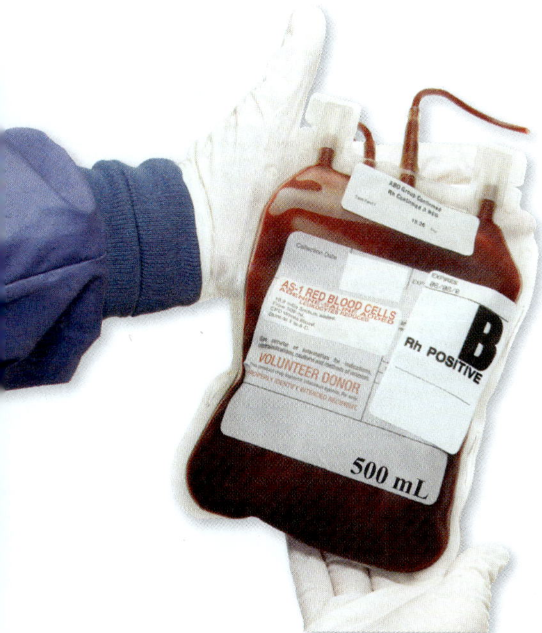

32. **MICROBES** A species of bacteria is 10 micrometers long. A virus is 10,000 times smaller than the bacteria.
    a. Using the table above, find the length of the virus in meters.
    b. Is the answer to part (a) *less than*, *greater than*, or *equal to* one nanometer?

33. **BLOOD DONATION** Every 2 seconds, someone in the United States needs blood. A sample blood donation is shown. ($1 \text{ mm}^3 = 10^{-3} \text{ mL}$)
    a. One cubic millimeter of blood contains about 10^4 white blood cells. How many white blood cells are in the donation? Write your answer in words.
    b. One cubic millimeter of blood contains about 5×10^6 red blood cells. How many red blood cells are in the donation? Write your answer in words.
    c. Compare your answers for parts (a) and (b).

34. **OPEN-ENDED** Write two different powers with negative exponents that have the same value.

35. **Reasoning** The rule for negative exponents states that $a^{-n} = \dfrac{1}{a^n}$. Explain why this rule does not apply when $a = 0$.

Fair Game Review What you learned in previous grades & lessons

Simplify the expression. *(Section 9.2 and Section 9.3)*

36. $10^3 \cdot 10^6$

37. $10^2 \cdot 10$

38. $\dfrac{10^8}{10^4}$

39. **MULTIPLE CHOICE** Which data display best shows the variability of a data set? *(Section 7.2)*

(A) bar graph
(B) circle graph
(C) scatter plot
(D) box-and-whisker plot

9.5 Reading Scientific Notation

Essential Question How can you read numbers that are written in scientific notation?

STATE STANDARDS
MA.8.A.6.1
MA.8.A.6.4

1 ACTIVITY: Very Large Numbers

Work with a partner.

- Use a calculator. Experiment with multiplying large numbers until your calculator gives an answer that is *not* in standard form.
- When the calculator at the right was used to multiply 2 billion by 3 billion, it listed the result as

  6.0 E+18.

- Multiply 2 billion by 3 billion by hand. Use the result to explain what 6.0 E+18 means.
- Check your explanation using products of other large numbers.
- Why didn't the calculator show the answer in standard form?
- Experiment to find the maximum number of digits your calculator displays. For instance, if you multiply 1000 by 1000 and your calculator shows 1,000,000, then it can display 7 digits.

2 ACTIVITY: Very Small Numbers

Work with a partner.

- Use a calculator. Experiment with multiplying very small numbers until your calculator gives an answer that is *not* in standard form.
- When the calculator at the right was used to multiply 2 billionths by 3 billionths, it listed the result as

  6.0 E−18.

- Multiply 2 billionths by 3 billionths by hand. Use the result to explain what 6.0 E–18 means.
- Check your explanation using products of other very small numbers.

3 ACTIVITY: Reading Scientific Notation

Work with a partner.

Each description gives an example of a number written in scientific notation. Answer the question in the description. Write your answer in standard form.

a. Nearly 1.0×10^5 dust mites can live in 1 square yard of carpet.

How many dust mites can live in 100 square yards of carpet?

b. A micron is about 4.0×10^{-5} inch. The length of a dust mite is 250 microns.

How long is a dust mite in inches?

c. About 1.0×10^{15} bacteria live in a human body.

How many bacteria are living in the humans in your classroom?

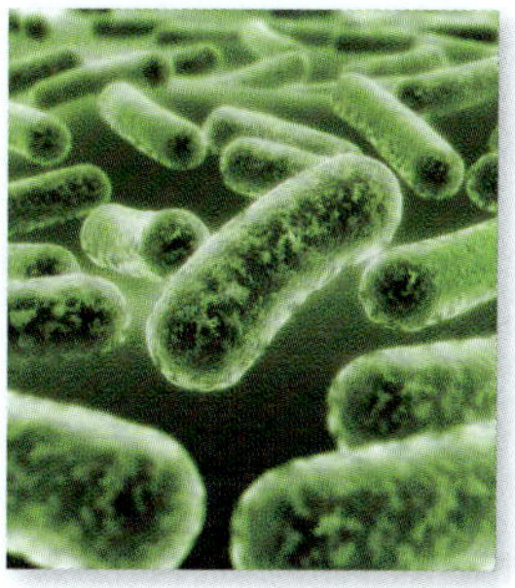

d. A micron is about 4.0×10^{-5} inch. The length of a bacterium is about 0.5 micron.

How many bacteria could lie end-to-end on your finger?

e. Earth has only about 1.5×10^8 kilograms of gold. Earth has a mass of 6.0×10^{24} kilograms.

What percent of Earth's mass is gold?

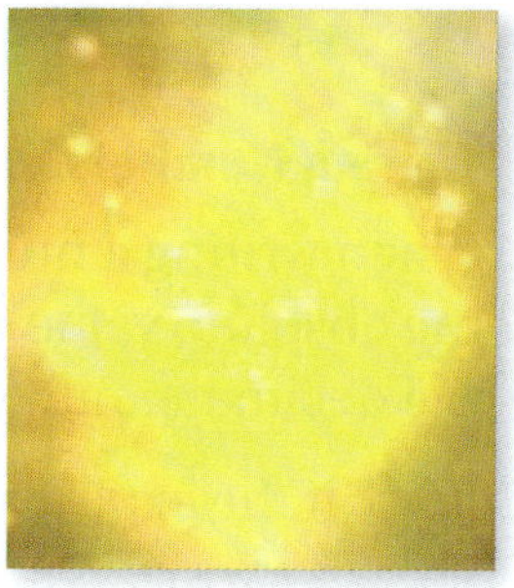

f. A gram is about 0.035 ounce. An atom of gold weighs about 3.3×10^{-22} gram.

How many atoms are in an ounce of gold?

What Is Your Answer?

4. **IN YOUR OWN WORDS** How can you read numbers that are written in scientific notation? Why do you think this type of notation is called "scientific notation?" Why is scientific notation important?

Practice Use what you learned about reading scientific notation to complete Exercises 3–5 on page 380.

9.5 Lesson

Check It Out
Lesson Tutorials
BigIdeasMath.com

Key Vocabulary
scientific notation, *p. 378*

Study Tip
Scientific notation is used to write very small and very large numbers.

Key Idea

Scientific Notation

A number is written in **scientific notation** when it is represented as the product of a factor and a power of 10. The factor must be at least 1 and less than 10.

The factor is at least 1 and less than 10. → 8.3×10^{-7} ← The power of 10 has an integer exponent.

EXAMPLE 1 Identifying Numbers Written in Scientific Notation

Tell whether the number is written in scientific notation. Explain.

a. 5.9×10^{-6}

∴ The factor is at least 1 and less than 10. The power of 10 has an integer exponent. So, the number is written in scientific notation.

b. 0.9×10^{8}

∴ The factor is less than 1. So, the number is not written in scientific notation.

Key Idea

Writing Numbers in Standard Form

When writing a number from scientific notation to standard form, the absolute value of the exponent tells you how many places to move the decimal point.

- If the exponent is negative, move the decimal point to the left.
- If the exponent is positive, move the decimal point to the right.

EXAMPLE 2 Writing Numbers in Standard Form

a. Write 3.22×10^{-4} in standard form.

$3.22 \times 10^{-4} = 0.000322$ (4) — Move decimal point $|-4| = 4$ places to the left.

b. Write 7.9×10^{5} in standard form.

$7.9 \times 10^{5} = 790{,}000$ (5) — Move decimal point $|5| = 5$ places to the right.

On Your Own

Now You're Ready
Exercises 6–23

1. Is 12×10^4 written in scientific notation? Explain.

Write the number in standard form.

2. 6×10^7
3. 9.9×10^{-5}
4. 1.285×10^4

EXAMPLE 3 Comparing Numbers in Scientific Notation

An object with a lesser density than water will float. An object with a greater density than water will sink. Use each given density (in kilograms per cubic meter) to explain what happens when you place a brick and an apple in water.

Water: 1.0×10^3 **Brick:** 1.84×10^3 **Apple:** 6.41×10^2

Write each density in standard form.

Water	Brick	Apple
$1.0 \times 10^3 = 1000$	$1.84 \times 10^3 = 1840$	$6.41 \times 10^2 = 641$

∴ The apple is less dense than water, so it will float. The brick is denser than water, so it will sink.

EXAMPLE 4 Real-Life Application

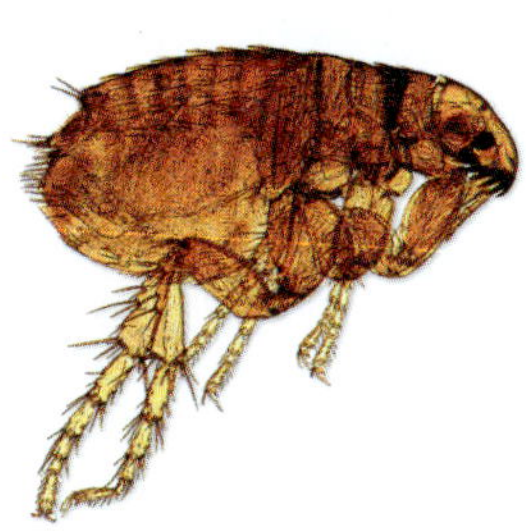

A female flea consumes about 1.4×10^{-5} liter of blood per day.

A dog has 100 female fleas. How many milliliters of blood do the fleas consume per day?

$$1.4 \times 10^{-5} \cdot 100 = 0.000014 \cdot 100 \quad \text{Write in standard form.}$$
$$= 0.0014 \quad \text{Multiply.}$$

∴ The fleas consume about 0.0014 liter, or 1.4 milliliters of blood per day.

On Your Own

Now You're Ready
Exercise 27

5. **WHAT IF?** In Example 3, the density of lead is 1.14×10^4 kilograms per cubic meter. What happens when lead is placed in water?
6. **WHAT IF?** In Example 4, a dog has 75 female fleas. How many milliliters of blood do the fleas consume per day?

9.5 Exercises

Check It Out
Help with Homework
BigIdeasMath.com

Vocabulary and Concept Check

1. **WRITING** Describe the difference between scientific notation and standard form.

2. **WHICH ONE DOESN'T BELONG?** Which number does *not* belong with the other three? Explain.

2.8×10^{15}	4.3×10^{-30}	1.05×10^{28}	10×9.2^{-13}

Practice and Problem Solving

Write your answer in standard form.

3. A micrometer is 1.0×10^{-6} meter. How long is 150 micrometers in meters?

4. An acre is about 4.05×10^{7} square centimeters. How many square centimeters are in 4 acres?

5. A cubic millimeter is about 6.1×10^{-5} cubic inches. How many cubic millimeters are in 1.22 cubic inches?

Tell whether the number is written in scientific notation. Explain.

1
6. 1.8×10^{9}
7. 3.45×10^{14}
8. 0.26×10^{-25}
9. 10.5×10^{12}
10. 46×10^{-17}
11. 5×10^{-19}
12. 7.814×10^{-36}
13. 0.999×10^{42}
14. 6.022×10^{23}

Write the number in standard form.

2
15. 7×10^{7}
16. 8×10^{-3}
17. 5×10^{2}
18. 2.7×10^{-4}
19. 4.4×10^{-5}
20. 2.1×10^{3}
21. 1.66×10^{9}
22. 3.85×10^{-8}
23. 9.725×10^{6}

24. **ERROR ANALYSIS** Describe and correct the error in writing the number in standard form.

✗ $4.1 \times 10^{-6} = 4100000$

2.7×10^{8} platelets per milliliter

25. **PLATELETS** Platelets are cell-like particles in the blood that help form blood clots.

    a. How many platelets are in 3 milliliters of blood? Write your answer in standard form.

    b. An adult body contains about 5 liters of blood. How many platelets are in an adult body?

26. **REASONING** A googol is 1.0×10^{100}. How many zeros are in a googol?

3 27. **STARS** The table shows the surface temperatures of five stars.

a. Which star has the highest surface temperature?

b. Which star has the lowest surface temperature?

Star	Betelgeuse	Bellatrix	Sun	Aldebaran	Rigel
Surface Temperature (°F)	6.2×10^3	3.8×10^4	1.1×10^4	7.2×10^3	2.2×10^4

28. **CORAL REEF** The area of the Florida Keys National Marine Sanctuary is about 9.6×10^3 square kilometers. The area of the Florida Reef Tract is about 16.2% of the area of the sanctuary. What is the area of the Florida Reef Tract in square kilometers?

29. **REASONING** A gigameter is 1.0×10^6 kilometers. How many square kilometers are in 5 square gigameters?

30. **WATER** There are about 1.4×10^9 cubic kilometers of water on Earth. About 2.5% of the water is fresh water. How much fresh water is on Earth?

31. **Critical Thinking** The table shows the speed of light through five media.

a. In which medium does light travel the fastest?

b. In which medium does light travel the slowest?

Medium	Speed
Air	6.7×10^8 mi/h
Glass	6.6×10^8 ft/sec
Ice	2.3×10^5 km/sec
Vacuum	3.0×10^8 m/sec
Water	2.3×10^{10} cm/sec

Fair Game Review What you learned in previous grades & lessons

Simplify. Write the expression using only positive exponents. *(Section 9.2 and Section 9.4)*

32. $6^3 \cdot 6^5$

33. $10^2 \cdot 10^5$

34. $8^{-1} \cdot 8^{-11}$

35. $10^{-6} \cdot 10^{-10}$

36. **MULTIPLE CHOICE** What is the length of the hypotenuse of the right triangle? *(Section 6.5)*

Ⓐ $\sqrt{18}$ in. Ⓑ $\sqrt{41}$ in.

Ⓒ 18 in. Ⓓ 41 in.

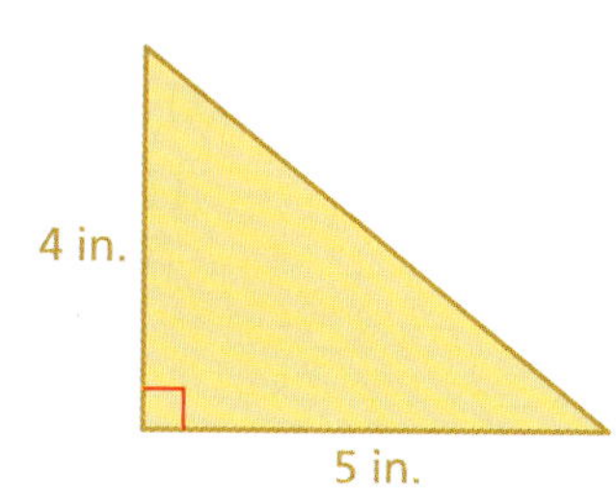

9.6 Writing Scientific Notation

STATE STANDARDS
MA.8.A.6.1
MA.8.A.6.4

Essential Question How can you write a number in scientific notation?

1 ACTIVITY: Finding pH Levels

Work with a partner. In chemistry, pH is a measure of the activity of dissolved hydrogen ions (H^+). Liquids with low pH values are called acids. Liquids with high pH values are called bases.

Find the pH of each liquid. Is the liquid a base, neutral, or an acid?

a. Lime juice:
$[H^+] = 0.01$

b. Egg:
$[H^+] = 0.00000001$

c. Distilled water:
$[H^+] = 0.0000001$

d. Ammonia water:
$[H^+] = 0.00000000001$

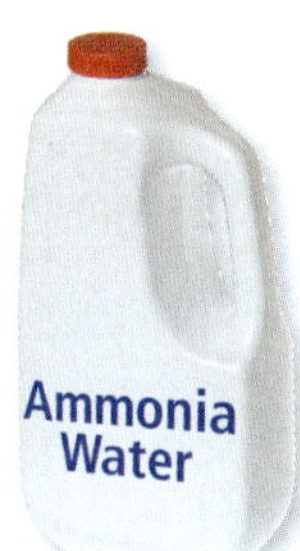

e. Tomato juice:
$[H^+] = 0.0001$

f. Hydrochloric acid:
$[H^+] = 1$

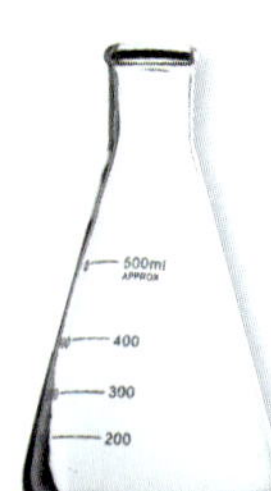

pH	$[H^+]$	
14	1×10^{-14}	Bases
13	1×10^{-13}	
12	1×10^{-12}	
11	1×10^{-11}	
10	1×10^{-10}	
9	1×10^{-9}	
8	1×10^{-8}	
7	1×10^{-7}	Neutral
6	1×10^{-6}	
5	1×10^{-5}	
4	1×10^{-4}	
3	1×10^{-3}	
2	1×10^{-2}	
1	1×10^{-1}	
0	1×10^{0}	Acids

2 ACTIVITY: Writing Scientific Notation

Work with a partner. Match each planet with its description. Then write each of the following in scientific notation.

- **Distance from the Sun (in miles)**
- **Distance from the Sun (in feet)**
- **Mass (in kilograms)**

a. Distance: 1,800,000,000 miles
Mass: 87,000,000,000,000,000,000,000,000 kg

b. Distance: 67,000,000 miles
Mass: 4,900,000,000,000,000,000,000,000 kg

c. Distance: 890,000,000 miles
Mass: 570,000,000,000,000,000,000,000,000 kg

d. Distance: 93,000,000 miles
Mass: 6,000,000,000,000,000,000,000,000 kg

e. Distance: 140,000,000 miles
Mass: 640,000,000,000,000,000,000,000 kg

f. Distance: 2,800,000,000 miles
Mass: 100,000,000,000,000,000,000,000,000 kg

g. Distance: 480,000,000 miles
Mass: 1,900,000,000,000,000,000,000,000,000 kg

h. Distance: 36,000,000 miles
Mass: 330,000,000,000,000,000,000,000 kg

3 ACTIVITY: Making a Scale Drawing

Work with a partner. The illustration in Activity 2 is not drawn to scale. Make a scale drawing of the distances in our solar system.

- **Cut a sheet of paper into three strips of equal width. Tape the strips together.**
- **Draw a long number line. Label the number line in hundreds of millions of miles.**
- **Locate each planet's position on the number line.**

What Is Your Answer?

4. **IN YOUR OWN WORDS** How can you write a number in scientific notation?

Practice Use what you learned about writing scientific notation to complete Exercises 3–5 on page 386.

Check It Out
Lesson Tutorials
BigIdeasMath.com

Key Idea

Writing Numbers in Scientific Notation

Step 1: Move the decimal point to the right of the first nonzero digit.

Step 2: Count the number of places you moved the decimal point. This determines the exponent of the power of 10.

Number greater than or equal to 10

Use a positive exponent when you move the decimal point to the left.

$$8600 = 8.6 \times 10^3$$

3

Number between 0 and 1

Use a negative exponent when you move the decimal point to the right.

$$0.0024 = 2.4 \times 10^{-3}$$

3

EXAMPLE 1 Writing Large Numbers in Scientific Notation

Google purchased YouTube for \$1,650,000,000. Write this number in scientific notation.

The number is greater than 1. So, move the decimal point 9 places to the left.

$$1{,}650{,}000{,}000 = 1.65 \times 10^9$$

9

The exponent is positive.

EXAMPLE 2 Writing Small Numbers in Scientific Notation

The 2004 Indonesian earthquake slowed the rotation of Earth, making the length of a day 0.00000268 second shorter. Write this number in scientific notation.

The number is between 0 and 1. So, move the decimal point 6 places to the right.

$$0.00000268 = 2.68 \times 10^{-6}$$

6

The exponent is negative.

On Your Own

Now You're Ready
Exercises 3–11

Write the number in scientific notation.

1. 50,000 **2.** 25,000,000 **3.** 683

4. 0.005 **5.** 0.00000033 **6.** 0.000506

EXAMPLE 3 Standardized Test Practice

An album receives an award when it sells 10,000,000 copies.

An album has sold 8,780,000 copies. How many more copies does it need to sell to receive the award?

Ⓐ 1.22×10^{-7} Ⓑ 1.22×10^{-6}

Ⓒ 1.22×10^{6} Ⓓ 1.22×10^{7}

Use a model to solve the problem.

Remaining sales needed for award = Sales required for award − Current sales total

$= 10{,}000{,}000 - 8{,}780{,}000$

$= 1{,}220{,}000$

$= 1.22 \times 10^{6}$

∴ The album must sell 1.22×10^{6} more copies to receive the award. The correct answer is Ⓒ.

On Your Own

7. An album has sold 955,000 copies. How many more copies does it need to sell to receive the award? Write your answer in scientific notation.

EXAMPLE 4 Multiplying Numbers in Scientific Notation

Find $(3 \times 10^{-5}) \times (5 \times 10^{-2})$. Write your answer in scientific notation.

$(3 \times 10^{-5}) \times (5 \times 10^{-2})$

$= 3 \times 5 \times 10^{-5} \times 10^{-2}$ Commutative Property of Multiplication

$= (3 \times 5) \times (10^{-5} \times 10^{-2})$ Associative Property of Multiplication

$= 15 \times 10^{-7}$ Simplify.

$= 1.5 \times 10^{1} \times 10^{-7}$ Write factor in scientific notation.

$= 1.5 \times 10^{-6}$ Simplify.

Study Tip

You can check your answer using standard form.

(3×10^{-5})
$\times (5 \times 10^{-2})$
$= 0.00003 \times 0.05$
$= 0.0000015$
$= 1.5 \times 10^{-6}$

On Your Own

Now You're Ready
Exercises 14–19

Multiply. Write your answer in scientific notation.

8. $(2.5 \times 10^{8}) \times (2 \times 10^{3})$

9. $(2 \times 10^{-4}) \times (1 \times 10^{-4})$

10. $(5 \times 10^{-4}) \times (5.4 \times 10^{-9})$

11. $(7 \times 10^{2}) \times (3 \times 10^{5})$

9.6 Exercises

Check It Out
Help with Homework
BigIdeasMath.com

Vocabulary and Concept Check

1. **REASONING** How do you know whether a number written in standard form will have a positive or negative exponent when written in scientific notation?

2. **WRITING** Describe how to write a number in scientific notation.

Practice and Problem Solving

Write the number in scientific notation.

1 2 **3.** 0.0021 **4.** 5,430,000 **5.** 321,000,000

6. 0.00000625 **7.** 0.00004 **8.** 10,700,000

9. 45,600,000,000 **10.** 0.000000000009256 **11.** 840,000

ERROR ANALYSIS **Describe and correct the error in writing the number in scientific notation.**

12. ✗ 0.000036 → 5 → 3.6×10^5

13. ✗ 72,500,000 → 6 → 72.5×10^6

Multiply. Write your answer in scientific notation.

4 **14.** $(4 \times 10^4) \times (2 \times 10^6)$ **15.** $(3 \times 10^{-8}) \times (3 \times 10^{-2})$

16. $(5 \times 10^{-7}) \times (3 \times 10^6)$ **17.** $(8 \times 10^3) \times (2 \times 10^4)$

18. $(6 \times 10^8) \times (1.4 \times 10^{-5})$ **19.** $(7.2 \times 10^{-1}) \times (4 \times 10^{-7})$

20. **HAIR** What is the diameter of a human hair in scientific notation?

Diameter: 0.000099 meter

21. **EARTH** What is the circumference of Earth in scientific notation?

Circumference at the equator: about 40,100,000 meters

22. **SPRINGS** Every day, more than 250,000,000 gallons of water flow from Wakulla Spring near Tallahassee, Florida. Write this number in scientific notation.

Find the area of the figure. Write your answer in scientific notation.

23.

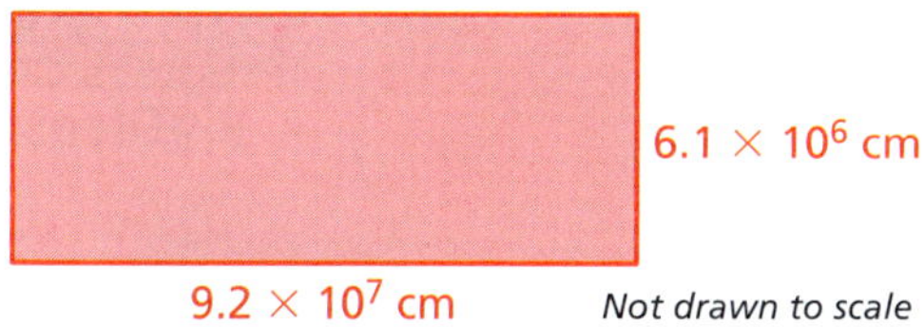

24.

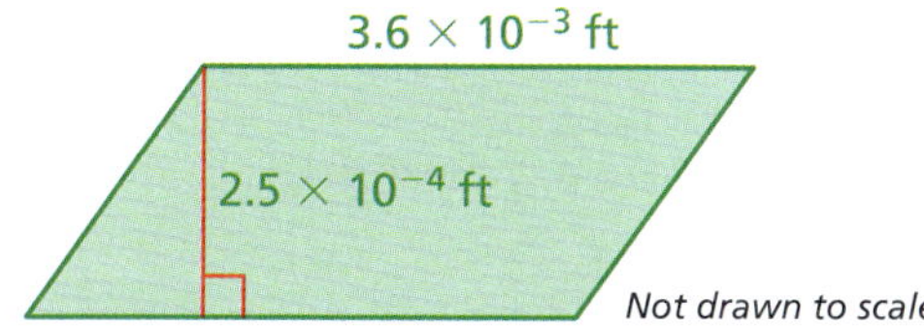

25. SPACE SHUTTLE The power of a space shuttle during launch is the force of the solid rocket boosters multiplied by the velocity. The velocity is 3.75×10^2 meters per second. What is the power (in newton-meters per second) of the shuttle shown during launch?

26. NUMBER SENSE Write 670 million in three ways.

27. PROJECT Use the Internet or some other reference to find the populations of India, China, Argentina, the United States, and Egypt. Round each population to the nearest million.

a. Write each population in scientific notation.

b. Use the Internet or some other reference to find the population density for each country.

c. Use the results of parts (a) and (b) to find the area of each country.

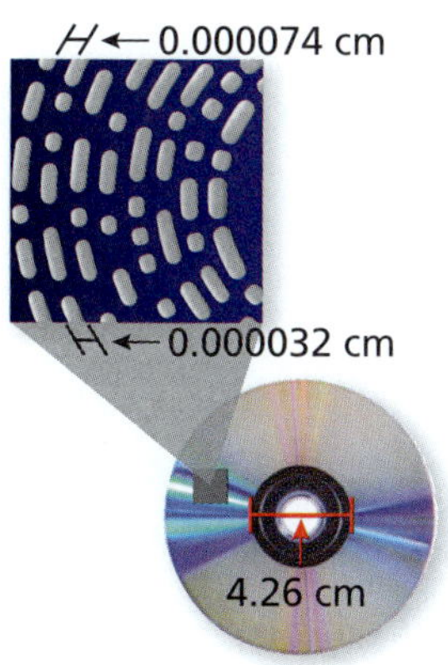

28. DVDS On a DVD, information is stored on bumps that spiral around the disk. There are 73,000 ridges (with bumps) and 73,000 valleys (without bumps) across the diameter of the DVD. What is the diameter of the DVD in centimeters?

29. Number Sense Simplify. Write your answer in scientific notation.

a. $\dfrac{(53{,}000{,}000)(0.002)}{(0.0004)}$

b. $\dfrac{(0.33)(60{,}000)}{(90{,}000{,}000)}$

Fair Game Review What you learned in previous grades & lessons

Write and solve an equation to answer the question. *(Skills Review Handbook)*

30. 15% of 60 is what number?

31. 85% of what number is 170?

32. MULTIPLE CHOICE What is the domain of the function represented by the table? *(Section 4.1)*

x	−2	−1	0	1	2
y	−6	−2	2	6	10

Ⓐ −2, −1, 0, 1, 2 Ⓑ −6, −2, 2, 6, 10

Ⓒ all integers Ⓓ all whole numbers

9.4–9.6 Quiz

Check It Out
Progress Check
BigIdeasMath.com

Evaluate the expression. *(Section 9.4)*

1. $(-4.8)^{-9} \cdot (-4.8)^9$

2. $\frac{5^4}{5^7}$

Simplify. Write the expression using only positive exponents. *(Section 9.4)*

3. $8d^{-6}$

4. $\frac{12x^5}{4x^7}$

Tell whether the number is written in scientific notation. Explain. *(Section 9.5)*

5. 23×10^9

6. 0.6×10^{-7}

Write the number in standard form. *(Section 9.5)*

7. 8×10^6

8. 1.6×10^{-2}

Write the number in scientific notation. *(Section 9.6)*

9. 0.00524

10. 892,000,000

Multiply. Write your answer in scientific notation. *(Section 9.6)*

11. $(9 \times 10^3) \times (4 \times 10^4)$

12. $(2 \times 10^{-5}) \times (3.1 \times 10^{-2})$

13. **PLANETS** The table shows the equatorial radii of the eight planets in our solar system. *(Section 9.5)*

a. Which planet has the second smallest equatorial radius?

b. Which planet has the second greatest equatorial radius?

Planet	Equatorial Radius (km)
Mercury	2.44×10^3
Venus	6.05×10^3
Earth	6.38×10^3
Mars	3.4×10^3
Jupiter	7.15×10^4
Saturn	6.03×10^4
Uranus	2.56×10^4
Neptune	2.48×10^4

14. **OORT CLOUD** The Oort cloud is a spherical cloud that surrounds our solar system. It is about 2×10^5 astronomical units from the Sun. An astronomical unit is about 1.5×10^8 kilometers. How far is the Oort cloud from the Sun in kilometers? *(Section 9.6)*

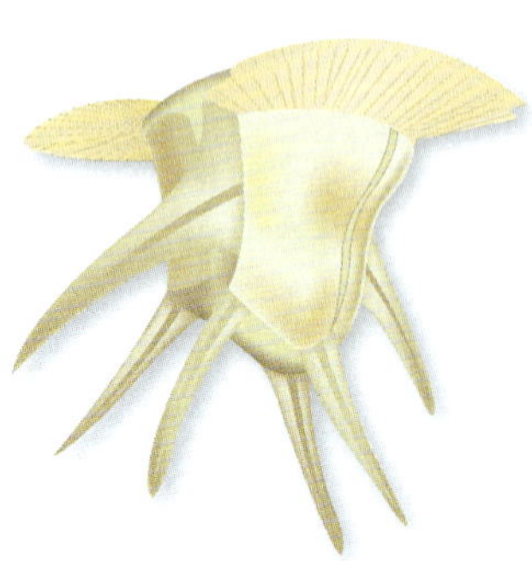

15. **ORGANISM** A one-celled, aquatic organism called a dinoflagellate is 1000 micrometers long. *(Section 9.4)*

a. One micrometer is 10^{-6} meter. What is the length of the dinoflagellate in meters?

b. Is the length of the dinoflagellate equal to 1 millimeter or 1 kilometer? Explain.

9 Chapter Review

Check It Out
Vocabulary Help
BigIdeasMath.com

Review Key Vocabulary

power, *p. 352*
base, *p. 352*
exponent, *p. 352*
scientific notation, *p. 378*

Review Examples and Exercises

9.1 Exponents *(pp. 350–355)*

Write $(-4) \cdot (-4) \cdot (-4) \cdot y \cdot y$ using exponents.

Because -4 is used as a factor 3 times, its exponent is 3. Because y is used as a factor 2 times, its exponent is 2.

So, $(-4) \cdot (-4) \cdot (-4) \cdot y \cdot y = (-4)^3y^2$.

Exercises

Write the product using exponents.

1. $(-9) \cdot (-9) \cdot (-9) \cdot (-9) \cdot (-9)$
2. $2 \cdot 2 \cdot 2 \cdot n \cdot n$

Evaluate the expression.

3. 6^3
4. $-\left(\frac{1}{2}\right)^4$
5. $\left|\frac{1}{2}(16 - 6^3)\right|$

9.2 Product of Powers Property *(pp. 356–361)*

a. $\left(-\frac{1}{8}\right)^7 \cdot \left(-\frac{1}{8}\right)^4 = \left(-\frac{1}{8}\right)^{7+4}$ The base is $-\frac{1}{8}$. Add the exponents.

$= \left(-\frac{1}{8}\right)^{11}$ Simplify.

b. $(3m)^2 = 3m \cdot 3m$ Write as repeated multiplication.

$= (3 \cdot 3) \cdot (m \cdot m)$ Use properties of multiplication.

$= 3^{1+1} \cdot m^{1+1}$ The bases are 3 and m. Add the exponents.

$= 3^2 \cdot m^2 = 9m^2$ Simplify.

Exercises

Simplify the expression.

6. $p^5 \cdot p^2$
7. $(n^{11})^2$
8. $(5y)^3$
9. $(-2k)^4$

9.3 Quotient of Powers Property (pp. 362–367)

a. $\dfrac{(-4)^9}{(-4)^6} = (-4)^{9-6}$ — The base is −4. Subtract the exponents.

$= (-4)^3$ — Simplify.

b. $\dfrac{x^4}{x^3} = x^{4-3}$ — The base is *x*. Subtract the exponents.

$= x^1$

$= x$ — Simplify.

Exercises

Simplify the expression. Write your answer as a power.

10. $\dfrac{8^8}{8^3}$ **11.** $\dfrac{5^2 \cdot 5^9}{5}$ **12.** $\dfrac{w^8}{w^7} \cdot \dfrac{w^5}{w^2}$

Simplify the expression.

13. $\dfrac{2^2 \cdot 2^5}{2^3}$ **14.** $\dfrac{(6c)^3}{c}$ **15.** $\dfrac{m^8}{m^6} \cdot \dfrac{m^{10}}{m^9}$

9.4 Zero and Negative Exponents (pp. 370–375)

a. $10^{-3} = \dfrac{1}{10^3}$ — Definition of negative exponent

$= \dfrac{1}{1000}$ — Evaluate power.

b. $(-0.5)^{-5} \cdot (-0.5)^5 = (-0.5)^{-5+5}$ — Add the exponents.

$= (-0.5)^0$ — Simplify.

$= 1$ — Definition of zero exponent

Exercises

Evaluate the expression.

16. 2^{-4} **17.** 95^0 **18.** $\dfrac{8^2}{8^4}$

19. $(-12)^{-7} \cdot (-12)^7$ **20.** $\dfrac{1}{7^9} \cdot \dfrac{1}{7^{-6}}$ **21.** $\dfrac{9^4 \cdot 9^{-2}}{9^2}$

9.5 Reading Scientific Notation (pp. 376–381)

a. Write 5.9×10^4 in standard form.

$$5.9 \times 10^4 = 59{,}000$$

4

Move decimal point 4 places to the right.

b. Write 7.31×10^{-6} in standard form.

$$7.31 \times 10^{-6} = 0.00000731$$

6

Move decimal point 6 places to the left.

Exercises

Tell whether the number is written in scientific notation. Explain.

22. 0.9×10^9 **23.** 3.04×10^{-11} **24.** 15×10^{26}

Write the number in standard form.

25. 2×10^7 **26.** 4.8×10^{-3} **27.** 6.25×10^5

9.6 Writing Scientific Notation (pp. 382–387)

a. In 2010, the population of the United States was about 309,000,000. Write this number in scientific notation.

The number is greater than 1. So, move the decimal point 8 places to the left.

$$309{,}000{,}000 = 3.09 \times 10^8$$

8

The exponent is positive.

b. The cornea of an eye is 0.00056 meter thick. Write this number in scientific notation.

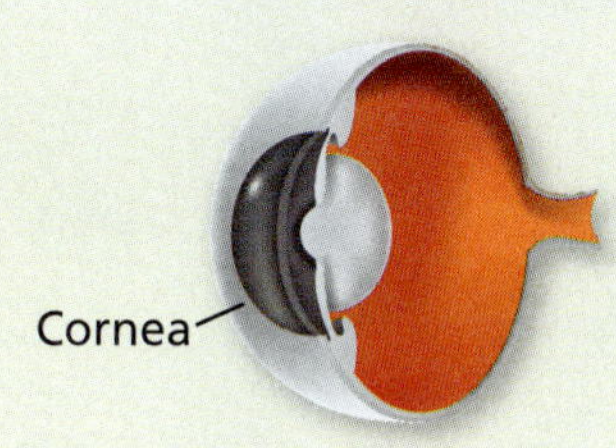

The number is between 0 and 1. So, move the decimal point 4 places to the right.

$$0.00056 = 5.6 \times 10^{-4}$$

4

The exponent is negative.

Exercises

Write the number in scientific notation.

28. 0.00036 **29.** 800,000 **30.** 79,200,000

Multiply. Write your answer in scientific notation.

31. $(4 \times 10^3) \times (2 \times 10^2)$ **32.** $(1.5 \times 10^{-9}) \times (8 \times 10^{-3})$

9 Chapter Test

Check It Out
Test Practice
BigIdeasMath.com

Write the product using exponents.

1. $(-15) \cdot (-15) \cdot (-15)$
2. $\left(\frac{1}{12}\right) \cdot \left(\frac{1}{12}\right) \cdot \left(\frac{1}{12}\right) \cdot \left(\frac{1}{12}\right) \cdot \left(\frac{1}{12}\right)$

Evaluate the expression.

3. -2^3
4. $10 + 3^3 \div 9$

Simplify the expression. Write your answer as a power.

5. $9^{10} \cdot 9$
6. $\frac{(-3.5)^{13}}{(-3.5)^9}$

Evaluate the expression.

7. $5^{-2} \cdot 5^2$
8. $\frac{-8}{(-8)^3}$

Write the number in standard form.

9. 3×10^7
10. 9.05×10^{-3}

Multiply. Write your answer in scientific notation.

11. $(7 \times 10^3) \times (5 \times 10^2)$
12. $(3 \times 10^{-5}) \times (2 \times 10^{-3})$

13. **HAMSTER** A hamster toy is in the shape of a sphere. The volume V of a sphere is represented by $V = \frac{4}{3}\pi r^3$, where r is the radius of the sphere. What is the volume of the toy? Round your answer to the nearest cubic centimeter. Use 3.14 for π.

14. **CRITICAL THINKING** Is $(xy^2)^3$ the same as $(xy^3)^2$? Explain.

15. **RICE** A grain of rice weighs about 3^3 milligrams. About how many grains of rice are in one scoop?

16. **TASTE BUDS** There are about 10,000 taste buds on a human tongue. Write this number in scientific notation.

17. **LEAD** From 1978 to 2008, the amount of lead allowed in the air in the United States was 1.5×10^{-6} gram per cubic meter. In 2008, the amount allowed was reduced by 90%. What is the new amount of lead allowed in the air?

One scoop of rice weighs about 3^9 milligrams.

9 Standarized Test Practice

Test-Taking Strategy
Use Intelligent Guessing

"It can't be 40 or 5000 because they aren't divisible by 3. So, you can intelligently guess between 3000 and 3072."

1. Mercury's distance to the Sun is approximately 5.79×10^7 kilometers. Write this distance in standard form.

   A. 5,790,000,000 km **C.** 57,900,000 km

   B. 579,000,000 km **D.** 5,790,000 km

2. The steps Jim took to answer the question are shown below. What should Jim change to correctly answer the question?

   How many degrees are in the largest angle in the triangle below?

   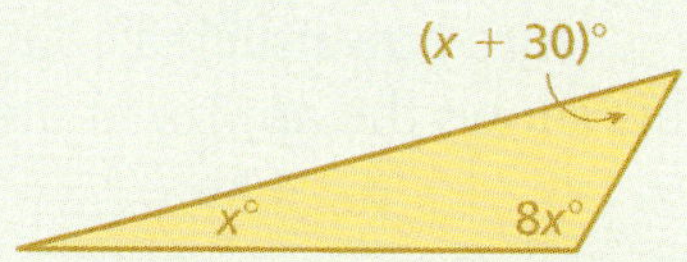

   $$x + 8x + x + 30 = 180$$
   $$10x = 150$$
   $$x = 15$$

   F. The left side of the equation should equal 360° instead of 180°.

   G. The sum of the acute angles should equal 90°.

   H. Evaluate the smallest angle when $x = 15$.

   I. Evaluate the largest angle when $x = 15$.

3. Which expression is equivalent to the expression below?

   $$2^4 2^3$$

   A. 2^{12} **C.** 48

   B. 4^7 **D.** 128

4. Your mean score for four rounds of golf was 71. Your scores on the first three rounds were 76, 70, and 70. What was your score on the fourth round?

5. The temperature in Frostbite Falls has never been above 38 degrees Fahrenheit. Let t represent the temperature, in degrees Fahrenheit. Write this as an inequality.

   F. $t < 38$ **H.** $t > 38$

   G. $t \leq 38$ **I.** $t \geq 38$

6. A bank account pays interest so that the amount in the account doubles every 10 years. The account started with \$5,000 in 1940. How much would be in the account in the year 2010?

A. \$40,000

B. \$320,000

C. \$640,000

D. \$1,280,000

7. Which expression is equivalent to $5\sqrt{5} + 2\sqrt{5}$?

F. $7\sqrt{5}$

G. $10\sqrt{5}$

H. $7\sqrt{10}$

I. $10\sqrt{10}$

8. The gross domestic product (GDP) is a way to measure how much a country produces economically in a year. The table below shows the approximate population and GDP for the United States.

Think Solve Explain

United States 2008	
Population	300 million (300,000,000)
GDP	14.4 trillion dollars (\$14,400,000,000,000)

Part A Find the GDP per person for the United States. Show your work and explain your reasoning.

Part B Write the population and GDP using scientific notation.

Part C Find the GDP per person for the United States using your answers from Part B. Write your answer in scientific notation. Show your work and explain your reasoning.

9. What is the equation of the line shown in the graph?

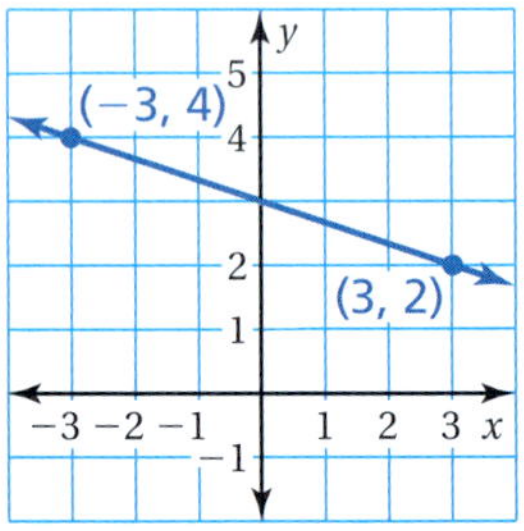

A. $y = -\frac{1}{3}x + 3$

B. $y = \frac{1}{3}x + 1$

C. $y = -3x + 3$

D. $y = 3x - \frac{1}{3}$

10. Which graph represents the inequality shown below?

$$x - 1.5 \le -1$$

F. 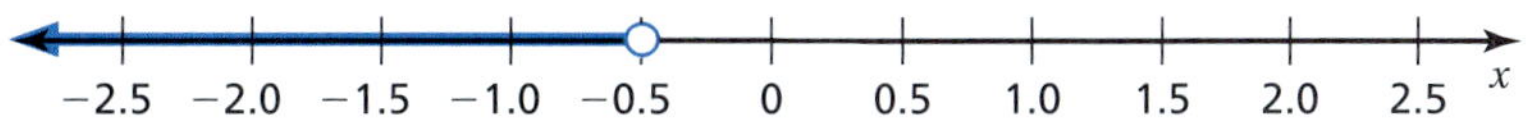

G. 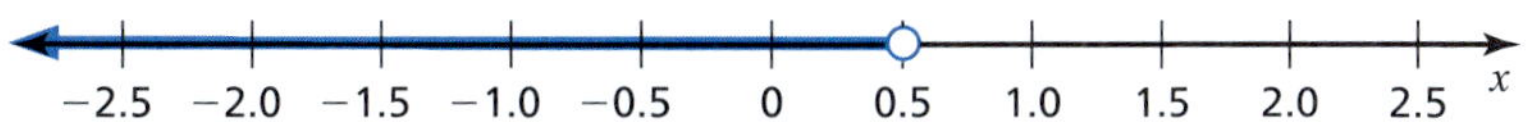

H. 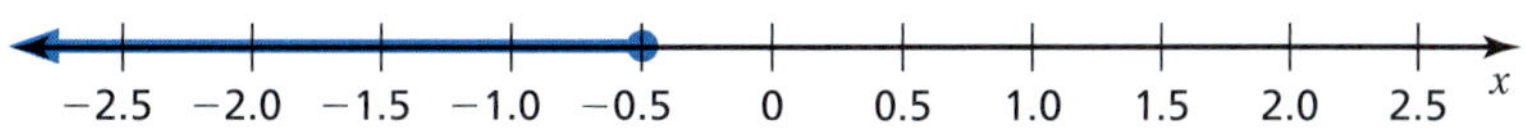

I. 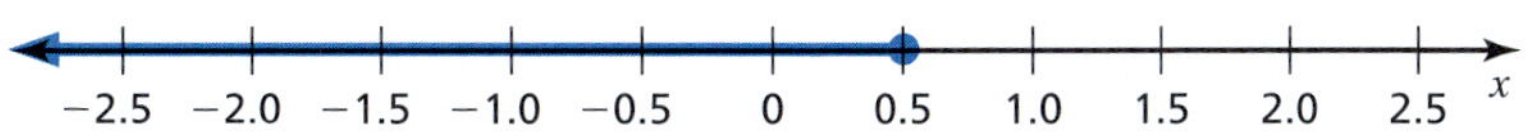

11. Find $(-2.5)^{-2}$.

12. The director of a research lab wants to present data to donors, showing how a great deal of donated money is used for research and how only a small amount of money is used for other expenses. Which type of display is best suited for showing this data?

A. box-and-whisker plot

B. circle graph

C. line graph

D. scatter plot

13. You earn \$14.75 per hour at your job. Your goal is to earn more than \$2000 next month. If you work h hours next month, which inequality represents this situation algebraically?

F. $14.75 + h > 2000$

G. $14.75 + h \ge 2000$

H. $14.75h > 2000$

I. $14.75h \ge 2000$

Appendix A
My Big Ideas Projects

"I'm nervous. The IRS dog is coming today to see if I have buried any bones that I didn't report."

"It's really simple, Descartes."

"If you simply earn interest on simple principal, that's simple interest."

"But if the simple interest earns more simple interest, that's compound interest."

Appendix B
Financial Literacy

A.1 Literature Project

Share Your Work at... My.BigIdeasMath.com

Swiss Family Robinson

STATE STANDARDS
MA.8.A.1.1–6
MA.8.G.2.1–4
MA.8.S.3.1–2
MA.8.A.4.1–2
MA.8.G.5.1
MA.8.A.6.1–4

1 Getting Started

Swiss Family Robinson is a novel about a Swiss family who was shipwrecked in the East Indies. The story was written by Johann David Wyss, and was first published in 1812.

Essential Question How does the knowledge of mathematics provide you and your family with survival tools?

Read *Swiss Family Robinson*. As you read the exciting adventures, think about the mathematics the family knew and used to survive.

Sample: The tree house built by the family was accessed by a long rope ladder. The ladder was about 30 feet long with a rung every 10 inches. To make the ladder, the family had to plan how many rungs were needed. They decided the number was $1 + 12(30) \div 10$. Why?

2 Things to Include

- Suppose you lived in the 18th century. Plan a trip from Switzerland to Australia. Describe your route. Estimate the length of the route and the number of miles you will travel each day. About how many days will the entire trip take?
- Suppose that your family is shipwrecked on an island that has no other people. What do you need to do to survive? What types of tools do you hope to salvage from the ship? Describe how mathematics could help you survive.
- Suppose that you are the oldest of four children in a shipwrecked family. Your parents have made you responsible for the education of your younger siblings. What type of mathematics would you teach them? Explain your reasoning.

3 Things to Remember

- You can download each part of the book at *BigIdeasMath.com*.
- Add your own illustrations to your project.
- Organize your math stories in a folder, and think of a title for your report.

A.2 History Project

Mathematics in Ancient China

Share Your Work at... My.BigIdeasMath.com

STATE STANDARDS
MA.8.A.1.1–6
MA.8.G.2.1–4
MA.8.S.3.1–2
MA.8.A.4.1–2
MA.8.G.5.1
MA.8.A.6.1–4

1 Getting Started

Mathematics was developed in China independently of the mathematics that was developed in Europe and the Middle East. For example, the Pythagorean Theorem and the computation of pi were used in China prior to the time when China and Europe began communicating with each other.

Essential Question How have tools and knowledge from the past influenced modern day mathematics?

Sample: Here are the names and symbols that were used in ancient China to represent the digits from 1 through 10.

1	yi	一
2	er	二
3	san	三
4	si	四
5	wu	五
6	liu	六
7	qi	七
8	ba	八
9	jiu	九
10	shi	十

Life-size Terra-cotta Warriors

A Chinese Abacus

2 Things to Include

- Describe the ancient Chinese book *The Nine Chapters on the Mathematical Art* (c. 100 B.C.). What types of mathematics are contained in this book?
- How did the ancient Chinese use the abacus to add and subtract numbers? How is the abacus related to base 10?
- How did the ancient Chinese use mathematics to build large structures, such as the Great Wall and the Forbidden City?
- How did the ancient Chinese write numbers that are greater than 10?

Ancient Chinese Teapot

- Describe how the ancient Chinese used mathematics. How does this compare with the ways in which mathematics is used today?

The Great Wall of China

3 Things to Remember

- Add your own illustrations to your project.
- Organize your math stories in a folder, and think of a title for your report.

Chinese Guardian Fu Lions

A.3 Art Project

STATE STANDARDS
MA.8.A.1.1–6
MA.8.G.2.1–4
MA.8.S.3.1–2
MA.8.A.4.1–2
MA.8.G.5.1
MA.8.A.6.1–4

Share Your Work at... My.BigIdeasMath.com

Polyhedra in Art

1 Getting Started

Polyhedra is the plural of *polyhedron*. Polyhedra have been used in art for many centuries, in cultures all over the world.

Essential Question Do polyhedra influence the design of games and architecture?

Some of the most famous polyhedra are the five Platonic solids. They have faces that are congruent, regular, convex polygons.

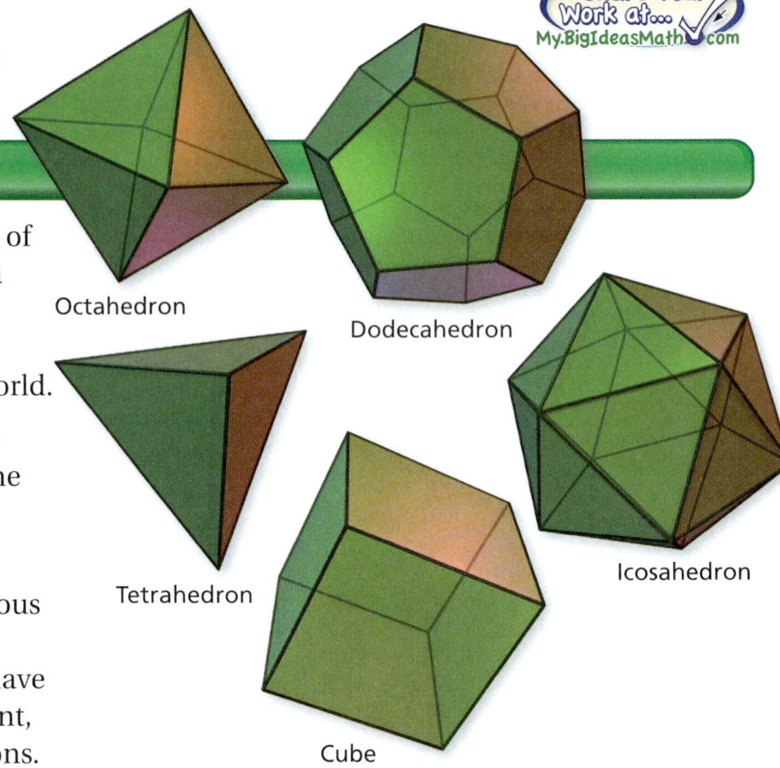

Sample: Three of the Platonic solids have faces that are equilateral triangles. One of these is the octahedron. Here is a net for an octahedron.

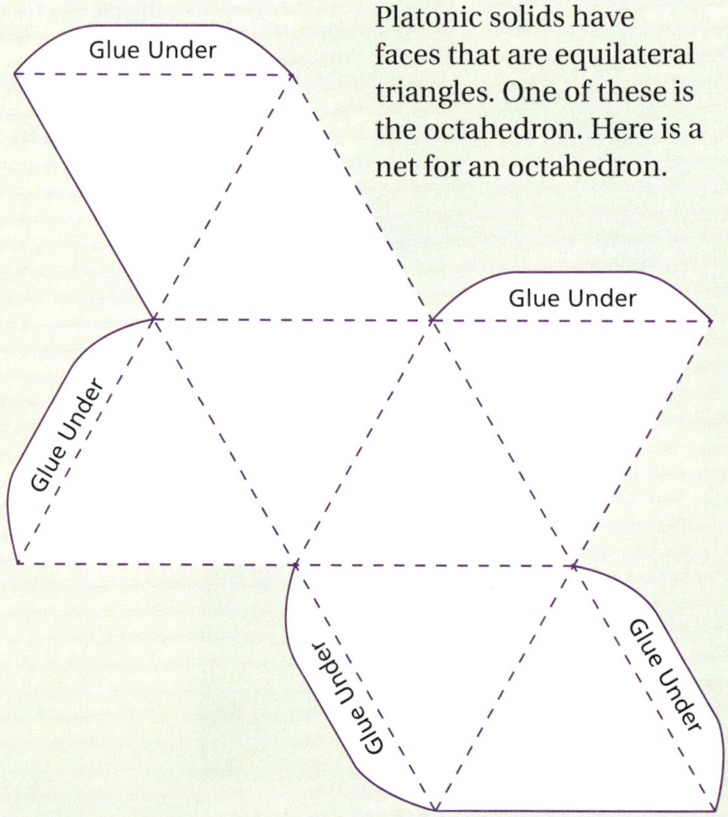

Mosaic by Paolo Uccello, 1430 A.D.

2 Things to Include

- Explain why the platonic solids are sometimes referred to as the cosmic figures.
- Draw a net for an icosahedron or a dodecahedron. Cut out the net and fold it to form the polyhedron.
- Describe the 13 polyhedra that are called Archimedean solids. What is the definition of this category of polyhedra? Draw a net for one of them. Then cut out the net and fold it to form the polyhedron.
- Find examples of polyhedra in games and architecture.

Faceted Cut Gem

Origami Polyhedron

3 Things to Remember

- Add your own illustrations or paper creations to your project.
- Organize your report in a folder, and think of a title for your report.

Concrete Tetrahedrons by Ocean

Bulatov Sculpture

A.4 Science Project

Our Solar System

STATE STANDARDS
MA.8.A.1.1–6
MA.8.G.2.1–4
MA.8.S.3.1–2
MA.8.A.4.1–2
MA.8.G.5.1
MA.8.A.6.1–4

Share Your Work at... MyBigIdeasMath.com

1 Getting Started

Our solar system consists of four inner planets, four outer planets, dwarf planets such as Pluto, several moons, and many asteroids and comets.

Essential Question How do the characteristics of a planet influence whether or not it can sustain life?

Sample: The average temperatures of the eight planets in our solar system are shown in the graph.

The average temperature tends to drop as the distance between the Sun and the planet increases.

An exception to this rule is Venus. It has a higher average temperature than Mercury, even though Mercury is closer to the Sun.

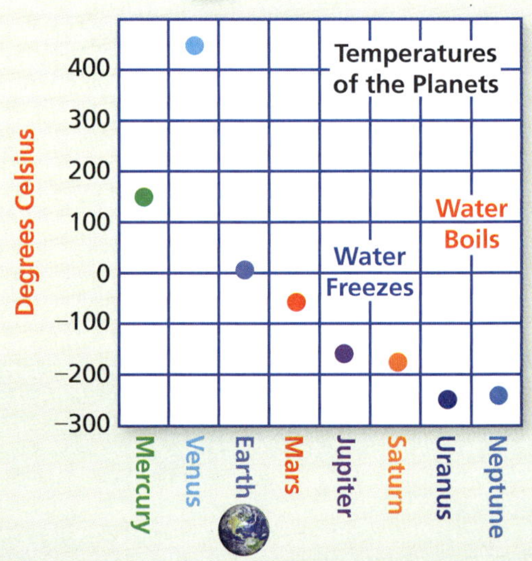

2 Things to Include

- Compare the masses of the planets.
- Compare the gravitational forces of the planets.
- How long is a "day" on each planet? Why?
- How long is a "year" on each planet? Why?
- Which planets or moons have humans explored?
- Which planets or moons could support human life? Explain your reasoning.

Mars Rover

3 Things to Remember

- Add your own drawings or photographs to your report. You can download photographs of the solar system and space travel at *NASA.gov*.
- Organize your report in a folder, and think of a title for your report.

Hubble Image of Space

Hubble Spacecraft

B.1 Simple and Compound Interest

STATE STANDARDS
MA.8.A.1.1
MA.8.A.1.6

Essential Question How can you find the balance in an account that earns simple interest or compound interest?

1 ACTIVITY: Comparing Simple and Compound Interest

Work with a partner. Interest that is calculated only on principal is simple interest. Interest that is calculated on principal *and* previously earned interest is compound interest.

You deposit $1000 in a savings account that earns 6% interest per year.

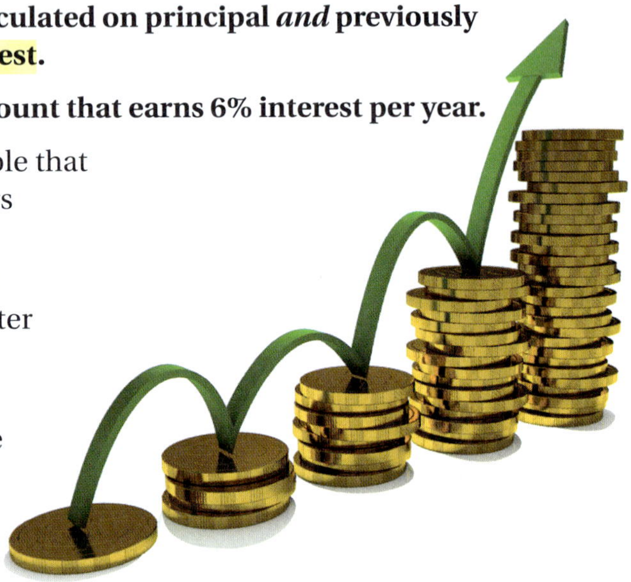

a. Copy and complete the first table that shows the balance after 10 years with simple interest.

b. Copy and complete the second table that shows the balance after 10 years with interest that is compounded annually.

c. Which type of interest gives the greater balance?

$I = Prt$
$= 1000(0.06)(1)$

Simple Interest			
t	Principal	Annual Interest	Balance at End of Year
1	$1000.00	$60.00	$1060.00
2	$1000.00	$60.00	$1120.00
3			
4			
5			
6			
7			
8			
9			
10			

Compound Interest			
t	Principal and Interest	Annual Interest	Balance at End of Year
1	$1000.00	$60.00	$1060.00
2	$1060.00	$63.60	$1123.60
3			
4			
5			
6			
7			
8			
9			
10			

2 ACTIVITY: Comparing Simple and Compound Interest

Work with a partner.

a. Graph the end-of-year balances for each type of interest in Activity 1.

b. Which graph is linear? Explain your reasoning.

c. For the linear graph, write a linear function that represents the balance after t years.

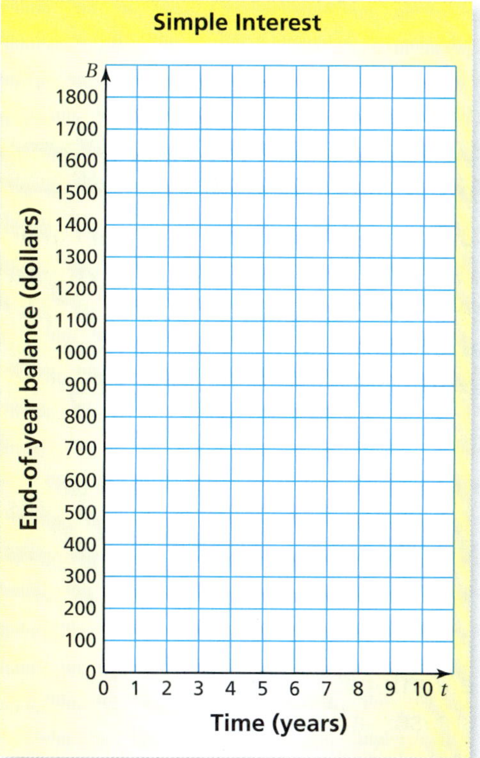

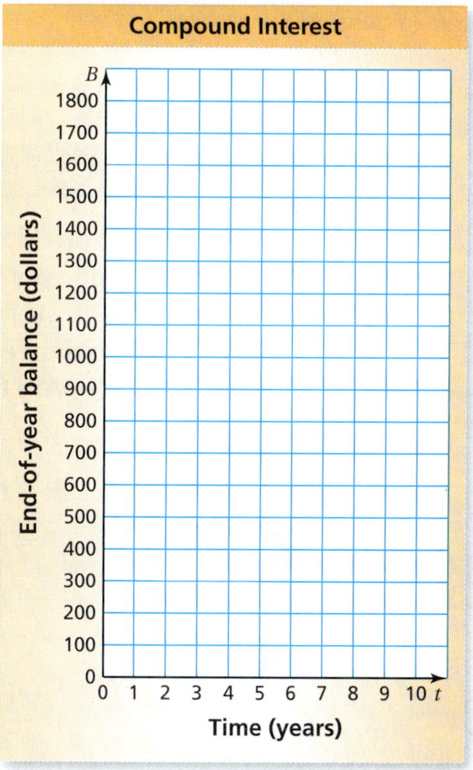

What Is Your Answer?

3. **IN YOUR OWN WORDS** How can you find the balance in an account that earns simple interest or compound interest?

4. Use what you learned in Activity 2. About how many years will it take for the balance to double with simple interest? with compound interest?

Practice Use what you learned about simple and compound interest to complete Exercise 3 on page A14.

Check It Out
Lesson Tutorials
BigIdeasMath.com

Key Vocabulary
compound interest, *p. A13*

Key Idea

Balance in a Simple Interest Account

The balance B of an account that earns simple interest is

$$B = P(1 + rt).$$

Balance: B
Principal: P
Annual interest rate (in decimal form): r
Time (in years): t

Remember

Interest is money paid or earned for the use of money. The *principal* is the amount of money borrowed or deposited.

EXAMPLE 1 Comparing Simple Interest Accounts

You deposit \$275 in a savings account that earns 4% simple interest per year. Your friend deposits \$300 in a savings account that earns 2% simple interest per year. (a) Write and graph two equations for the balance B in each account after t years. Describe the equations. (b) Are the account balances ever equal? Explain.

a. ***You***

You		Your Friend
$B = P(1 + rt)$	Write balance formula.	$B = P(1 + rt)$
$= 275(1 + 0.04t)$	Substitute values.	$= 300(1 + 0.02t)$
$= 275 + 11t$	Simplify.	$= 300 + 6t$

Reading

Notice that both equations are of the form
$y = mx + b$.

Graph the equations. Your account balance increases at a constant rate of \$11 per year. Your friend's account balance increases at a constant rate of \$6 per year.

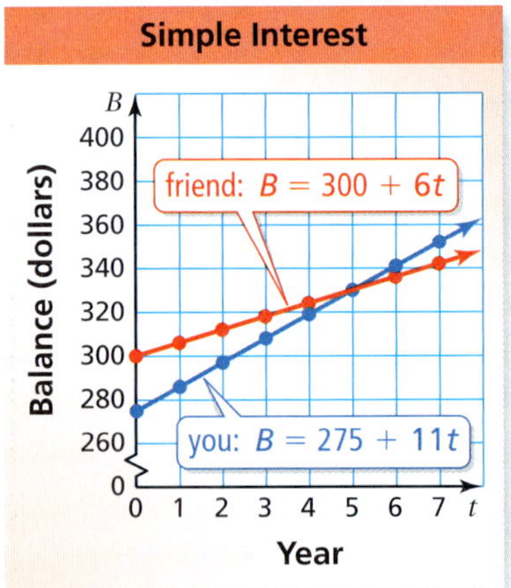

b. Yes, the graphs appear to intersect when $t = 5$. So, after 5 years, the accounts are equal.

On Your Own

Now You're Ready
Exercise 4

1. **WHAT IF?** In Example 1, your friend deposits \$600 in the account. Are the account balances ever equal? Explain.

Multi-Language Glossary at BigIdeasMath.com.

Compound interest is interest earned on the principal *and* on the previously earned interest.

EXAMPLE 2 Calculating Compound Interest

You deposit $400 in an account that earns 3.5% interest compounded annually. What is the account balance after 2 years?

Study Tip

The simple interest formula $I = Prt$ can be used to find the balance in an account that earns compound interest.

Interest for 1st year: $I = Prt$ — Write simple interest formula.

$= 400(0.035)(1)$ — Substitute.

$= 14$ — Simplify.

The account balance after 1 year is \$400 + \$14 = \$414.

Interest for 2nd year: $I = Prt$ — Write simple interest formula.

$= 414(0.035)(1)$ — Substitute. Use 414 for P.

$= 14.49$ — Simplify.

The account balance after 2 years is \$414 + \$14.49 = \$428.49.

EXAMPLE 3 Comparing Simple and Compound Interest

You want to invest $8000 for 3 years. Which account should you choose?

- **Account A earns 4% simple interest per year.**
- **Account B earns 4% interest compounded annually.**

Make two tables that show the account balances for 3 years.

Account A			
Year	Principal	4% Simple Interest	Balance at End of Year
1	\$8000	\$320	\$8320
2	\$8000	\$320	\$8640
3	\$8000	\$320	\$8960

Account B			
Year	Principal and Interest	4% Compound Interest	Balance at End of Year
1	\$8000.00	\$320.00	\$8320.00
2	\$8320.00	\$332.80	\$8652.80
3	\$8652.80	\$346.11	\$8998.91

Account B earns \$8998.91 − \$8960 = \$38.91 more than Account A after 3 years. So, you should choose Account B.

On Your Own

Now You're Ready
Exercises 5, 6, and 8

2. In Example 2, what is the account balance after 5 years?
3. In Example 3, how much more does Account B earn than Account A after 2 more years?

B.1 Exercises

Check It Out
Help with Homework
BigIdeasMath.com

Vocabulary and Concept Check

1. **VOCABULARY** What type of interest is money earned only on the principal?

2. **WRITING** How are simple interest and compound interest similar? How are they different?

Practice and Problem Solving

3. You deposit \$500 in a savings account that earns 3% interest per year.

   a. Copy and complete the tables that show the balances after 5 years with simple interest and compound interest.

   b. Which type of interest gives the greater balance?

Simple Interest			
t	Principal	Annual Interest	Balance at End of Year
1	\$500	\$15	\$515
2	\$500	\$15	\$530
3			
4			
5			

Compound Interest			
t	Principal and Interest	Annual Interest	Balance at End of Year
1	\$500	\$15.00	\$515.00
2	\$515	\$15.45	\$530.45
3			
4			
5			

1 4. You deposit \$600 in a savings account that earns 4% simple interest per year. Your friend deposits \$400 in a savings account that earns 5% simple interest per year.

   a. Write and graph two equations for the balance B in each account after t years.

   b. Are the account balances ever equal? Explain.

2 5. You deposit \$1200 in a savings account that earns 5.4% interest compounded annually. What is the account balance after 3 years?

6. You deposit \$300 in a savings account that earns 7.2% interest compounded annually. What is the account balance after 2 years?

7. **ERROR ANALYSIS** Describe and correct the error in finding the balance of the simple interest account after two years.

✗ Principal: \$700 Rate: 3% per year

t	Principal	Annual Interest	Balance at End of Year
1	\$700.00	\$21.00	\$721.00
2	\$721.00	\$21.63	\$742.63

③ 8. **INVESTMENT** The owners of a company want to invest \$12,000 for 4 years.

a. Which account should they choose? Explain.
- Account A earns 5% simple interest per year.
- Account B earns 5% interest compounded annually.

b. How much more do they earn by choosing the better account?

9. **GAMES** After 7 years at 3% simple interest per year, your savings account earns \$63.

a. What is the principal?

b. How much money do you have left after buying the video game system?

10. **DIRT BIKE** Your friend borrows \$1050 from you to buy a new dirt bike. Your friend pays you back the principal plus 7.25% simple interest per year in 3 years. How much money do you earn?

11. **Critical Thinking** You want to deposit \$600 in a savings account. Account A earns 3.5% simple interest per year. Account B earns 3.3% interest compounded annually.

a. Which account should you choose if you invest your money for 3 years? 6 years? Explain.

b. After how many years is the balance in Account B greater than the balance in Account A?

c. After how many years is the difference between the account balances greater than \$5?

Fair Game Review What you learned in previous grades & lessons

Simplify the expression. *(Section 9.2)*

12. $12^2 \cdot 12^5$ 13. $(1.5x^2)^3$ 14. $3p^4 \cdot 2.4p^8$ 15. $(5.4d^9)^2$

16. **MULTIPLE CHOICE** The ratio of the number of students with Attention Deficit Hyperactivity Disorder to the number of students in an entire school is 1 : 20. What percent of students in the school have Attention Deficit Hyperactivity Disorder? *(Skills Review Handbook)*

Ⓐ 5% Ⓑ 8%

Ⓒ 20% Ⓓ 25%

B.2 Formula for Compound Interest

STATE STANDARDS
MA.8.A.6.3
MA.8.A.6.4

Essential Question How can you use a formula to find the balance in an account that earns compound interest?

1 ACTIVITY: Finding a Formula for Compound Interest

Work with a partner. You deposit \$1000 in a savings account that earns 6% interest compounded annually. Explain why parts (a)–(d) in the table are true.

a. Sample: Balance at End of Year = Principal + Annual Interest

$$= \$1000 + \$1000(0.06)$$

$$= \$1000(1 + 0.06)$$

$$= \$1000(1.06)$$

t	Principal and Interest	Annual Interest	Balance at End of Year
1	\$1000.00	\$1000(0.06)	**a.** $B = \$1000(1.06)$
2	\$1000(1.06)	**b.** \$1000(1.06)(0.06)	**c.** $B = \$1000(1.06)^2$
3	$\$1000(1.06)^2$	$\$1000(1.06)^2(0.06)$	$B = \$1000(1.06)^3$
4	$\$1000(1.06)^3$	$\$1000(1.06)^3(0.06)$	$B = \$1000(1.06)^4$
5	$\$1000(1.06)^4$	$\$1000(1.06)^4(0.06)$	**d.** $B = \$1000(1.06)^5$
6	$\$1000(1.06)^5$	$\$1000(1.06)^5(0.06)$	$B = \$1000(1.06)^6$
7	$\$1000(1.06)^6$	$\$1000(1.06)^6(0.06)$	$B = \$1000(1.06)^7$
8	$\$1000(1.06)^7$	$\$1000(1.06)^7(0.06)$	$B = \$1000(1.06)^8$
9	$\$1000(1.06)^8$	$\$1000(1.06)^8(0.06)$	$B = \$1000(1.06)^9$
10	$\$1000(1.06)^9$	$\$1000(1.06)^9(0.06)$	$B = \$1000(1.06)^{10}$

e. Use the pattern in the table to find the balance after 20 years.

2 ACTIVITY: Writing a Formula

Work with a partner. Use the pattern in Activity 1 to write a formula for the balance in an account that earns interest compounded annually after *t* years.

P = Principal (initial deposit)

r = Annual interest rate (in decimal form)

t = Time (in years)

B = Balance after t years

B =

3 ACTIVITY: A Penny Saved

Work with a partner. In his will, Benjamin Franklin gave $4000 to the state of Pennsylvania. He instructed that the money be invested for 200 years. After 200 years, the money should be used to do good. Franklin died in 1790. In 1990, his gift had grown to over $2 million.

"A penny saved is a penny earned." Benjamin Franklin

Use your formula from Activity 2 to copy and complete the table. Then approximate the annual interest rate that Benjamin Franklin's gift earned.

Annual Interest Rate	Balance After 200 years
3.0%	
3.1%	
3.2%	
3.3%	
3.4%	

What Is Your Answer?

4. **IN YOUR OWN WORDS** How can you use a formula to find the balance in an account that earns compound interest? How much would Benjamin Franklin's account have been worth if it had earned 6% interest each year?

Practice Use what you learned about the formula for compound interest to complete Exercises 3 and 4 on page A20.

B.2 Lesson

Check It Out
Lesson Tutorials
BigIdeasMath.com

Key Idea

Interest Compounded Annually

$$B = P(1 + r)^t$$

Balance (B); Principal (P); Annual interest rate (in decimal form) (r); Time (in years) (t)

EXAMPLE 1 Finding an Account Balance

You deposit \$1000 in a savings account. The account earns 3% interest compounded annually. (a) What is the balance after 2 years? (b) What is the interest earned?

a. $B = P(1 + r)^t$ — Write compound interest formula.

$= 1000(1 + 0.03)^2$ — Substitute.

$= 1000(1.03)^2$ — Add.

$= 1060.9$ — Simplify.

The balance is \$1060.90 after 2 years.

b. The interest earned is the difference of the balance and the principal.

So, the interest earned is \$1060.90 − \$1000.00 = \$60.90.

Study Tip

For small values of t, you can use the simple interest formula to estimate the interest.

$I = Prt$

$= 1000(0.03)(2)$

$= 60$

EXAMPLE 2 Finding a Principal

An account opened on January 1, 2010 earns 5% interest compounded annually. The only change in the account is interest earned. What is the principal?

Descartes the Cat
9 Lives Lane
Meow Town, FL

Account No. 3-141592654

Date	Description	Deposits	Balance
01/01/2012	Previous Balance		\$8820.00
12/31/2012	Interest Earned	\$441.00	\$9261.00

The balance is \$9261.00 after 3 years. Use the formula to find the principal.

$B = P(1 + r)^t$ — Write formula.

$9261.00 = P(1 + 0.05)^3$ — Substitute.

$9261.00 = P(1.05)^3$ — Add.

$8000 = P$ — Solve for P.

The principal is \$8000.

On Your Own

Now You're Ready
Exercises 5 and 6

1. You deposit \$2500 in a savings account. The account earns 2% interest compounded annually. (a) What is the balance after 3 years? (b) What is the interest earned?

2. An account earns 1% interest compounded annually. The balance after 2 years is \$10,201. What is the principal?

EXAMPLE 3 Choosing an Account

A business owner wants to invest \$10,000 for 10 years. Which account should the business owner choose? Explain.

Account A

Account B

Account A

$B = P(1 + rt)$	Write simple interest balance formula.
$= 10{,}000(1 + 0.05(10))$	Substitute.
$= 15{,}000$	Simplify.

The balance after 10 years is \$15,000.

Account B

$B = P(1 + r)^t$	Write compound interest formula.
$= 10{,}000(1 + 0.05)^{10}$	Substitute.
$= 10{,}000(1.05)^{10}$	Add.
$\approx 16{,}288.95$	Use a calculator.

The balance after 10 years is \$16,288.95.

∴ The business owner should choose Account B because its balance is greater than the balance of Account A after 10 years.

On Your Own

Now You're Ready
Exercises 8 and 9

3. **WHAT IF?** In Example 3, which account should the business owner choose if the money is being invested for 5 years? Explain.

B.2 Exercises

Check It Out
Help with Homework
BigIdeasMath.com

Vocabulary and Concept Check

1. **WRITING** For a given interest rate, how does compound interest differ from simple interest?

2. **DIFFERENT WORDS, SAME QUESTION** You deposit \$1000 in an account that earns 4% interest compounded annually. Which question is different? Find "both" answers.

   What is the balance after 1 year?

   What is the interest earned after 1 year?

   By how much money did your account increase after 1 year?

   What is the difference of the balance and the principal after 1 year?

Practice and Problem Solving

Copy and complete the table for an account that earns interest compounded annually.

3. Principal: \$500

Annual Interest Rate	Balance After 5 Years
3.0%	
3.5%	
4.0%	
4.5%	

4. Principal: \$1000

Annual Interest Rate	Balance After 10 Years
3.5%	
4.0%	
4.5%	
5.0%	

1 5. You deposit \$600 in a savings account. The account earns 3.5% interest compounded annually.

   a. What is the balance after 4 years?

   b. What is the interest earned?

2 6. An account earns 1.5% interest compounded annually. The balance after 2 years is \$8241.80. What is the principal?

7. **ERROR ANALYSIS** Describe and correct the error in finding the balance of a savings account with interest compounded annually.

   ✗

   Principal: \$2000
   Annual Interest Rate: 4%
   Time: 2 years

   $$B = P(1 + r)^t$$
   $$= 2000(1 + 4)^2$$
   $$= 2000(5)^2$$
   $$= 50{,}000$$

A business owner wants to invest \$7000 for 8 years. Which account should the business owner choose? Explain.

3 **8.**

Account A

Account B

9.

Account A

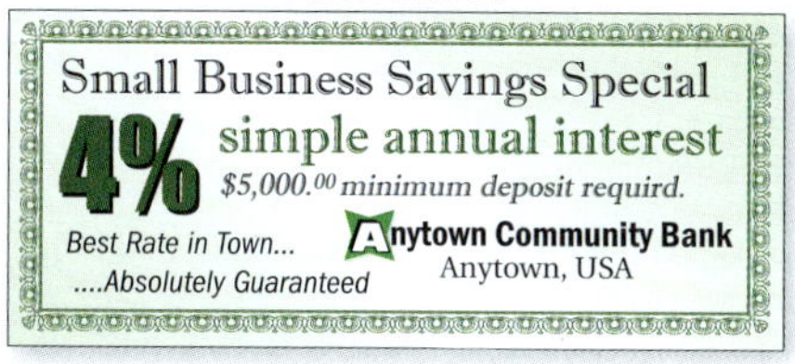

Account B

10. **COLLEGE** You want to save \$13,000 in 4 years for college tuition. About how much money should you deposit in a savings account that earns 5% interest compounded annually?

11. **CRITICAL THINKING** Interest can be compounded more than once per year. The formula for interest compounded more than once per year is $B = P\left(1 + \frac{r}{n}\right)^{nt}$, where n is the number of times the balance is compounded each year.

a. You deposit \$1000 in a savings account. The account earns 3% interest compounded 4 times per year. What is the balance after 2 years?

b. What happens to the balance after 2 years as n increases?

12. Critical Thinking Two accounts have the same interest rate, but one account earns interest compounded annually and the other earns interest compounded monthly. Which account earns more interest? Explain.

Fair Game Review What you learned in previous grades & lessons

Simplify the expression. *(Skills Review Handbook)*

13. $465.25 \cdot 60$ **14.** $\frac{1}{12} \cdot 0.036$ **15.** $\frac{4.8}{12}$ **16.** $268.68 \cdot 72$

17. **MULTIPLE CHOICE** Which graph represents the solution of the inequality $-\frac{c}{4} \geq -5$? *(Section 8.3)*

Ⓐ 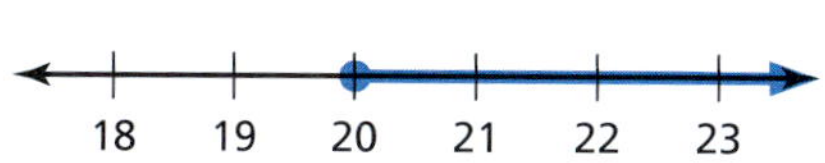

Ⓑ 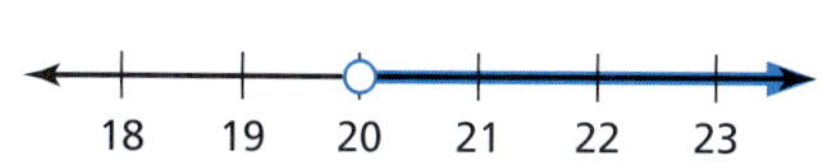

Ⓒ 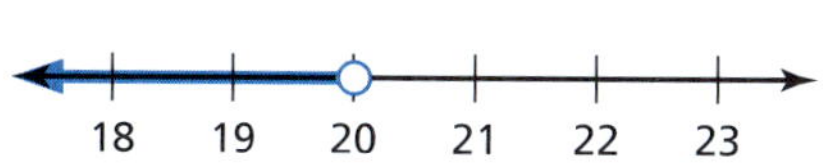

Ⓓ 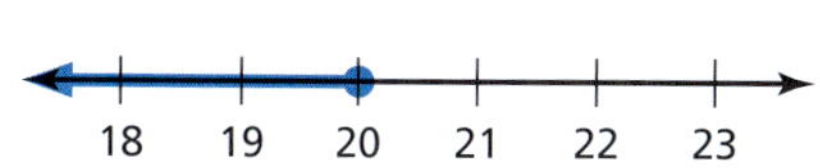

B.3 Installment Loans

STATE STANDARDS
MA.7.A.1.2
MA.8.A.6.4

Essential Question How can you find the total amount of interest paid on an installment loan?

1 ACTIVITY: Monthly Loan Payments

Work with a partner.

You borrow $1000 for 6 months. Your annual interest rate is 12%, which means that your monthly interest rate is

$$\text{Monthly Interest Rate} = \frac{\text{Annual Interest Rate}}{\text{Months in a Year}} = \frac{0.12}{12} = 0.01.$$

LOAN STATEMENT	
Amount of Loan	$1000.00
Monthly Payment	$172.55
Annual Interest Rate	12.00%
Monthly Interest Rate	1.00%
Payment Due Date	1st of Month
Term	6 Months

a. Copy and complete the table.

t	Balance Before Payment	Monthly Payment	Monthly Interest	Balance After Payment
1	$1000.00	$172.55	$1000.00(0.01) = $10.00	$1000.00 − 172.55 + 10.00 = $837.45
2	$837.45	$172.55	$837.45(0.01) = $8.37	$837.45 − 172.55 + 8.37 = $673.27
3		$172.55		
4		$172.55		
5		$172.55		
6		$172.53		

b. How much interest do you pay during the 6 months?

c. How is your answer to part (b) related to the expression 6(172.55) − 1000?

2 ACTIVITY: Finding the Total Interest Paid

Work with a partner. Use the result of Activity 1.

a. **CAR LOAN** You borrow $18,000 for a car. Make a table that shows about how much interest you pay for each annual interest rate and term.

Monthly Payments					
Annual Interest Rate / Term	6%	8%	10%	12%	14%
36 months	$547.59	$564.05	$580.81	$597.86	$615.20
48 months	$422.73	$439.43	$456.53	$474.01	$491.88
60 months	$347.99	$364.98	$382.45	$400.40	$418.83
72 months	$298.31	$315.60	$333.47	$351.90	$370.90

b. **HOME MORTGAGE** You borrow $180,000 to buy a house. Make a table that shows about how much interest you pay for each annual interest rate and term.

Monthly Payments					
Annual Interest Rate / Term	4%	6%	8%	10%	12%
20 years	$1090.76	$1289.58	$1505.59	$1737.04	$1981.96
30 years	$859.35	$1079.19	$1320.78	$1579.63	$1851.50
40 years	$752.29	$990.38	$1251.56	$1528.46	$1815.30
50 years	$694.27	$947.53	$1222.69	$1510.39	$1804.61

What Is Your Answer?

3. **IN YOUR OWN WORDS** How can you find the total amount of interest paid on an installment loan? Why is it important to search for low interest rates and short terms when you are applying for an installment loan?

Practice Use what you learned about installment loans to complete Exercise 3 on page A26.

B.3 Lesson

Check It Out
Lesson Tutorials
BigIdeasMath.com

Key Vocabulary
loan, *p. A24*
term, *p. A24*
installment loan, *p. A24*
mortgage, *p. A25*

A **loan** is money borrowed for a specified amount of time, called the **term** of the loan. The lender may add interest to the principal, increasing the total amount due. When the total amount due is paid back in equal increments over the term of the loan, it is called an **installment loan**.

EXAMPLE 1 Paying Interest on a Loan

A student borrows money for college. The loan summary is shown below. How much interest is paid on the loan?

LOAN AMOUNT	\$1450.00
MONTHLY PAYMENT AMOUNT	\$250.20
ANNUAL INTEREST RATE	12% annual interest
TERM	6 months

Study Tip
Due to rounding, the last payment of an installment loan may be a few cents more or less than the regular monthly payment.

The annual interest rate is 12%. So, the monthly interest rate is $\frac{0.12}{12} = 0.01$.

Use a table to track the payment and the balance each month.

Balance Before Payment	Monthly Payment	Monthly Interest	Balance After Payment
\$1450.00	\$250.20	\$1450.00(0.01) = \$14.50	\$1450.00 − 250.20 + 14.50 = \$1214.30
\$1214.30	\$250.20	\$1214.30(0.01) ≈ \$12.14	\$1214.30 − 250.20 + 12.14 = \$976.24
\$976.24	\$250.20	\$976.24(0.01) ≈ \$9.76	\$976.24 − 250.20 + 9.76 = \$735.80
\$735.80	\$250.20	\$735.80(0.01) ≈ \$7.36	\$735.80 − 250.20 + 7.36 = \$492.96
\$492.96	\$250.20	\$492.96(0.01) ≈ \$4.93	\$492.96 − 250.20 + 4.93 = \$247.69
\$247.69	\$250.17	\$247.69(0.01) ≈ \$2.48	\$247.69 − 250.17 + 2.48 = \$0.00

Find the sum of the monthly interest totals in the table.

$$\$14.50 + \$12.14 + \$9.76 + \$7.36 + \$4.93 + \$2.48 = \$51.17$$

So, \$51.17 in interest is paid on the loan.

On Your Own

Now You're Ready
Exercises 4–7

1. In Example 1, the loan amount is \$1750 and the monthly payment amount is \$301.96. How much interest is paid on the loan?

A **mortgage** is an installment loan used to buy a house. The term of a mortgage is usually greater than the terms of other types of loans.

EXAMPLE 2 Paying a Mortgage

The table shows loan options for a $100,000 mortgage. (a) How much interest is paid on each option? (b) Discuss an advantage and a disadvantage of the 15-year mortgage.

Monthly Payments

Term \ Annual Interest Rate	5%	7%
15 years	$790.79	$898.83
20 years	$659.96	$775.30
30 years	$536.82	$665.30

a. Find the number of monthly payments for each term.

15 years	*20 years*	*30 years*
$15 \cdot 12 = 180$	$20 \cdot 12 = 240$	$30 \cdot 12 = 360$

Use a table to find the interest I paid on each option. The total amount paid for the mortgage is the product of the monthly payment and the number of payments. The interest paid is the difference of this product and the principal.

Total Interest Paid

Term \ Annual Interest Rate	5%	7%
15 years	$I = 180(790.79) - 100{,}000$ $= \$42{,}342.20$	$I = 180(898.83) - 100{,}000$ $= \$61{,}789.40$
20 years	$I = 240(659.96) - 100{,}000$ $= \$58{,}390.40$	$I = 240(775.30) - 100{,}000$ $= \$86{,}072.00$
30 years	$I = 360(536.82) - 100{,}000$ $= \$93{,}255.20$	$I = 360(665.30) - 100{,}000$ $= \$139{,}508.00$

b. You pay less interest with the 15-year mortgage than with the 20-year and 30-year mortgages, but the monthly payment for the 15-year mortgage is higher than the monthly payments for the 20-year and 30-year mortgages.

On Your Own

Now You're Ready
Exercises 9 and 10

2. The principal on a home mortgage is $125,000, the monthly payment is $749.44, and the term is 30 years. How much interest is paid on the loan?

B.3 Exercises

Check It Out
Help with Homework
BigIdeasMath.com

Vocabulary and Concept Check

1. **VOCABULARY** Compare the principal of a loan and the total amount due.

2. **REASONING** As the term of a loan increases, what happens to the monthly payment? What happens to the total amount of interest paid?

Practice and Problem Solving

3. Copy and complete the table. How much interest do you pay during the 5 months? The annual interest rate is 12%

Balance Before Payment	Monthly Payment	Monthly Interest	Balance After Payment
\$1135.00	\$233.86		
	\$233.86		
	\$233.86		
	\$233.86		
	\$233.84		

A loan summary is shown. How much interest is paid on the loan?

1 4.

LOAN AMOUNT	\$1020.00
MONTHLY PAYMENT	\$176.00
ANNUAL INTEREST RATE	12% annual interest
TERM	6 months

5.

LOAN AMOUNT	\$1635.00
MONTHLY PAYMENT	\$277.29
ANNUAL INTEREST RATE	6% annual interest
TERM	6 months

6.

LOAN AMOUNT	\$1500.00
MONTHLY PAYMENT	\$129.10
ANNUAL INTEREST RATE	6% annual interest
TERM	12 months

7.

LOAN AMOUNT	\$920.00
MONTHLY PAYMENT	\$81.74
ANNUAL INTEREST RATE	12% annual interest
TERM	12 months

8. **ERROR ANALYSIS** Describe and correct the error in finding the balance after payment.

✗
Balance Before Payment: \$1000
Monthly Payment: \$172.55
Monthly Interest: \$10
Balance After Payment:
\$1000 − \$172.55 − \$10 = \$817.45

The table shows loan options for a mortgage. How much interest is paid on each option?

9.

Monthly Payments $120,000 Mortgage		
Term \ Annual Interest Rate	5%	7%
20 years	$791.95	$930.36
25 years	$701.51	$848.14
30 years	$644.19	$798.36

10.

Monthly Payments $135,000 Mortgage		
Term \ Annual Interest Rate	4%	6%
20 years	$818.07	$967.18
25 years	$712.58	$869.81
30 years	$644.51	$809.39

11. **TUITION** You pay off a $15,000 college tuition loan in 10 years. The monthly payment is $166.53. What percent of the total amount paid is the interest on the loan?

12. **CONDOMINIUM** A luxury condominium in Miami Beach costs $13 million. The monthly payment for a 30-year mortgage on the condominium is $77,941.57. What percent of the total amount paid is the principal of the loan?

13. **Critical Thinking** You have $5000 in a savings account that earns 3.5% interest compounded annually. You need $16,000 to buy a car. What should you do? Explain.

- Let your savings earn interest for 3 years and get Loan A.
- Use your savings to get Loan B.

Loan A

APPLY NOW!
$16,000 for 36 months
7% annual interest rate
$494.03 monthly payment

Loan B

New Loan Special
$11,000 for 24 months
7% annual interest rate
$492.50 monthly payment

Fair Game Review What you learned in previous grades & lessons

Find the sum or difference. *(Skills Review Handbook)*

14. 666.35 + 485.21

15. 872.38 − 900.44

16. **MULTIPLE CHOICE** Which data display is most appropriate for the information given in the table? *(Section 7.4)*

Ⓐ box-and-whisker plot
Ⓑ stem-and-leaf plot
Ⓒ circle graph
Ⓓ line graph

Year	Enrollment
2006	5042 students
2007	4921 students
2008	5198 students
2009	5003 students
2010	5117 students

B.4 Checking Accounts

STATE STANDARDS
MA.7.A.1.2
MA.8.A.6.4

Essential Question How do checking accounts work?

Share Your Work at... My.BigIdeasMath.com

1 ACTIVITY: Writing and Entering a Check

Work with a partner. You receive a bill for three pairs of jeans.

a. Copy and complete the check for the bill.

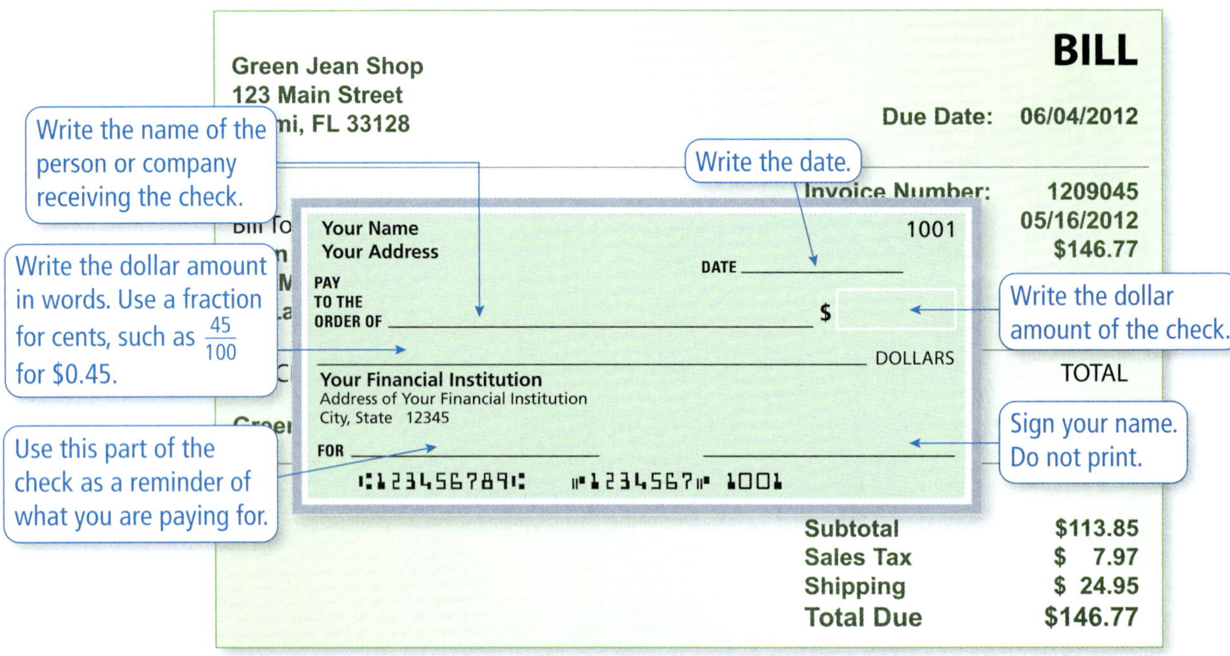

b. Enter the check in your checkbook registry below.

c. Copy and complete your checkbook registry.

Date	Check #	Transaction	Credit	Debit	Balance
		Balance Forward			100.00
5/11/12	996	Grocery Store		29.55	
5/11/12		Deposit Paycheck	482.75		
5/16/12		ATM Withdrawal		100.00	
5/17/12	997	Gym Membership		22.35	
5/17/12	998	Cell Phone Company		58.00	
5/17/12	999	Car Payment		82.66	
5/18/12		Deposit Paycheck	501.50		
5/18/12	1000	Birthday Gift		41.28	

2 ACTIVITY: Reconciling Your Account

Work with a partner. Your monthly bank statement shows that the balance of your account is $602.96. This balance does not agree with your balance. You notice that you did not enter the two transactions shown below in your checkbook registry.

- **5/16/12 ATM Fee $2.50 (debit)**
- **5/24/12 Interest $1.82 (credit)**

Enter these transactions and the check you wrote in Activity 1 in your checkbook registry. Does your balance now agree with the bank statement?

Date	Check #	Transaction	Credit	Debit	Balance
5/17/12	998	Cell Phone Company		58.00	372.85
5/17/12	999	Car Payment		82.66	
5/18/12		Deposit Paycheck	501.50		
5/18/12	1000	Birthday Gift		41.28	

3 ACTIVITY: Choosing a Checking Account

RESEARCH PROJECT **Research the checking accounts offered by three different banks. (Do this online or by visiting banks and collecting brochures.)**

Write a report that compares the different features of the checking accounts. Then decide which account is best for you.

- **Does the account pay interest?**
- **Is there a minimum initial deposit?**
- **Are the checks free?**
- **What is the overdraft fee?**
- **Is the ATM card free to use?**
- **Is online banking free?**

What Is Your Answer?

4. **IN YOUR OWN WORDS** How do checking accounts work?

Practice Use what you learned about checking accounts to complete Exercise 3 on page A32.

B.4 Lesson

Check It Out
Lesson Tutorials
BigIdeasMath.com

Key Vocabulary
checking account, *p. A30*
checkbook registry, *p. A30*
debit card, *p. A30*
bad check, *p. A31*

A **checking account** is a service provided by a financial institution (such as a bank or credit union) that allows you to deposit money in an account and then write personal checks using money from the account. A **checkbook registry** is used to record credits to an account and debits from an account.

EXAMPLE 1 Completing a Checkbook Registry

You balance your checkbook at the end of the month.

a. Complete the balance column. What is the ending balance?

Date	Check #	Transaction	Credit	Debit	Balance
		Balance Forward			275.58
9/2/12	1308	Grocery Store		70.89	204.69
9/5/12		Deposit Paycheck	402.98		607.67
9/5/12	1309	Cable Company		35.60	572.07
9/7/12	1310	Cell Phone Company		65.40	506.67
9/9/12	1311	Car Payment		239.17	267.50
9/12/12	1312	Electric Company		72.54	194.96
9/12/12		Deposit Paycheck	402.95		597.91
9/16/12	1313	Birthday Gift		25.00	572.91
9/19/12		Deposit Paycheck	402.57		975.48
9/23/12	1314	Dinner		42.00	933.48
9/25/12	1315	Gas Company		45.75	887.73
9/26/12		Deposit Paycheck	403.08		1290.81
9/29/12	1316	Rent		450.00	840.81

Calculations
275.58 − 70.89
204.69 + 402.98
607.67 − 35.60
572.07 − 65.40
506.67 − 239.17
267.50 − 72.54
194.96 + 402.95
597.91 − 25.00
572.91 + 402.57
975.48 − 42.00
933.48 − 45.75
887.73 + 403.08
1290.81 − 450.00

∴ The ending balance is $840.81.

b. Your monthly bank statement shows a balance of $1290.81. Explain why your balance does not agree with the bank statement.

∴ The balance on the bank statement is equal to your balance on 9/26/12. Because Check #1316 is close to the end of the month, it likely was not processed in time to appear on your statement.

On Your Own

Now You're Ready
Exercise 4

1. You write a check on 9/30/12 for $74.95 to the auto shop, and another on 10/1/12 for $120.85 to the plumber. Complete two more rows in the registry for these checks. What is your new balance?

Reading
ATM stands for Automated Teller Machine.

Most banks include a **debit card**, or check card, with a checking account. This card allows you to withdraw cash from ATMs *and* pay for purchases in stores. The money is taken directly from your account.

Multi-Language Glossary at BigIdeasMath.com.

EXAMPLE 2 Choosing a Checking Account

New Customers

Basic Checking
- $20 monthly fee
- free ATM usage

Regular Checking
- no monthly fee
- $2 charge for each ATM withdrawal

How many ATM withdrawals must you make each month for the cost of *Regular Checking* to be greater than the cost of *Basic Checking*?

Use a model to write and solve an inequality.

Words	Cost per withdrawal	times	the number of withdrawals	is greater than	the monthly fee.
Variable	Let w be the number of withdrawals.				
Inequality	2	$\cdot$	w	$>$	20

$2w > 20$ Write the inequality.

$w > 10$ Solve for w.

You must make more than 10 withdrawals each month.

A **bad check**, or "bounced" check, occurs when the amount written on the check is greater than the balance available in the checking account. Many banks charge a fee for a bad check.

EXAMPLE 3 Finding a Bad Check

Which check is the bad check?

Date	Check #	Transaction	Credit	Debit	Balance
		Balance Forward			190.60
6/2/12	417	Clothing Store		69.95	120.65
6/3/12		Deposit Paycheck	370.54		491.19
6/3/12	418	Gas Company		52.37	438.82
6/5/12	419	Cell Phone Company		75.89	362.93
6/7/12	420	Car Payment		275.10	87.83
6/9/12	421	Water Company		93.75	−5.92
6/13/12		Deposit	100.00		94.08
6/14/12	422	Birthday Gift		25.00	69.08

The amount of Check #421 is greater than the available balance.

So, Check #421 is the bad check.

On Your Own

Now You're Ready
Exercises 5 and 7

2. **WHAT IF?** In Example 2, there is a $15 monthly fee for *Basic Checking*. How many ATM withdrawals must you make each month for the cost of *Regular Checking* to be greater than the cost of *Basic Checking*?

3. Write an inequality that describes the possible car payments for Check #420 in Example 3 that may result in a bad check. Explain.

B.4 Exercises

Check It Out
Help with Homework
BigIdeasMath.com

Vocabulary and Concept Check

1. **VOCABULARY** What information do you need to know to enter a transaction in your checkbook registry?

2. **VOCABULARY** How do you know if a check is a bad check?

Practice and Problem Solving

3. Your checkbook registry is shown.

   a. Copy and complete your checkbook registry.

Date	Check #	Transaction	Credit	Debit	Balance
		Balance Forward			300.00
3/4/12	506	Dinner		45.32	
3/7/12		ATM Withdrawal		80.00	
3/8/12	507	Electric Company		39.76	
3/14/12		Deposit	150.00		
3/18/12	508	Credit Card Payment		129.50	

   b. Your monthly bank statement shows that your account balance is $81.12. You notice that you did not enter the two transactions shown below in your checkbook registry. After entering these transactions in your checkbook registry, does your balance now agree with the bank statement?

   - 3/7/12 ATM Fee $2.00
   - 3/10/12 Department Store $72.30

1 4. You balance your checkbook at the end of the month.

   a. Complete the balance column. What is the ending balance?

Date	Check #	Transaction	Credit	Debit	Balance
		Balance Forward			756.90
1/2/12	1563	Cell Phone Company		78.83	
1/5/12	1564	Grocery Store		132.70	
1/8/12		Deposit Paycheck	1567.00		
1/11/12	1565	Mortgage Payment		876.00	
1/13/12	1566	Credit Card Payment		651.25	
1/18/12		ATM Withdrawal		40.00	
1/20/12	1567	Grocery Store		143.78	
1/24/12		Deposit Paycheck	1567.00		
1/29/12	1568	Gas Company		68.23	

   b. Your monthly bank statement shows a balance of $1968.34. Explain why your balance does not agree with the bank statement.

2 **5.** How many ATM withdrawals must you make each month for the cost of *Basic Checking* to be greater than the cost of *Advanced Checking*?

BASIC CHECKING
- $10 monthly fee
- $0.50 charge for each ATM withdrawal

Anytown Community Bank
Anytown, USA

ADVANCED CHECKING
- $20 monthly fee
- free ATM usage

Anytown Community Bank
Anytown, USA

6. ERROR ANALYSIS Describe and correct the error in the checkbook registry.

Date	Check #	Transaction	Credit	Debit	Balance
		Balance Forward			760.00
4/5/12	1462	Gas Company		34.00	726.00
4/6/12		Deposit		154.00	572.00

3 **7.** Complete the balance column. Which check is the bad check?

Date	Check #	Transaction	Credit	Debit	Balance
		Balance Forward			230.50
11/3/12	1021	Electric Company		37.29	
11/8/12	1022	Shoe Store		49.21	
11/9/12	1023	Credit Card Payment		173.80	
11/12/12		Deposit Paycheck	275.00		
11/17/12	1024	Car Payment		133.61	
11/26/12		Deposit Paycheck	275.00		
11/29/12	1025	Rent		375.00	

8. Critical Thinking When you write a bad check, a bank charges a fee of $30.

a. How will the bad check fee change the checkbook registry in Exercise 7?

b. What mistakes can lead to writing a bad check? How can you avoid writing a bad check?

Fair Game Review What you learned in previous grades & lessons

Write the number in scientific notation. *(Section 9.6)*

9. 0.000000079

10. 1,500,000

11. 360,000,000,000

12. 0.000024

13. MULTIPLE CHOICE An account earns 4% interest compounded annually. The balance after 3 years is $618.68. What is the principal? *(Section B.2)*

Ⓐ $500 Ⓑ $525 Ⓒ $550 Ⓓ $600

B Study Help

You can use an **information frame** to help you organize and remember topics. Here is an example of an information frame for simple interest.

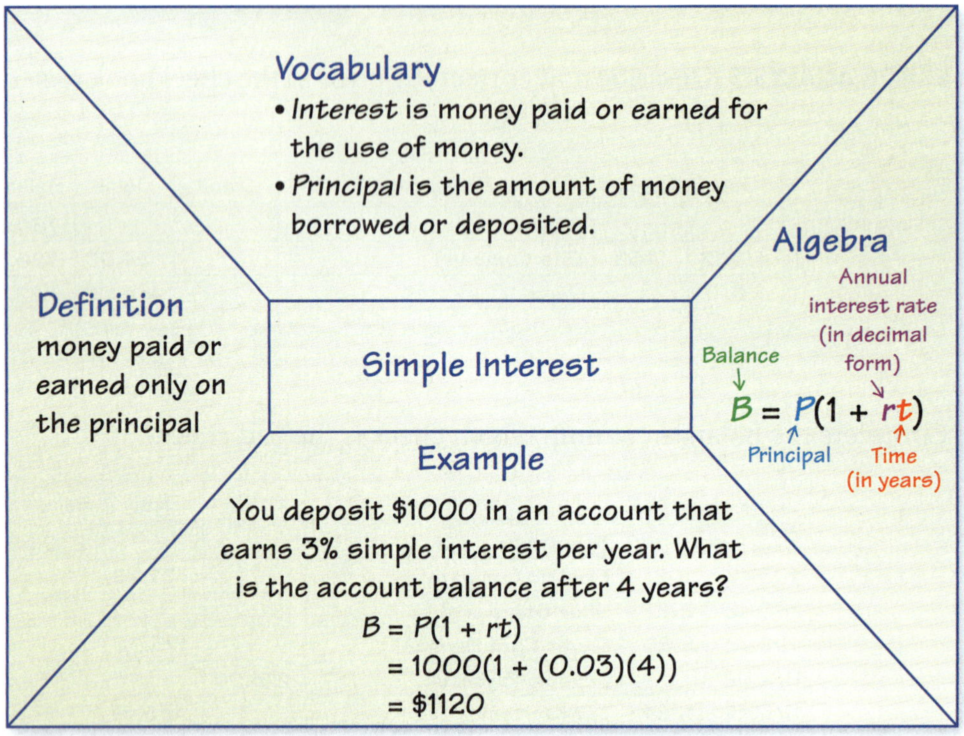

On Your Own

Make an information frame to help you study these topics.

1. compound interest
2. loan
3. mortgage
4. checking account

After you complete this chapter, make information frames for the following topics.

5. credit card
6. payroll statement

"Dear Teacher, I am emailing my information frame showing the characteristics of circles."

B.1–B.4 Quiz

Check It Out
Progress Check
BigIdeasMath.com

1. You deposit \$1200 in a savings account that earns 2.5% simple interest per year. Your friend deposits \$1000 in a savings account that earns 5% simple interest per year. *(Section B.1)*

   a. Write and graph two equations for the balance B in each account after t years.

   b. Are the account balances ever equal? Explain.

2. You deposit \$1200 in a savings account that earns 4% interest compounded annually. *(Section B.2)*

   a. What is the balance after 2 years?

   b. What is the interest earned?

3. An account earns 2% interest compounded annually. The balance after 3 years is \$5306.04. What is the principal? *(Section B.2)*

4. A loan summary is shown. How much interest is paid on the loan? *(Section B.3)*

LOAN AMOUNT	\$2135.00
MONTHLY PAYMENT	\$368.39
ANNUAL INTEREST RATE	12%
TERM	6 months

5. You balance your checkbook at the end of the month. *(Section B.4)*

   a. Complete the balance column. What is the ending balance? (Assume there is no fee for a bad check.)

Date	Check #	Transaction	Credit	Debit	Balance
		Balance Forward			98.12
11/1/12		ATM Withdrawal		40.00	
11/5/12	312	Grocery Store		64.53	
11/15/12		Deposit Paycheck	623.50		
11/16/12	313	Credit Card Payment		150.00	
11/21/12	314	Electric Company		42.79	
11/28/12	315	Rent		325.00	

   b. Your monthly bank statement shows a balance of \$424.30. Explain why your balance does not agree with the bank statement.

   c. Which check is the bad check?

6. **MORTGAGE** The principal on a home mortgage is \$150,000, the monthly payment is \$997.95, and the term is 30 years. How much interest is paid on the loan? *(Section B.3)*

7. **KEYBOARD** Your friend borrows \$450 from you for a new electronic keyboard. Your friend pays you back the principal plus 9% simple interest per year in 1.5 years. How much money do you earn? *(Section B.1)*

B.5 Credit Cards

Essential Question How do credit cards work?

STATE STANDARDS
MA.7.A.1.2
MA.8.A.6.4

1 ACTIVITY: Running a Business

Work with a partner. You own a shop that sells greeting cards. A customer uses a credit card to pay the total shown on the receipt.

a. How much will the credit card company deposit in your bank account?

b. Of the amount in part (a), how much do you have to send to the state and county?

Polly Greetings
123 Main Street
Somewhere, FL 33456
(905) 555-1212

Credit Card Sale

Card #xxxx xxxx xxxx 4321
Date: 05/26/2012 10:16 AM

Auth. $025873

SALE:	$34.95
STATE TAX:	$ 2.10
COUNTY TAX:	$ 0.55
TOTAL:	$37.60

X *Maria Lopez*

- - THANK YOU - -

Send the state this amount for sales tax

Send the county this amount for sales tax

The credit card company keeps 6% of the total

2 ACTIVITY: Credit Rating

RESEARCH PROJECT **Your credit score is a number from 300 to 850 that rates your credit history. Write a report on the types of things that increase your credit score and also the types of things that decrease it.**

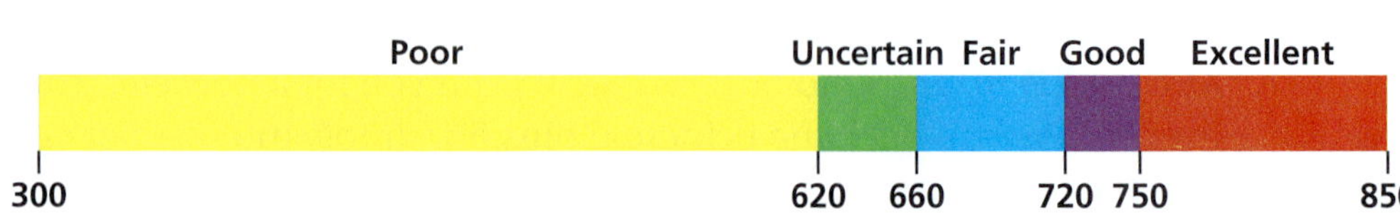

3 ACTIVITY: Credit Card Bills

Work with a partner. Maria paid \$140.00 of her previous month's credit card balance of \$490.00. Her credit card company charged her interest on the average daily balance for the current month.

The average daily balance for the first week is

$$\frac{490.00 + 490.00 + 490.00 + 350.00 + 350.00 + 350.00 + 350.00}{7} = 410.00.$$

a. Copy and complete the table for the entire month of May to find the average daily balance for the month.

Date	Activity Since Last Statement	Amount	Balance
05/01/2012			490.00
05/02/2012			490.00
05/03/2012			490.00
05/04/2012	Payment	−140.00	350.00
05/05/2012			350.00
⋮	⋮	⋮	⋮
05/25/2012			350.00
05/26/2012	Polly Greetings	37.60	
05/27/2012			
05/28/2012			
05/29/2012			
05/30/2012			
05/31/2012			

b. Maria's credit card company charges her \$5.56 interest on her average daily balance for the month. What is her monthly interest rate?

c. Find the average daily balance for the month without the purchase at Polly Greetings. Use the interest rate you found in part (b) to calculate the interest charge.

d. What does the difference between the interest charges in parts (b) and (c) represent?

What Is Your Answer?

4. **IN YOUR OWN WORDS** How do credit cards work?

Practice Use what you learned about credit cards to complete Exercises 3–5 on page A40.

Check It Out
Lesson Tutorials
BigIdeasMath.com

Key Vocabulary
credit card, *p. A38*

A **credit card** is used to postpone payment on a purchase. To allow this, the credit card company keeps a percentage of the total and sends the remaining amount to the seller. The seller pays the taxes due on the purchase.

EXAMPLE 1 Analyzing a Credit Card Transaction

Louis uses his credit card to buy CDs and guitar strings at a music store. The credit card company keeps 6% of the total.

Caja de Música

Louis A. Cardholder
Credit card # xxxx xxxx xxxx 8080
5:19 P.M. 08/10/12

ITEM	AMOUNT
2 CDs	22.98
Pkg Guitar Strings	12.05
SUBTOTAL	35.03
7% State Tax	2.45
TOTAL	**37.48**

-THANK YOU-
Customer Copy

a. How much money does the credit card company keep from the transaction?

The credit card company keeps 6% of the total. Find 6% of \$37.48.

$$6\% \text{ of } \$37.48 = 0.06 \cdot 37.48$$
$$\approx 2.25$$

∴ The credit card company keeps \$2.25.

b. How much money does the music store receive from the credit card company?

The music store receives the remaining amount.

∴ So, the music store receives \$37.48 − \$2.25 = \$35.23.

c. How much money does the music store keep from the transaction?

The music store sends the state sales tax to the state.

∴ So, the music store keeps \$35.23 − \$2.45 = \$32.78.

On Your Own

Now You're Ready
Exercises 6–9

1. **WHAT IF?** In Example 1, Louis charges \$90.00, including sales tax, to his credit card. How much money does the music store receive from the credit card company?

Multi-Language Glossary at BigIdeasMath.com.

Credit card owners are sent a monthly bill. The new balance shown on the bill may consist of unpaid balances, new purchases, interest, or other fees.

Many credit card companies charge interest based on the average daily balance on the account during a statement period. The interest charge is the product of the monthly interest rate and the average daily balance.

EXAMPLE 2 Paying a Credit Card Bill

Louis receives a credit card bill in the mail. His monthly interest rate is 1.5%. Find the interest charge and the new balance.

Credit Card Statement

Account Number XXXX XXXX XXXX 8080

Date	Activity Since Last Statement	Amount
08/10/12	Online Payment	-600.00
	- - - - Purchases - - - -	
08/02/12	Gas Station	59.61
08/06/12	Cell Phone Company	58.86
08/07/12	Restaurant	28.65
08/10/12	Caja de Música	37.48
08/15/12	Electric Bill	50.47
08/20/12	Car Repair Shop	164.93

Average Daily Balance	$522.47
Statement Closing Date	09/01/12
Previous Balance	$700.00
Payments	-600.00
Unpaid Balance	$100.00
New Purchases	$400.00
Interest Charge	
New Balance	

Find the product of the monthly interest rate and the average daily balance.

$$0.015 \cdot 522.47 \approx \$7.84$$

The previous balance was not fully paid. So, the new balance is the sum of the unpaid balance, the new purchases, and the interest charge.

$$\$100.00 + \$400.00 + \$7.84 = \$507.84$$

So, the interest charge is $7.84 and the new balance is $507.84.

On Your Own

Now You're Ready
Exercise 10

2. **WHAT IF?** In Example 2, if Louis had paid only $50.00 on last month's bill, the average daily balance would be $912.79. Find the interest charge and the new balance.

B.5 Exercises

Vocabulary and Concept Check

1. **VOCABULARY** How is a credit card similar to a debit card? How is a credit card similar to a loan? Explain.
2. **REASONING** Why is it important to know the annual interest rate of a credit card?

Practice and Problem Solving

You own a bookstore. A customer uses a credit card to pay the total shown on the receipt.

3. How much money is sent to the state?
4. How much money is sent to the county?
5. The credit card company keeps 5% of the total. How much money does the credit card company keep?

SALE:	$45.00
STATE TAX:	$ 2.70
COUNTY TAX:	$ 0.72
TOTAL:	$48.42

In Exercises 6–9, a credit card company keeps the given percent of the total. Find the amount of money (a) the credit card company keeps from the transaction, (b) the store receives from the credit card company, and (c) the store keeps from the transaction.

1 6. 4%

Office Supply Store
Maria Lopez
Credit card # xxxx xxxx xxxx 4321
11:24 A.M. 05/13/11

SALE:	$62.00
SALES TAX:	$ 4.03
TOTAL:	**$66.03**

-THANK YOU-

7. 3%

ROY'S HARDWARE Store
Michael Jones
Credit card # xxxx xxxx xxxx 7914
9:45 A.M. 06/02/12

SALE:	$105.80
SALES TAX:	$ 5.29
TOTAL:	**$111.09**

THANK YOU

8. 3.5%

Pretzel Shop
Gustav Tanner
Credit card # xxxx xxxx xxxx 5059
1:24 P.M. 08/24/11

SALE:	$8.25
SALES TAX:	$0.50
TOTAL:	**$8.75**

THANK YOU

9. 5%

Arts and Crafts Outlet
Gloria Moore
Credit card # xxxx xxxx xxxx 6338
3:12 P.M. 11/30/12

SALE:	$75.00
SALES TAX:	$ 3.75
TOTAL:	**$78.75**

-THANK YOU-

② **10.** A credit card bill arrives in the mail. The monthly interest rate is 1.25%. Find the interest charge and the new balance.

Credit Card Statement

Account Number XXXX XXXX XXXX 0947

Date	Activity Since Last Statement	Amount
02/15/11	Payment	250.00
	- - - - Purchases - - - -	
02/03/11	Town Mall	75.08
02/07/11	Grocery Store	92.65
02/13/11	Costume Shop	18.33
02/20/11	Camping Store	48.50
02/25/11	Shoe Outlet	65.44

Average Daily Balance	$552.93
Statement Closing Date	03/01/11
Previous Balance	$500.00
Payments	-250.00
Unpaid Balance	$250.00
New Purchases	$300.00
Interest Charge	
New Balance	

11. **CLOTHING** You use a credit card to buy $100 worth of merchandise at a clothing store. The sales tax is 6%. The credit card company keeps 2% of the total.

a. How much money does the store keep from the transaction?

b. About what percent of the transaction does the store keep?

c. The clothing store decides to charge a 2% fee for using a credit card. The fee is added to the purchase price before taxes are calculated. Does the store keep $100 from the sale now? Explain.

12. **LAYAWAY** *Layaway* is a payment method used to purchase an item without paying the entire cost at one time. A customer pays a deposit to put an item "on layaway" for a given period of time. The store reserves the item until it is paid for in full. What is an advantage of putting an item "on layaway" compared to using a credit card?

13. **Number Sense** The statement closing date on your credit card is the first of each month. Your new balance on January 1 is $800. You stop making purchases with the card and pay the minimum payment of $20 on the first of each month, starting January 1. The monthly interest rate is 2%. What is the balance one year later? How much have you paid after one year? How much did the amount due decrease? Explain why you should pay more than the minimum payment.

Fair Game Review What you learned in previous grades & lessons

Simplify the expression. *(Section 6.4)*

14. $6\sqrt{3} - 2\sqrt{3}$ **15.** $\sqrt{8} \cdot \sqrt{18}$ **16.** $\sqrt{\frac{10}{49}}$ **17.** $5\sqrt{19} + \sqrt{19}$

18. **MULTIPLE CHOICE** What is $\frac{30}{800}$ as a percent? *(Skills Review Handbook)*

Ⓐ 0.0375% Ⓑ 0.375% Ⓒ 3.75% Ⓓ 37.5%

B.6 Payroll Deductions

STATE STANDARDS
MA.7.A.1.2
MA.8.A.6.4

Essential Question How can you read a payroll earnings statement?

1 ACTIVITY: Reading a Payroll Earnings Statement

Work with a partner. Talk about each entry on the payroll statement. Do you understand each entry?

a. How much does John get paid per hour?

b. How often does John get paid?

c. How many paychecks has John received so far in 2012?

d. How is the gross pay related to the net pay?

e. How much does John pay for:

- federal income tax?
- Social Security tax?
- Medicare tax?

NAME
John Doe

EARNINGS STATEMENT

HOURS	RATE	AMOUNT
80	17.50	1400.00

YTD GROSS PAY	GROSS PAY
14,000.00	1400.00

PAY ENDING	PAY DATE	DEPARTMENT	NET PAY
05/18/12	05/18/12	312	983.87

	FEDERAL	SOC. SEC.	HEALTH INS.	MEDICARE	401K
PAYROLL TAX WITHHELD	192.85	83.70	50.00	19.58	70.00
YEAR TO DATE	1928.50	837.00	500.00	195.80	700.00

PAYROLL CHECK

EMPLOYEE ID NO.
876 3210

DEPT.	CHECK NO.	PAY DATE
312	12789	05/18/12

PAY EXACTLY ***Nine Hundred Eighty Three and 87/100*** DOLLARS

PAY TO THE ORDER OF
John Doe
Street Address
City, FL 33045

BANK NAME
CITY, STATE, ZIP CODE

AMOUNT
***983.87

Susan Florencia

2 ACTIVITY: Federal Income Tax Rates

Work with a partner. Use the table to find the federal income tax for 2008 for the given taxable income.

a. $7982.00 **b.** $26,500.00 **c.** $54,780.00

d. $125,348.40 **e.** $235,893.70 **f.** $356,789.00

Filing Status: Single		
If taxable income is between. . .	**Tax bracket is:**	**Federal income tax is:**
\$0 and \$8,024	10%	10% of the income over \$0
\$8,025 and \$32,549	15%	\$802.50 plus 15% of the income over \$8,025
\$32,550 and \$78,849	25%	\$4,481.25 plus 25% of the income over \$32,550
\$78,850 and \$164,549	28%	\$16,056.25 plus 28% of the income over \$78,850
\$164,550 and \$357,699	33%	\$40,052.25 plus 33% of the income over \$164,550
\$357,700 and Above	35%	\$103,791.75 plus 35% of the income over \$357,700

3 ACTIVITY: Federal Income Tax Rates

Work with a partner.

a. Make a scatter plot of the data in Activity 2.

b. Federal income tax is called a *graduated tax.* What does this mean?

c. Graduated income tax graphs are not linear. Does the graph show a linear pattern?

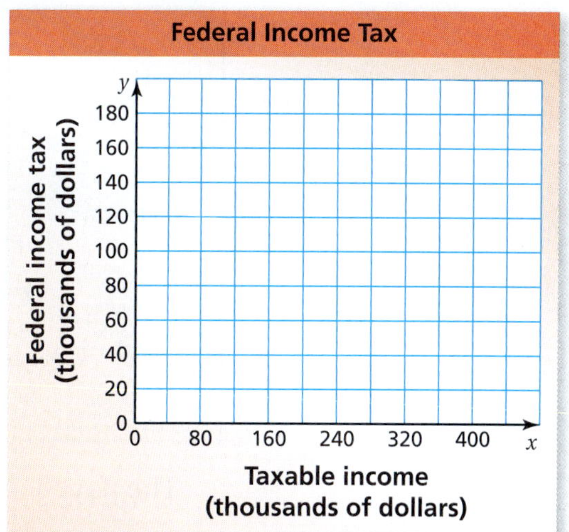

What Is Your Answer?

4. IN YOUR OWN WORDS How can you read a payroll earnings statement?

Practice Use what you learned about payroll deductions to complete Exercises 3–5 on page A46.

B.6 Lesson

Check It Out
Lesson Tutorials
BigIdeasMath.com

Key Vocabulary
gross pay, *p. A44*
net pay, *p. A44*
payroll statement, *p. A44*

Study Tip
Other deductions, such as health insurance, may appear on a payroll statement.

The amount of money earned by an employee during a pay period is called **gross pay**. Several deductions are made from an employee's gross pay. Here are some of the deductions that are required by U.S. law.

- ***Federal income tax*** A tax based on an employee's earned income that is paid directly to the U.S. federal government.
- ***Social Security tax*** A tax used to fund the Social Security system, which provides benefits mainly for retired workers.
- ***Medicare tax*** A tax used to fund Medicare, which is a medical plan administered by the federal government mainly for people age 65 and older.

The amount of money that an employee receives after all the deductions have been made is called **net pay**. A **payroll statement** shows the gross pay, the deductions, and the net pay.

EXAMPLE 1 Reading a Payroll Statement

According to the payroll statement, how often does Cara get paid?

NAME
Cara Jimenez

EARNINGS STATEMENT

HOURS	RATE	AMOUNT
168	14.50	2436.00

YTD GROSS PAY	GROSS PAY
14,616.00	2436.00

PAY ENDING	PAY DATE	DEPARTMENT	NET PAY
06/30/12	06/30/12	M	1845.39

PAYROLL TAX WITHHELD

FEDERAL	SOC. SEC.	HEALTH INS.	MEDICARE	401K
170.94	143.55	120.75	33.57	121.80

YEAR TO DATE

FEDERAL	SOC. SEC.	HEALTH INS.	MEDICARE	401K
1025.64	861.30	724.50	201.42	730.80

PAYROLL CHECK

EMPLOYEE ID NO.
987 4320

DEPT.	CHECK NO.	PAY DATE
M	10259	06/30/12

PAY EXACTLY ***ONE THOUSAND EIGHT HUNDRED FORTY FIVE and 39/100*** DOLLARS

PAY TO THE ORDER OF
Cara Jimenez
314 Pi Circle
Mathville, FL 31416

BANK NAME
CITY, STATE, ZIP CODE

AMOUNT
***1845.39

Matthew Pascal

The paycheck is for 168 hours and is dated for the end of June.

So, you can conclude that Cara is paid once per month.

On Your Own

Now You're Ready
Exercises 6–11

1. What is Cara's hourly wage?
2. What is the difference between Cara's gross pay and net pay?
3. How much does Cara pay annually for health insurance?

Multi-Language Glossary at BigIdeasMath.com.

A 401(k) plan is an optional retirement savings plan offered by some companies to their employees. The amount an employee can contribute toward a 401(k) plan is limited by U.S. law. The amount of money in a 401(k) is not taxed until it is withdrawn.

EXAMPLE 2 Analyzing a Payroll Statement

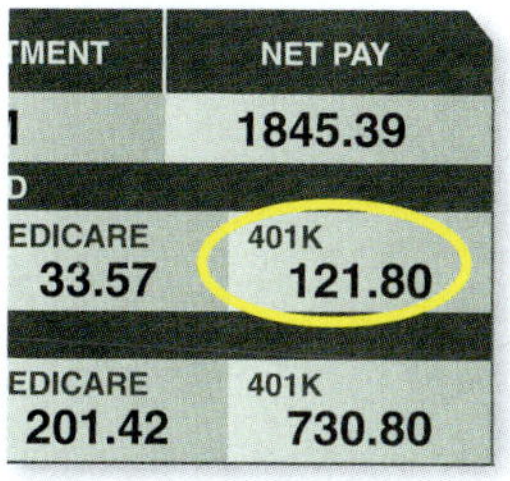

In Example 1, what percent of Cara's gross pay is put in her 401(k)?

Divide the amount put in her 401(k) by her gross pay.

$$\frac{\text{amount put in 401(k)}}{\text{gross pay}} = \frac{121.80}{2436.00} = 0.05 = 5\%$$

∴ 5% of her gross pay is put in her 401(k).

Social Security and Medicare taxes are calculated after the health insurance amount is subtracted from the gross pay. Federal income taxes are calculated after the health insurance amount *and* the 401(k) contribution are subtracted from the gross pay.

EXAMPLE 3 Finding Tax Percents

a. What percent does Cara pay for Social Security tax?

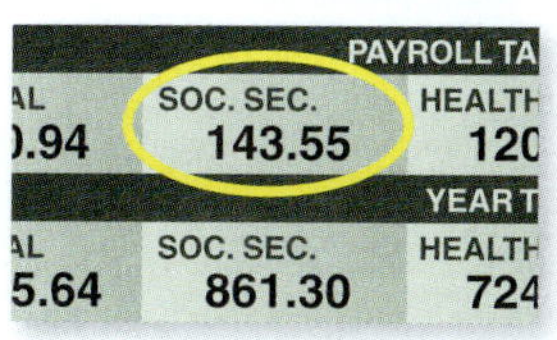

Divide the amount of Social Security tax by the difference of her gross pay and health insurance amount.

$$\frac{\text{Social Security tax}}{\text{gross pay} - \text{health insurance}} = \frac{143.55}{2436.00 - 120.75} = \frac{143.55}{2315.25}$$

$$\approx 0.062 = 6.2\%$$

∴ Cara pays 6.2% for Social Security tax.

b. What percent does Cara pay for federal income tax?

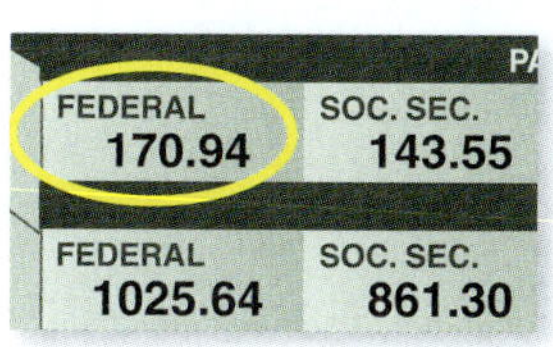

Divide the amount of federal income tax by the difference of her gross pay and the sum of the health insurance amount and 401(k) contribution.

$$\frac{\text{federal income tax}}{\text{gross pay} - [\text{health insurance} + \text{401(k)}]} = \frac{170.94}{2436.00 - 242.55}$$

$$\approx 0.078 = 7.8\%$$

∴ Cara pays 7.8% for federal income tax.

On Your Own

Now You're Ready
Exercises 13–15

4. Cara changes her 401(k) contribution to 8% of her gross pay. How much will Cara put in her 401(k) each month?
5. What percent does Cara pay for Medicare tax?

B.6 Exercises

Check It Out
Help with Homework
BigIdeasMath.com

Vocabulary and Concept Check

1. **REASONING** Is the net pay *sometimes*, *always*, or *never* greater than the gross pay? Explain.

2. **REASONING** Why do you think the amount of money an employee can contribute toward a 401(k) plan is limited by U.S. law?

Practice and Problem Solving

In Exercises 3–5, use the table in Activity 2 to find the federal income tax for a person who had the given amount of taxable income.

3. $10,500.00
4. $58,435.00
5. $6120.75

In Exercises 6–12, use the payroll statement.

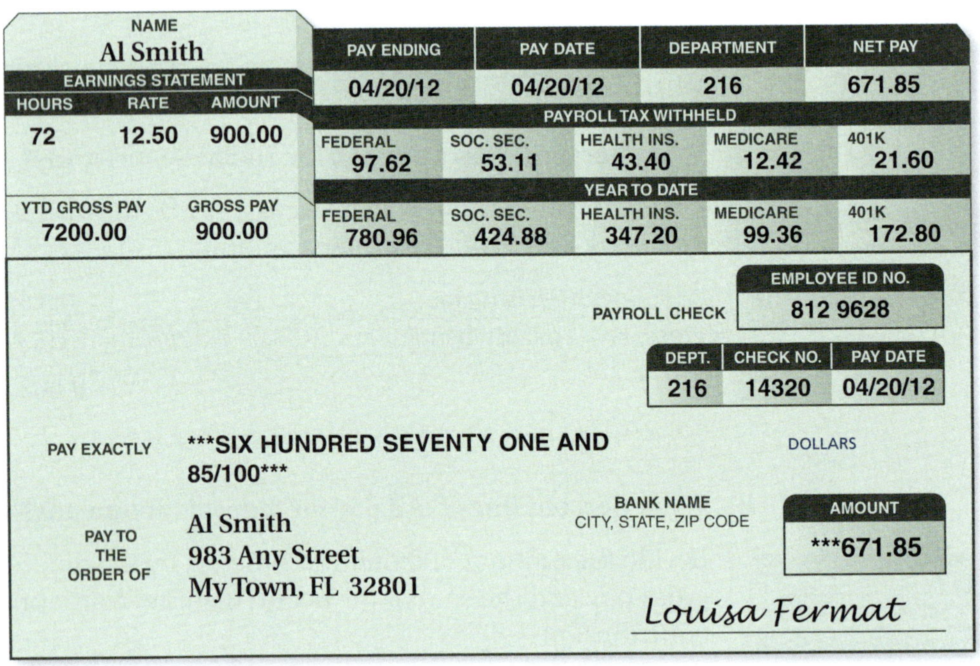

NAME: Al Smith

EARNINGS STATEMENT

HOURS	RATE	AMOUNT
72	12.50	900.00

YTD GROSS PAY	GROSS PAY
7200.00	900.00

PAY ENDING	PAY DATE	DEPARTMENT	NET PAY
04/20/12	04/20/12	216	671.85

PAYROLL TAX WITHHELD

FEDERAL	SOC. SEC.	HEALTH INS.	MEDICARE	401K
97.62	53.11	43.40	12.42	21.60

YEAR TO DATE

FEDERAL	SOC. SEC.	HEALTH INS.	MEDICARE	401K
780.96	424.88	347.20	99.36	172.80

PAYROLL CHECK

EMPLOYEE ID NO.
812 9628

DEPT.	CHECK NO.	PAY DATE
216	14320	04/20/12

PAY EXACTLY ***SIX HUNDRED SEVENTY ONE AND 85/100*** DOLLARS

PAY TO THE ORDER OF
Al Smith
983 Any Street
My Town, FL 32801

BANK NAME
CITY, STATE, ZIP CODE

AMOUNT ***671.85

Louisa Fermat

6. What is Al's hourly wage?
7. How often does Al get paid?
8. What is Al's gross pay for April?
9. What is Al's annual net pay?
10. How many paychecks has Al received so far in 2012?
11. What is the difference between Al's gross pay and net pay?
12. **ERROR ANALYSIS** Describe and correct the error in finding Al's annual gross pay.

> ✗ Gross Pay: \$900.00
> \$900.00 • 12 = \$10,800
> His annual gross pay is \$10,800.

In Exercises 13–15, use the payroll statement on page A46.

2 3 **13.** What percent does Al pay for Social Security tax?

14. What percent does Al pay for Medicare tax?

15. What percent does Al pay for federal income tax?

16. **PROJECT** The table shows the federal income tax brackets for 2008.

Filing Status: Head of Household		
If taxable income is between...	**Tax bracket is:**	**Federal income tax is:**
\$0 and \$11,449	10%	10% of the income over \$0
\$11,450 and \$43,649	15%	\$1,145 plus 15% of the income over \$11,450
\$43,650 and \$112,649	25%	\$5,975 plus 25% of the income over \$43,650
\$112,650 and \$182,399	28%	\$23,225 plus 28% of the income over \$112,650
\$182,400 and \$357,699	33%	\$42,755 plus 33% of the income over \$182,400
\$357,700 and Above	35%	\$100,604 plus 35% of the income over \$357,700

a. Use the Internet to research the federal income tax brackets for the current year. Make a table similar to the one above.

b. Describe the change in the federal income tax from 2008 to the current year for each tax bracket.

c. Compare the federal income tax for 2008 and the federal income tax for the current year for a person whose taxable income was \$30,000.

17. **STATE TAXES** In addition to federal income tax, you may have to pay state income tax. Use the Internet to research the income taxes for Colorado, Florida, and Massachusetts. Which state has the lowest income tax? the highest income tax?

18. **Critical Thinking** Your annual gross pay is \$32,400. Five percent of your gross pay is put in your 401(k) and 7% pays for health insurance. You pay 15% for federal income tax. What amount of federal income tax do you pay?

Fair Game Review What you learned in previous grades & lessons

Evaluate the expression. *(Section 9.4)*

19. $3^{-1} \cdot 3^2$ **20.** $5^0 \cdot 5^3$ **21.** $\dfrac{7^2 \cdot 7^{-4}}{7^{-5}}$ **22.** $\dfrac{6^{-1} \cdot 6^2}{6^3}$

23. **MULTIPLE CHOICE** Which inequality represents "20 is less than a number x"? *(Section 8.1)*

Ⓐ $x < 20$ Ⓑ $20 \geq x$ Ⓒ $20 \leq x$ Ⓓ $20 < x$

B.5–B.6 Quiz

Check It Out
Progress Check
BigIdeasMath.com

1. A credit card bill arrives in the mail. The annual interest rate is 15%. Find the interest charge and the new balance. *(Section B.5)*

Credit Card Statement
Account Number xxxx xxxx xxxx 6003

Date	Activity Since Last Statement	Amount
07/11/12	Payment	50.00
	- - - - Purchases - - - -	
07/02/12	Hair Salon	26.18
07/12/12	Farmer's Market	35.05
07/24/12	Office Supply Store	63.42
07/30/12	Gas Station	25.35

Average Daily Balance	$132.08
Statement Closing Date	08/01/12
Previous Balance	$100.00
Payments	-50.00
Unpaid Balance	$50.00
New Purchases	$150.00
Interest Charge	
New Balance	

In Exercises 2–5, use the payroll statement. *(Section B.6)*

NAME
Sally Wong

EARNINGS STATEMENT

HOURS	RATE	AMOUNT
72	11.25	810.00

YTD GROSS PAY	GROSS PAY
7290.00	810.00

PAY ENDING	PAY DATE	DEPARTMENT	NET PAY
05/04/12	05/04/12	103	623.58

	FEDERAL	SOC. SEC.	HEALTH INS.	MEDICARE	401K
PAYROLL TAX WITHHELD	75.90	48.21	32.40	11.28	18.63
YEAR TO DATE	683.10	433.89	291.60	101.52	167.67

2. What is Sally's hourly wage?

3. What would you expect Sally's net pay to be for the entire year?

4. How much does Sally put in her 401(k) annually?

5. What percent does Sally pay for Medicare tax?

In Exercises 6 and 7, a credit card company keeps the given percent of the total. Find the percent of the transaction the store keeps. *(Section B.5)*

6. 3%

Ice Cream Shoppe
Pablo Rodriguez
Credit card # xxxx xxxx xxxx 5631
8:10 P.M. 07/22/12

SALE:	$18.75
SALES TAX:	$ 0.94
TOTAL:	**$19.69**

-THANK YOU-

7. 4%

Appliance Outlet — #1 IN CUSTOMER SATISFACTION
Celia Danielson
Credit card # xxxx xxxx xxxx 9103
2:08 P.M. 09/30/12

SALE:	$845.15
SALES TAX:	$ 50.71
TOTAL:	**$895.86**

THANK YOU

B Appendix Review

Check It Out
Vocabulary Help
BigIdeasMath.com

Review Key Vocabulary

compound interest, *p. A13*
loan, *p. A24*
term, *p. A24*
installment loan, *p. A24*
mortgage, *p. A25*
checking account, *p. A30*
checkbook registry, *p. A30*
debit card, *p. A30*
bad check, *p. A31*
credit card, *p. A38*
gross pay, *p. A44*
net pay, *p. A44*
payroll statement, *p. A44*

Review Examples and Exercises

B.1 Simple and Compound Interest *(pp. A10–A15)*

You want to invest $400 for 2 years. Which account should you choose?

- **Account A earns 4% simple interest per year.**
- **Account B earns 3% interest compounded annually.**

Make two tables that show the account balances for 2 years.

Account A			
Year	Principal	4% Simple Interest	Balance at End of Year
1	$400	$16	$416
2	$400	$16	$432

Account B			
Year	Principal and Interest	3% Compound Interest	Balance at End of Year
1	$400.00	$12.00	$412.00
2	$412.00	$12.36	$424.36

Account A earns $432 − $424.36 = $7.64 more than Account B after 2 years. So, you should choose Account A.

Exercises

1. You deposit $500 in an account that earns 4% interest compounded annually. What is the account balance after 2 years?

B.2 Formula for Compound Interest *(pp. A16–A21)*

You deposit $1800 in a savings account. The account earns 3% interest compounded annually. What is the balance after 7 years?

$B = P(1 + r)^t$ — Write compound interest formula.

$= 1800(1 + 0.03)^7$ — Substitute.

$= 1800(1.03)^7$ — Add.

≈ 2213.77 — Simplify.

The balance is $2213.77 after 7 years.

Exercises

2. An account earns 4% interest compounded annually. The balance after 5 years is \$7421.58. What is the principal?

B.3 Installment Loans *(pp. A22–A27)*

The principal on a home mortgage is \$145,000, the monthly payment is \$1081.08, and the term is 20 years. How much interest is paid on the loan?

The total amount paid for the mortgage is the product of the monthly payment and the number of payments. The interest paid on the loan is the difference of the total amount paid for the mortgage and the principal.

$1081.08 \cdot 240 = 259{,}459.20$ — Multiply the monthly payment by number of payments.

$259{,}459.20 - 145{,}000 = 114{,}459.20$ — Subtract the principal from the total amount paid.

∴ \$114,459.20 in interest is paid on the loan.

Exercises

3. The principal on a home mortgage is \$250,000, the montly payment is \$1580.17, and the term is 30 years. How much interest is paid on the loan?

B.4 Checking Accounts *(pp. A28–A33)*

Which check is the bad check?

Date	Check #	Transaction	Credit	Debit	Balance
		Balance Forward			89.37
11/1/12	528	Cell Phone Company		65.11	24.26
11/5/12	529	Grocery Store		52.08	-27.82
11/15/12		Deposit Paycheck	418.94		391.12
11/16/12	530	Credit Card Payment		200.00	191.12
11/30/12		Charity Donation		30.00	161.12
11/30/12		Magazine Subscription		20.00	141.12

The amount of Check #529 is greater than the available balance.

∴ So, Check #529 is the bad check.

Exercises

4. Your monthly bank statement shows a balance of \$191.12. Explain why your balance does not agree with the bank statement.

5. Your next two transactions are shown below. What is your new balance?

- 12/1/12 Deposit \$300.00
- 12/2/12 ATM Withdrawal \$60.00

B.5 Credit Cards (pp. A36–A41)

Rocco receives a credit card bill in the mail. The monthly interest rate is 1.1%. Find the interest charge.

Average Daily Balance	$645.79
Statement Closing Date	06/01/12
Previous Balance	$800.00
Payments	-400.00
Unpaid Balance	$400.00
New Purchases	$200.00
Interest Charge	
New Balance	

Find the product of the monthly interest rate and the average daily balance.

$$0.011 \cdot 645.79 \approx \$7.10$$

∴ The interest charge is $7.10.

Exercises

6. What is the new balance?

7. If Rocco had paid $600 on last month's bill, the average daily balance would be $497.41. Find the interest charge and the new balance.

B.6 Payroll Deductions (pp. A42–A47)

What percent does Raj pay for health insurance?

NAME	Raj Singh	
EARNINGS STATEMENT		
HOURS	RATE	AMOUNT
80	13.50	1080.00
YTD GROSS PAY	GROSS PAY	
4320.00	1080.00	

PAY ENDING	PAY DATE	DEPARTMENT	NET PAY	
02/24/12	02/24/12	P	744.21	
PAYROLL TAX WITHHELD				
FEDERAL	SOC. SEC.	HEALTH INS.	MEDICARE	401K
115.37	61.33	90.75	14.34	54.00
YEAR TO DATE				
FEDERAL	SOC. SEC.	HEALTH INS.	MEDICARE	401K
461.48	245.32	363.00	57.36	216.00

$$\frac{\text{amount for health insurance}}{\text{gross pay}} = \frac{90.75}{1080.00}$$

$$\approx 0.084 = 8.4\%$$

∴ Raj pays 8.4% for health insurance.

Exercises

8. What percent does Raj pay for federal income tax?

9. How much will Raj put in his 401(k) for the entire year?

10. What is Raj's annual gross pay?

B Appendix Test

Check It Out
Test Practice
BigIdeasMath.com

1. You deposit $800 in a savings account that earns 4% simple interest per year. Your friend deposits $900 in a savings account that earns 3% simple interest per year.

   a. Write and graph two equations for the balance *B* in each account after *t* years.

   b. Are the account balances ever equal? Explain.

2. An account earns 3% interest compounded annually. The balance after 2 years is $4774.05. What is the principal?

3. A bank offers two types of checking accounts. How many transactions must you make each month for the cost of *Basic Checking* to be greater than the cost of *Advanced Checking*?

   Basic Checking
   - no monthly fee
   - $0.15 per transaction

   Anytown Savings and Loan
   Anytown, USA

   Advanced Checking
   - $12 monthly fee
   - no fee per transaction

   Anytown Savings and Loan
   Anytown, USA

4. You use a credit card to pay for a $186.90 purchase at a sporting goods store. The credit card company keeps 3% of the total. Find the amount of money the store receives from the credit card company for the transaction.

5. **Use the payroll statement.**

NAME	Karen Walker		
EARNINGS STATEMENT			
HOURS	RATE	AMOUNT	
168	15.60	2620.80	
YTD GROSS PAY	15,724.80	GROSS PAY	2620.80

PAY ENDING	PAY DATE	DEPARTMENT	NET PAY	
06/30/12	06/30/12	J	1990.98	
PAYROLL TAX WITHHELD				
FEDERAL	SOC. SEC.	HEALTH INS.	MEDICARE	401K
189.37	155.18	117.94	36.29	131.04
YEAR TO DATE				
FEDERAL	SOC. SEC.	HEALTH INS.	MEDICARE	401K
1136.22	931.08	707.64	217.74	786.24

   a. How often does Karen get paid?

   b. What percent does Karen pay for federal income tax?

   c. How much will Karen put in her 401(k) for the entire year?

6. **USED CAR** The average sale price of a used car at a car dealership is $6580. A savings account earns 4% interest compounded annually. About how much should you deposit in the account to have enough to buy a car in 3 years?

7. **MORTGAGE** The principal on a home mortgage is $128,000, the monthly payment is $1097.50, and the term is 15 years. How much interest is paid on the loan?

Photo Credits

inside front cover ©iStockphoto.com/Aldo Murillo; **iv** Big Ideas Learning, LLC; **viii** *top* ©iStockphoto.com/Lisa Thornberg, ©iStockphoto.com/ Ann Marie Kurtz; *bottom* Evok20; **ix** *top* ©iStockphoto.com/Jonathan Larsen; *bottom* Apollofoto; **x** *top* ©iStockphoto.com/Angel Rodriguez, ©iStockphoto.com/Ann Marie Kurtz; *bottom* ©iStockphoto.com/daaronj; **xi** ©iStockphoto.com/Ron Sumners, ©iStockphoto.com/Ann Marie Kurtz; *bottom* Apollofoto; **xii** *top* ©iStockphoto.com/Stefan Klein; *bottom* ©iStockphoto.com/ronen; **xiii** *top* ©iStockphoto.com/Michael Flippo, ©iStockphoto.com/Ann Marie Kurtz; *bottom* Heather Prosch-Jensen; **xiv** *top* ©iStockphoto.com/Alistair Cotton; *bottom* Peter Close; **xv** *top* ©iStockphoto.com/Angel Rodriguez, ©iStockphoto.com/Ann Marie Kurtz; *bottom* ©iStockphoto.com/ranplett; **xvi** *top* ©iStockphoto.com/Varina and Jay Patel, ©iStockphoto.com/Ann Marie Kurtz; *bottom* Jane Norton; **xvii** *top* ©iStockphoto.com/Björn Kindler; *bottom* ©iStockphoto.com/ Thomas Perkins; **xviii** ©iStockphoto.com/Ekaterina Monakhova **1** ©iStockphoto.com/Lisa Thornberg, ©iStockphoto.com/Ann Marie Kurtz; **6** ©iStockphoto.com/David Freund; **7** ©iStockphoto.com/shapecharge; **8** NASA; **9** ©iStockphoto.com/Judson Lane; **12** ©iStockphoto.com/Roman Dekan; **13** ©iStockphoto.com/Jacom Stephens; **14** ©iStockphoto.com/ Harry Hu; **15** ©Paul Slaughter; **20** ©iStockphoto.com/Andrey Krasnov; **29** *top right* ©iStockphoto.com/Alan Crawford; *center left* ©iStockphoto.com/ Julio Yeste; *center right* ©iStockphoto.com/Mark Stay; **31** ©iStockphoto.com/ Michel de Nijs; **33** Steve MacAulay Photography; **34** *top* Benedict S. Gibson; *bottom* Frantz Petion; **35** Jeff Kinsey; **36** *bottom right* ©iStockphoto.com/ Paul Tessier; *center left* Tom Uhlenberg; *top right* ©iStockphoto.com/ polarica; **37** ©iStockphoto.com/Winston Davidian; **38** ©iStockphoto.com/ Steve Mcsweeny; **41** ©iStockphoto.com/machinim; **46** ©iStockphoto.com/ Jonathan Larsen; **51** NASA; **52** ©iStockphoto.com/David Morgan; **53** *center left* ©iStockphoto.com/Jakub Semeniuk; *top right* NASA; **60** ©iStockphoto.com/ Amanda Rohde; **61** Julián Rovagnati; **66** ©iStockphoto.com/whitechild; **67** *top right* Jerry Horbert; *bottom left* ©iStockphoto.com/Chris Schmidt; **69** ©iStockphoto.com/Peter Finnie; **72** ©iStockphoto.com/Stephen Pothier; **73** *bottom* Dewayne Flowers; *top* Gina Smith; **75** Philip Lange; **76** Howard Sandler, ©iStockphoto.com/Dori O'Connell; **79** Yuri Arcurs; **80** ©iStockphoto.com/Kathy Hicks; **81** ©iStockphoto.com/webphotographeer; **82** ©iStockphoto.com/walik; **86** ©iStockphoto.com/Corina Estepa; **87** ©iStockphoto.com/Tomislav Forgo; **91** Kateryna Potrokhova; **92** ©iStockphoto.com/Jason Stitt; **93** ©iStockphoto.com/Gabor Izso, ©iStockphoto.com/Dawn Jagroop; **94** ©iStockphoto.com/ZanyZeus; **98** ©iStockphoto.com/Justin Horrocks; **104** ©iStockphoto.com/Angel Rodriguez, ©iStockphoto.com/Ann Marie Kurtz; **109** Printed courtesy of the Florida Department of Transportation; **110** ©iStockphoto.com/Adam Mattel; **111** *center right* ©iStockphoto.com/Pawel Liprec, ©iStockphoto.com/beetle8; *top left* ©iStockphoto.com/Gene Chutka; **113** ©iStockphoto.com/Kirsty Pargeter, **115** ©iStockphoto.com/Connie Maher; **116** ©iStockphoto.com/ Jacom Stephens; **117** *center left* ©iStockphoto.com/Andrea Krause; *top right* ©iStockphoto.com/Petr Podzemny; **121** ©iStockphoto.com/Michael Chen; **123** *bottom* ©iStockphoto.com/adrian beesley; *top* ©iStockphoto.com/Brian McEntire; **125** *center right* ©iStockphoto.com/John Kounadeas; *bottom right* ©iStockphoto.com/Rich Yasick; **128** ©iStockphoto.com/Ryan Putnam; **129** ©iStockphoto.com/ iLexx; **130** ©iStockphoto.com/Jeremy Edwards; **131** ©iStockphoto.com/Robert Kohlhuber; **136** *center left* ©iStockphoto.com/ George Peters; *center right* ©iStockphoto.com/Olga Shelego; *bottom left* ©iStockphoto.com/Valerie Loiseleux; **137** ©iStockphoto.com/Duncan Walker; **140** ©iStockphoto.com/ Marcio Silva; **142** ©iStockphoto.com/Carmen Martínez Banús; **146** ©iStockphoto.com/Ron Sumners, ©iStockphoto.com/ Ann Marie Kurtz; **149** *right* ©2008 Zappos.com, Inc.; *left* ©iStockphoto.com/ Andrew Johnson; **151** ©iStockphoto.com/alohaspirit; **152** ©iStockphoto.com/ Timur Kulgarin; **153** *center right* Primo Ponies Photography; *center left* ©iStockphoto.com/Wayne Johnson; **155** Digital Vision Royalty Free Photograph/Getty Images; **158** ©iStockphoto.com/Hannu Liivaar; **159** *center left* ©iStockphoto.com/LoopAll; *center right* ©iStockphoto.com/ Justin Horrocks, ©iStockphoto.com/Ana Abejon, ©iStockphoto.com/Huchen Lu; **161** ©iStockphoto.com/Lisa F. Young; **165** ©iStockphoto.com/technotr; **167** *top* ©iStockphoto.com/Alexander Hafemann; *bottom* ©iStockphoto.com/ medobear; **169** *left* ©iStockphoto.com/PeskyMonkey; *right* ©iStockphoto.com/shapecharge; **173** *top right* ©iStockphoto.com/ Dean Turner; *bottom* ©iStockphoto.com/Mladen Mladenov; *center* ©iStockphoto.com/Tom Buttle; **176** Junial Enterprises;**178** ©iStockphoto.com/ Louis Aguinaldo; **182** ©iStockphoto.com/Stefan Klein; **189** ©iStockphoto.com/Jorgen Jacobsen; **194** *exercise 9* ©iStockphoto.com/ Chih-Feng Chen; *exercise 10* ©iStockphoto.com/Andreas Gradin; *exercise 11* ©iStockphoto.com/Jim Lopes; **195** ©iStockphoto.com/Zoran Kolundzija; **198** *center* Booker Middle School; *bottom left* ©iStockphoto.com/ Black Jack 3D; *bottom right* ©iStockphoto.com/Vadym Volodin; **199** NASA; **202** ©iStockphoto.com/Evelyn Peyton; **203** *top left* ©iStockphoto.com/Lora Clark; *top right* ©iStockphoto.com/Terraxplorer; **212** Estate Craft Homes, Inc.; **228** ©iStockphoto.com/Michael Flippo, ©iStockphoto.com/Ann Marie Kurtz; **234** ©iStockphoto.com/Benjamin Lazare; **235** *center* ©iStockphoto.com/ Jill Chen; *top right* ©iStockphoto.com/Sheldon Kralstein; **236** ©ImageState **240** ©iStockphoto.com/Melissa Carroll; **241** *center right* ©iStockphoto.com/ Cathy Keifer; *center left* ©iStockphoto.com/Alex Slobodkin, ©iStockphoto.com/ Sebastian Duda; **243** *bottom right* ©iStockphoto.com/MACIEJ NOSKOWSKI; *center left* ©iStockphoto.com/Yvan Dubé **244** ©iStockphoto.com/Kais Tolmats; **248** *top left* ©iStockphoto.com/Don Bayley; *bottom right* ©iStockphoto.com/ iLexx; **251** ©iStockphoto.com/Marcio Silva; **253** Luminis; **256** *exercise 3* Joshua Haviv; *exercise 4* ©iStockphoto.com/William D Fergus McNeill; *exercise 5* ©iStockphoto.com/Klaas Jan Schraa; **257** ©iStockphoto.com/ Nikontiger; **267** *center right* Hasan Shaheed; *bottom left* ©iStockphoto.com/ JenDen2005; *bottom right* Orla; **268** *center right* CD Lenzen; *bottom left* red06; **272** ©iStockphoto.com/Alistair Cotton **278** ©iStockphoto.com/Jan Will; **279** Elena Elisseeva; **282** Frederic J. Brown/Getty Images; **284** ©iStockphoto.com/Mike Elliott; **285** Monkey Business Images; **287** ©iStockphoto.com/David15; **288** Gina Brockett; **293** ©iStockphoto.com/ Jill Fromer; **294** ©iStockphoto.com/Janis Litavnieks; **295** DLD; **296** *top right* Florida Park Service; *bottom right* ©iStockphoto.com/Eric Isselée; *center left* ©iStockphoto.com/Tony Campbell; **297** *top right* Larry Korhnak; *center right* Photo by Andy Newman; **301** *center left* ©iStockphoto.com/ Jane norton; *center right* ©iStockphoto.com/Krzysztof Zmij; **302** Dwight Smith; **304** Nikola Bilic; **305** WizData, inc.; **310** ©iStockphoto.com/Angel Rodriguez, ©iStockphoto.com/Ann Marie Kurtz; **313** ©iStockphoto.com/ Floortje; **317** *top* The ESRB rating icons are registered trademarks of the Entertainment Software Association; *center right* ©iStockphoto.com/Richard Goerg; **318** ©iStockphoto.com/George Peters; **319** ©iStockphoto.com/ George Peters; **322** pandapaw; **323** ©iStockphoto.com/Daniel Van Beek; **325** ©iStockphoto.com/Eric Simard; **331** Alexander Kalina; **332** *top right* ©iStockphoto.com/Martin Firus; *bottom* Robert Pernell; **333** ©iStockphoto.com/Trevor Fisher; **338** ©iStockphoto.com/fotoVoyager; **339** *exercise 26* ©iStockphoto.com/Jill Chen; *exercise 27* ©iStockphoto.com/ 7unit; *exercise 28* ©iStockphoto.com/George Clerk; **340** Tatiana Popova; **344** ©iStockphoto.com/Heather Shimmin, ©iStockphoto.com/itographer; **348** ©iStockphoto.com/Varina and Jay Patel, ©iStockphoto.com/Ann Marie Kurtz; **350** ©iStockphoto.com/Franck Boston; **351** *activity 3a* NASA; *activity 3b* NASA/JPL-Caltech/R. Hurt (SSC); *activity 3c* NASA; *activity 3d* NASA; *center right;* Stevyn Colgan; **353** ©iStockphoto.com/Philippa Banks; **354** ©iStockphoto.com/Clotilde Hulin; **355** ©iStockphoto.com/Boris Yankov; **356** ©iStockphoto.com/John Tomaselli; **360** ©iStockphoto.com/ Viktoriia Kulish; **361** *top right* ©iStockphoto.com/Paul Tessier; *center left* ©iStockphoto.com/Marie-france Bélanger, ©iStockphoto.com/Valerie Loiseleux, ©iStockphoto.com/Linda Steward; **366** ©iStockphoto.com/ Andrey Volodin; **367** *top right* Dash; **367** *center left* NASA/JPL-Caltech/ L. Cieza (UT Austin); **369** ©iStockphoto.com/Dan Moore; **373** ©iStockphoto.com/Aliaksandr Autayeu; **374** EugeneF; **375** ©iStockphoto.com/Nancy Louie; **376** ©iStockphoto.com/Kais Tolmats; **377** *center* ©iStockphoto.com/Henrik Jonsson; *top* ©iStockphoto.com/ Sebastian Kaulitzki; *bottom 7artscreensavers.com*; **379** *top center* ©iStockphoto.com/Frank Wright; *top left* ©iStockphoto.com/Mark Stay; *bottom left* ©iStockphoto.com/Oliver Sun Kim; *top right* ©iStockphoto.com/ Evgeniy Ivanov; **380** ©iStockphoto.com/ChristianAnthony; **381** Microgen; **382** ©iStockphoto.com/camilla wisbauer, ©iStockphoto.com/David Freund, ©iStockphoto.com/Joe Belanger, ©iStockphoto.com/thumb, ©iStockphoto.com/Marie-france Bélanger, ©iStockphoto.com/Susan Trigg; **383** NASA; **384** *center left* Google logo ©Google Inc., 2008. Reprinted with Permission.; *center right* YouTube logo ©Google Inc., 2008. Reprinted with Permission.; **385** Elaine Barker; **386** *bottom left* ©iStockphoto.com/angelhell; *bottom right* ©iStockphoto.com/Jan Rysavy; **387** BORTEL Pavel; **392** *bottom right* ©iStockphoto.com/Eric Holsinger; *center left* Eric Isselée; **A1** ©iStockphoto.com/Björn Kindler; **A4** *top right* ©iStockphoto.com/ Hsing-WenHsu; *center right* ©iStockphoto.com/blackred; *bottom left* ©iStockphoto.com/Thomas Kuest; *bottom right* ©iStockphoto.com/ Lim ChewHow; **A5** *top right* ©iStockphoto.com/Richard Cano; *bottom left* ©iStockphoto.com/best-photo; *bottom right* ©iStockphoto.com/Mika Makkonen; **A6** Emmer, Michele, ed., The Visual Mind: Art and Mathematics, Plate 2, ©1993 Massachusetts Institute of Technology, by permission of The MIT Press.; **A7** *top right* Elena Borodynkina; *center right* ©iStockphoto.com/ Andrew Cribb; *bottom left* ©iStockphoto.com/Matthew Okimi; *bottom right* Vladimir Bulatov; **A8** NASA; **A9** NASA; **A10** Koksharov Dmitry; **A11** Valva Vitaly; **A15** *top right* Dmitry Terentjev; *center left* ©iStockphoto.com/James Ferrie; *bottom right* ©iStockphoto.com/Nicholas Monu; **A17** patrimonio; **A19** ©iStockphoto.com/Roel Smart; **A21** ©iStockphoto.com/Jacom Stephens; **A23** *top right* ©iStockphoto.com/Brian sullivan; *center left* ©iStockphoto.com/ Justin Horrocks; **A25** Denis Miraniuk; **A27** ©iStockphoto.com/Rich Vintage; **A29** Andresr; **A36** ©iStockphoto.com/Roberto A Sanchez, ©iStockphoto.com/ Frank Leung; **A38** Elnur; **A39** Laurent Renault; **A41** Yuri Arcurs; **A42** ©iStockphoto.com/Feng Yu

Cartoon illustrations Tyler Stout

Cover image Lechner & Benson Design

Selected Answers

Section 1.1 Solving Simple Equations
(pages 7–9)

1. $+$ and $-$ are inverses. $\times$ and $\div$ are inverses.

3. $x - 3 = 6$; It is the only equation that does not have $x = 6$ as a solution.

5. $x = 57$ **7.** $x = -5$ **9.** $p = 21$ **11.** $x = 9\pi$ **13.** $d = \frac{1}{2}$ **15.** $n = -4.9$

17. a. $105 = x + 14$; $x = 91$

b. no; Because $82 + 9 = 91$, you did not knock down the last pin with the second ball of the frame.

19. $n = -5$ **21.** $m = 7.3\pi$ **23.** $k = 1\frac{2}{3}$ **25.** $p = -2\frac{1}{3}$

27. They should have added 1.5 to each side.

$-1.5 + k = 8.2$

$k = 8.2 + 1.5$

$k = 9.7$

29. $6.5x = 42.25$; \$6.50 per hour

31. $420 = \frac{7}{6}b$, $b = 360$; \$60

33. $h = -7$ **35.** $q = 3.2$ **37.** $x = -1\frac{4}{9}$

39. greater than; Because a negative number divided by a negative number is a positive number.

41. 3 mg **43.** 8 in. **45.** $7x - 4$ **47.** $\frac{25}{4}g - \frac{2}{3}$

Section 1.2 Solving Multi-Step Equations
(pages 14 and 15)

1. $2 + 3x = 17$; $x = 5$ **3.** $k = 45$; $45°, 45°, 90°$ **5.** $b = 90$; $90°, 135°, 90°, 90°, 135°$

7. $c = 0.5$ **9.** $h = -9$ **11.** $x = -\frac{2}{9}$ **13.** 20 watches

15. $4(b + 3) = 24$; 3 in. **17.** $\frac{129 + 147.7 + x}{3} = 150$; 173.3 million boxes

19. $<$ **21.** $>$

Section 1.3 Solving Equations with Variables on Both Sides
(pages 20 and 21)

1. no; When 3 is substituted for x, the left side simplifies to 4 and the right side simplifies to 3.

3. $x = 13.2$ in. **5.** $x = 7.5$ in. **7.** $k = -0.75$

9. $p = -48$ **11.** $n = -3.5$ **13.** $x = -4$

15. The 4 should have been added to the right side.

$$\begin{aligned} 3x - 4 &= 2x + 1 \\ 3x - 2x - 4 &= 2x + 1 - 2x \\ x - 4 &= 1 \\ x - 4 + 4 &= 1 + 4 \\ x &= 5 \end{aligned}$$

17. $15 + 0.5m = 25 + 0.25m$; 40 mi

19. 7.5 units

21. Remember that the box is with priority mail and the envelope is with express mail.

23. 10 mL

25. square: 12 units; triangle: 10 units, 19 units, 19 units

27. 24 in.^{3}

29. C

Section 1.4 Rewriting Equations and Formulas
(pages 28 and 29)

1. no; The equation only contains one variable.

3. **a.** $A = \frac{1}{2}bh$ **b.** $b = \frac{2A}{h}$ **c.** $b = 12$ mm

5. $y = 4 - \frac{1}{3}x$

7. $y = \frac{2}{3} - \frac{4}{9}x$

9. $y = 3x - 1.5$

11. The y should have a negative sign in front of it.

$$\begin{aligned} 2x - y &= 5 \\ -y &= -2x + 5 \\ y &= 2x - 5 \end{aligned}$$

13. **a.** $t = \frac{I}{Pr}$

b. $t = 3$ yr

15. $m = \frac{e}{c^2}$

17. $\ell = \frac{A - \frac{1}{2}\pi w^2}{2w}$

19. $w = 6g - 40$

21. **a.** $F = 32 + \frac{9}{5}(K - 273.15)$

b. 32°F

c. liquid nitrogen

23. $r^3 = \frac{3V}{4\pi}$; $r = 4.5$ in.

25. $6\frac{2}{5}$

27. $1\frac{1}{4}$

Section 1.5 Converting Units of Measure
(pages 35–37)

1. yes; Because 1 centimeter is equal to 10 millimeters, the conversion factor equals 1.

3. 6.25 ft; The other three represent the same length.

5. 11 yd, 33 ft

7. 12.63

9. 1.22

11. 0.19

13. 37.78

15. 14.4

17. **a.** 57.93 m

b. 8.85 km

19. 1320

21. 112.5

23. 0.001

25. 0.99 mL/sec

27. 80

29. **a.** spine-tailed swift; mallard

b. yes, It is faster than all of the other birds in the table. Its dive speed is about 201.25 miles per hour.

31. 34,848

33. 3,000,000,000

35. 0.00042

Section 1.5 Converting Units of Measure *(continued)*
(pages 35–37)

37. **a.** 120 in.^{3} **b.** 138 tissues

39. 113,000 mm^3

41–43.

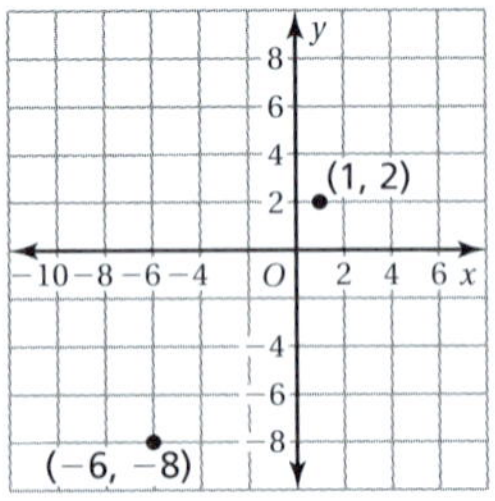

45. B

Section 2.1 Graphing Linear Equations
(pages 52 and 53)

1. a line

3. *Sample answer:*

x	0	1
$y = 3x - 1$	−1	2

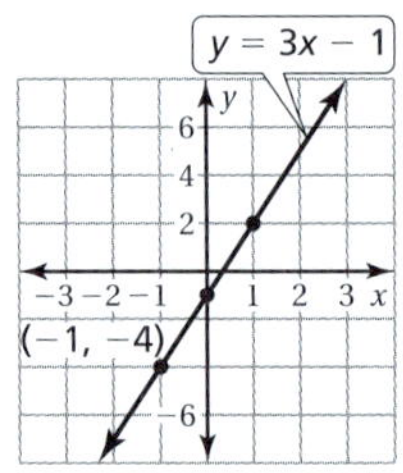

5.

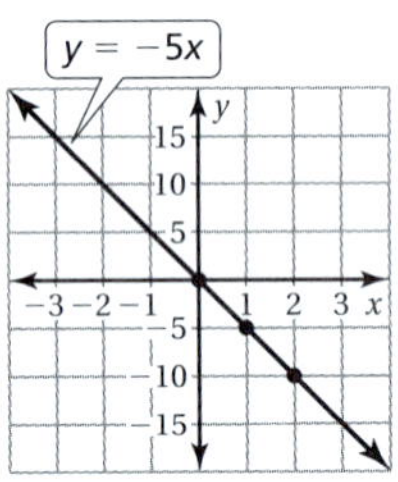

7.

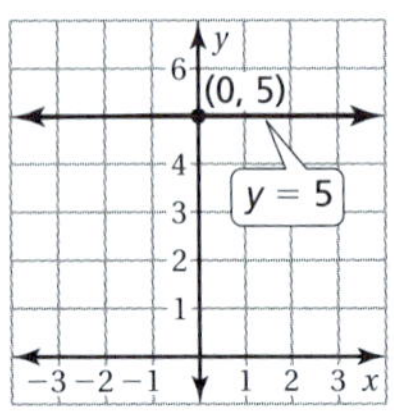

9.

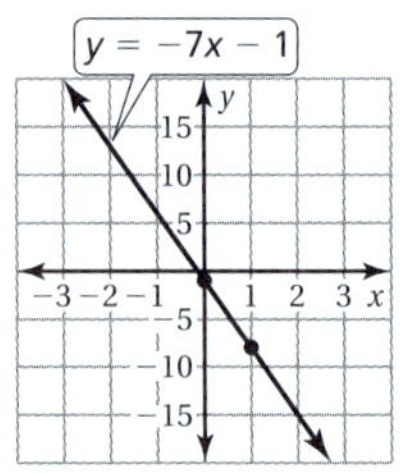

11.

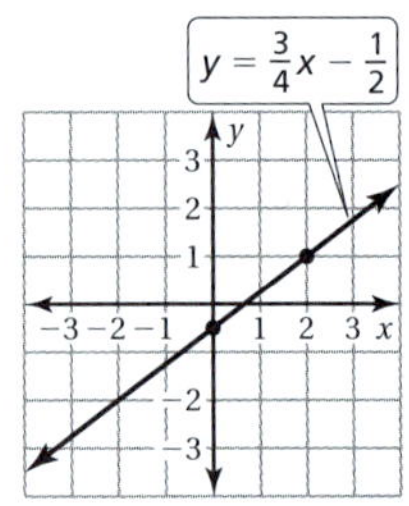

13.

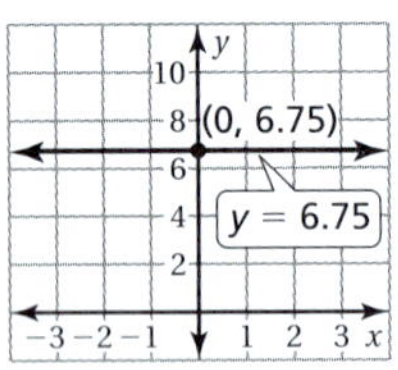

15.

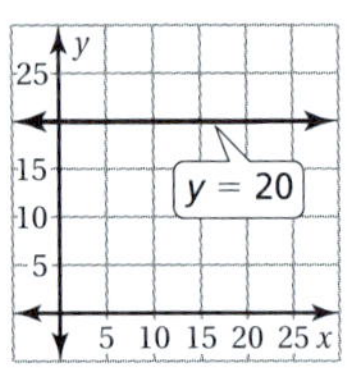

17. $y = 3x + 1$

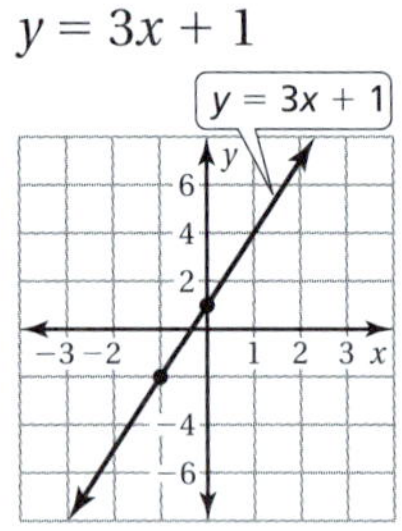

19. $y = 12x - 9$

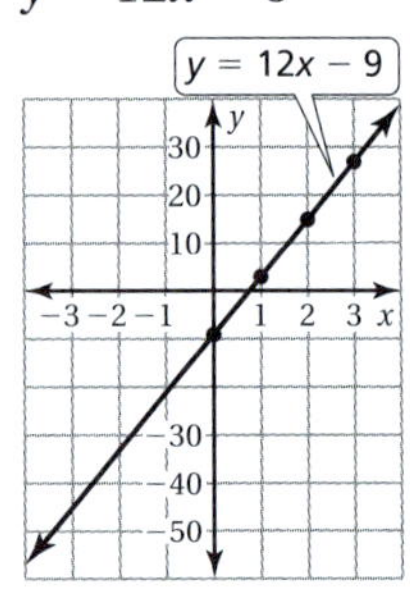

21. **a.** $y = 100 + 12.5x$

b. 6 mo

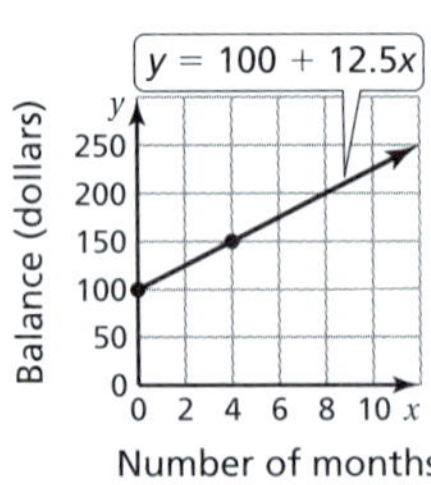

23. **a.** $y = 3x$

b. *Sample answer:* If you are 13 years old, the sea level has risen 39 millimeters since you were born.

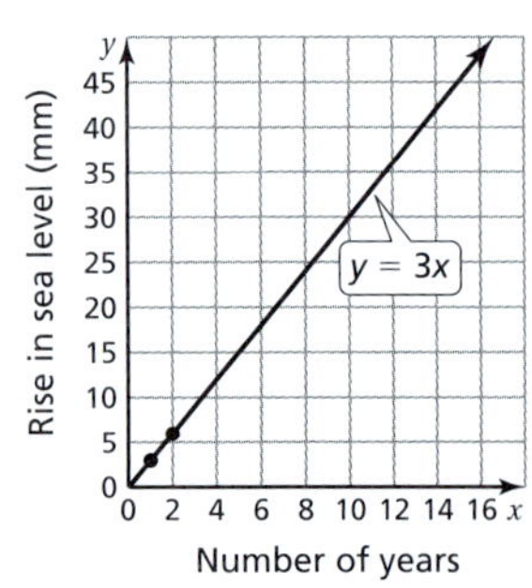

25. (5, 3) **27.** (2, −2) **29.** B

Section 2.2 Slope of a Line (pages 59–61)

1. **a.** B and C

b. A

c. no; All of the slopes are different.

3. The line is horizontal.

5.

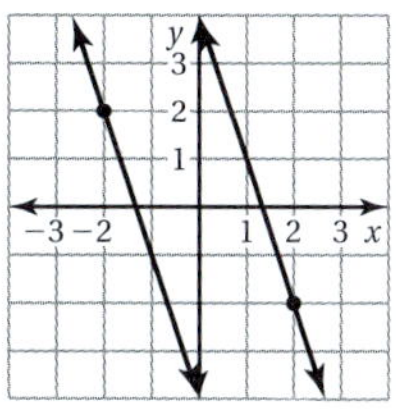

The lines are parallel.

7. $\frac{3}{4}$

9. $-\frac{3}{5}$

11. 0

13. The 2 should be -2 because it goes down.

Slope $= -\frac{2}{3}$

15. 4

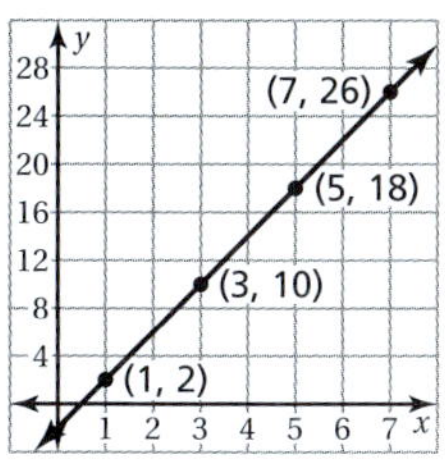

17. $-\frac{3}{4}$

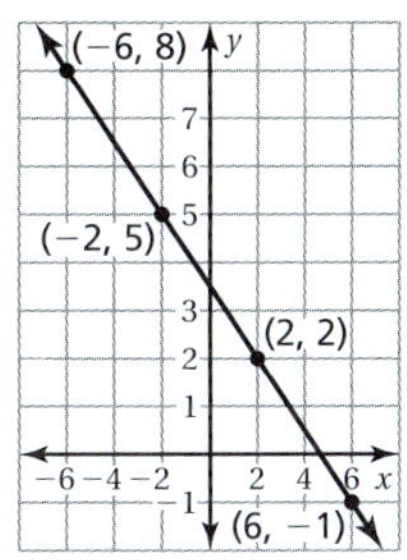

19. $\frac{1}{3}$

21. red and green; They both have a slope of $\frac{4}{3}$.

23. no; Opposite sides have different slopes.

25. **a.** $\frac{3}{40}$

b. The cost increases by \$3 for every 40 miles you drive, or the cost increases by \$0.075 for every mile you drive.

27. You can draw the slide in a coordinate plane and let the x-axis be the ground to find the slope.

29.

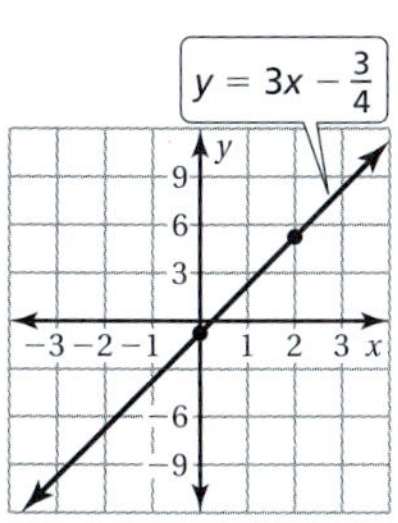

31. B

Section 2.3 Graphing Linear Equations in Slope-Intercept Form *(pages 66 and 67)*

1. Find the x-intercept of the point where the graph crosses the x-axis.

3. *Sample answer:* The amount of gasoline y (in gallons) left in your tank after you travel x miles is $y = \frac{1}{-20}x + 20$. The slope of $-\frac{1}{20}$ means the car uses 1 gallon of gas for every 20 miles driven. The y-intercept of 20 means there is originally 20 gallons of gas in the tank.

5. A; slope: $\frac{1}{3}$; y-intercept: -2

7. slope: 4; y-intercept: -5

Section 2.3 Graphing Linear Equations in Slope-Intercept Form *(continued) (pages 66 and 67)*

9. slope: $-\frac{4}{5}$; y-intercept: -2

11. slope: $\frac{4}{3}$; y-intercept: -1

13. slope: -2; y-intercept: 3.5

15. slope: 1.5; y-intercept: 11

17. a.

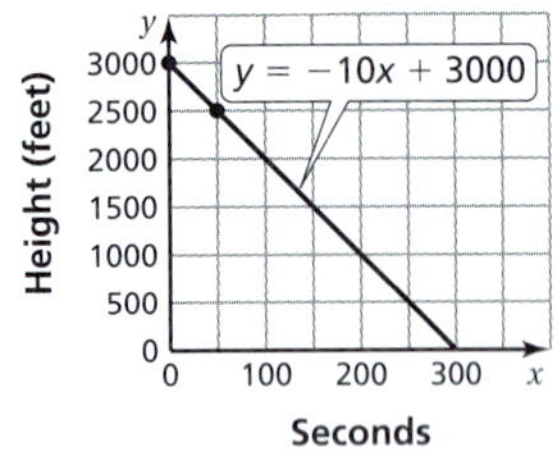

b. The x-intercept of 300 means the skydiver lands on the ground after 300 seconds. The slope of -10 means that the skydiver falls to the ground at a rate of 10 feet per second.

19.

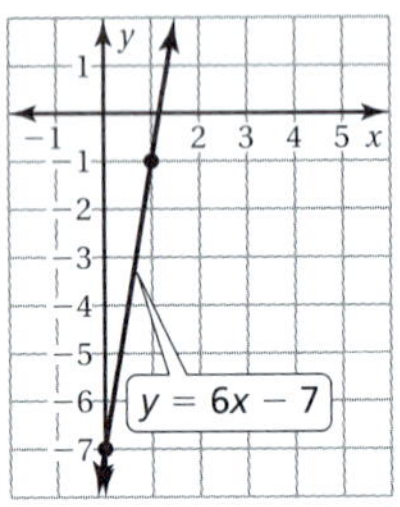

x-intercept: $\frac{7}{6}$

21.

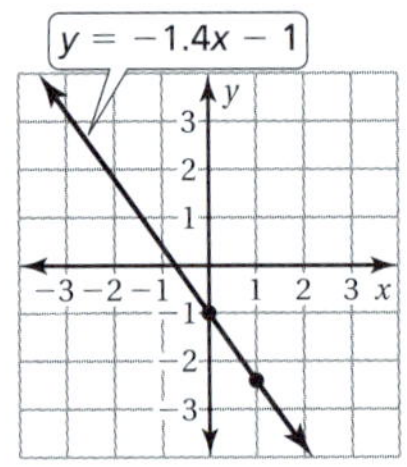

x-intercept: $-\frac{5}{7}$

23.

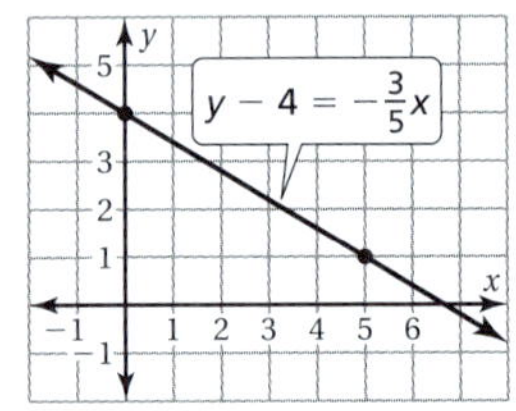

x-intercept: $\frac{20}{3}$

25. $y = 0.75x + 5$

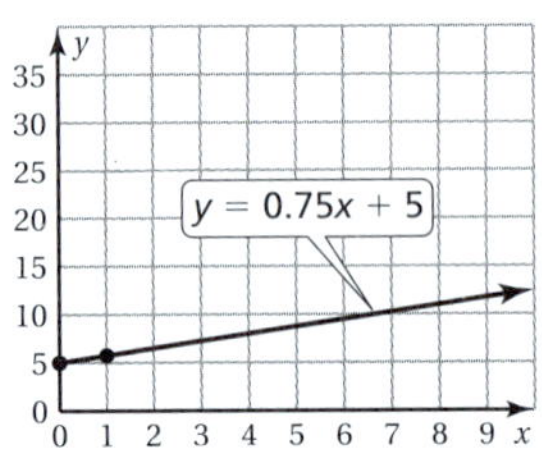

27. $y = 0.15x + 35$

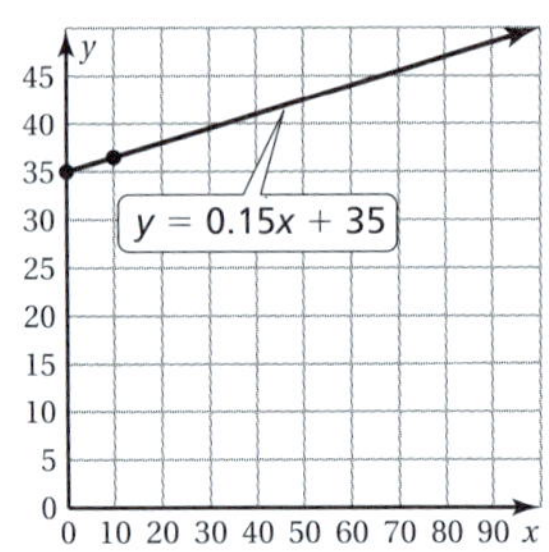

29. $y = 2x + 3$

31. $y = \frac{2}{3}x - 2$

33. B

Section 2.4 Graphing Linear Equations in Standard Form *(pages 72 and 73)*

1. no; The equation is in slope-intercept form.

3. x = pounds of peaches

y = pounds of apples

$y = -\frac{4}{3}x + 10$

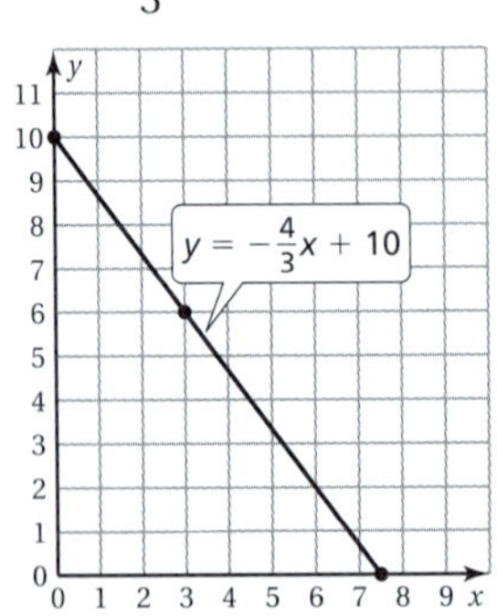

5. $y = -2x + 17$

7. $y = \frac{1}{2}x + 10$

9.

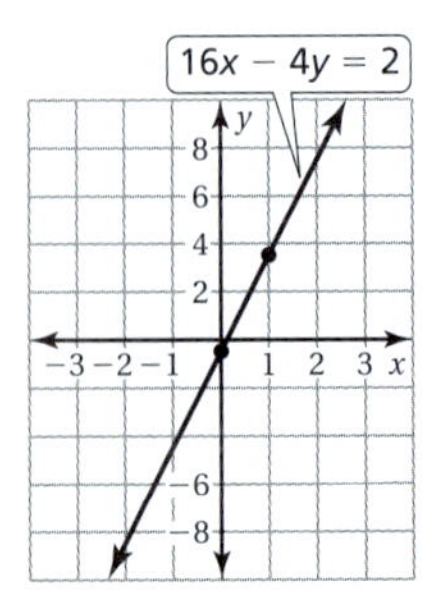

11. x-intercept: -6

y-intercept: 3

13. x-intercept: none

y-intercept: -3

15. a. $y - 25x = 65$

b. $390

17. 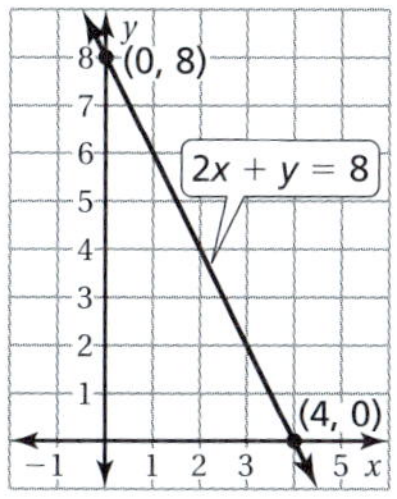

19. x-intercept: 9

y-intercept: 7

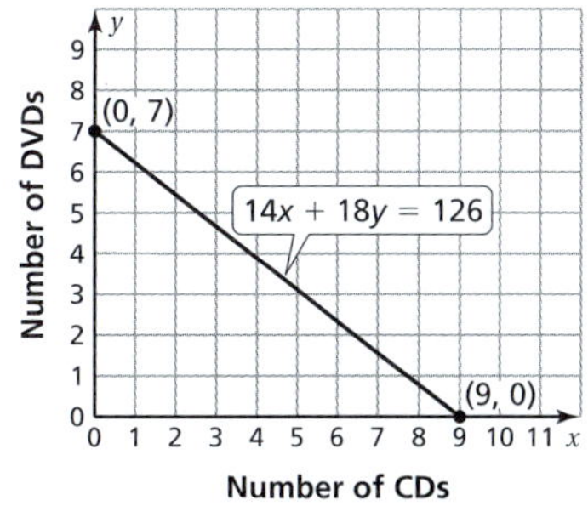

21. **a.** $9.45x + 7.65y = 160.65$

b. 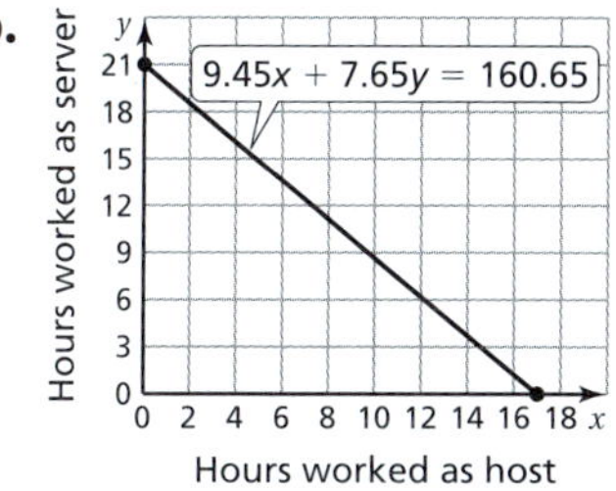

23. **a.** $y = 40x + 70$

b. x-intercept: $-\frac{7}{4}$; It will not be on the graph because you cannot have a negative time.

c. 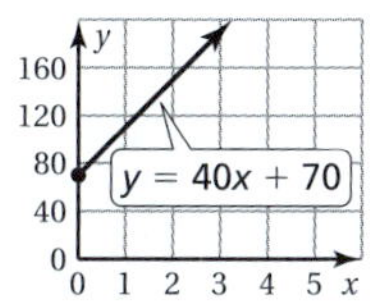

25.

x	-2	-1	0	1	2
$-5 - 3x$	1	-2	-5	-8	-11

Section 2.5 Systems of Linear Equations
(pages 80 and 81)

1. yes; The set of equations are linear and in the same variables.

3.

x	0	1	2	3	4	5	6
C	150	165	180	195	210	225	240
R	0	45	90	135	180	225	257

(5, 225)

5. (2.5, 6.5)

7. (3, −1)

9. **a.** $R = 35x$ **b.** 100 rides

11. (−5, 1)

13. (12, 15)

15. (8, 1)

17. **a.** 6 h **b.** 49 mi

19. yes

21. no

Section 2.6 Special Systems of Linear Equations
(pages 86 and 87)

1. The graph of the system with no solution has two parallel lines, and the graph of a system with infinitely many solutions is one line.

3. one solution; because the lines are not parallel and will not be the same equation

5. no solution

7. infinitely many solutions; all points on the line $y = -\frac{1}{6}x + 5$

9. one solution; (2, −3)

11. no solution

13. no; because they are running at the same speed and your pig had a head start

15. no solution

17. **a.** 6 h

b. You both work the same number of hours.

Section 2.6 Special Systems of Linear Equations *(continued) (pages 86 and 87)*

19. **a.** *Sample answer:* $y = -7$

b. *Sample answer:* $y = 3x$

c. *Sample answer:* $2y - 6x = -2$

21. $x = -3$

23. B

Section 2.7 Solving Equations by Graphing *(pages 92 and 93)*

1. algebraic method; Graphing fractions is harder than solving the equation.

3. $x = 6$

5. $x = 6$

7. $x = 3$

9. yes; Because a solution to $3x + 2 = 4x$ exists ($x = 2$).

11. $x = 2$

13. The two lines are parallel, which means there is no solution. Using an algebraic method, you obtain $-5 = 8$, which is not true and means that there is no solution.

15. Organize the home and away games for last year and this year in a table before solving.

17. 4

19. -3

21. A

Section 3.1 Writing Equations in Slope-Intercept Form *(pages 110 and 111)*

1. *Sample answer:* Find the ratio of the rise to the run of the intercepts.

3. $y = 3x + 2$; $y = 3x - 10$; $y = 5$; $y = -1$

5. $y = x + 4$

7. $y = \frac{1}{4}x + 1$

9. $y = \frac{1}{3}x - 3$

11. The x-intercept was used instead of the y-intercept. $y = \frac{1}{2}x - 2$

13. $y = 5$

15. $y = -2$

17. **a–b.**

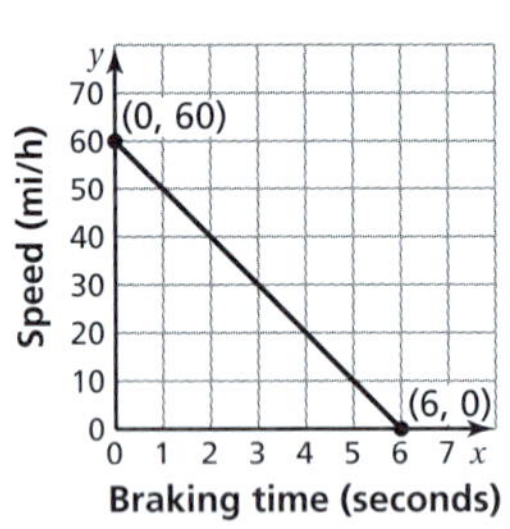

(0, 60) represents the speed of the automobile before braking. (6, 0) represents the amount of time it takes to stop. The line represents the speed y of the automobile after x seconds of braking.

c. $y = -10x + 60$

19. Be sure to check that your rate of growth will not lead to a 0-year-old tree with a negative height.

21–23.

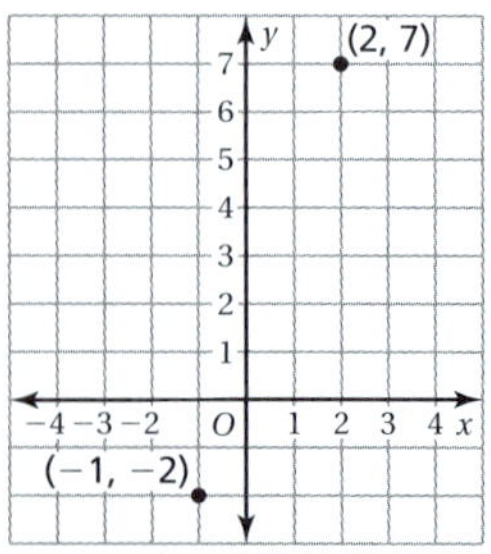

Section 3.2 Writing Equations Using a Slope and a Point (pages 116 and 117)

1. *Sample answer:* slope and a point

3. $y = \frac{1}{2}x + 1$ **5.** $y = -3x + 8$ **7.** $y = \frac{3}{4}x + 5$

9. $y = -\frac{1}{7}x - 4$ **11.** $y = -2x - 6$ **13.** $V = \frac{2}{25}T + 22$

15. The rate of change is 0.25 degree per chirp.

17. **a.** $y = -0.03x + 2.9$

b. 2 g/cm^2

c. *Sample answer:* Eventually $y = 0$, which means the astronaut's bones will be very weak.

19. B

Section 3.3 Writing Equations Using Two Points (pages 122 and 123)

1. Plot both points and draw the line that passes through them. Use the graph to find the slope and y-intercept. Then write the equation in slope-intercept form.

3. slope $= -1$; y-intercept: 0; $y = -x$

5. slope $= \frac{1}{3}$; y-intercept: -2; $y = \frac{1}{3}x - 2$

7. $y = 2x$ **9.** $y = \frac{1}{4}x$ **11.** $y = x + 1$ **13.** $y = \frac{3}{2}x - 10$

15. They switched the slope and y-intercept in the equation. $y = 2x - 4$

17. **a.**

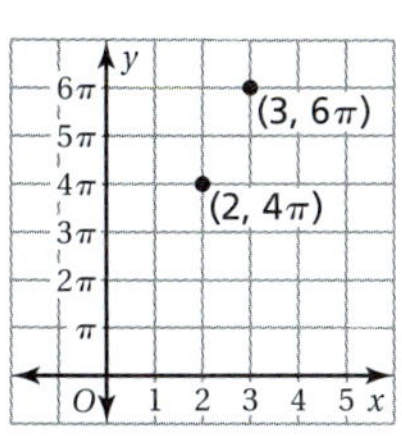

b. $y = 2\pi x$

19. **a.** $y = -2000x + 21{,}000$

b.

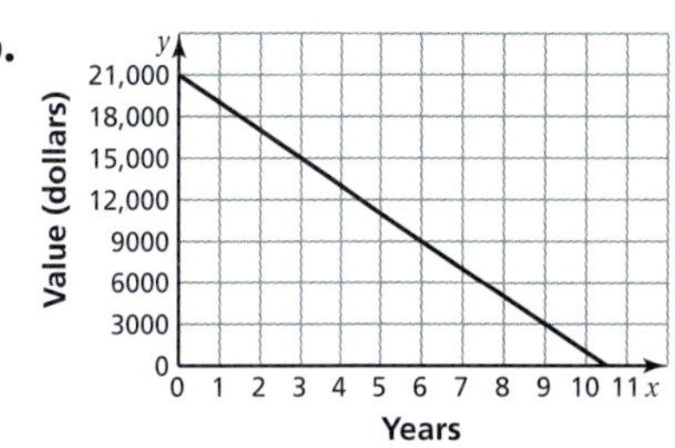

c. \$21,000; the original price of the car

21. **a.** $y = 14x - 108.5$

b. 4 m

23. 175 **25.** D

Section 3.4 Solving Real-Life Problems
(pages 130 and 131)

1. The y-intercept is -6 because the line crosses the y-axis at the point $(0, -6)$. The x-intercept is 2 because the line crosses the x-axis at the point $(2, 0)$. You can use these two points to find the slope.

   $\text{Slope} = \frac{\text{change in } y}{\text{change in } x} = \frac{6}{2} = 3$

3. *Sample answer:* the rate at which something is happening

5. *Sample answer:* On a visit to Mexico, you spend 45 pesos every week. After 4 weeks, you have no pesos left.

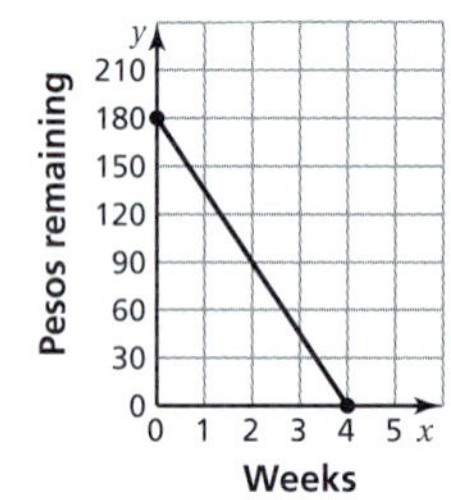

7. **a.** slope: -3.6; y-intercept: 59 **b.** $y = -3.6x + 59$

   c. 59°F

9. **a.** Antananarivo: 19°S, 47°E; Denver: 39°N, 105°W; Brasilia: 16°S, 48°W; London: 51°N, 0°W; Beijing: 40°N, 116°E

   b. $y = \frac{1}{221}x + \frac{8724}{221}$

   c. a place that is on the prime meridian

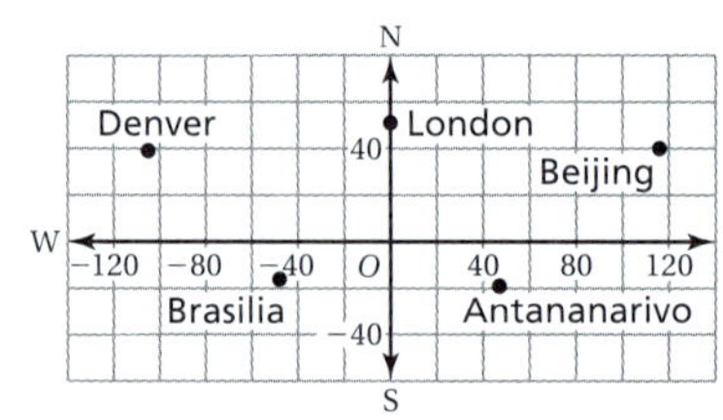

11. infinitely many solutions

13. no solution

Section 3.5 Writing Systems of Linear Equations
(pages 136 and 137)

1. because its graph is a line

3. You can use a table to see when the two equations are equal. You can use a graph to see whether or not the two lines intersect. You can use algebra and set the equations equal to each other to see when they have the same value.

5. **a.** $x + y = 12$

   $3x + 2y = 32$

   b.

x	0	1	2	3	4	5	6	7	8
$y = 12 - x$	12	11	10	9	8	7	6	5	4
$y = 16 - \frac{3}{2}x$	16	14.5	13	11.5	10	8.5	7	5.5	4

   8 lilies and 4 tulips

   c.

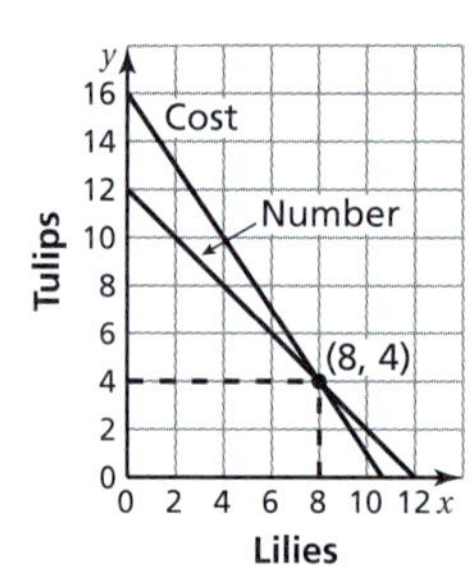

   8 lilies and 4 tulips

   d. $12 - x = 16 - \frac{3}{2}x$; $x = 8$; 8 lilies and 4 tulips

7. **a.** no; You need to know how many more dimes there are than nickels or how many coins there are total.

   b. *Sample answer:* 9 dimes and 1 nickel

9. no; A linear system must have either one, none, or infinitely many solutions. Lines cannot intersect at exactly two points.

11. Each equation is the same. So, the graph of the system is the same line.

13. $(1, 0), (-2, 3), (-6, 1)$

15. $y = \frac{1}{4}x - 2$

17. B

Section 4.1 Domain and Range of a Function
(pages 152 and 153)

1. no; The equation is not solved for *y*.

3. **a.** $y = 6 - 2x$ **b.** domain: 0, 1, 2, 3; range: 6, 4, 2, 0

   c. $x = 6$ is not in the domain because it would make *y* negative, and it is not possible to buy a negative number of headbands.

5. domain: −2, −1, 0, 1, 2; range: −2, 0, 2

7. The domain and range are switched. The domain is −3, −1, 1, and 3. The range is −2, 0, 2, and 4.

9.

x	−1	0	1	2
y	−4	2	8	14

domain: −1, 0, 1, 2

range: −4, 2, 8, 14

11.

x	−1	0	1	2
y	1.5	3	4.5	6

domain: −1, 0, 1, 2

range: 1.5, 3, 4.5, 6

13. Rewrite the percent as a fraction or decimal before writing an equation.

15. 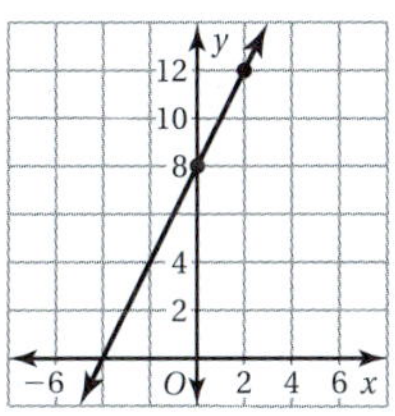

17. 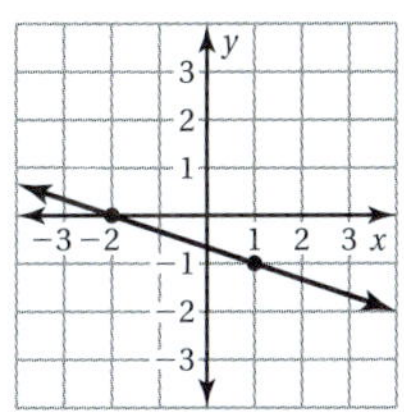

19. D

Section 4.2 Discrete and Continuous Domains
(pages 158 and 159)

1. A discrete domain consists of only certain numbers in an interval, whereas a continuous domain consists of all numbers in an interval.

3. domain: $x \geq 0$ and $x \leq 6$,

   range: $y \geq 0$ and $y \leq 6$;

   continuous

5. 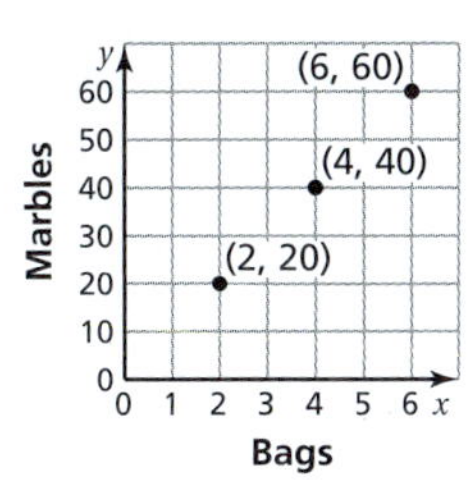

discrete

7. 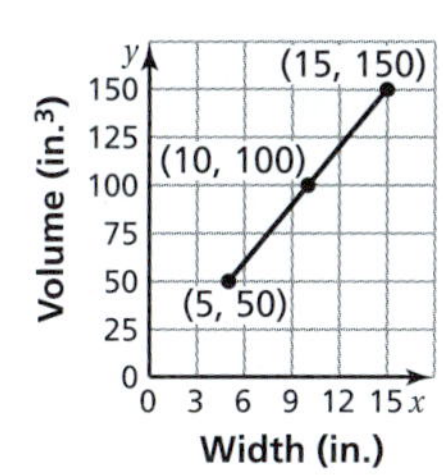

continuous

9. The domain is discrete because only certain numbers are inputs.

11. The function with an input of length has a continuous domain because you can use any length, but you cannot have half a shirt.

13. 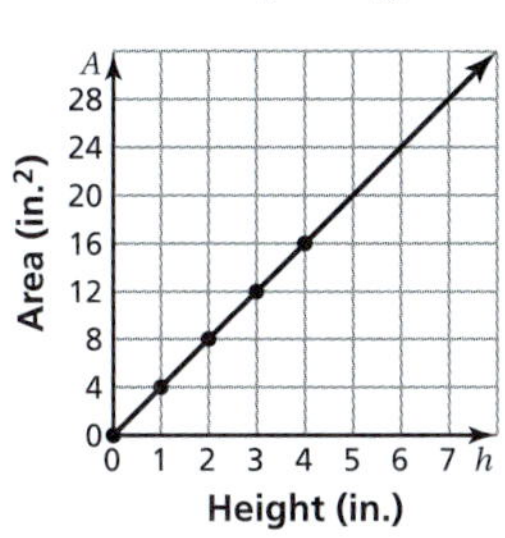

continuous

15. Before writing a function, draw one possible arrangement to understand the problem.

17. $-\frac{5}{2}$

19. C

Selected Answers

Section 4.3 Linear Function Patterns
(pages 166 and 167)

1. words, equation, table, graph

3. $y = \pi x$; x is the diameter; y is the circumference.

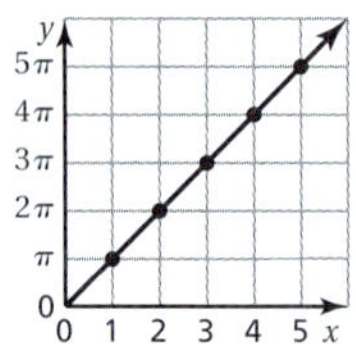

5. $y = \frac{4}{3}x + 2$

7. $y = 3$

9. $y = -\frac{1}{4}x$

11. a.

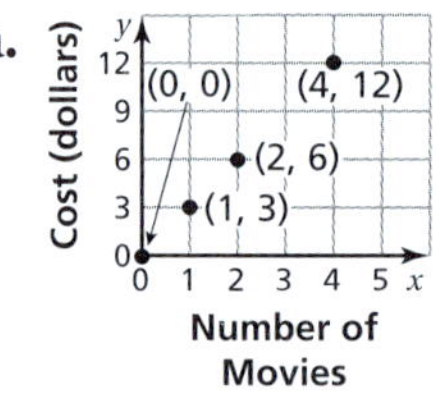

discrete

b. $y = 3x$

c. $9

13. Substitute 8 for t in the equation.

15. 5%

17. B

Section 4.4 Comparing Linear and Nonlinear Functions
(pages 172 and 173)

1. A linear function has a constant rate of change. A nonlinear function does not have a constant rate of change.

3.

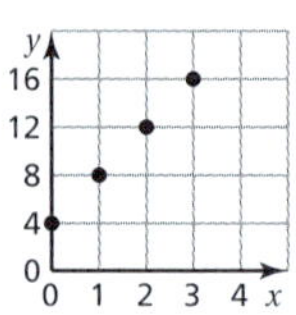

linear

5.

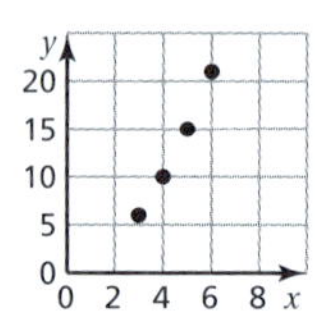

nonlinear

7. linear; The graph is a line.

9. linear; As x increases by 6, y increases by 4.

11. nonlinear; As x increases by 1, V increases by different amounts.

13. linear; The equation can be written in slope-intercept form.

15. Because you want the table to represent a linear function and 3 is half-way between 2 and 4, the missing value is half-way between 2.80 and 5.60.

17. nonlinear; The graph is not a line.

19. linear

21. straight

23. right

Section 5.1 Classifying Angles
(pages 188 and 189)

1. The sum of two complementary angles is 90°. The sum of two supplementary angles is 180°.

3. sometimes; Either x or y may be obtuse.

5. never; Because x and y must both be less than 90° and greater than 0°.

7. complementary

9. supplementary

11. neither

13. 128

15. Vertical angles are congruent. The value of x is 35.

17. 37

19. 20

21. **a.** $\angle CBD$ and $\angle DBE$; $\angle ABF$ and $\angle FBE$

b. $\angle ABE$ and $\angle CBE$; $\angle ABD$ and $\angle CBD$; $\angle CBF$ and $\angle ABF$

23. 54°

25. $7x + y + 90 = 180$; $5x + 2y = 90$; $x = 10$; $y = 20$

27. 29.3

29. B

Section 5.2 Angles and Sides of Triangles
(pages 194 and 195)

1. An equilateral triangle has three congruent sides. An isosceles triangle has at least two congruent sides. So, an equilateral triangle is a specific type of isosceles triangle.

3. right isosceles triangle

5. obtuse isosceles triangle

7. 94; obtuse triangle

9. 67.5; acute isosceles triangle

11. 24; obtuse isosceles triangle

13. **a.** 70 **b.** acute isosceles triangle

15. no; 39.5°

17. yes

19. If two angles of a triangle are greater than or equal to 90°, the sum of those two angles will be greater than or equal to 180°. The sum of the three angles will be greater than 180°, which is not possible.

21. $x + 2x + 2x + 8 + 5 = 48$; 7

23. $4x - 4 + 3\pi = 25.42$ or $2x - 4 = 6$; 5

Section 5.3 Angles of Polygons
(pages 201–203)

1. 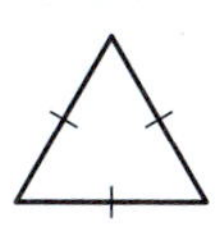

3. What is the measure of an angle of a regular pentagon?; 108°; 540°

5. 1260°

7. 720°

9. 1080°

11. no; The angles given add up to 535°, but the sum of the angles of a pentagon is 540°.

13. 135

15. 140°

17. 140°

19. The sum of the angle measures should have been divided by the number of angles, 20. $3240° \div 20 = 162°$; The measure of each angle is 162°.

21. 24 sides

23. convex; No line segment connecting two vertices lies outside the polygon.

25. no; All of the angles would not be congruent.

27. 135°

29. 120°

31. You can determine if it is a linear function by writing an equation or by graphing the points.

33. 9

35. 3

37. D

Section 5.4 Using Similar Triangles
(pages 210 and 211)

1. Write a proportion that uses the missing measurement because the ratios of corresponding side lengths are equal.

3. Student should draw a triangle with the same angle measures as the textbook. The ratio of the corresponding side lengths, $\frac{\text{student's triangle length}}{\text{book's triangle length}}$, should be greater than one.

5. yes; The triangles have the same angle measures, 107°, 39°, and 34°.

7. no; The triangles do not have the same angle measures.

9. The numerators of the fractions should be from the same triangle.

$$\frac{18}{16} = \frac{x}{8}$$
$$16x = 144$$
$$x = 9$$

11. 65

13. no; Each side increases by 50%, which is a factor of $\frac{3}{2}$. The area is $\frac{3}{2}\left(\frac{3}{2}\right) = \frac{9}{4}$ or 225% of the area of the original, which is a 125% increase.

15. When two triangles are similar, the ratios of corresponding sides are equal.

17. linear; The equation can be rewritten in slope-intercept form.

19. nonlinear; The equation cannot be rewritten in slope-intercept form.

Section 5.5 Parallel Lines and Transversals
(pages 217–219)

1. *Sample answer:*

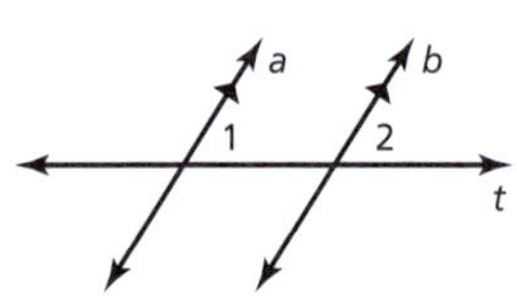

3. m and n

5. 8

7. $\angle 1 = 107°$, $\angle 2 = 73°$

9. $\angle 5 = 49°$, $\angle 6 = 131°$

11. 60°; Corresponding angles are congruent.

13. $\angle 1$, $\angle 3$, $\angle 5$ and $\angle 7$ are congruent; $\angle 2$, $\angle 4$, $\angle 6$ and $\angle 8$ are congruent.

15. $\angle 6 = 61°$; $\angle 6$ and the given angle are vertical angles. $\angle 5 = 119°$ and $\angle 7 = 119°$; $\angle 5$ and $\angle 7$ are supplementary to the given angle.
$\angle 1 = 61°$; $\angle 1$ and the given angle are corresponding angles.
$\angle 3 = 61°$; $\angle 1$ and $\angle 3$ are vertical angles.
$\angle 2 = 119°$ and $\angle 4 = 119°$; $\angle 2$ and $\angle 4$ are supplementary to $\angle 1$.

17. $\angle 2 = 90°$; $\angle 2$ and the given angle are vertical angles.
$\angle 1 = 90°$ and $\angle 3 = 90°$; $\angle 1$ and $\angle 3$ are supplementary to the given angle.
$\angle 4 = 90°$; $\angle 4$ and the given angle are corresponding angles.
$\angle 6 = 90°$; $\angle 4$ and $\angle 6$ are vertical angles.
$\angle 5 = 90°$ and $\angle 7 = 90°$; $\angle 5$ and $\angle 7$ are supplementary to $\angle 4$.

19. 132°; *Sample answer:* $\angle 2$ and $\angle 4$ are alternate interior angles and $\angle 4$ and $\angle 3$ are supplementary.

21. 120°; *Sample answer:* $\angle 6$ and $\angle 8$ are alternate exterior angles.

23. 61.3°; *Sample answer:* $\angle 3$ and $\angle 1$ are alternate interior angles and $\angle 1$ and $\angle 2$ are supplementary.

25. They are all right angles because perpendicular lines form 90° angles.

27. 130

29. **a.** no; They look like they are spreading apart. **b.** Check students' work.

31. 13

33. 51

35. B

Section 6.1 Finding Square Roots
(pages 234 and 235)

1. no; There is no integer whose square is 26.

3. $\sqrt{256}$ represents the positive square root because there is not a – or a ± in front.

5. 1.3 km

7. 3 and −3

9. 2 and −2

11. 25

13. $\frac{1}{31}$ and $-\frac{1}{31}$

15. 2.2 and −2.2

17. The positive and negative square roots should have been given.
$\pm\sqrt{\frac{1}{4}} = \frac{1}{2}$ and $-\frac{1}{2}$

19. 9

21. 25

23. 40

25. because a negative radius does not make sense

27. =

29. 9 ft

31. 8 m/sec

33. 2.5 ft

35. 25

37. 144

39. B

Section 6.2 The Pythagorean Theorem
(pages 240 and 241)

1. The hypotenuse is the longest side and the legs are the other two sides.

3. 24 cm

5. 9 in.

7. 12 ft

9. The length of the hypotenuse was substituted for the wrong variable.

$$
\begin{aligned}
a^2 + b^2 &= c^2 \\
7^2 + b^2 &= 25^2 \\
49 + b^2 &= 625 \\
b^2 &= 576 \\
b &= 24
\end{aligned}
$$

11. 16 cm

13. 10 ft

15. 8.4 cm

17. **a.** *Sample answer:*

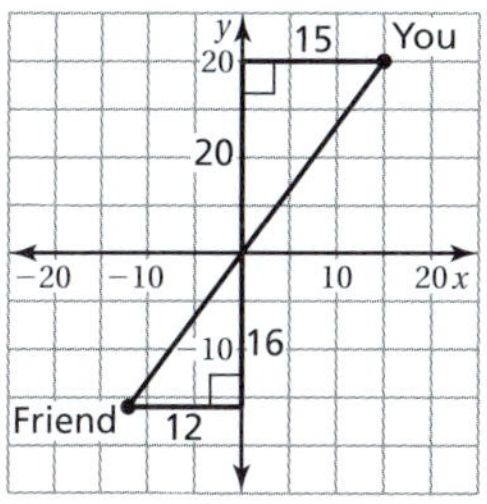

b. 45 ft

19. 6 and −6

21. 13

23. C

Section 6.3 Approximating Square Roots
(pages 249–251)

1. A rational number can be written as the ratio of two integers. An irrational number cannot be written as the ratio of two integers.

3. all rational and irrational numbers; *Sample answer:* $-2, \frac{1}{8}, \sqrt{7}$

5. yes

7. no

9. rational; $3.\overline{6}$ is a repeating decimal.

11. irrational; 7 is not a perfect square.

13. rational; $-3\frac{8}{9}$ can be written as the ratio of two integers.

15. 144 is a perfect square. So, $\sqrt{144}$ is rational.

17. **a.** natural number **b.** irrational number **c.** irrational number

19. 26

21. -10

23. -13

25. 10; 10 is to the right of $\sqrt{20}$.

27. $\sqrt{133}$; $\sqrt{133}$ is to the right of $10\frac{3}{4}$.

29. -0.25; -0.25 is to the right of $-\sqrt{0.25}$.

31. 8 ft

33. *Sample answer:* $a = 82$, $b = 97$

35. 1.1

37. 30.1 m/sec

39. Falling objects do not fall at a linear rate. Their speed increases with each second they are falling.

41. $-3x + 3y$

43. $40k - 9$

Section 6.4 Simplifying Square Roots
(pages 256 and 257)

1. *Sample answer:* The square root is like a variable. So, you add or subtract the number in front to simplify.

3. about 1.62; yes

5. about 1.11; no

7. $\frac{\sqrt{7}+1}{3}$

9. $6\sqrt{3}$

11. $2\sqrt{5}$

13. $-7.7\sqrt{15}$

15. You do not add the radicands. $4\sqrt{5} + 3\sqrt{5} = 7\sqrt{5}$

17. $10\sqrt{2}$

19. $4\sqrt{3}$

21. $\frac{\sqrt{23}}{8}$

23. $\frac{\sqrt{17}}{7}$

25. $10\sqrt{2}$ in.

27. $6\sqrt{6}$

29. 210 ft^3

31. **a.** $38.8\sqrt{7}$ mi **b.** 638.4 mi^2

33. Remember to take the square root of each side when solving for r.

35. 24 in.

37. C

Section 6.5 Using the Pythagorean Theorem
(pages 262 and 263)

1. *Sample answer:* You can plot a point at the origin and then draw lengths that represent the legs. Then, you can use the Pythagorean Theorem to find the hypotenuse of the triangle.

3. 27.7 m 5. 11.3 yd 7. 7.2 units 9. 27.5 ft 11. 15.1 m

13. yes 15. no 17. yes 19. 12.8 ft

21. a. *Sample answer:* 5 in., 7 in., 3 in.

  b. *Sample answer:* $BC \approx 8.6$ in.; $AB \approx 9.1$ in.

  c. Check students' work.

23. mean: 13; median: 12.5; mode: 12 25. mean: 58; median: 59; mode: 59

Section 7.1 Measures of Central Tendency
(pages 278 and 279)

1. no; The definition of an outlier means that it is not in the center of the data.

3. If the outlier is larger than the mean, removing it will lower the mean. If the outlier is smaller than the mean, removing it will increase the mean.

5. mean: 1; median: 1; mode: -1 7. mean: $1\frac{29}{30}$ h; median: 2 h; mode: $1\frac{2}{3}$ h and 2 h

9. They calculated the mean, not the median. Test scores: 80, 80, 90, 90, 90, 98

$$\text{Median} = \frac{90 + 90}{2} = \frac{180}{2} = 90$$

11. 4 13. 16 15. a. 105°F b. mean

17. The mean and median both decrease by \$0.05. There is still no mode.

19. $-8, -5, -3, 1, 4, 7$ 21. B

Section 7.2 Box-and-Whisker Plots
(pages 284 and 285)

1. 25%; 50%

3. The length gives the range of the data set. This tells how much the data vary.

5. 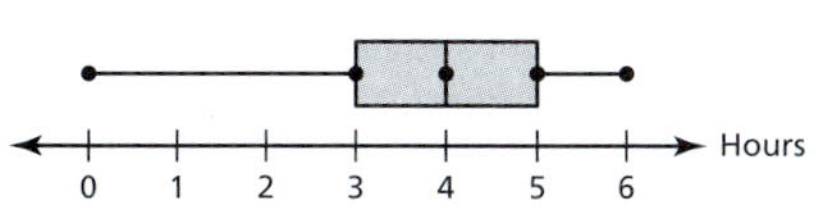

7. 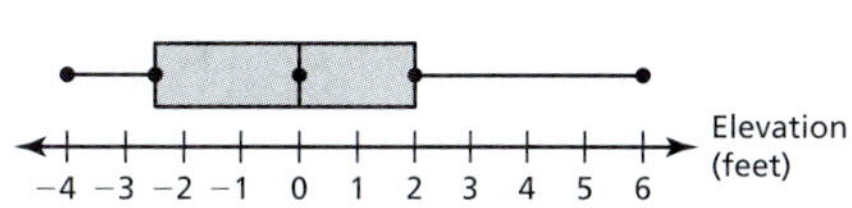

9. 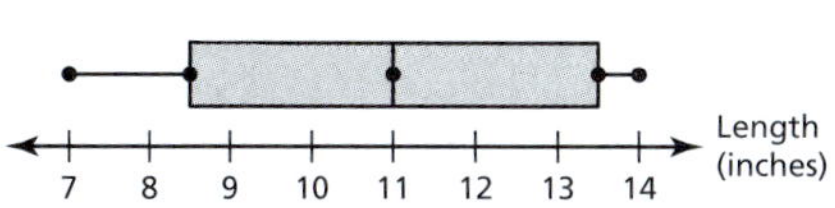

range = 7

Section 7.2 Box-and-Whisker Plots *(continued)*

(pages 284 and 285)

11. a. 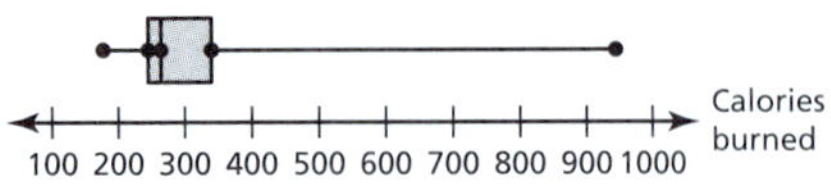

b. 944 calories **c.** 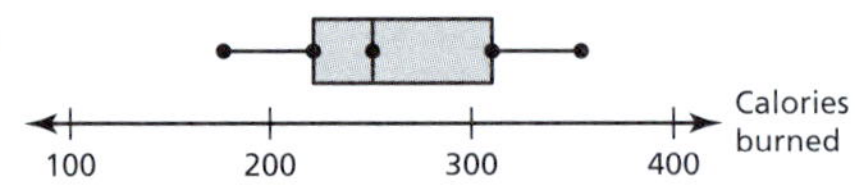

d. The outlier makes the right whisker longer, increases the length of the box, increases the third quartile, and increases the median. In this case, the first quartile and the left whisker were not affected.

13. *Sample answer:* 0, 5, 10, 10, 10, 15, 20

15. *Sample answer:* 1, 7, 9, 10, 11, 11, 12

17. $y = 3x + 2$

19. $y = -\frac{1}{4}x$

21. B

Section 7.3 Scatter Plots and Lines of Best Fit

(pages 293–295)

1. They must be ordered pairs so there are equal amounts of x- and y-values.

3. a–b. 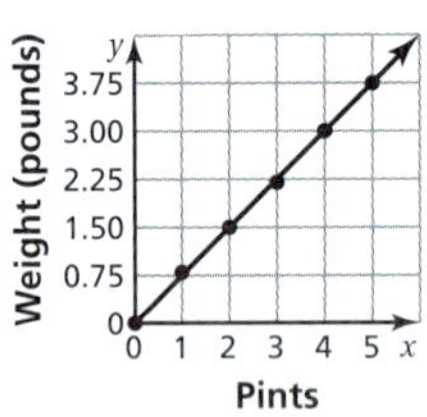

c. *Sample answer:* $y = 0.75x$

d. *Sample answer:* 7.5 lb

e. *Sample answer:* \$16.88

5. a. 3.5 h **b.** \$85

c. There is a positive relationship between hours worked and earnings.

7. positive relationship

9. negative relationship

11. a–b. 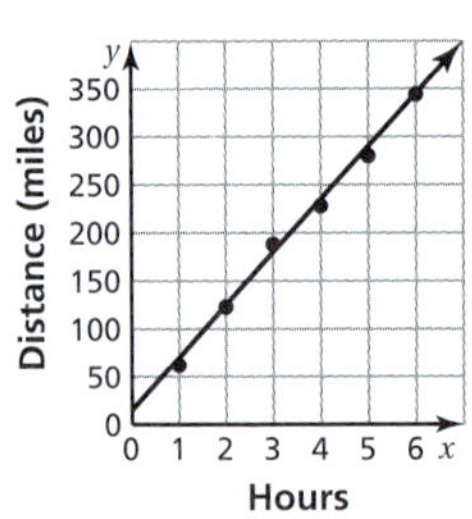

c. *Sample answer:* $y = 55x + 15$

d. *Sample answer:* 400 mi

13. a. positive relationship

b. The more time spent studying, the better the test score.

15. The slope of the line of best fit should be close to 1.

17. 2

19. -4

Section 7.4 Choosing a Data Display
(pages 300 and 301)

1. yes; Different displays may show different aspects of the data.

3. *Sample answer:*

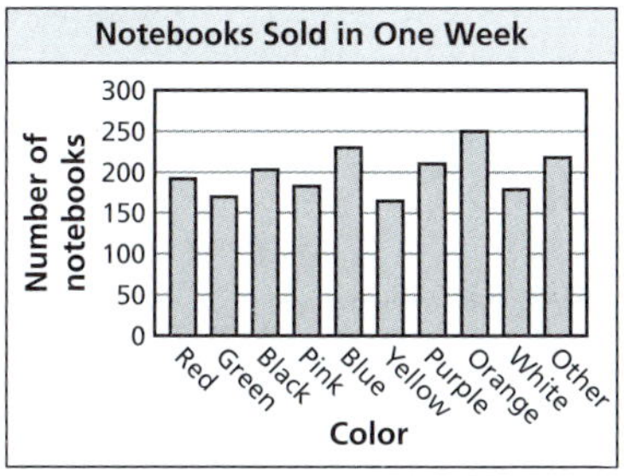

A bar graph shows the data in different color categories.

5. *Sample answer:* Line graph: shows changes over time.

7. *Sample answer:* Line graph: shows changes changes over time.

9. The pictures of the bikes are larger on Monday, which makes it seem like the distance is the same each day.

11. The intervals are not the same size.

13. *Sample answer:* bar graph; Each bar can represent a different vegetable.

15. *Sample answer:* line plot

17. Does one display better show the differences in digits?

19. $8x = 24$

Section 8.1 Writing and Graphing Inequalities
(pages 316 and 317)

1. An open circle would be used because 250 is not a solution.

3. no; $x \geq -9$ is all values of x greater than or equal to -9. $-9 \geq x$ is all values of x less than or equal to -9.

5. $x < -3$; all values of x less than -3

7. $y + 5.2 < 23$

9. $k - 8.3 > 48$

11. yes

13. yes

15. no

17. 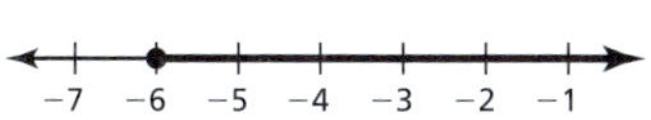

19. 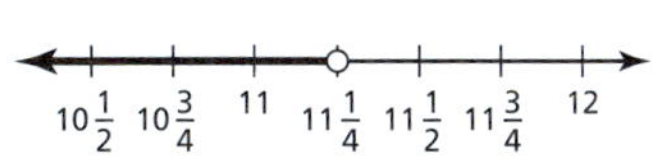

21. $x \geq 21$

23. yes

25. a. $a \geq 10$; 

$s \geq 200$; 

$t \geq 10$; 

b. yes; You satisfy the swimming requirement of the course because $10(25) = 250$ and $250 \geq 200$.

27. a. $m < n$; $n \leq p$ b. $m < p$

c. no; Because n is no more than p and m is less than n, m cannot be equal to p.

29. -1.7

31. D

Section 8.2 Solving Inequalities Using Addition or Subtraction *(pages 322 and 323)*

1. no; The solution of $r - 5 \le 8$ is $r \le 13$ and the solution of $8 \le r - 5$ is $r \ge 13$.

3. *Sample answer:* $A = 350$, $C = 275$, $Y = 3105$, $T = 50$, $N = 2$

5. *Sample answer:* $A = 400$, $C = 380$, $Y = 6510$, $T = 83$, $N = 0$

7. $t > 4$; 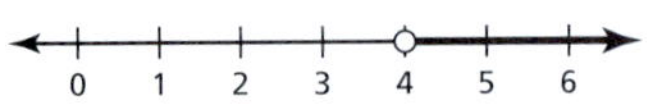

9. $a > -8$; 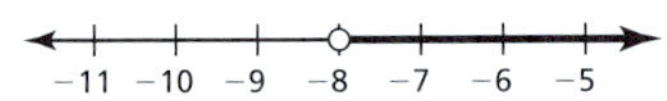

11. $-\frac{3}{5} > d$; 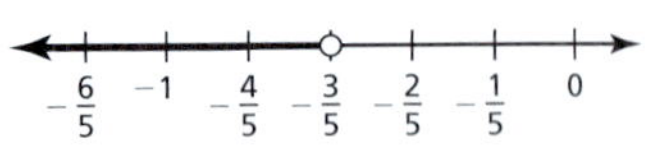

13. $m \le 1$; 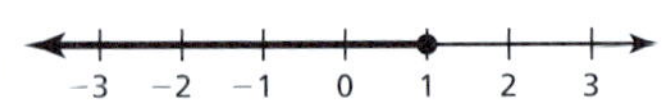

15. $h < -1.5$; 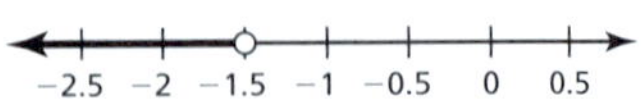

17. $9.5 \ge u$; 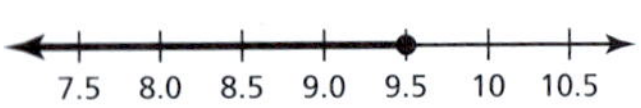

19. **a.** $100 + V \le 700$; $V \le 600$ in.^{3} **b.** $V \le \frac{700}{3}$ in.^{3}

21. $x + 2 > 10$; $x > 8$

23. 5

25. **a.** $4500 + x \ge 12{,}000$; $x \ge 7500$ points

b. This changes the number added to x by 60%, so the inequality becomes $7200 + x \ge 12{,}000$. So, you need less points to advance to the next level.

27. $2\pi h + 2\pi \le 15\pi$; $h \le 6.5$ mm

29. 10

31. 12

Section 8.3 Solving Inequalities Using Multiplication or Division *(pages 331–333)*

1. Multiply each side of the inequality by 6.

3. *Sample answer:* $-3x < 6$

5. $x \ge -1$

7. $x \le -3$

9. $x \le \frac{3}{2}$

11. $c \le -36$; 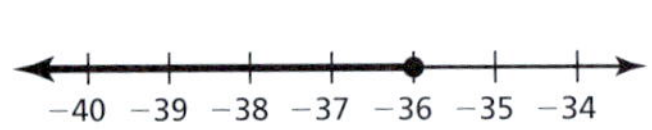

13. $x < -28$; 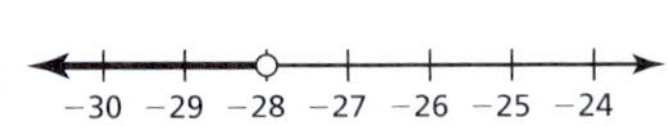

15. $k > 2$; 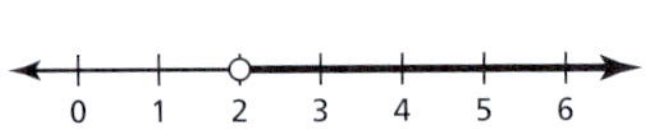

17. $y \le -4$; 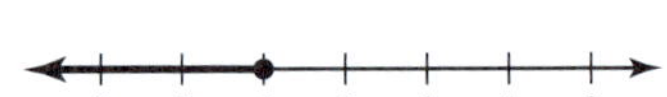

19. The inequality sign should not have been reversed.

$$\frac{x}{2} < -5$$
$$2 \cdot \frac{x}{2} < 2 \cdot (-5)$$
$$x < -10$$

21. $\frac{x}{8} < -2$; $x < -16$

23. $5x > 20$; $x > 4$

25. $0.25x \le 3.65$; $x \le 14.6$; You can make **at** most 14 copies.

27. $n \geq -5$; 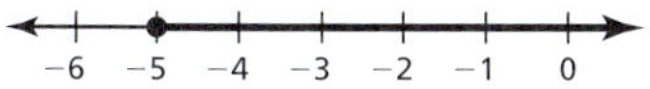

29. $h \leq -42$; 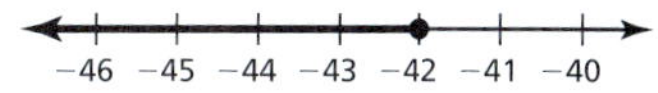

31. $y > \frac{11}{2}$; 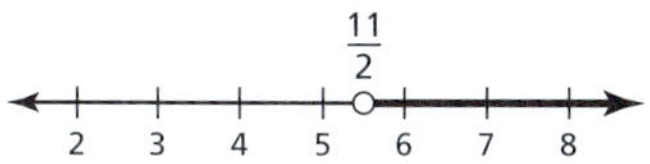

33. $m > -12$; 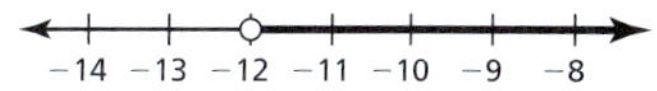

35. $b > 4$; 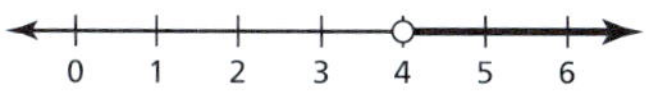

37. no; You need to solve the inequality for x. The solution is $x < 0$. Therefore numbers greater than 0 are not solutions.

39. $12x \geq 102$; $x \geq 8.5$ cm

41. $\frac{x}{4} < 80$; $x < \$320$

43. *Answer should include, but is not limited to:* Make sure students use the correct number of months that the CD has been out. In part (d), an acceptable answer could be never because the top selling CD could have a higher monthly average.

45. $n \geq -6$ and $n \leq -4$;

47. $m < 20$;

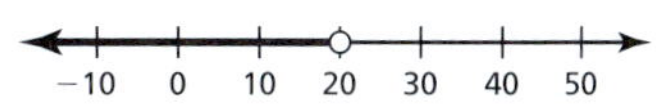

49. $8\frac{1}{4}$

51. 84

Section 8.4 Solving Multi-Step Inequalities
(pages 338 and 339)

1. *Sample answer:* They use the same techniques, but when solving an inequality you must be careful to reverse the inequality symbol when you multiply or divide by a negative number.

3. $k > 0$ and $k \leq 16$ units

5. $b \geq 1$; 

7. $m \geq -15$; 

9. $p < -1$; 

11. They did not perform the operations in proper order.

$$\frac{x}{4} + 6 \geq 3$$
$$\frac{x}{4} \geq -3$$
$$x \geq -12$$

13. $y \leq 13$; 

15. $u < -17$; 

17. $z > -0.9$; 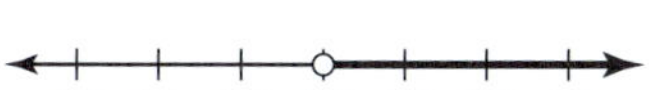

19. $x \leq 6$; 

21. $\frac{3}{16}x + 2 \leq 11$; $x > 0$ and $x \leq 48$ lines

23. Remember to add the height of the truck to find the height the ladder can reach.

25. $r \geq 3$ units

27. 625π in.^{2}

29. A

Section 9.1 Exponents
(pages 354 and 355)

1. An exponent describes the number of times the base is used as a factor. A power is the entire expression (base and exponent). A power tells you the value of the factor and the number of factors. No, the two cannot be used interchangeably.

3. 3^4

5. $\left(-\frac{1}{2}\right)^3$

7. $\pi^3 x^4$

9. $8^4 b^3$

11. 25

13. 1

15. $\frac{1}{144}$

17. The exponent 3 describes how many times the base 6 should be used as a factor. Three should not appear as a factor in the product. $6^3 = 6 \cdot 6 \cdot 6 = 216$

19. $-\left(\frac{1}{4}\right)^4$

21. 29

23. 5

25. 66

27.

h	1	2	3	4	5
$2^h - 1$	1	3	7	15	31
2^{h-1}	1	2	4	8	16

$2^h - 1$; The option $2^h - 1$ pays you more money when $h > 1$.

29. Remember to add the black keys when finding how many notes you travel.

31. Associative Property of Multiplication

33. B

Section 9.2 Product of Powers Property
(pages 360 and 361)

1. When multiplying powers with the same base

3. 3^4

5. $(-4)^{12}$

7. h^7

9. $\left(-\frac{5}{7}\right)^{17}$

11. 5^{12}

13. 3.8^{12}

15. The bases should not be multiplied. $5^2 \cdot 5^9 = 5^{2+9} = 5^{11}$

17. $216g^3$

19. $\frac{1}{25}k^2$

21. $r^{12}t^{12}$

23. no; $3^2 + 3^3 = 9 + 27 = 36$ and $3^5 = 243$

25. 496

27. 78,125

29. **a.** $16\pi \approx 50.24$ in.^{3}

  b. $192\pi \approx 602.88$ in.^{3} Squaring each of the dimensions causes the volume to be 12 times larger.

31. Use the Commutative and Associative Properties of Multiplication to group the powers.

33. 4

35. 3

37. B

Section 9.3 Quotient of Powers Property

(pages 366 and 367)

1. To divide powers means to divide out the common factors of the numerator and denominator. To divide powers with the same base, write the power with the common base and an exponent found by subtracting the exponent in the denominator from the exponent in the numerator.

3. 6^6

5. $(-3)^3$

7. 5^6

9. $(-17)^3$

11. $(-6.4)^2$

13. b^{13}

15. You should subtract the exponents instead of dividing them. $\dfrac{6^{15}}{6^5} = 6^{15-5} = 6^{10}$

17. 2^9

19. π^8

21. k^{14}

23. $64x$

25. $125a^3b^2$

27. x^7y^6

29. You are checking to see if there is a constant rate of change in the prices, not if it is a linear function.

31. 10^{13} galaxies

33. -9

35. 61

37. B

Section 9.4 Zero and Negative Exponents

(pages 374 and 375)

1. no; Any nonzero base raised to the zero power is always 1.

3. $5^{-5}, 5^0, 5^4$

5.

n	4	3	2	1
$\dfrac{5^n}{5^2}$	$5^2 = 25$	$5^1 = 5$	$5^0 = 1$	$5^{-1} = \dfrac{1}{5}$

7. One-fifth of 5^1; $5^0 = \dfrac{1}{5}(5^1) = 1$

9. $\dfrac{1}{36}$

11. $\dfrac{1}{16}$

13. 1

15. $\dfrac{1}{125}$

17. The negative sign goes with the exponent, not the base. $(4)^{-3} = \dfrac{1}{4^3} = \dfrac{1}{64}$

19. 2^0; 10^0

21. $\dfrac{a^7}{64}$

23. $5b$

25. 12

27. $\dfrac{w^6}{9}$

29. 10,000 micrometers

31. 1,000,000 micrometers

33. Convert the blood donation to cubic millimeters before answering the parts.

35. If $a = 0$, then $0^n = 0$. Because you can not divide by 0, the expression $\dfrac{1}{0}$ is undefined.

37. 10^3

39. D

Section 9.5 Reading Scientific Notation

(pages 380 and 381)

1. Scientific notation uses a factor of at least one but less than 10 multiplied by a power of 10. A number in standard form is written out with all the zeros and place values included.

3. 0.00015 m

5. 20,000 mm^3

7. yes; The factor is at least 1 and less than 10. The power of 10 has an integer exponent.

9. no; The factor is greater than 10.

11. yes; The factor is at least 1 and less than 10. The power of 10 has an integer exponent.

13. no; The factor is less than 1.

15. 70,000,000

17. 500

19. 0.000044

21. 1,660,000,000

23. 9,725,000

25. **a.** 810,000,000 platelets

b. 1,350,000,000,000 platelets

27. **a.** Bellatrix

b. Betelgeuse

29. 5×10^{12} km^2

31. Be sure to convert some of the speeds so that they all have the same units.

33. 10^7

35. $\frac{1}{10^{16}}$

Section 9.6 Writing Scientific Notation

(pages 386 and 387)

1. If the number is greater than or equal to 10, the exponent will be positive. If the number is less than 1 and greater than 0, the exponent will be negative.

3. 2.1×10^{-3}

5. 3.21×10^8

7. 4×10^{-5}

9. 4.56×10^{10}

11. 8.4×10^5

13. 72.5 is not less than 10. The decimal point needs to move one more place to the left. 7.25×10^7

15. 9×10^{-10}

17. 1.6×10^8

19. 2.88×10^{-7}

21. 4.01×10^7 m

23. 5.612×10^{14} cm^2

25. 9.75×10^9 N•m per sec

27. *Answer should include, but is not limited to:* Make sure calculations using scientific notation are done correctly.

29. **a.** 2.65×10^8

b. 2.2×10^{-4}

31. 200

Section B.1 Simple and Compound Interest

(pages A14 and A15)

1. simple interest

3. a.

Simple Interest			
t	Principal	Annual Interest	Balance at End of Year
3	\$500	\$15	\$545
4	\$500	\$15	\$560
5	\$500	\$15	\$575

Compound Interest			
t	Principal and Interest	Annual Interest	Balance at End of Year
3	\$530.45	\$15.91	\$546.36
4	\$546.36	\$16.39	\$562.75
5	\$562.75	\$16.88	\$579.63

   b. compound interest

5. \$1405.09

7. The table shows compound interest.

t	Principal	Annual Interest	Balance at End of Year
1	\$700.00	\$21	\$721.00
2	\$700.00	\$21	\$742.00

9. a. \$300 b. \$13

11. One account will not always be greater than the other account.

13. $3.375x^6$

15. $29.16d^{18}$

Section B.2 Formula for Compound Interest

(pages A20 and A21)

1. Compound interest earns more than simple interest for a given interest rate because it is interest earned on the principal and on the previously earned interest.

3.

Annual Interest Rate	Balance after 5 Years
3.0%	\$579.64
3.5%	\$593.84
4.0%	\$608.33
4.5%	\$623.09

5. a. \$688.51

   b. \$88.51

7. The interest rate should be in decimal form.

$$\begin{aligned} B &= P(1 + r)^t \\ &= 2000(1 + 0.04)^2 \\ &= 2000(1.04)^2 \\ &= \$2163.20 \end{aligned}$$

9. Account B; Its balance is greater after 8 years.

11. Compounding interest four times a year is compounding it quarterly.

13. 27,915

15. 0.4

17. D

Section B.3 Installment Loans
(pages A26 and A27)

1. The principal of a loan is less than the total amount due because interest is included in the amount due.

3.

Balance Before Payment	Monthly Payment	Monthly Interest	Balance After Payment
\$1135.00	\$233.86	\$11.35	\$912.49
\$912.49	\$233.86	\$9.12	\$687.75
\$687.75	\$233.86	\$6.88	\$460.77
\$460.77	\$233.86	\$4.61	\$231.52
\$231.52	\$233.84	\$2.32	\$0.00

\$34.28

5. \$28.74

7. \$60.88

9.

Term \ Annual Interest Rate	Interest Paid 5%	Interest Paid 7%
20	\$70,068.00	\$103,286.40
25	\$90,453.00	\$134,442.00
30	\$111,908.40	\$167,409.60

11. about 24.9%

13. For Loan A, you will use the amount of money in your savings account after 3 years to pay off the loan.

15. −28.06

Section B.4 Checking Accounts
(pages A32 and A33)

1. date, check number (if one is written), transaction description, credit or debit amount

3. a.

Date	Check #	Transaction	Credit	Debit	Balance
		Balance Forward			300.00
3/4/12	506	Dinner		45.32	254.68
3/7/12		ATM Withdrawal		80.00	174.68
3/8/12	507	Electric Company		39.76	134.92
3/14/12		Deposit	150.00		284.92
3/18/12	508	Credit Card Payment		129.50	155.42

b. yes

5. more than 20 withdrawals each month

7. You should balance your checkbook after every transaction so that you do not write any bad checks.

9. 7.9×10^{-8}

11. 3.6×10^{11}

13. C

Section B.5 Credit Cards
(pages A40 and A41)

1. *Sample answer:* You can use both cards to make a purchase, instead of using cash or a check. *Sample answer:* A credit card is similar to a loan in that payment is postponed and interest is charged.

3. \$2.70 **5.** \$2.42

7. **a.** \$3.33 **b.** \$107.76 **c.** \$102.47

9. **a.** \$3.94 **b.** \$74.81 **c.** \$71.06

11. To find the percent the store keeps, divide the amount the store keeps by the total amount with the sales tax.

13. \$740.98; \$240.00; \$59.02; The amount you paid is more than four times the decrease in the amount due. You end up decreasing your balance by about \$5 each month because of the interest. You should pay more than the minimum so that more money goes toward decreasing the balance, rather than paying mostly interest.

15. 12 **17.** $6\sqrt{19}$

Section B.6 Payroll Deductions
(pages A46 and A47)

1. never; At most, net pay can equal gross pay if there are no deductions, but it will never be greater than gross pay.

3. \$1173.75 **5.** \$612.08 **7.** every 2 weeks

9. \$17,468.10 **11.** \$228.15

13–15. Make sure to look back at page A45 and divide by the correct amounts.

17. *Sample answer:* Colorado: 4.63%; Florida: none; Massachusetts: 5.3%; Florida has the lowest income tax and Massachusetts has the highest income tax.

19. 3 **21.** 343 **23.** D

Student Index

This student-friendly index will help you find vocabulary, key ideas, and concepts. It is easily accessible and designed to be a reference for you whether you are looking for a definition, real-life application, or help with avoiding common errors.

A

D

E

F

G

H

I

Student Index

L

M

N

Q

R

S

T

U

V

W

X

Y

Z

Conversions

U.S. Customary

1 foot = 12 inches
1 yard = 3 feet
1 mile = 5280 feet
1 acre ≈ 43,560 square feet
1 cup = 8 fluid ounces
1 pint = 2 cups
1 quart = 2 pints
1 gallon = 4 quarts
1 gallon = 231 cubic inches
1 pound = 16 ounces
1 ton = 2000 pounds
1 cubic foot ≈ 7.5 gallons

U.S. Customary to Metric

1 inch ≈ 2.54 centimeters
1 foot ≈ 0.3 meter
1 mile ≈ 1.6 kilometers
1 quart ≈ 0.95 liter
1 gallon ≈ 3.79 liters
1 cup ≈ 237 milliliters
1 pound ≈ 0.45 kilogram
1 ounce ≈ 28 grams
1 gallon ≈ 3785 cubic centimeters

Time

1 minute = 60 seconds
1 hour = 60 minutes
1 hour = 3600 seconds
1 year = 52 weeks

Temperature

$C = \frac{5}{9}(F - 32)$

$F = \frac{9}{5}C + 32$

Metric

1 centimeter = 10 millimeters
1 meter = 100 centimeters
1 kilometer = 1000 meters
1 liter = 1000 milliliters
1 kiloliter = 1000 liters
1 milliliter = 1 cubic centimeter
1 liter = 1000 cubic centimeters
1 cubic millimeter = 0.001 milliliter
1 gram = 1000 milligrams
1 kilogram = 1000 grams

Metric to U.S. Customary

1 centimeter ≈ 0.39 inch
1 meter ≈ 3.28 feet
1 kilometer ≈ 0.6 mile
1 liter ≈ 1.06 quarts
1 liter ≈ 0.26 gallon
1 kilogram ≈ 2.2 pounds
1 gram ≈ 0.035 ounce
1 cubic meter ≈ 264 gallon

Equations of Lines

Slope-intercept form

$y = mx + b$

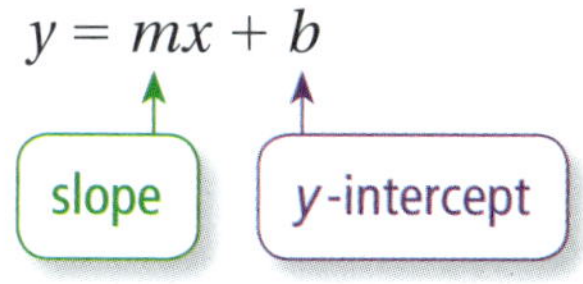

Standard form

$ax + by = c,\ a, b \neq 0$

Pythagorean Theorem

$a^2 + b^2 = c^2$

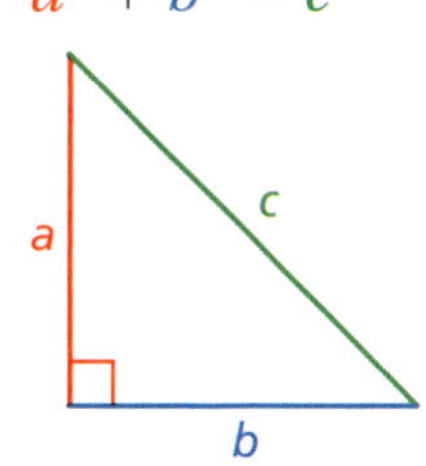

Rules of Exponents

Product of Powers Property	$a^m \cdot a^n = a^{m+n}$
Quotient of Powers Property	$\frac{a^m}{a^n} = a^{m-n}$, where $a \neq 0$
Zero Exponents	$a^0 = 1$, where $a \neq 0$
Negative Exponents	$a^{-n} = \frac{1}{a^n}$, where $a \neq 0$

K

BIG Idea 1 Represent, compare, and order whole numbers and join and separate sets.

BIG Idea 2 Describe shapes and space.

BIG Idea 3 Order objects by measurable attributes.

Supporting Ideas

Algebra • Geometry and Measurement

1

BIG Idea 1 Develop understandings of addition and subtraction strategies for basic addition facts and related subtraction facts.

BIG Idea 2 Develop an understanding of whole number relationships, including grouping by tens and ones.

BIG Idea 3 Compose and decompose two-dimensional and three-dimensional geometric shapes.

Supporting Ideas

Algebra • Geometry and Measurement • Number and Operations

2

BIG Idea 1 Develop an understanding of base-ten numerations system and place-value concepts.

BIG Idea 2 Develop quick recall of addition facts and related subtraction facts and fluency with multi-digit addition and subtraction.

BIG Idea 3 Develop an understanding of linear measurement and facility in measuring lengths.

Supporting Ideas

Algebra • Geometry and Measurement • Number and Operations